"BEST BOOK"

"BEST BOOK"

Physics

"Best Book"

Jonathan S. Wolf

Formerly Department of Physics
Queens College
City University of New York

All inquiries should be addressed to:
Barron's Educational Series, Inc.
250 Wireless Boulevard
Hauppauge, New York 11788

Library of Congress Catalog Card No. 95-51483
International Standard Book No. 0-8120-9522-7

Library of Congress Catologing-in-Publication Data

Wolf, Jonathan S.
Physics / Jonathan S. Wolf.
p. cm. — (College review series)
Includes index.
ISBN 0-8120-9522-7
1. Physics—Outlines, syllabi, etc. I. Title. II. Series.
QC31.W74 1996
530—dc20 95-51483
CIP

PRINTED IN THE UNITED STATES OF AMERICA
6789 8800 987654321

CONTENTS

PREFACE

This book is concerned with a systematic review of the material contained in a standard course in college physics. The course of study is typically offered in the freshman year to non-science majors or science majors who are not focusing on physics or engineering. The course is therefore developed algebraically *without* the use of calculus. It is not required that the student have taken a course in physics previously in high school. It is expected that the student has had a thorough exposure of mathematics through trigonometry.

Reviewing means much more than going over course material or doing extra problems. Reviewing means developing a systematic approach to studying, understanding, and applying course material. This includes identifying key concepts, looking for cross links and associations from the various sub-divisions of physics, developing a scheme for organizing the acquired knowledge, and enhancing problem-solving skills.

From this list of items, problem solving occupies a major goal of this review book and any science course. Students learn by doing problems. In the "constructivist" view of learning, a student brings to a physics problem a large amount of prior knowledge based on life experience and past courses. When new knowledge is acquired, the student actively constructs meaning to that knowledge based on processes that analyze, synthesize, and otherwise associate that knowledge with the student's prior knowledge. Learning is an "active" as opposed to a "passive" process. Problem solving is a learning process also.

It has been well documented that beginning physics students often find physics difficult. One of the reasons for this is the way in which physics deals with everyday experiences in an abstract way. Life experiences have given the student an intuitive idea about how the universe works and how to make inferences or draw conclusions. These concepts are often based directly upon what the senses detect. As such, the student often arrives into a physics class for the first time with misconceptions about the nature of physical knowledge that is counter-intuitive to what they are being exposed to in the course.

Additionally, beginning students have difficulty solving problems that are not simply "plug-and-chug," meaning the direct substitution of numerical values into an algorithmic formula (usually an algebraic equation). Thus, the student is faced with an often new way of thinking and looking at the world around them.

A major goal of this book is to enhance the problem solving and study skills of the student as reflected in the research performed in the field of physics education. Therefore, in addition to a full review of content material and extra problems to solve and think about, the book will present ideas throughout the review that provide learning and problem-solving strategies.

Each review chapter will contain a summary of the relevant content material followed by a series of review questions. These questions will take the form of "thought questions" which are short answer essay type, multiple choice, and solution based problems. Following the review chapters are additional problem solving sets that the student can use to assess their understanding of the course material. There is also an appendix, glossary, and index of all material at the end of the book.

I am grateful to the many people at Barron's who helped in the preparation of this manuscript. Specific thanks go to Warren Bratter, Project Editor, for all of his guidance and suggestions throughout the production process. Additionally, I would like to thank Todd Gordon, Richard New, and Pat Jablonowski for all of their advice and suggestions during all phases of this project. Finally, I would like to thank my wife Karen and my daughters Marissa and Ilana for all of their love, support, and understanding. This book is dedicated to them.

Jonathan S. Wolf

1
THE NATURE OF PHYSICS

WHAT IS PHYSICS?

Physics is a branch of knowledge that attempts to establish relationships among natural phenomena. These phenomena are a part of our sensory experiences from everyday life. As human beings, we use these sensory inputs to make decisions about the world around us, decisions that are based on recognized patterns of orderly events (like the rising and setting of the sun, the changes in seasons, the migrations of animals, and the movements of objects).

Occasionally, we come across situations that seem to occur randomly. These discrepant events challenge the routine of predictability that is the foundation of the scientific method. Many of these naturally occurring random events display orderly tendencies on average, which means that while individual events might be random, the occurrence of a large number of events may indeed be predictable. This *statistical* behavior of natural phenomena will come in handy when we consider the interactions of large numbers of interacting particles such as those that make up a gas.

In order to understand how these causal relationships are established, we must first adopt a set of rules by which all observers might be led to similar conclusions. There is a caveat, however, that if we restrict our decisions too rigidly, then we may be forced to make conclusions that are contrary to information obtained from our senses and the real world. In physics, the former happens all the time, however, the nature of science demands that we must always be describing real events. It is to these *rules of reasoning* that we now proceed.

RULES OF REASONING WHEN DOING PHYSICS

The branch of knowledge that develops these rules of reasoning about the world is called *philosophy.* Philosophy differs from science in that the scientific method demands that experimental evidence verify a prediction made using the rules of reasoning (or logic) to be discussed here. Thus, the ancient school of Greek philosophy called φισικοσ (*physikos*) or natural philosophy developed from abstract thinking about how the universe works into a systematic experimental discipline.

In physics we usually employ two modes of reasoning called inductive and deductive. *Inductive reasoning* begins with an accumulation of specific phenomena, and then we make conclusions about a general concept. This is the basis of experimental science. In a controlled experiment, we vary one quantity *independently* and then observe what happens to another quantity called the *dependent* quantity.

The relationship is established by comparing the effects of the changes. If the dependent quantity does not change at all, the relationship is referred to as being *constant.* If the dependent quantity increases as the independent quantity increases, the relationship is referred to as being *direct.* If the dependent quantity decreases as the independent quantity increases, the relationship is referred to as being *inverse.*

Deductive reasoning begins with a general premise, and then we draw conclusions about particular aspects of that concept. Syllogistic reasoning is one type of deductive reasoning. Consider this standard example:

All men are mortal.
Socrates is a man.
Therefore, Socrates is mortal.

If the truth of the first two statements is valid, then the conclusion is judged valid. Notice that the first statement is an all-inclusive general statement: All men are mortal.

An example from physics might be something like this:

Gravity accelerates all free bodies near the earth at the same rate.
A sheet of paper and a pencil are both bodies near the earth.
Therefore, if the paper and pencil are free, they will both fall with the same acceleration.

If you try this example, you will find that the sheet of paper floats down at a constant speed because of air resistance. Therefore, in light of the counter-example, we are forced to either discount or modify the original premise. In science, we often choose to do the latter.

In his treatise, *Principia,* outlining much of what we now call *classical Mechanics,* Sir Isaac Newton presented, more comprehensively, these rules of reasoning which we can keep in the backs of our minds as we proceed to study physics:

1. We are to admit no more causes of natural things than such as are both true and sufficient to explain their appearances.
2. Therefore, to the same natural effects, we will assign the same natural causes.
3. The qualities of bodies that do not increase or decrease in amount and are found to belong to all bodies, as far as we can determine from our experimental instruments, will be called *universal qualities* of all bodies.
4. In experimental science we look at conclusions inferred by general induction from phenomena as being accurate and very nearly true, notwithstanding any contrary hypotheses we may consider, unless some new phenomena occurs which causes us to modify these conclusions.

ESTABLISHING RELATIONSHIPS IN PHYSICS

Relationships are based on quantitative comparisons using the rules of reasoning just described. To begin with, we ascribe to quantities of matter *fundamental characteristics* that can be used to distinguish them. Among these are mass, length, and electric current. And since the universe undergoes constant change, we add time to this list.

Quantitative comparisons are based on assigning a magnitude and a scale to each of these quantities. The magnitude indicates the relative size of the quantity, and the scale provides the rule for measurement. Another term we could use instead of *scale* is *unit* or *dimension.* Specifically, the units assigned to mass, length, and time (and several others) are called *fundamental units.* In physics, we use the International System of Units (SI, for *Système International d'Unités*) and some of the more common units are presented in Tables 1 and 2.

TABLE 1. FUNDAMENTAL SI UNITS USED IN PHYSICS

Quantity	Unit Name	Symbol
Length	meter	m
Mass	kilogram	kg
Time	second	s
Electric current	ampere	A
Temperature	kelvin	K
Amount of substance	mole	mol
Plane angle	radian	rad

The units in Table 2 are called *derived* since they can be expressed in terms of the fundamental units shown in Table 1.

TABLE 2. SOME DERIVED SI UNITS USED IN PHYSICS

Quantity	Unit Name	Symbol	Expression in Other SI Units
Force	newton	N	kg-m/s^2
Frequency, cyclic	hertz	Hz	cycles/s
Frequency, angular	—	—	rad/s = 1/s
Pressure	pascal	Pa	N/m^2 = kg/m-s^2
Energy, work and heat	joule	J	N-m = kg-m^2/s^2
Power	watt	W	J/s = kg-m^2/s^3
Electric charge	coulomb	C	As
Electric potential (EMF)	volt	V	J/C = J/As = kg-m^2/As^3
Capacitance	farad	F	C/V = A^{2}s^4/kg-m^2
Resistance	ohm	Ω	V/A = kg-m^2/A^{2}s^3
Magnetic flux	weber	Wb	Vs = kg-m^2/As^2
Magnetic flux density	tesla	T	Wb/m^2 = kg/As^2
Inductance	henry	H	Wb/A = kg-m^2/A^{2}s^2

One way to establish a relationship between two physical quantities is by means of a graph. If the graph indicates that for one value of the independent quantity we assign one and only one value for the dependent quantity, then we have a *functional relationship.* This usually means that the relationship can be represented by an algebraic equation.

Certain relationships produce graphs that are immediately recognizable. Usually, we plot a graph using *Cartesian coordinates* (see Chapter 3) and represent the independent quantity along the horizontal or *x axis.* When there is no change in the dependent quantity for any change in the independent quantity, then we obtain a horizontal line as in Figure 1.

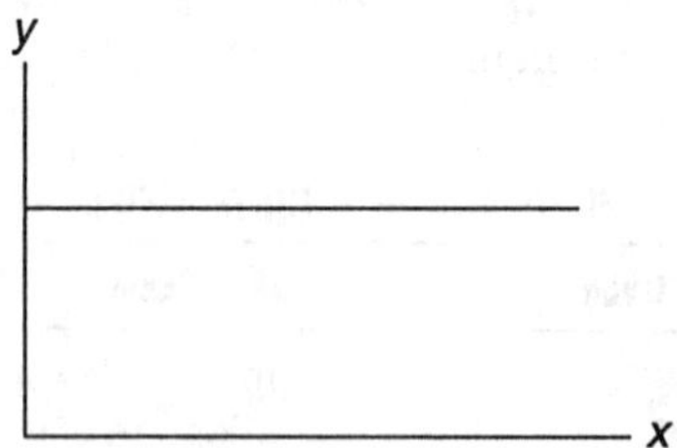

FIGURE 1. *Constant relationship.*

If there is a constant change in the dependent quantity proportional to the independent quantity such that both increase (or decrease) simultaneously, then we say that the relationship is a *direct relationship.* When the dependent quantity decreases while the independent quantity increases (or vice versa), we have an *inverse relationship.* Now these relationships can be either linear or nonlinear. Examples of linear direct and inverse relationships are presented in Figures 2 and 3.

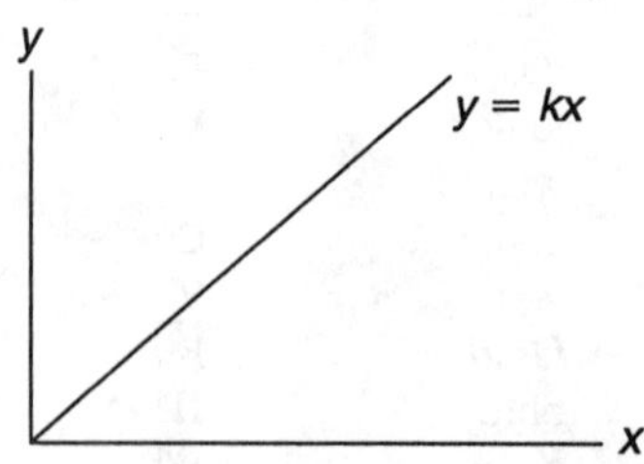

FIGURE 2. *Linear direct relationship.*

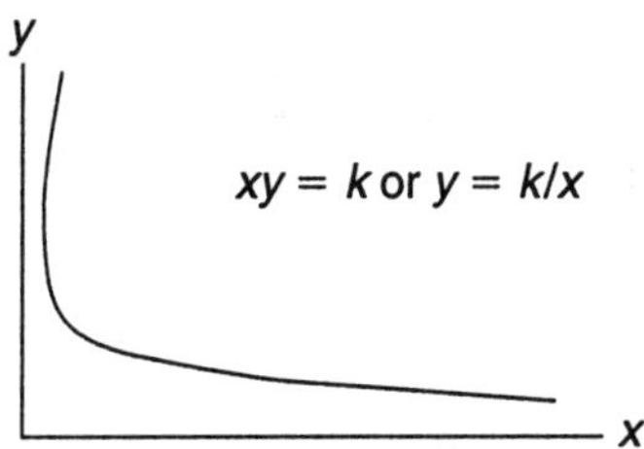

FIGURE 3. *Linear inverse relationship.*

The graph of a linear direct relationship is a diagonal straight line that starts at the origin. This is desirable since it makes the form of the equation simpler. This equation can be expressed as $y = kx$, where the constant k is called the *slope* of the line. Direct relationships that contain *y intercepts* other than the origin can be rescaled to fit this form. One example is the absolute temperature (Kelvin) scale. Instead of plotting Celsius temperature versus pressure for an ideal gas (and getting −273°C as the lowest possible temperature), we rescale the temperature axis such that the lowest possible temperature is defined to be 0 K and then obtain the familiar gas law relationships defined for absolute temperature only.

There are two typical nonlinear direct relationships that appear often in physics. The first is the *parabolic* or *squared relationship.* For example, in freely falling motion, the displacement of a mass is directly proportional to the square of the elapsed time. A graph of this relationship is shown in Figure 4.

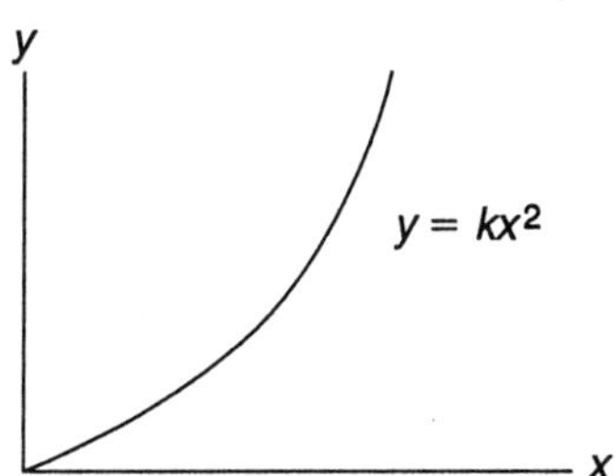

FIGURE 4. *Parabolic relationship.*

The other nonlinear direct relationship is called the *square root relationship.* For example, the period of a swinging pendulum is directly proportional to the square root of its length. A graph of this relationship is shown in Figure 5.

Having looked at how we establish relationships in physics, we are then left with the question of how to identify key concepts when doing physics. This

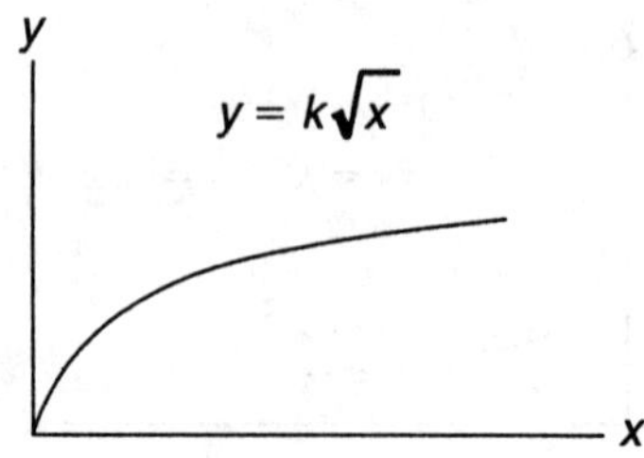

FIGURE 5. *Square root relationship.*

is an important aspect of learning, studying, reviewing, and doing physics. The remaining chapters in this book will be devoted to answering this question as well as to reviewing most of the content material covered in a standard college physics course. Let's get started!

2 LEARNING STRATEGIES AND PROBLEM SOLVING IN PHYSICS

INTRODUCTION

Have you ever tried to solve a physics problem only to quit in frustration because you didn't know how to get started? Many students find themselves in this situation. Often students say, after doing poorly on a test, that they "really studied hard for the test" or "really knew the stuff." When asked how they studied, most say that they reviewed their notes and looked at old homework assignments. A few industrious students may even have tried to do some extra problems.

When interviewing students about their study habits, I have also discovered that they often do not really understand how all the material is connected. They memorize formulas and gain experience solving certain types of problems, but when faced with a more challenging problem, many students who study in this way may not even know how to begin.

In physics, success and achievement require more than just being able to memorize and use formulas, and there is also more to learning physics than just learning how to solve problems. First, you must understand what the problem is asking you to do. Then, you must access from your memory all the information you feel is related to the ideas being discussed in the problem. Then, you must determine which information is relevant to the problem. Finally, you must decide on a solution path that will hopefully lead you to the correct answer.

To succeed in problem solving, you must first make sure that you understand and can apply a large body of information that may seem at first glance to be unrelated. The ability to do this requires an organizational hierarchy of ideas called concepts.

WHAT ARE CONCEPTS?

If I were to say the word *big* to you, what picture would you think of? If I were to say the word *digging* to you, what picture would you think? What image comes to mind when you hear (or read) the word *atom*? Even though everyone hears or uses the same words, each of us develops different mental images or interpretations for them using our own past experiences and ideas.

The mental images or ideas we create about words are called **concepts**. Words are merely the labels we assign to concepts, and a great deal of confusion and misunderstanding arises when we mix words with concepts. In physics, this is especially true in attempting to understand the meaning of terms and ideas about the world. In the branch of physics known as mechanics, there are many words that are also often used in ordinary colloquial speech.

These are words you grew up with and have come to associate with daily living experiences. Some of these words are *speed, force, momentum, strength, weight,* and *work.* Often a student who has conceptual difficulties in physics has a difficult time understanding the different meanings of these colloquially used terms.

HOW ARE CONCEPTS LINKED?

In this section, you will begin to see how concepts can be listed in a hierarchical, organized way. The goal is to help you study and learn more meaningfully. For example, if *energy* is the key concept being discussed, some terms associated with energy might be *kinetic, potential, work, motion, heat,* or even *photosynthesis.* This list can be made more meaningful with the inclusion of so-called linking words. Examples of linking words are *is, as, are, can, be,* and *with.* Action phrases such as *contains, is composed of, are made from,* and so forth, can also be used. For example, we might start with *energy* and then say, "Energy is the capacity to do work." We can also say, "Energy can depend on velocity or position." When concepts are linked together in a hierarchical way, a better view of what you do and do not understand can be made more apparent.

We began this chapter by considering the nature of concepts and how important they are in understanding and studying physics. When you study, it is very important to pay attention to the organization of concepts. Since the goal of any course is to assess how well you understand the concepts and their applications, it is only natural to think about the nature of problem solving and how you can improve your ability to solve novel and difficult physics problems.

WHAT IS A PROBLEM?

Consider the following situation. You are driving in your car at 25 miles per hour (mi/h). A truck with a constant velocity of 20 mi/h is 100 feet (ft) in front of you. When you are about 10 ft behind the truck, you decide to pass it by accelerating to 45 mi/h. The highway has only two lanes, and so you have to cross into the oncoming traffic lane. By the time you have reached the back of the truck after starting this maneuver, you notice a car coming at you with a velocity of 45 mi/h and about $\frac{1}{4}$ mi away. Do you have enough time to safely pass the truck?

Think for a moment about what is going on here. You have been given a situation that is somewhat realistic. A question has been asked prompted by the phrase "Do you have . . . ?" What kind of answer is expected from you? Clearly, the answer prompted by the word *do* is either yes or no. But, what information do you need to have in front of you before you can decide whether the answer is yes or no? In other words, you have been given a "problem" to solve, which is implicit from the question previously stated.

A **problem** can be defined as a situation in which you want to achieve a goal but are unsure of how to go about it. There are many elements in the problem presented to you. Some information is explicit (like how fast you and the truck are going), and some information is implicit (it is never stated explicitly that you and the truck are going in the same direction, no mention is made of any acceleration by the truck, and it is assumed that your rate of acceleration is constant). In addition, the goal of the problem is implicit since in order to answer yes or no to the question, you must first ascertain whether the simultaneous distances covered by you and the oncoming car produce a number that can be interpreted as meaning either a successful or an unsuccessful completion of the maneuver.

In the next few sections, you will explore the nature of problem solving to better understand where you succeed and where you have difficulty in solving physics problems. Effective studying and reviewing means that you must develop an instinct for certain familiar problem types and techniques and be prepared for more novel problems that may be difficult to approach.

WHAT ARE THE STRUCTURAL ELEMENTS OF A PROBLEM?

The first element of a typical problem consists of the **givens**. This is information that is explicitly provided in the problem statement. As you read a problem, the words that are associated with concepts begin to access information from your memory. What kind of information is retrieved and in what form depend on what you already know about the subject, the type of problem, your experience, and your expertise.

"Difficult" problems are difficult because sometimes the information given is implicit or not well defined. If you are used to solving certain types of problems, you come to expect that a problem will look and sound a certain way. Even so, identifying the givens is the first step after accepting the problem in your mind. That is, after you initially read the problem, you begin to form a **representation** of it in your mind. This representation might begin with a redescription of the problem (perhaps as a mental image or mental model of the situation), followed by translating the problem into symbols that link the concepts already known with the given information in the present problem.

The representation might be similar to a concept map in which associated knowledge is linked by "operators" that tell your brain how to deal with particular concepts using formulas or learned rules of thumb for problem solving. These representations can vary from solver to solver across a wide domain of problem types.

Forming a suitable representation for the problem is the key to getting started. The second issue concerns the degree of relevancy (choosing which information to pay attention to and incorporate into your solution path). As you learn, you actively interact with the given information and prior knowledge. You construct meanings based on your interpretations of circumstances.

As a simple example, consider the problem of a mass sliding down an inclined plane that is frictionless (see Figure 1). The length of the incline is ℓ, and a mass M is starting from the top a height h above the ground. The problem is to determine the velocity of the mass when it reaches the bottom of the incline.

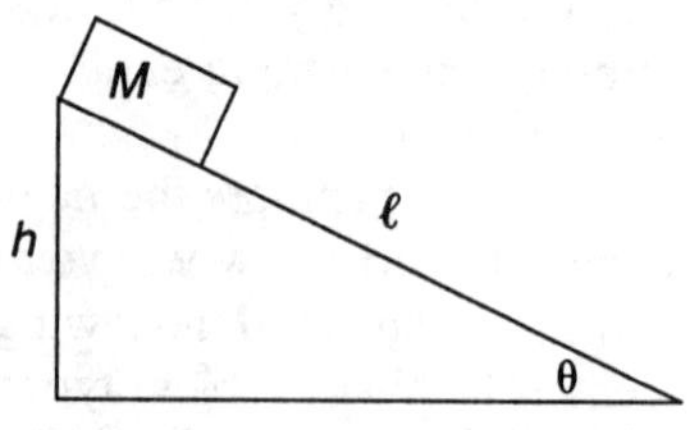

FIGURE 1.

If the mass starts from rest, then there are two ways to approach this problem. Suppose your first idea is to determine the acceleration of the block down the incline. Because gravity acts on the block, the acceleration down the incline can be shown to be equal to

$$\mathbf{a} = \mathbf{g} \sin \theta$$

where $\mathbf{g}$ is the acceleration of gravity and is equal to 9.8 m/s^2.

Now, since the time required to go down the incline is not known and we assume that the acceleration is constant (always ask yourself questions like What assumptions am I making about this problem? or What representation have I chosen?), we could try to determine the final velocity by using a known expression from linear motion (kinematics) that does not explicitly involve time. (Does that mean that the concept of time is not relevant to this problem?):

$$\mathbf{v}^2 = 2\mathbf{a}\ell$$

which means that the final velocity at the bottom of the incline is given by (after substituting for the acceleration)

$$\mathbf{v} = \sqrt{2\mathbf{g}\ell \sin \theta}$$

Notice that both the mass and the height of the incline are absent from the final expression. Are these quantities (like the time) also not relevant to the problem? Clearly, what we use is dictated by how we choose to interpret and represent the problem. If we designate this problem as one of a class of kinematics problems, then certain expressions are independent of given quantities. However, did you observe that the height is implicit since $h = \ell \sin \theta$?

There is another way to do this example problem. If we consider the law of conservation of energy (since no friction is present), then we can represent the problem in terms of the loss of potential energy balanced by the gain in kinetic energy. Mathematically this is stated as

$$Mgh = \frac{1}{2} Mv^2$$

Now, we simply cancel out the mass from both sides and solve for velocity to obtain

$$v^2 = 2gh$$

Since h and ℓ are related, the answers are equivalent. Notice that in this representation the final expression is independent of the mass and the length of the incline. Thus, two equivalent representations of the same problem lead to the same final answer (as we would expect). The example chosen happened to be a very simple one. When faced with a really difficult problem, choosing the correct representation is the key to solving it correctly.

The second element in a problem consists of the **obstacles**. These are the factors that prevent you from immediately achieving your goal. In the first example, answering yes or no was hindered by the obstacle of needing to know how long the maneuver will take and how much additional distance will be covered by all three objects (your car, the truck, and the oncoming car). In the second example, involving the inclined plane, the obstacles depended on the representation chosen for the problem. In the motion representation, perhaps not knowing how long it takes to slide down the incline was an obstacle. For some students, however, this may not have been the case since different solvers may encounter different obstacles.

Deciding on the proper method (or path) for solving the problem most efficiently is the main focus of problem solving as a process. In the next section, we will discuss several methods that have been identified by researchers as typical of students solving physics problems. In studying them, you might recognize the way you solve problems, and if you are not satisfied with your results, you can begin to modify your procedures.

WHAT IS PROBLEM SOLVING?

Problem solving consists of the mental and behavioral operations necessary to reach the goal of a problem. When you begin working on a problem, you rely on stored knowledge about the particular problem, the subject area, and your past problem solving experience.

A set procedure that has been shown to solve a problem successfully is known as an *algorithm.* The algorithm may or may not make use of a formula. For example, the procedure for locating the image formed by a convex lens using a ray diagram is an example of an algorithm. On the other hand, given the mass and acceleration of an object, use of the formula $F = ma$ to calculate the net force acting on the object is also an example of an algorithm. Your reliance on algorithms, however, can sometimes hinder your ability to answer qualitative questions that require explanations.

When an algorithm is not used, several other rules of thumb or **heuristics** have been identified as being useful. These heuristics do not guarantee a successful conclusion, but they are often powerful and useful in moving toward the solution of a problem.

PROBLEM-SOLVING HEURISTICS

One simple heuristic that aids in the development of a successful representation of a problem is to redescribe it mentally. This means that when you read a problem for the first time, you use your knowledge of reading comprehension to understand the sentences and interpret them mentally in terms of the conceptual knowledge required. Often, if a problem is poorly worded or if you have difficulty understand the meaning of some words (called *lexical* difficulty), a redescription is often useful.

One way to redescribe a problem is to rephrase it. On an exam, your teacher will not be able to help you do this, but when reviewing, learning how to rephrase a problem may help you handle a similar situation better on a test. Another thing you can do is make a sketch of the situation. The old phrase, "A picture is worth a thousand words," is certainly true in problem solving. If the problem is especially theoretical, putting in simple numbers sometimes helps you to understand the relationship better.

A heuristic often employed in mathematical proofs is called *reductio ad absurdum* or "reducing to an absurdity." In this heuristic, you try to establish the truth of something by first assuming that it is false. Following the rules of deductive logic, you proceed to analyze the situation until you reach an absurdity. You can then conclude that the original assumption about the falsity of the proposition was wrong and that it is indeed true. Reductio ad absurdum is not often employed in the solution of physics problems, but it is a method of logic that can be useful if needed. In free response questions, if something needs to be proved correct, you might on occasion find this heuristic useful.

Beginning physics students often use a heuristic called *working backward.* This means that they start with the goals and try to associate them with the givens. Usually, this means searching for a formula that contains the desired quantity and seeing if the givens fit into it. For example, in the inclined plane problem, if the goal is to determine the final velocity at the bottom, then in the motion representation we establish a formula for final velocity that is independent of time and then use algebraic manipulation to obtain an answer. This method of operation is alright for simple problems, but for solving more difficult problems greater conceptual knowledge is required since there may not be a prior explicit relationship among the quantities allowing an equation to be written down immediately.

A heuristic commonly used by more experienced solvers is the method of *working forward.* With this method, you start with the givens and choose a representation that will link the givens together under the conditions of the laws of physics. From this broad viewpoint, mathematical relationships are established that can then be manipulated to obtain the solution. The energy representation of the same problem is an example of this methodology. By stating that the law of conservation of energy is basically responsible for the actions, we wrote down this condition mathematically and then solved for the velocity as a consequence of this law of physics.

Another heuristic to consider is called *hill climbing.* When you use hill climbing, you plan steps that will take you closer to your goal. In chess, for example, hill climbing is used when a move is made to put the opponent's king in check. In a physics problem, hill climbing can be used to answer a difficult question.

Suppose, for example, you are given a problem involving a suspension bridge. The goal is to determine if a given load on the supported roadway will hold. Starting with a sketch, looking at incremental force elements on a given portion of the cable, and then stating that the sum of all forces must remain zero is an example of hill climbing.

The name *hill climbing* derives from the technique used in hiking to find the way out of a mountainous forest. If you know that the top of the mountain is above the tree line, where you will be able to see your camp or a town, then you walk only in a direction that will lead you uphill (hence the name *hill climbing*).

Related to hill climbing is a heuristic known as *means-ends analysis.* In this heuristic, you break the problem down into smaller subgoals and proceed to solve these easier problems. By taking the solutions to the subgoals together, the hope is that they will allow the problem to be solved more efficiently. Problems that have sequential steps labeled a, b, c, and so on, use means-ends analysis explicitly. Writing a problem in which the three subgoals are suppressed is an implicit use of means-ends analysis since these steps must be solved first in order to reach the final goal.

The final heuristic we will discuss is called the *method of analogies.* Often, a difficult problem has some similarity to a previously solved problem. It is also possible to use an analogy when attempting to understand the behavior of a system. For example, certain types of electric circuits using alternating current behave analogously to a mechanical oscillating system. In fact, the equations governing these systems are remarkably similar. The equations for rotational kinematics are similar to those used for linear kinematics. The use of an analogy can often lead to a better understanding of the system. Caution must be used, however, since an analogy cannot replace the actual system.

3
VECTORS

COORDINATE SYSTEMS AND FRAMES OF REFERENCE

We begin our review of physics with the idea that all observations and measurements are made relative to a suitably chosen frame of reference. That is, when observations are made that will be the basis of future predictions, we must be careful to note from what point of view they are being made. As an example, if you are standing on the street and see a car driving by, then you observe the car and all of its occupants moving relative to you. However, the driver and occupants of the car may not consider themselves in motion relative to one another, but you appear to be moving backward relative to them.

If you were to get into a car, drive out to meet them, and travel with the same speed in the same direction right next to them, there would be no relative motion between the two cars. These different points of view are known as **frames of reference**, and they are a very important aspect of physics.

A **coordinate system** within a frame of reference is defined to be a set of reference lines that intersect at an arbitrarily chosen fixed point called the *origin*. In the **Cartesian coordinate system,** the reference lines are three mutually perpendicular lines designated x, y, and z (see Figure 1). The coordinate system must provide a set of rules for locating objects within this frame of reference. In the three-dimensional Cartesian system, if we define a plane containing the x, y coordinates (let's say a horizontal plane), then the z axis specifies a direction up or down.

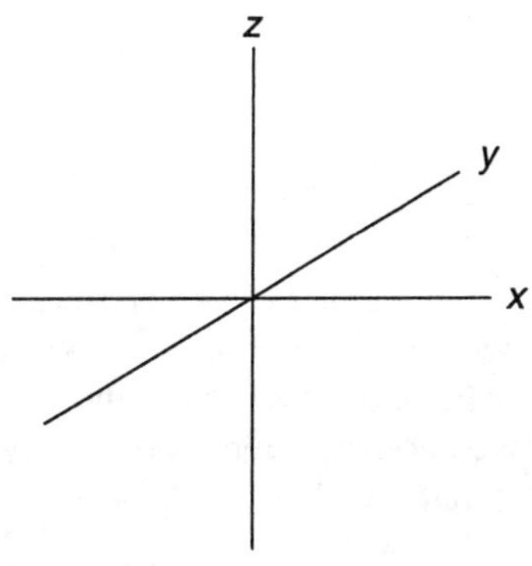

FIGURE 1.

It is often useful to compare observations made in two different frames of reference. In the example just mentioned, the motion of the occupants of the car was reduced to zero when we transformed our coordinate system to the car moving at constant velocity. In this case we say that the car is an **inertial frame of reference**. In such a frame, it is impossible to observe whether or not the reference frame is in motion if the observers are moving with it.

VECTORS

Another way of locating the position of an object in the Cartesian system is with a directed line segment or vector. If you draw an arrow, starting from the origin, to a point in space (see Figure 2), you have defined a position vector **R** whose magnitude is given by | **R** | and is equal to the linear distance between the origin and the point (x,y). The direction of the vector is given by the acute angle that the arrow makes with the positive x axis, which is designated by the Greek letter θ. Any quantity that has both magnitude and direction is called a **vector** quantity, and any quantity that has only magnitude is called a **scalar** quantity. Examples of vector quantities are force, velocity, weight, and displacement. Examples of scalar quantities are mass, distance, speed, and energy.

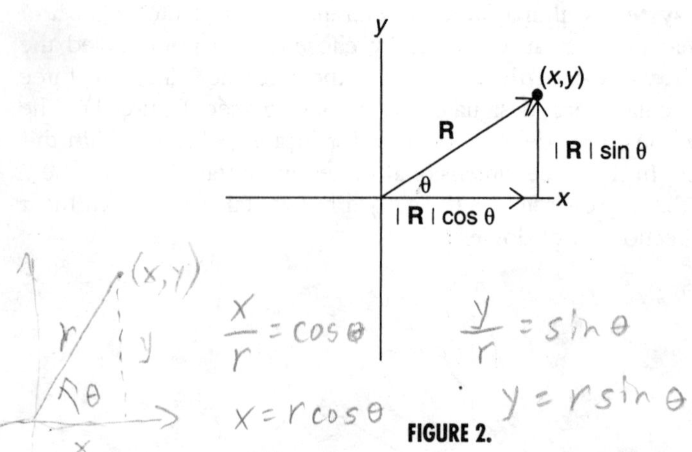

FIGURE 2.

If we designate the magnitude of the vector **R** to be r and the direction angle is given by θ, then we have an alternative coordinate system called the **polar coordinate** system. In two dimensions, the location of point (x,y) in the Cartesian system involves going x units horizontally and y units vertically. Additionally, we can show that (see Figure 2)

$$x = r \cos \theta$$
$$y = r \sin \theta$$

Having established this polar form for vectors, we can easily show that r and θ can be expressed in terms of x and y as

$$r^2 = x^2 + y^2$$

$$\tan\theta = \frac{y}{x}$$

The magnitude of **R** is then given by

$$|\mathbf{R}| = r = \sqrt{x^2 + y^2}$$

and the direction angle θ is given by

$$\theta = \arctan\left(\frac{y}{x}\right)$$

ADDITION OF VECTORS

Geometric Considerations

The ability to combine vectors is a very important tool of physics. From a geometric standpoint, the addition of two vectors is not the same as the addition of two numbers. When we state that given two vectors **A** and **B,** we wish to form a third vector **C** such that **A** + **B** = **C**, we must be careful to preserve the directions of the vectors relative to our chosen frame of reference.

One way to do this addition is by the construction of what is called a **vector diagram**. Vectors are identified geometrically by a directed arrow that has a "head" and a "tail." As a series of examples, consider a girl walking from her house a distance of 5 blocks east and then an additional 2 blocks east. How can we represent these displacements vectorially? First, we must choose a suitable scale to represent the magnitude lengths of the vectors. In this case, let us just call the scale 1 vector unit or just 1 unit to be equal to one block of distance. Let us also agree that east is to the right (and hence north is directed upward). Figure 3 is a vector diagram for this set of displacements.

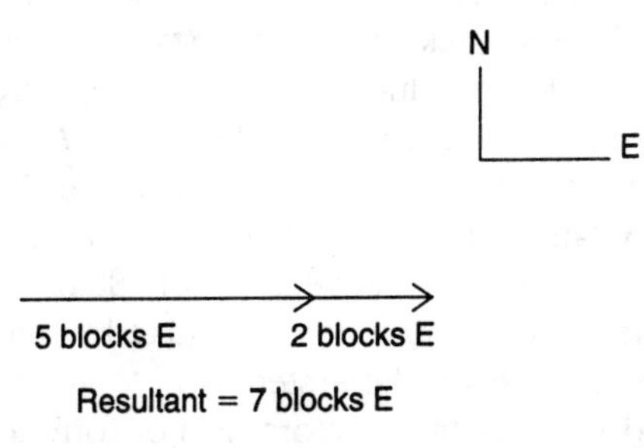

FIGURE 3.

Now suppose the girl walks 5 blocks east and then 2 blocks northeast (i.e., 45 degrees north of east). Since our intuition tells us that the final displacement will be in the general direction of north and east, we draw our vector diagram so that the two vectors are connected head to tail. For a similar scale, the vector diagram would look like Figure 4. The resultant is drawn from the tail of the first to the head of the second, forming a triangle (some texts use the *parallelogram method* of construction, which is equivalent). The direction of the resultant is measured from the tail of the vector inside the triangle.

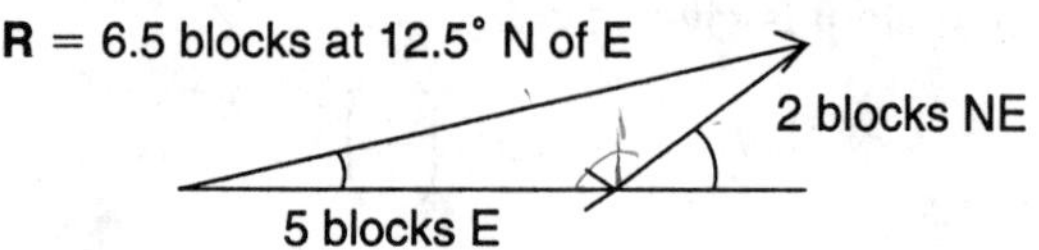

FIGURE 4. *Vector construction—45° displacement NE.*

Note that when such a construction is used, the magnitudes are in slight error from the actual calculated values (to be discussed in a later section).

Continuing our example, suppose the girl walks 5 blocks east and then 2 blocks north. Figure 5 is the vector construction diagram for this set of displacements.

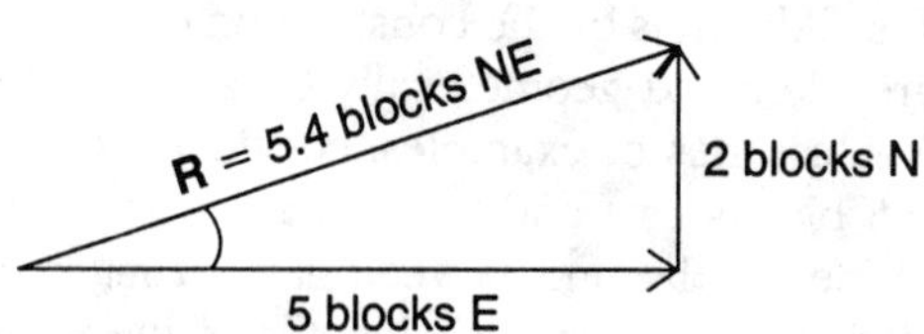

FIGURE 5. *Vector construction—right angle displacement.*

Finally, suppose the girl walks 5 block east and then 2 blocks west. Her final displacement will be 3 blocks east of the starting point.

From these examples, it is concluded that as the angle between two vectors increases, the magnitude of their resultant decreases. Also, the magnitude of the resultant between two vectors is a maximum when the vectors are in the same direction (at a relative angle of 0 degrees) and a minimum when they are in the opposite direction (at a relative angle of 180 degrees). It is important to remember that the vectors must be constructed head to tail and that a suitable scale must be chosen for the system.

It should be noted that the two vectors do not form a closed figure. The "missing link" is the resultant. The suggests an important concept. If a third vector were given that was equal in magnitude but opposite in direction to the predicted resultant, the vector sum of that resultant and the third vector would be zero. Stated another way, the vector sum of these three vectors would form a closed triangle of zero resultant. This would mean that the girl returned to her starting point, leaving a zero displacement. In physics, that third vector, equal and opposite to the resultant, is called the **equilibrant.** Figure 6 provides an illustration of this concept.

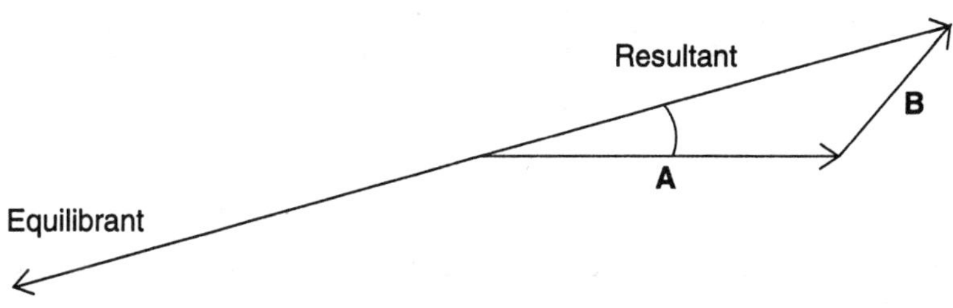

FIGURE 6.

Algebraic Considerations

In the preceding examples, the resultant between two vectors was constructed using a vector diagram. The magnitude of the resultant was the measured length of the vector drawn from the tail of the first to the head of the last. If we sketch such a situation, we form a triangle whose sides are related by the law of cosines and whose angles are related by the law of sines.

Consider the vector triangle in Figure 7 with arbitrary sides a, b, and c and corresponding angles A, B, and C.

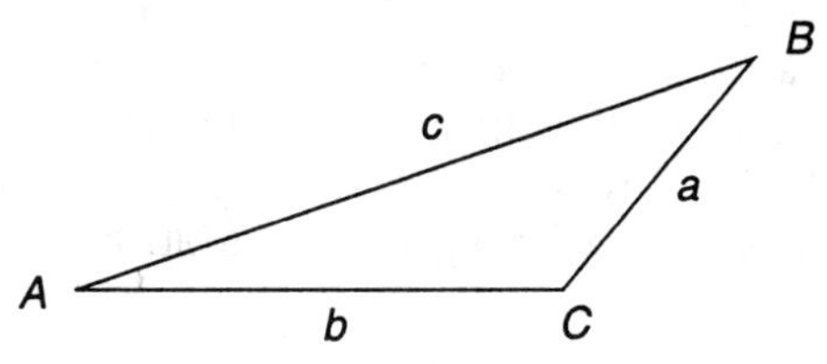

FIGURE 7.

The **law of cosines** states that

$$c^2 = a^2 + b^2 - 2ab \cos C$$

and the **law of sines** states that

$$\frac{a}{\sin A} = \frac{b}{\sin B} = \frac{c}{\sin C}$$

If we use the information given in the second example, we see that the vectors have the magnitudes and directions shown in Figure 8.

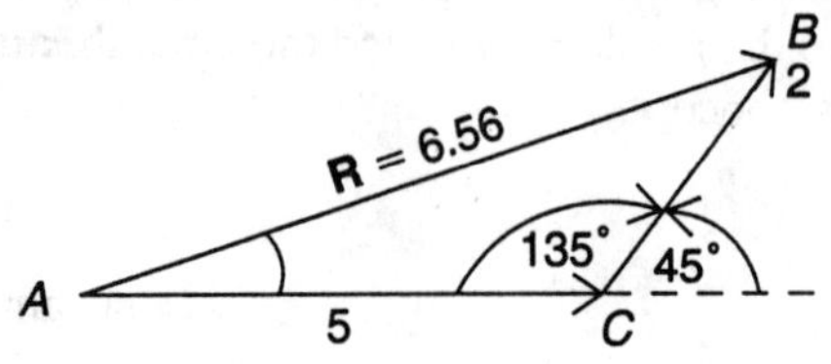

FIGURE 8.

Using the law of cosines, we obtain for the magnitude of the resultant (in "blocks"):

$$c = \sqrt{(2)^2 + (5)^2 - (2)(2)(5)(\cos 135)} = 6.56$$

Using the law of sines, the direction of the resultant is angle A:

$$\frac{6.56}{\sin 135} = \frac{2}{\sin A}$$

Angle A turns out to be equal to 12.45 degrees north of east.

Addition of Multiple Vectors

The discussion about the equilibrant leads to the idea of the addition of multiple vectors. As long as the vectors are constructed head to tail, multiple vectors can be added in any order. If the resultant is zero, the diagram constructed will be a closed geometric figure. This was shown in the case of three vectors forming a closed triangle. Thus, if the girl walks 5 blocks east, 2 blocks north, and 5 blocks west, the vector diagram will look like Figure 9, with the resultant displacement being equal to 2 blocks north.

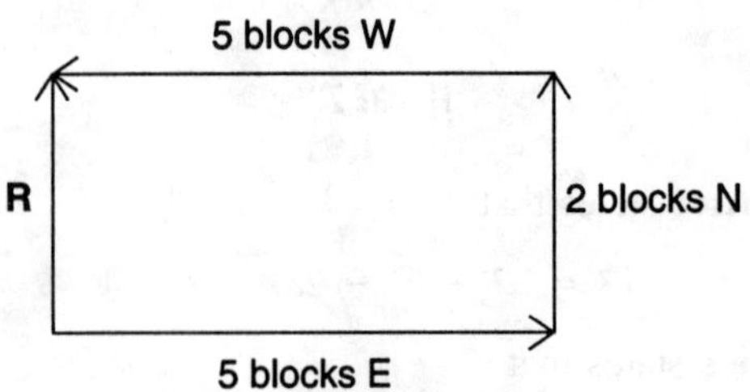

FIGURE 9.

SUBTRACTION OF VECTORS

The difference between two vectors can be understood by considering the process of displacement. Suppose we identify two points (x,y) and (u,v) in a coordinate system as shown in Figure 10. Each point can be located within the coordinate system by a position vector drawn from the origin to that point. The displacement of an object from point A to point B is represented by the vector drawn from A to B. This vector is defined to be the difference between the two position vectors and is sometimes called $\Delta\mathbf{R}$ such that $\Delta\mathbf{R} = \mathbf{R}_B - \mathbf{R}_A$.

From our understanding of vector diagrams, you can see that $\mathbf{R}_1$ and $\Delta\mathbf{R}$ are connected head to tail. Thus, we can see that $\mathbf{R}_A + \Delta\mathbf{R} = \mathbf{R}_B$.

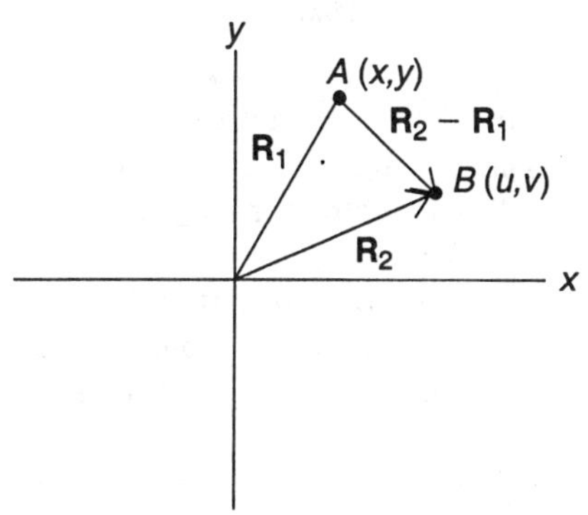

FIGURE 10.

ADDITION METHODS USING THE COMPONENTS OF VECTORS

Any single vector in space can be **resolved** into two perpendicular components in a suitably chosen coordinate system. The methods of vector resolution were actually discussed on page 16, but we can review them here. Given a vector (see Figure 11) representing a displacement of 100 m northeast (i.e., making a 45-degree angle with the positive x axis), we can observe that it is equivalent to the sum of x and y (or horizontal and vertical) components that can be geometrically or algebraically determined.

Geometrically, if we were to draw the 100-m vector to scale at the correct angle, then projecting a perpendicular line down from the head of the given vector to the x axis would allow us to construct the two perpendicular components. Algebraically, we see that if we have a given vector $\mathbf{R}$, then in the x direction we have $\mathbf{R}_x = \mathbf{R}\cos\theta$ and $\mathbf{R}_y = \mathbf{R}\sin\theta$.

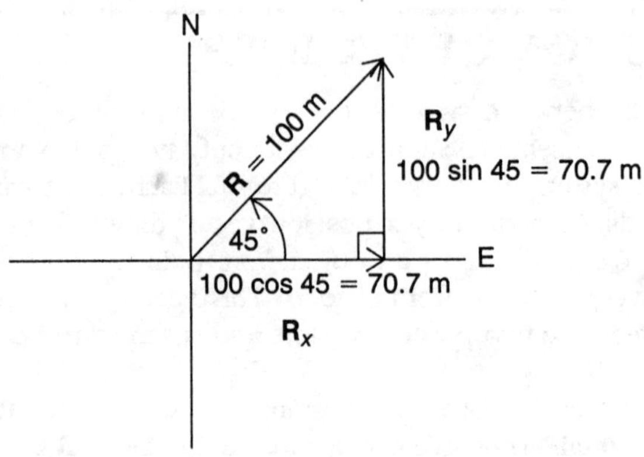

FIGURE 11.

This method of vector resolution can be useful when adding two or more vectors. Since all vectors in the same direction add up numerically and all vectors in opposite directions subtract numerically, if we are given two vectors, then the resultant between them can be found from the addition and subtraction of the respective components. In other words, if vector **A** has components $\mathbf{A}_x$ and $\mathbf{A}_y$ and if vector **B** has components $\mathbf{B}_x$ and $\mathbf{B}_y$, then the resultant vector **C** will have components $\mathbf{C}_x$ and $\mathbf{C}_y$ such that $\mathbf{C}_x = \mathbf{A}_x + \mathbf{B}_x$ and $\mathbf{C}_y = \mathbf{A}_y + \mathbf{B}_y$.

An example of this method is shown in Figure 12. Suppose we have two vectors **A** and **B** that represent tensions in two ropes applied at the origin of the coordinate system (let's say it's a box). The resultant force of these tensions (which are vector quantities) can be found by determining the respective x and y components of the given vectors.

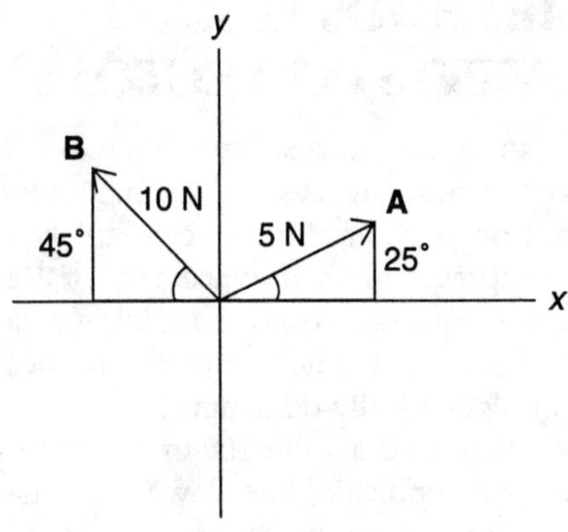

FIGURE 12.

From the diagram, we see that $\mathbf{A}_x = 5 \cos 25 = 4.53$ N and $\mathbf{A}_y = 5 \sin 25 = 2.11$ N. For the second vector, we have $\mathbf{B}_x = -10 \cos 45 = -7.07$ N (since it is pointing left) and $\mathbf{B}_y = 10 \sin 45 = 7.07$ N. Therefore, $\mathbf{C}_x = 4.53 \text{ N} - 7.07 \text{ N} = -2.54$ N and $\mathbf{C}_y = 2.11 \text{ N} + 7.07 \text{ N} = 9.18$ N. The magnitude of the resultant vector **C** is given by $|\mathbf{C}| = \sqrt{(-2.54)^2 + (9.18)^2} = 9.52$ N. The angle that vector **C** makes with the negative x axis is obtained from arctan $\theta = \mathbf{C}_x/\mathbf{C}_y = -3.614$, and so $\theta = 74.53$ degrees.

It is interesting to point out that vectors obey the commutative and associative laws for addition. That is, given two vectors **A** and **B**, then **A** + **B** + **A**. Moreover, if we introduce a third vector **C**, then **A** + (**B** + **C**) = (**A** + **B**) + **C**.

PROBLEM-SOLVING STRATEGIES

Vectors are a fundamental tool of physics. They provide a representation for physical quantities that are changing in both magnitude and direction. It is therefore essential that you be able to understand the way in which vectors combine algebraically and geometrically. When solving a vector problem, you should be sure to

1. Understand the frame of reference for the situation.
2. Select an appropriate coordinate system. Cartesian systems do not necessarily have to be vertical and horizontal (a mass sliding down an inclined plane is an example of a rotated system with the "x axis" parallel to the incline).
3. Pick an appropriate scale for constructing your vector diagram. For example, in a problem with displacement, a scale of 1 cm = 1 m might be appropriate.
4. Recognize the given orientation of the vectors as they correspond to geographic directions (N, E, S, W). Be sure to maintain the same directions as you draw your vector triangle or parallelogram. Use a protractor and a metric ruler.
5. Connect vectors head to tail and connect your resultant from the tail of the first vector to the head of the last vector.
6. Identify and determine the components of vectors. In most cases, vector problems can be simplified using components.

PRACTICE PROBLEMS FOR CHAPTER 3

Thought Problems

1. Give a geometric explanation of the following statement: Three vectors that add up to zero must be coplanar (i.e., they must lie in the same plane).

2. (a) Is it necessary to specify a coordinate system when adding two vectors?
   (b) Is it necessary to specify a coordinate system when forming the components of a vector?

3. Can a vector have a zero magnitude if one of its components is nonzero?

Multiple-Choice Problems

1. A vector is given by its components $\mathbf{A}_x = 2.5$ and $\mathbf{A}_y = 7.5$. What angle does the vector **A** make with the positive x axis?
   (A) 72 degrees
   (B) 18 degrees
   (C) 25 degrees
   (D) 50 degrees
   (E) 75 degrees

2. Which pair of vectors can produce a resultant of 35?
   (A) 15 and 15
   (B) 20 and 20
   (C) 30 and 70
   (D) 20 and 60
   (E) 20 and 70

3. A vector has a magnitude of 17 units and makes an angle of 20 degrees with the positive x axis. The magnitude of the horizontal component of this vector is _____ units.
   (A) 16
   (B) 4.1
   (C) 5.8
   (D) 50
   (E) 12

4. As the angle between a given vector and the horizontal axis increases from 0 to 90 degrees, the magnitude of the vertical component of this vector _____.
   (A) decreases
   (B) increases and then decreases
   (C) decreases and then increases
   (D) increases
   (E) remains the same

5. Vector **A** has a magnitude of 10 units and makes an angle of 30 degrees with the positive x axis. Vector **B** has a magnitude of 25 units and makes an angle of 50 degrees with the negative x axis. What is the magnitude of the resultant between these two vectors?
   (A) 20 units
   (B) 35 units

(C) 15 units
(D) 45 units
(E) 25 units

6. Two concurrent vectors have magnitudes of 3 units and 8 units. The difference between these vectors is 8 units. The angle between these two vectors is _____ degrees.
   (A) 34
   (B) 56
   (C) 79
   (D) 113
   (E) 127

7. Which of the following sets of displacements have equal resultants when performed in the order given?
   I: 6 m east, 9 m north, 12 m west
   II: 6 m north, 9 m west, 12 m east
   III: 6 m east, 12 m west, 9 m north
   IV: 9 m north, 6 m west, 12 m west
   (A) I and IV
   (B) I and II
   (C) I, III, and IV
   (D) I, II, and IV
   (E) II and IV

8. Which vector represents the direction of the sum of the two concurrent vectors shown.

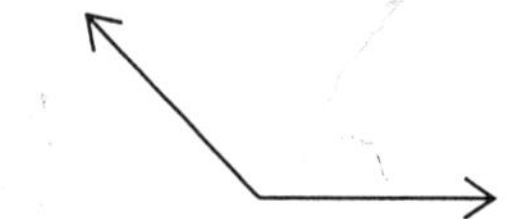

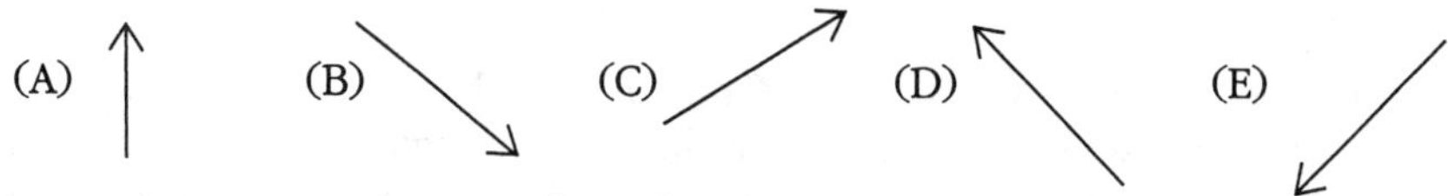

9. Three forces act concurrently on a point *P* as shown. Which of the choices represents the direction of the resultant force on point *P*?

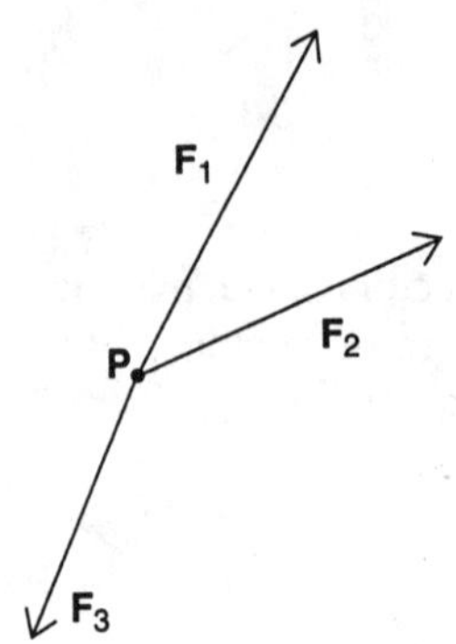

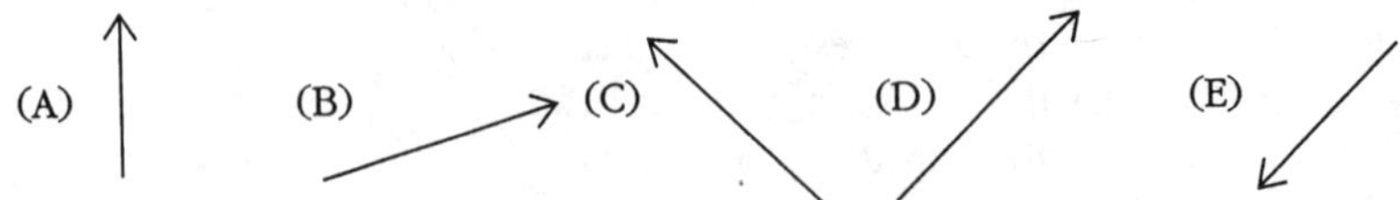

10. On a baseball field, first base is about 30 m away from home plate. Suppose a batter gets a hit and runs toward first base. He runs 3 m past the base and then runs back to stand on it. The magnitude of his final displacement from home plate is ____ m.
(A) 27
(B) 30
(C) 33
(D) 36
(E) 40

B

Free-Response Problems

1. Using vector methods, prove that the line segment joining the midpoints of a triangle is parallel to the third side and equal to one half its magnitude.
2. Two vectors **A** and **B** are concurrent and attached at the tails with an angle θ between them. Given that each vector has components $\mathbf{A}_x$, $\mathbf{A}_y$ and $\mathbf{B}_x$, $\mathbf{B}_y$, respectively, use the law of cosines to show that $\cos\theta = (\mathbf{A}_x\mathbf{B}_x + \mathbf{A}_y\mathbf{B}_y)/|\mathbf{A}|\,|\mathbf{B}|$.
3. Suppose we have a rectangular Cartesian coordinate system that is rotated counterclockwise about the origin through an angle θ. Show that a point that has coordinates (x,y) in the original coordinate system now has coordinates (x',y') in the rotated system such that

$$x' = x\cos\theta + y\sin\theta$$
$$y' = -x\sin\theta + y\cos\theta$$

SOLUTIONS TO PRACTICE PROBLEMS

Thought Problems

1. Any two vectors can be added geometrically using the parallelogram method. A parallelogram is a plane figure. The resultant of the two vectors is a diagonal of the parallelogram, and so the two original vectors and the resultant are all coplanar. Now, if we take our original two vectors and introduce a third vector, equal to and opposite the resultant of the first two, then the vector sum of the three vectors will be zero and they will all be coplanar.

2. (a) Using the logic of Problem 1, we can add two vectors geometrically regardless of how they are orientated in any coordinate system. Therefore, to find their resultant (i.e., to add two vectors), no coordinate system is necessary.
   (b) The components of a vector must lie along the principal axes of some coordinate system by definition. Therefore, a coordinate system must be specified in order to determine a vector's components.

3. The magnitude of a vector is a positive number formed by taking the square root of the sum of the squares of the magnitudes of its components. If one component is nonzero, the magnitude of the whole vector is necessarily nonzero also.

Multiple-Choice Problems

1. **A** The angle for the vector to the positive x axis is given by $\tan\theta = \mathbf{A}_y / \mathbf{A}_x = 7.5 / 2.5$. Thus, $\tan\theta = 3$ and $\theta = 71.5$ degrees or, rounded off, 72 degrees.

2. **B** Two vectors have a maximum resultant whose magnitude is equal to the numerical sum of the vector magnitudes when the angle between them is 0 degrees. The resultant is a minimum at 180 degrees, and the magnitude is equal to the numerical difference between the magnitudes. Using this logic, only the pair (20,20) produces a maximum and minimum set that could include the given value of 35 if the angle were specified. All the others have maximum and minimum resultants that do not include 35 in their range.

3. **A** The formula for the horizontal component of a vector **A** is $\mathbf{A}_x = A\cos\theta$ where A is the magnitude of **A**. Using the given values, $\mathbf{A}_x = 17\cos 20 = 15.9$ units.

4. **D** The vertical component of a vector is proportional to the sine of the angle which increases to a maximum at 90 degrees.

5. **E** Form the components of each vector and then add the respective components remembering the sign conventions; right and upward are positive values. Thus, we have $\mathbf{A}_x = 10\cos 30 = 8.66$ and $\mathbf{A}_y = 10\sin 30 = 5$. $\mathbf{B}_x = -25\cos 50 = -16.06$, and $\mathbf{B}_y = 25\sin 50 = 19.15$. The

components of the resultant are $\mathbf{C}_x = \mathbf{A}_x + \mathbf{B}_x = 8.66 - 16.06$ and so $\mathbf{C}_x = -7.4$ units and $\mathbf{C}_y = \mathbf{A}_y + \mathbf{B}_y = 5 + 19.15 = 24.15$. The magnitude of this resultant vector is given by $|\mathbf{C}| = \sqrt{\mathbf{C}_x^2 + \mathbf{C}_y^2} = \sqrt{(7.4)^2 + (24.15)^2}$. This implies that $\mathbf{C} = 25$ units.

6. **C** Use the law of cosines, where $a = 3$, $b = 8$, and $c = 8$, and solve for $\cos\theta$. We therefore have $(8)^2 = (3)^2 + (8)^2 - (2)(3)(8)\cos\theta$. Solving, $\cos\theta = 0.1875$ and $\theta = 79$ degrees.

7. **C** Vectors can be added in any order. The only requirement is that they have the same magnitude and direction as they are shuffled. A look at the four choices indicates that I, III, and IV are the same set of vectors listed in a different order. These three will produce the same resultant.

8. **C** Sketch the vectors head to tail as if forming a vector triangle in construction. The resultant is drawn from the tail of the first to the head of the second. Choosing the horizontal vector as the first one, choice (C) is the general direction of the resultant.

9. **B** Sketch the vectors in any order as shown. The resultant is drawn following the instructions in Problem 2 (from the tail of the first to the head of the last).

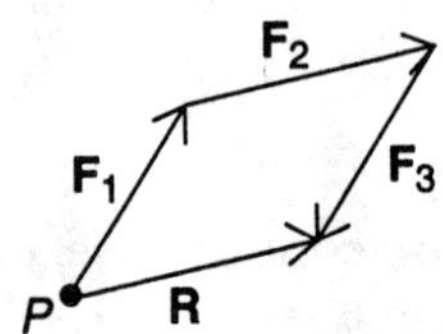

10. **B** The displacement is in the direction of first base and is equal in magnitude to the straight-line distance between home plate and first base (30 m).

Free-Response Problems

1. A typical triangle is shown here. Let the sides be represented by vectors such that **DB** + **BE** = **DE** and **AB** + **BC** = **AC**. Since our line joins the midpoints, we can see that $\frac{1}{2}$ (**AB**) + $\frac{1}{2}$ (**BC**) = **DE**, and therefore 2(**DE**) = **AB** + **BC** = **AC** and the proof is complete.

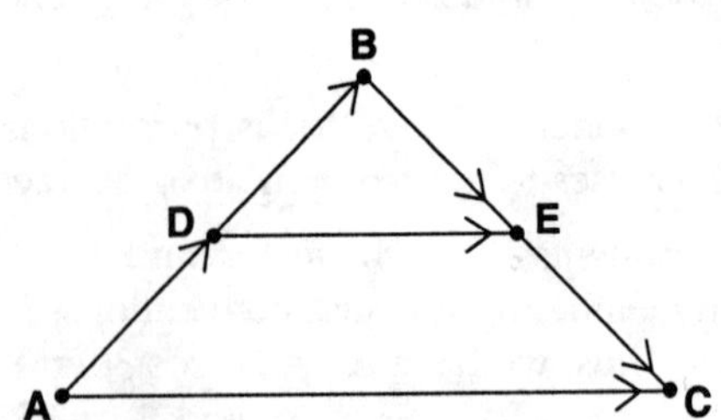

2. The situation is shown here. The third side of the vector triangle is **A** − **B**.

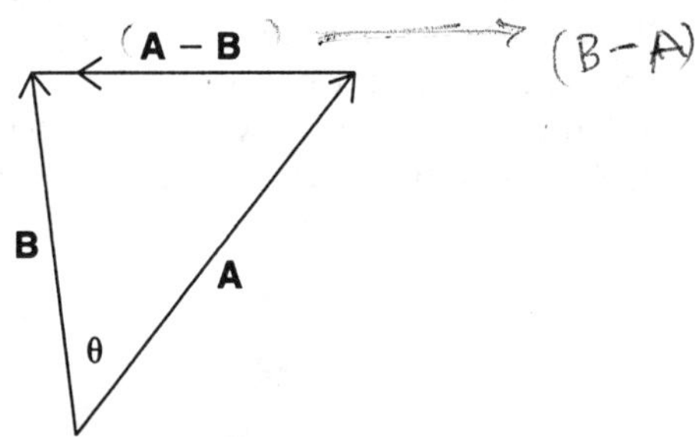

From the rules for vector subtraction, the components of **A** − **B** are $\mathbf{A}_x - \mathbf{B}_x$ and $\mathbf{A}_y - \mathbf{B}_y$. The law of cosines states that for the given triangle,

$$|\mathbf{A} - \mathbf{B}|^2 = |\mathbf{A}|^2 + |\mathbf{B}|^2 - 2|\mathbf{A}||\mathbf{B}|\cos\theta$$

Using our definition of magnitudes and components, we have

$$(\mathbf{A}_x - \mathbf{B}_x)^2 + (\mathbf{A}_y - \mathbf{B}_y)^2 = \mathbf{A}_x^2 + \mathbf{A}_y^2 + \mathbf{B}_x^2 + \mathbf{B}_y^2 - 2|\mathbf{A}||\mathbf{B}|\cos\theta$$

Expanding the left-hand side and canceling like terms on both sides, we are left with

$$-2\mathbf{A}_x\mathbf{B}_x - 2\mathbf{A}_y\mathbf{B}_y = -2\,|\mathbf{A}||\mathbf{B}|\cos\theta$$

Again, canceling like terms on both sides leaves us with the final expression as we solve for cos θ:

$$\cos\theta = \frac{\mathbf{A}_x\mathbf{B}_x + \mathbf{A}_y\mathbf{B}_y}{|\mathbf{A}||\mathbf{B}|}$$

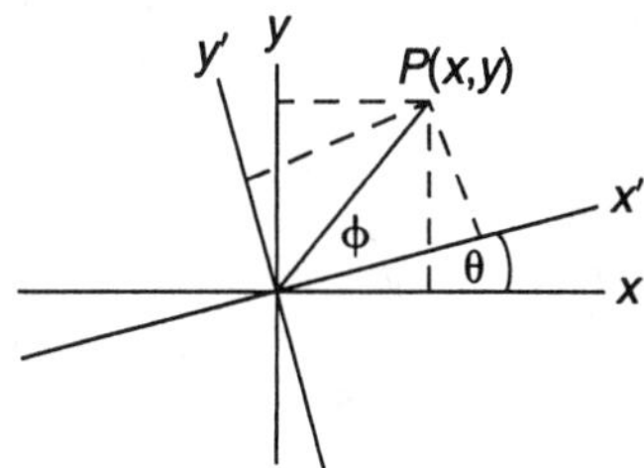

3. From the diagram for Problem 2, we can see that the vector drawn to the point $P(x, y)$ makes an angle ϕ with the original x axis. Let the magnitude of this vector **P** be R. It will have components $x = R\cos\phi$ and $y = R\sin\phi$. The coordinate system is rotated counterclockwise through an angle θ as shown. Let the vector **P** make an angle α with the new x axis, the x' axis, such that $\alpha = \phi - \theta$. Thus, in the new rotated system, the vector **P** has components given by $x' = R\cos\alpha$ and $y' = R\sin\alpha$. This means that

$x' = R\cos(\phi - \theta)$ and $y' = R\sin(\phi - \theta)$. Using the trigonometric identities for the difference between two angles, we have $x' = R\cos(\phi - \theta) = R\cos\phi\cos\theta + R\sin\phi\sin\theta$ and $y' = R\sin(\phi - \theta) = R\sin\phi\cos\theta - R\cos\phi\sin\theta$. But $x = R\cos\phi$ and $y = R\sin\phi$, and so making the substitution and rearranging, we arrive at our transformation equations: $x' = x\cos\theta + y\sin\theta$ and $y' = -x\sin\theta + y\cos\theta$.

4
ONE-DIMENSIONAL MOTION

INTRODUCTION

Motion involves the change in position of an object over time. When we observe an object moving, it is always with respect to a frame of reference. Since motion occurs in a particular direction, it is dependent on the choice of coordinate system for that frame of reference. This leads us to the conclusion that motion has a vector nature which must be taken into account when we analyze how something moves.

In this chapter, we shall confine our discussion to motion in one dimension only. In physics, the study of motion is called **kinematics**. No hypotheses are made in kinematics concerning *why* something moves. Kinematics is a completely descriptive study of *how* something moves.

AVERAGE AND INSTANTANEOUS MOTION

Motion can take place either uniformly or nonuniformly. By this we mean that if an object is moving relative to a frame of reference, then the displacement changes over equal periods of time might be equal or unequal.

If we consider the actual distance traveled by the object in Figure 1 along some arbitrary path, then we are dealing with a scalar quantity. The displacement vector **AB**, however, is directed along the line connecting points *A* and *B* (whether or not it is the actual route taken). Thus, when a baseball player hits a home run and runs around the bases, the player may travel a distance of 360 ft (since it is 90 ft between bases), but the final displacement is zero (because the player starts and ends up at the same place).

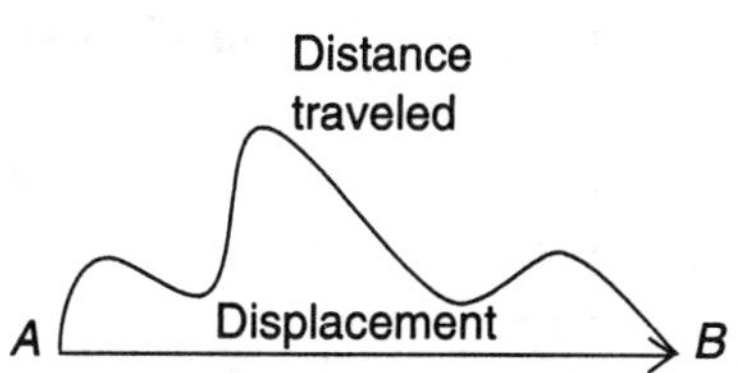

FIGURE 1. *Displacement versus distance.*

If we are given the displacement vector of an object for a period of time Δt, then we define the **average velocity** to be $\bar{\mathbf{v}} = \Delta\mathbf{x}/\Delta t$. This is a vector quantity since directions are specified. Numerically, we think of the average speed as being the ratio of the total distance traveled to the total elapsed time. The units of velocity are m/s.

If we are interested in the velocity at any instant in time, then we can define the **instantaneous velocity** to be the velocity $\mathbf{v}$ as determined at any precise time. In a car, the speedometer registers instantaneous speed, which can become velocity if we take into account the direction of motion. If the velocity is constant, then the average and instantaneous velocities are equal and we can simply write $\Delta\mathbf{x} = \mathbf{v}\,\Delta t$.

It is possible that the observer making the measurements of motion is likewise in motion relative to the earth. If this is true, then we must consider the relative velocity between the two systems. For example, if two objects, *a* and *b*, are moving in the same direction, then the relative velocity between them is given by $\mathbf{v}_{rel} = \mathbf{v}_a - \mathbf{v}_b$.

UNIFORMLY ACCELERATED MOTION

When an object is moving with a constant velocity such that its position is taken to be zero when it is first observed, then a graph of the expression $\mathbf{x} = \mathbf{v}t$ represents a direct relationship between position and time (see Figure 2). Since the line is straight, the constant slope (in which $\mathbf{v} = \Delta\mathbf{x}/\Delta t$) indicates that the velocity is constant throughout the time interval. If we were to plot the velocity versus time graph for this motion, it might look like Figure 3. Notice that for any time t the area under the graph equals the displacement.

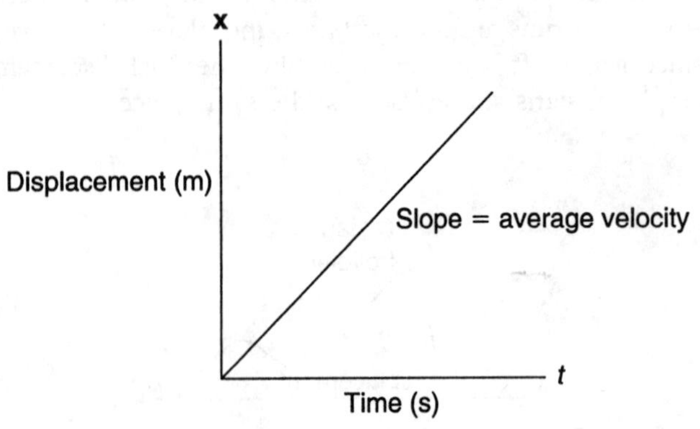

FIGURE 2.

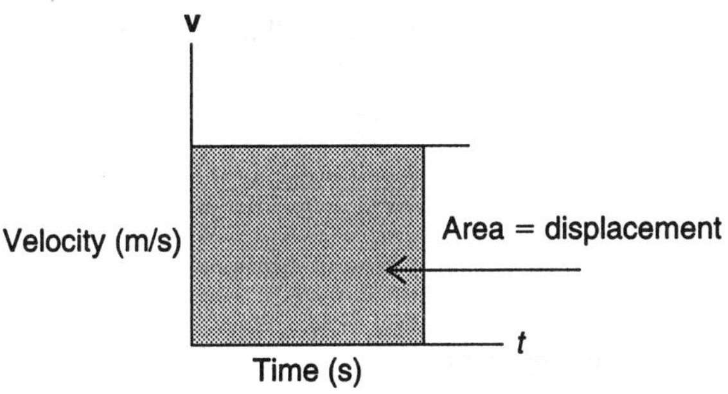

FIGURE 3.

It is possible that the object is changing direction while maintaining a constant speed (uniform circular motion is an example of this) or changing both speed and direction. In any case, if the velocity is changing, then we say that the object is **accelerating**. If the velocity is changing uniformly (at a constant rate), then we say that the object has *uniform acceleration.* In this case, a graph of velocity versus time would look like Figure 4.

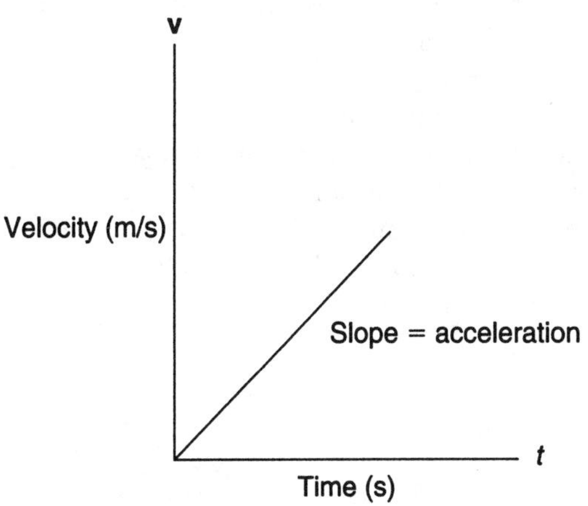

FIGURE 4.

The displacement from $t = 0$ to any other time is equal to the area of the triangle formed. However, between any two intermediary times the resulting figure is a trapezoid. If we make several measurements, then the displacement versus time graph for uniformly accelerated motion is a parabola starting from the origin (if we have the initial conditions that when $t = 0$, $\mathbf{x} = 0$ and $\mathbf{v} = 0$; see Figure 5).

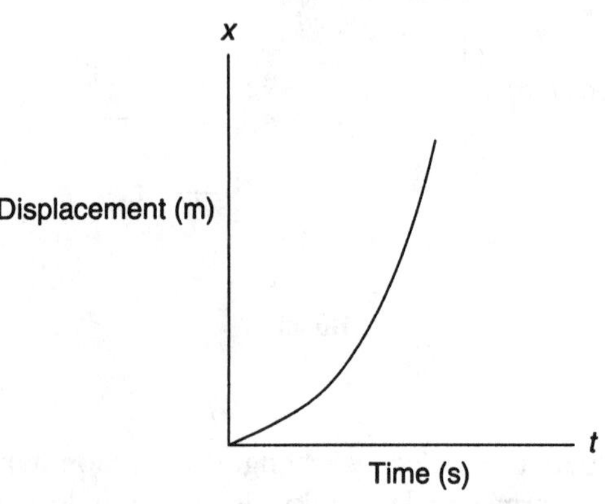

FIGURE 5.

The slope of the velocity time graph is defined to be the average acceleration in units of m/s^2. We can now write for the average acceleration $\bar{\mathbf{a}} = \Delta\mathbf{v}/\Delta t$. If the acceleration is taken to be constant in time, the expression for average acceleration can be written in a form that allows us to calculate the instantaneous final velocity after a period of acceleration has taken place. In other words, $\Delta\mathbf{v} = \mathbf{a}t$ (if we start our time interval from zero). If we define $\Delta\mathbf{v}$ to be equal to the difference between an initial and a final velocity, we can arrive at the fact that $\mathbf{v}_f = \mathbf{v}_i + \mathbf{a}t$. This means that if we were to plot velocity versus time for uniformly accelerated motion starting with a nonzero initial velocity, we would obtain a graph that looks like Figure 6.

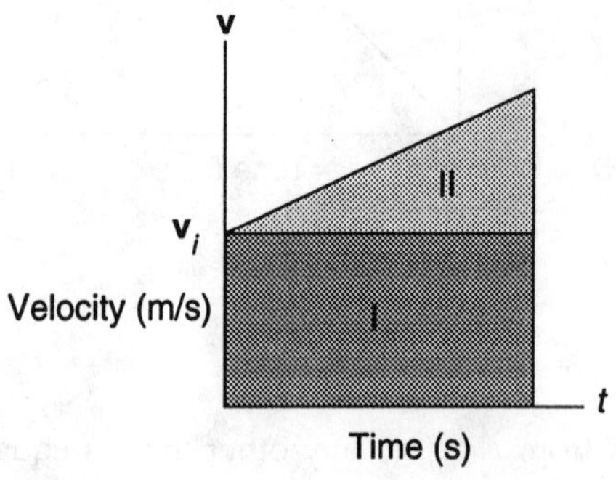

FIGURE 6.

The displacement during any period of time is equal to the total area under the line. In this case, the total area will be the sum of two areas: one a triangle and the other a rectangle. The area of the rectangle for some time t is just $\mathbf{v}_i t$. The area of the triangle is one half the base times the height. The "base" in this case is the time period t, and the "height" is the change in velocity $\Delta\mathbf{v}$. Therefore, the area of the triangle is $\frac{1}{2}\,\Delta\mathbf{v}\,t$. If we recall the definition of $\Delta\mathbf{v}$ and that $\mathbf{v}_f = \mathbf{v}_i + \mathbf{a}t$ then we have the following formula for the distance traveled during uniformly accelerated motion starting with an initial velocity (assuming we start from the origin):

$$\mathbf{x} = \mathbf{v}_i t + \frac{1}{2}\mathbf{a}t^2$$

This analysis suggests an alternative method of determining the average velocity of an object during uniformly accelerated motion. When we start, for example, with a velocity of 10 m/s and accelerate uniformly for 5 s at a rate of 2 m/s^2, then the average velocity during that interval is the average of 10, 12, 14, 16, 18, and 20 m/s, which is 15 m/s. Therefore, we can simply write

$$\overline{\mathbf{v}} = \frac{\mathbf{v}_i + \mathbf{v}_f}{2}$$

Occasionally, a problem in kinematics does not explicitly mention the time involved. It would be nice to have a formula for velocity that does not involve the time factor, and we can derive one from all the other formulas. Since the preceding equation relates the average velocity to the initial and final velocities (for uniformly accelerated motion), we can write our displacement formula as

$$\mathbf{x} = \frac{\mathbf{v}_i + \mathbf{v}_f}{2}\,t$$

Now, since $\mathbf{v}_f = \mathbf{v}_i + \mathbf{a}t$, we can express the time as $t = (\mathbf{v}_f - \mathbf{v}_i)/\mathbf{a}$. Therefore,

$$\mathbf{x} = \frac{\mathbf{v}_i + \mathbf{v}_f}{2}\,\frac{\mathbf{v}_f - \mathbf{v}_i}{a}$$

$$= \frac{\mathbf{v}_f^2 - \mathbf{v}_i^2}{2\mathbf{a}}$$

This expression is often given in the form $\mathbf{v}_f^2 = \mathbf{v}_i^2 + 2\,\mathbf{a}x$.

ACCELERATED MOTION DUE TO GRAVITY

Since velocity and acceleration are vector quantities, we need to consider the algebraic conventions accepted for dealing with various directions. For example, we usually agree to consider motion upward or to the right as positive, and motion downward or to the left as negative. In this way, negative velocity implies backward motion (since negative speed doesn't make sense). Negative acceleration, on the other hand, implies that the object is slowing down.

On page 32, we considered the case of uniformly accelerated motion. A naturally occurring situation in which acceleration is constant involves motion due to gravity. Since gravity acts in the downward direction, its acceleration, common to all masses, is likewise taken to be downward or negative. Notice that this may or may not imply deceleration. An object dropped from rest accelerates as time passes, but in the downward or negative direction.

In physics, the acceleration due to gravity is generally taken to be -9.8 m/s^2 and is represented by **g**. As a vector quantity, **g** is negative and the direction of motion is reflected in either the position or velocity equation. For example, if an object is dropped, the displacement in the y direction is given by

$$\Delta y = -\frac{1}{2}\mathbf{g}t^2$$

If, however, the object were thrown with some initial velocity, then we could write

$$\Delta \mathbf{y} = \mathbf{v}_i t + \frac{1}{2}\mathbf{g}t^2$$

GRAPHICAL ANALYSIS OF MOTION

We have seen that much information can be obtained if we consider the graphical analysis of motion. If complex changes in motion are taking place, then visualization may provide a better understanding of the physics than algebra. The techniques of graphical analysis are as simple as determining slopes and areas. For example, we already know that for uniformly accelerated motion, the graph of distance versus time is a parabola. Since the slope is changing, the instantaneous velocity can be approximated at a point by finding the slope of a tangent line drawn to a given point on the curve (see Figure 7).

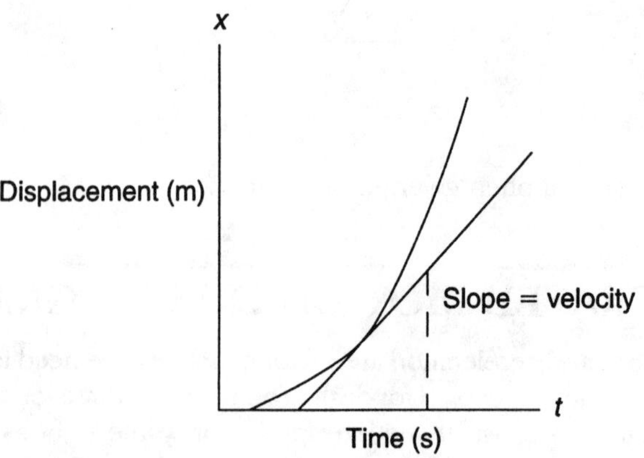

FIGURE 7.

What would happen if the object accelerated from rest, maintained a constant velocity for a while, and then slowed down to a stop? Using what we know about velocity and acceleration for displacement versus time graphs, the motion might look like Figure 8.

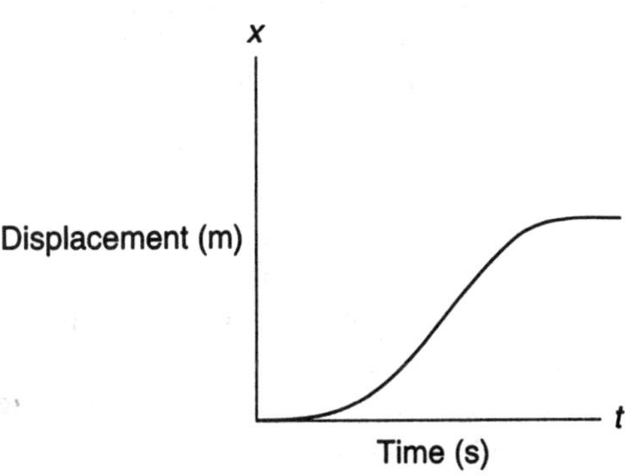

FIGURE 8.

We can apply many instances of motion to graphs. In the case of a changing velocity, consider the graph of velocity versus time for an object thrown upward into the air, reaching its highest point, changing direction, and then accelerating downward. This motion has a constant downward acceleration which at first acts to slow the object down but later acts to speed it up. Figure 9 is a graph of this motion.

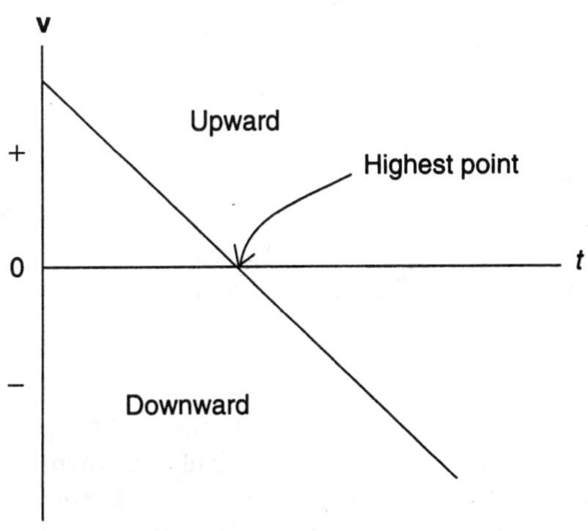

FIGURE 9.

PROBLEM-SOLVING STRATEGIES

Since all motion is relative, it is very important to always ask yourself the following question: From what frame of reference am I viewing the situation? Also, keep in mind that whenever you solve a physics problem, you consider the assumptions being made explicitly or implicitly about the moving object or objects. In this way, your ability to keep track of what is relevant to your solution path will be maintained. In addition, it is suggested that you

1. Identify all the goals and givens in the problem. Recall from Chapter 2 that the goals may be explicit or implicit. If a question is asked based on a decision or prediction, then be sure you understand the algebraic requirements necessary to make your decision.
2. Consider the "meaningfulness" of your solution and process. Does your answer or methodology make sense?
3. Choose a proper coordinate system and remember the sign conventions for algebraically treating vector quantities. In addition, make sure you understand the nature of the concepts being discussed (i.e., are they vectors or scalars?).
4. Make sure you carry proper international units throughout your calculations and that the units are included in your final answer. Try dimensional analysis to see if the answer makes sense. Perhaps your answer looks too large or small because you're using the wrong units.
5. Make a sketch of the situation if one is not provided.
6. If you are interpreting a graph, be sure you understand the interrelationship among all the kinematic variables discussed (slopes, areas, etc.).
7. If you are constructing a graph, be sure to label both axes completely, choose a proper scale for each axis, and draw neatly and clearly.
8. Try different problem-solving heuristics if you get stuck on a difficult problem.

Sample Problem

A ball is thrown straight up into the air and then caught by a person 5 m above the ground after it has been in the air for 9 s. What maximum height did the ball reach?

Solution

The equation $y = \mathbf{v}_i t - \frac{1}{2}\mathbf{g}t^2$ represents the vertical position of the ball above the ground for any time t. Since the ball is thrown upward, the initial velocity is positive while the acceleration of gravity is always directed downward (and hence algebraically negative). Thus, we can use this equation to find the initial velocity when $y = 5$ m and $t = 9$ s. Substituting these values, as well as the magnitude of $\mathbf{g} = 9.8\ \text{m/s}^2$, we have $\mathbf{v}_i = 44.66$ m/s.

Now to find the maximum height we note that when the ball reaches its maximum height, its speed becomes zero because its velocity is changing direction. Thus, since we do not know how long it takes to rise to its maximum height (we could determine this value if desired), we can use the formula $\mathbf{v}_f^2 - \mathbf{v}_i^2 = -2\,\mathbf{g}y$. When the final velocity equals zero, the value of y is equal to the maximum height. Using our answer for the initial velocity and the known value of **g**, we find that $y_{max} = 101.7$ m.

PRACTICE PROBLEMS FOR CHAPTER 4

Thought Problems

1. If the average velocity of an object is nonzero, does that mean that the instantaneous velocity can never be zero?

2. A girl standing on top of a roof throws a stone straight up into the air with a velocity of $+\mathbf{v}$. She then throws a similar stone straight down with a velocity of $-\mathbf{v}$. Compare the velocities of the two stones when they eventually strike the ground.

3. Explain how it is possible for an object to have zero average velocity with a nonzero average speed.

Multiple-Choice Problems

1. A ball is thrown upward with an initial velocity of 20 m/s. How long will it take it to reach its maximum height?
   (A) 19.6 s
   (B) 9.8 s
   (C) 6.3 s
   (D) 3.4 s
   (E) 2.04 s

2. An airplane lands on a runway with a velocity of 150 m/s. How far will it travel before it stops if its acceleration is constant at -3 m/s^2?
   (A) 525 m
   (B) 3,214 m
   (C) 6,235 m
   (D) 9,813 m
   (E) 10,435 m

3. A ball is thrown downward from the top of a roof with a speed of 25 m/s. After 2 s, its velocity will be _____ m/s.
   (A) 19.6
   (B) −5.4
   (C) −44.6

(D) 44.6
(E) −25

4. A rocket is propelled upward with an acceleration of 25 m/s^2 for 5 s. After that time, the engine is shut off and the rocket continues to move upward. The maximum height the rocket will reach is _____ m.
(A) 900
(B) 1,000
(C) 1,100
(D) 1,200
(E) 1,400

Problems 5 through 7 refer to the velocity versus time graph shown here.

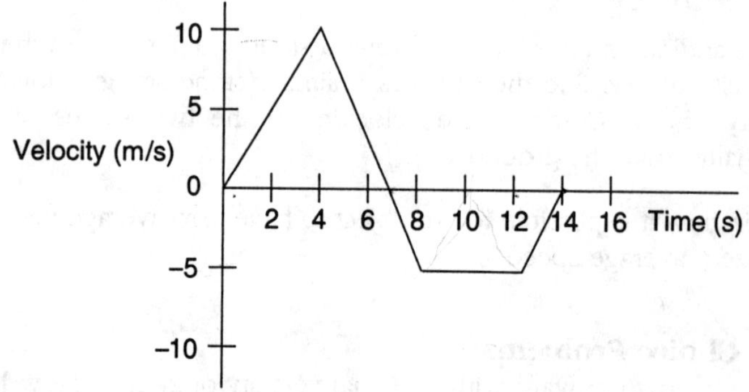

5. The total distance traveled by the object during the 14 s is _____ m.
(A) 7.5
(B) 25
(C) 62.5
(D) 77.5
(E) 82.1

6. The total displacement for the object during the 14 s is _____ m.
(A) 7.5
(B) 25
(C) 62.5
(D) 77.5
(E) 82.1

7. The average velocity for the object is _____ m/s.
(A) 0
(B) 0.5
(C) 2.5

(D) 4.5
(E) 5.6

8. What is the total change in velocity for an object whose acceleration versus time graph is given here.

A

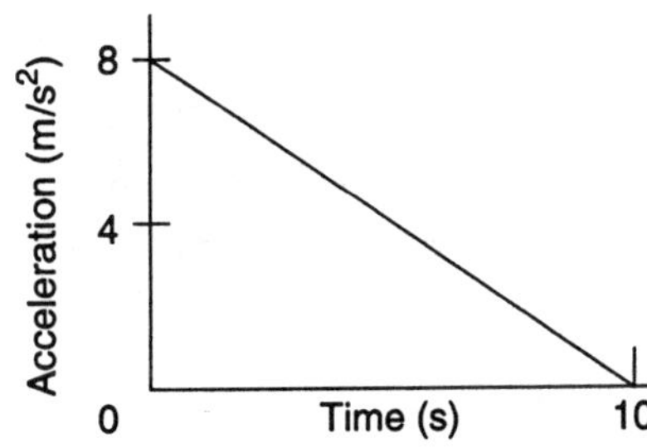

(A) 40 m/s
(B) −40 m/s
(C) 80 m/s
(D) −80 m/s
(E) 0

9. A particle moves along the x axis subject to the following position function: $x(t) = 2t^2 + 3t - 1$. What is its average velocity during the interval $t = 0$ to $t = 3$?
E
(A) 2.0 m/s
(B) 3.3 m/s
(C) 5 m/s
(D) 8.3 m/s
(E) 9.0 m/s

10. An object has an initial velocity of 15 m/s. How long must it accelerate at a constant rate of 3 m/s² before its average velocity is equal to twice its initial velocity?
B
(A) 5 s
(B) 10 s
(C) 15 s
(D) 20 s
(E) 25 s

Free-Response Problems

1. A particle has an acceleration versus time graph as shown. If the particle begins its motion at $t = 0$, with $\mathbf{v} = 0$ and $\mathbf{x} = 0$, draw a graph of velocity

versus time and position versus time for the particle's motion from $t = 0$ to $t = 13$ s.

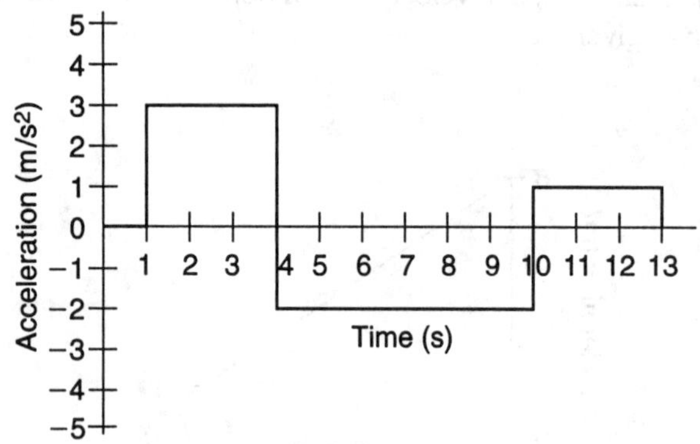

2. A stone is dropped from a 75-m-high building. When it has dropped 15 m, a second stone is thrown downward with an initial velocity such that the two stones hit the ground at the same time. What was the initial velocity of the second stone?

3. A car is stopped at a traffic light. The instant the light turns green, it is passed by another car traveling at 17 m/s. The first car begins to accelerate at a constant rate of 2.5 m/s^2 for a certain amount of time and then maintains a constant speed until it catches up to the second car. If the total time involved is 60 s, how fast is the first car going when it passes the second car and how far from the traffic light is it?

4. A particle is moving in one dimension along the x axis. The position of the particle at any given time is determined from the position function $x(t) = 3t^2 + 2$.
   (a) Evaluate the average velocity of the particle starting at $t = 2$ s for $\Delta t = 1, 0.5, 0.2, 0.1, 0.01$, and 0.001 s.
   (b) Interpret the physical meaning of your results for part (a).

SOLUTIONS TO PRACTICE PROBLEMS

Thought Problems

1. The average velocity is the ratio of the total displacement to the total elapsed time. It is quite possible that the object stopped for a while since only the final displacement counts in calculating the average velocity.

2. The two stones will have the same velocities when they strike the ground. When the first stone is thrown upward, it rises and momentarily stops. When it passes the level of the roof, it has the same speed as when it started, except in the opposite direction. This is because the magnitude of

the deceleration going up is equal to the magnitude of the acceleration going down. Since the time required to go up and down is the same, the velocity when the stone passes the roof is the same in magnitude as when it was thrown up. Thus, both stones are effectively "thrown" downward with the same initial velocity and hence land with the same velocity.

3. It is possible to have a zero average velocity with a nonzero average speed. Take the example of a baseball player hitting a home run and running around all the bases. The average speed is the total distance over the total time (a nonzero number). However, the total displacement is zero. Since average velocity is defined as total displacement over total time, the average velocity is technically zero for that motion.

Multiple-Choice Problems

1. **E** Given the initial velocity of 20 m/s, we know that the ball is decelerated by gravity at a rate of 9.8 m/s^2. That means we need to know how long it will take gravity to decelerate the ball to zero velocity. Clearly, from the definition of acceleration, dividing 20 m/s by 9.8 m/s^2 gives an answer of 2.04 s.

2. **B** We do not know the time needed to stop, but we do know that the final velocity is zero. If we use the formula $\mathbf{v}_f^2 - \mathbf{v}_i^2 = 2\,\mathbf{a}x$ and substitute for the initial velocity and the acceleration, we obtain an answer of 3,214 m.

3. **C** The initial velocity is $-$ 25 m/s, and so $\mathbf{v} = -(25) - (9.8)(2) = -44.6$ m/s.

4. **C** The rocket accelerates from rest, and so the first distance traveled in 5 s is given by $y_1 = \frac{1}{2}\,\mathbf{a}t^2$. Substituting the values for time and acceleration gives $y_1 = 312.5$ m. After 5 s of accelerating, the rocket has a velocity of 125 m/s. After this time, the engine stops and the rocket is decelerated by gravity as it continues to move upward. The time required to decelerate to zero is found by dividing 125 m/s by the acceleration of gravity. Thus, 125/9.8 = 12.76 s. The distance traveled during this time is added to the first accelerated distance. The second distance is given by $y_2 = (125)(12.76) - (4.9)(12.76)^2 = 796.6$ m. The total distance traveled is therefore equal to 1,108 m (or, rounded off, equal to 1,100 m).

5. **C** The total distance traveled in 14 s is equal to the total area under the curve. Breaking the figure up into triangles and rectangles adds up to 62.5 m (recall that distance is a scalar).

6. **A** The total displacement is the sum of the positive (forward) and negative (backward) areas representing the fact that the object moves away from the origin and then back (as indicated by the velocity time graph going below the x axis. Adding and subtracting the proper areas gives us 7.5 m for the final displacement.

7. **B** Average velocity is the displacement (not distance) over the change in time. Since from Question 6 we know the displacement to be 7.5 m in 14 s, the average velocity is therefore 0.54 m/s.

8. **A** The change in velocity is equal to the area under the graph, which is equal to $\frac{1}{2}$ (8)(10) = + 40 m/s.

9. **E** According to the formula, at $t = 0$, $x = -1$. When $t = 3$, $x = 26$. Thus, $\Delta x = 27$ m, and since the average velocity is $\Delta x/\Delta t$, the average velocity for 3 s is 9 m/s.

10. **B** For uniformly accelerated motion, the average velocity is $\bar{\mathbf{v}} = (\mathbf{v}_f + \mathbf{v}_i)/2$. It is desired that the average velocity be equal to twice the initial velocity. Thus if the average velocity is 30 m/s, the final velocity attained must be 45 m/s. The question now is How long must the object accelerate from 15 m/s to achieve a speed of 45 m/s². This implies a change in velocity of 30 m/s at a rate of 3 m/s² for 10 s.

Free-Response Problems

1. The acceleration graph indicates constant acceleration from $t = 1$ to $t = 3$. Therefore, the area represents a constant change in velocity from 0 m/s to 9 m/s. The second region corresponds to a negative acceleration, which slows the object down and turns it around. We know this since the area is -12 m/s. This brings the graph from 9 m/s to -3 m/s (a change in direction). The final area is a change of 3 m/s, which brings the velocity to 0 m/s. A graph of this motion is shown here, as desired.

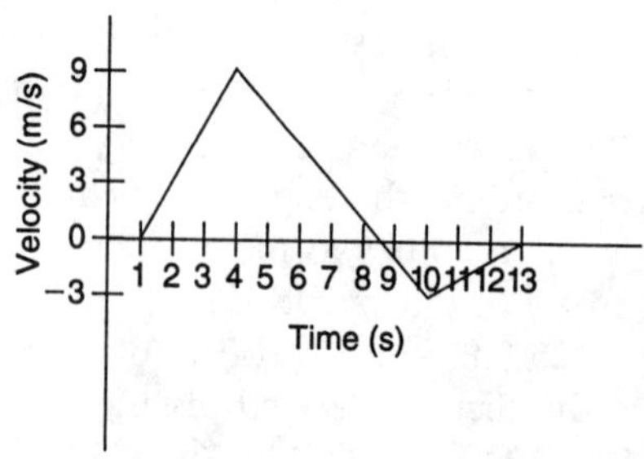

To determine the corresponding position time graph, we note that from $t = 0$ to $t = 1$ s no motion occurs. From $t = 1$ to $t = 4$ s, we have uniformly accelerated motion. The total distance traveled is equal to the area under the curve (which is a triangle of altitude 9 m/s and a base of 3 s). This area is equal to 13.5 m, and so we should sketch in a parabola starting at $t = 1$ and ending at $t = 4$. From $t = 4$ to about $t = 7.5$ s, the object is decelerating in the forward direction. Thus, it continues to move forward but at a slower rate. The total distance traveled in this segment is again equal to the area, which comes out to be 15.75 m. Since we have uniformly decelerated motion, we need an inverted parabola ending at about 30 m from the origin. At this time, the object reverses direction and begins to accelerate backward from $t = 7.5$ until $t = 10$ s. The area of this segment is -3.75 m, which is subtracted from the ending point. Since the object is accelerating backward, there is a reversal of the parabola to reflect an increasing negative slope until $t = 10$ s. From $t = 10$ to $t = 13$ s, the object slows down in the backward direction (the velocity graph is still below the x axis). This last area is equal to 4.5 m, and this is again subtracted off. However, for this last segment, we need to show the slopes approaching zero as the velocity is zero at $t = 13$ s. A completed version of the displacement versus time graph is shown.

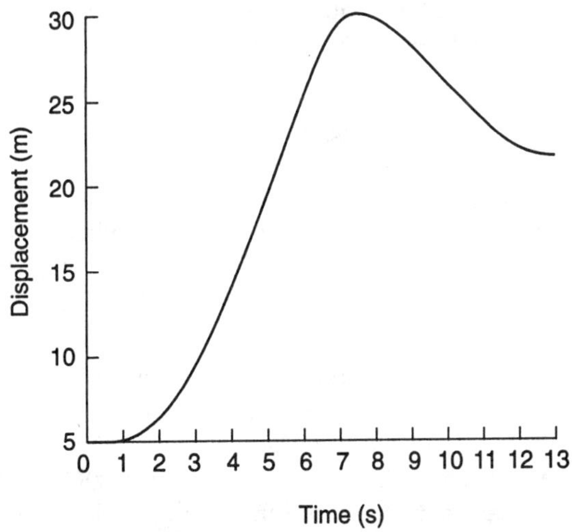

2. In this problem, we know that the first ball will be dropped from a height of 75 m and will be in free fall because of gravity. Therefore, the time required to fall is given by $t = \sqrt{2y/\mathbf{g}} = 3.91$ s. Since the first ball is to fall 15 m, we can likewise determine that $t = 1.75$ s to fall that distance. Thus, we can say that since the two balls must reach the ground at the same time, then we know that the second ball must reach the ground after the first ball has traveled for 3.91 s. Since the second ball is thrown when the distance fallen is 15 m, we subtract the 1.75 s from the total time of 3.9 s to obtain the second ball's duration. This time is equal to 2.16 s. Therefore, since the second ball is thrown downward, it has a negative displacement, and so we can write

$$-75 = -\mathbf{v}_i(2.16) - \frac{1}{2}(9.8)(2.16)^2$$

This gives us a downward initial velocity of 24.13 m/s.

3. In this problem when $t = 0$, the second car passes the first at a constant speed of 17 m/s. Since the entire episode lasts for 60 s, the second car travels a distance of 1,020 m. This is the distance the first car is from the

traffic light when it catches up to the second car. Now the first part of the car's motion is an acceleration of 2.5 m/s^2 for T seconds. After T seconds, the car travels at a constant speed for (60 − T) s until it catches up with the second car. Thus, there are two sets of distances: one for the accelerated phase and one for the constant-velocity phase.

The distance traveled from rest during the acceleration phase is given by $\frac{1}{2}(2.5)T^2$. The velocity attained after T seconds is of course 2.5T m/s, and this is the constant velocity for the second phase which lasts for (60 − T) seconds. This is because we know that the total elapsed time is 60 s. Therefore, the distance for the second phase is given by $x = \mathbf{v}t$, or in this case, the distance is equal to $2.5T(60 - T)$. To find the velocity attained when the first car catches up with the second, we must determine the time T. Since the total displacement must be 1,020 m, we therefore find, adding up both phases, that

$$1{,}020 = \frac{1}{2}(2.5)T^2 + (2.5T)(60 - T) = 1.25T^2 + 150T - 2.5T^2$$

The time T can be solved for using the quadratic formula after we collect terms and bring the 1,020 over to the other side. Of the two solutions, only 7.24 s is realistic (since the other term is greater than 60 s), which means that the velocity in question is (2.5)(7.24) = 18.1 m/s.

4. (a) Using the position function, we know that when t = 2 s, x = 14 m. For the first interval, we need the average velocity from t = 2 s to t = 3 s. When t = 3 s, x = 29 m. The displacement is 15 m, and the average velocity is 15 m/s (displacement over time interval: 15 m / 1 s). For the next interval, when t = 2.5 s, x = 20.75 m. The displacement in this case is 6.75 m, and the average velocity is 13.5 m/s (the time interval is 0.5 s). Continuing this process, we find that x (2.2) = 16.52 m, and the average velocity is 12.6 m/s. Next, x (2.1) = 15.23 m, and the average velocity is 12.3 m/s. Then, x (2.01) = 14.12 m, and the average velocity is 14.12 m/s.

   Finally, x (2.001) = 12.003 m, and the average velocity is 12.003 m/s.

   (b) It appears that as Δt approaches zero, the average velocity approaches 12 m/s, which we call the instantaneous velocity. In calculus, the method described above is called the *method of limits*.

5
TWO-DIMENSIONAL MOTION

RELATIVE MOTION

In Chapter 4, we reviewed the basic elements of one-dimensional rectilinear motion. In this chapter, we will consider only two-dimensional motion. In one sense, one-dimensional motion can be viewed as two-dimensional motion by a suitable transformation of coordinate systems.

For example, consider the definition of **displacement**. If we define a coordinate system for a reference frame, then the location of a point in that frame is determined by a position vector drawn from the origin of that coordinate system to the point. If the point is displaced, then the one-dimensional vector drawn from point A to point B is called the displacement vector and is designated as **B** − **A** (see Figure 1).

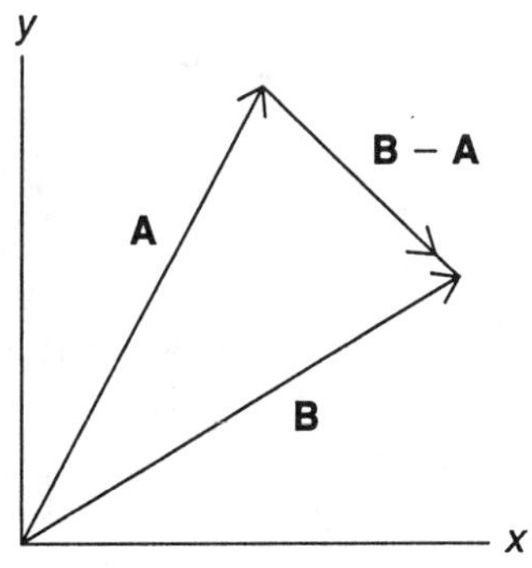

FIGURE 1.

We can resolve it into two components that represent mutually perpendicular and independent simultaneous motions. An example of this type of motion can be seen when considering a boat trying to cross a river or an airplane meeting a crosswind. In the case of the boat, its velocity relative to the river is based on the properties of the engine and is measured by the speedometer on board. However, to a person on the shore, its **relative velocity** (or effective velocity) is different from what the speedometer in the boat may report. In

Figure 2, we see such a situation with the river moving to the east at 4 m/s and the boat moving north at 10 m/s relative to the water.

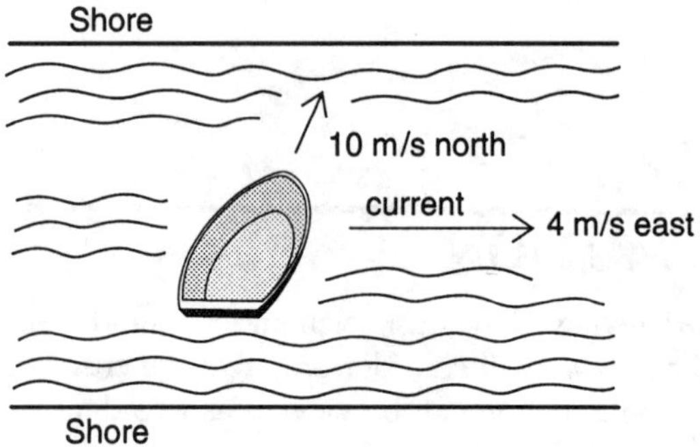

FIGURE 2.

By vector methods, the resultant velocity relative to the shore is given by the *Pythagorean theorem.* The direction is found by drawing a simple sketch connecting the vectors head to tail to preserve the proper orientation. Numerically, we can use the tangent function or the law of sines (see Figure 3).

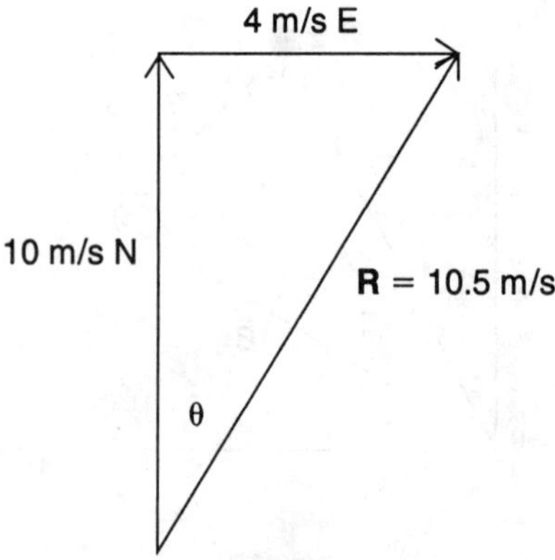

FIGURE 3.

The resultant velocity is 10.5 m/s at an angle of 22 degrees east of north.

In this chapter, we shall consider two-dimensional motion as viewed from the earth frame of reference. Typical of this kind of two-dimensional motion is that of a projectile. Galileo Galilei was one of the pioneers who studied the mechanics of flying projectiles and the first to discover that the path of a projectile is a parabola.

HORIZONTALLY LAUNCHED PROJECTILES

If you roll a ball off a smooth table, you will observe that it does not fall straight down. Using trial and error, you might find that how far it falls depends on how fast it is moving forward. Initially, however, it has no vertical velocity. The ability to "fall" is provided by gravity, and its acceleration is $-9.8\ \text{m/s}^2$. Since gravity acts in a direction perpendicular to the initial horizontal motion, the two motions are simultaneous and independent. Galileo demonstrated that the trajectory is a parabola.

We know that the distance fallen by a mass dropped from rest is given by the equation $y = -\frac{1}{2}\mathbf{g}t^2$. Since the ball is moving horizontally with some initial constant velocity, it covers a distance (called the **range**) of $x = \mathbf{v}_{ix}t$. Since the required time is the same for both motions, we can first solve for the time from the x equation and then substitute it into the y equation. In other words, $t = x/\mathbf{v}_{ix}$ and therefore

$$y = -\frac{1}{2}\mathbf{g}\left(\frac{x}{v_{ix}}\right)^2 = -\frac{\mathbf{g}x^2}{2\mathbf{v}_{ix}^2}$$

which is of course the equation of an inverted parabola. This equation of y in terms of x is called the **trajectory** of the projectile, while the two separate equations for x and y as functions of time are called **parametric equations**. Figure 4 illustrates this trajectory as well as a position vector **R** which locates a point in space at any given time.

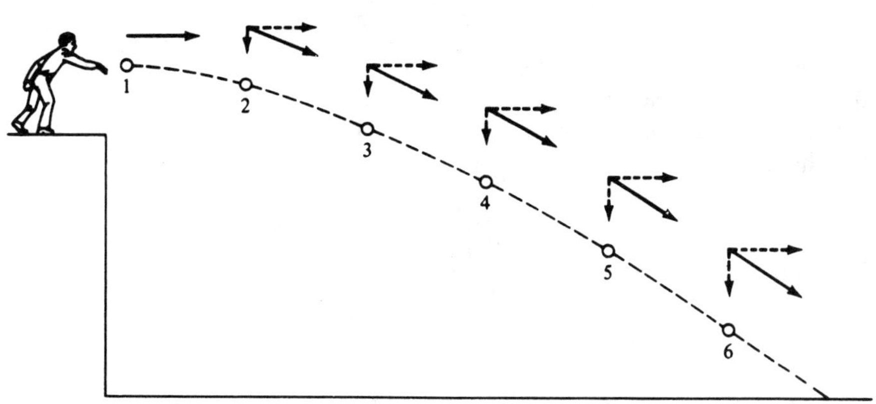

FIGURE 4.

If the height from which the projectile is launched is known, then the time required to fall can be calculated from the equation for free fall. For example, if the height is 49 m, then the time required to fall is 3.16 s. If the horizontal velocity is 10 m/s, then the maximum range will be 31.6 m. We can also follow the trajectory by determining how far the object has fallen when it is 10 m away from the cliff. Using the trajectory formula and the velocity given, this turns out to be −4.9 m.

PROJECTILES LAUNCHED AT AN ANGLE

Suppose we have a missile on the ground and that it is launched with some initial velocity at some angle θ. The vector nature of velocity allows us to immediately write the equations for the horizontal and vertical components of initial velocity:

$$\mathbf{v}_{ix} = \mathbf{v}_i \cos\theta$$
$$\mathbf{v}_{iy} = \mathbf{v}_i \sin\theta$$

Since each motion is independent, we can consider the fact that in the absence of friction, the horizontal velocity will be constant while the y velocity will decrease as the projectile rises. When the projectile reaches its maximum height, its vertical velocity will be zero and then gravity will accelerate the object back downward. The projectile will continue to move forward at a constant rate. How long will it take the object to reach its maximum height? From the definition of acceleration, and the equations in Chapter 4, this is the time needed for gravity to decelerate the vertical velocity to zero. That is,

$$\mathbf{v}_{fy} = 0 = \mathbf{v}_{iy} - \mathbf{g}t$$
$$t_{\text{up}} = \frac{\mathbf{v}_{iy}}{\mathbf{g}} = \frac{\mathbf{v}_i \sin\theta}{\mathbf{g}}$$

The total time of flight will be twice this time since gravity accelerates and decelerates the projectile by the same magnitude. Therefore, the range is the product of the initial horizontal velocity (which is constant) and the total time. In other words,

$$\text{Range} = 2\mathbf{v}_{ix}t_{\text{up}} = 2\mathbf{v}_i \cos\theta\, t_{\text{up}}$$

If the time required to go up is now substituted into this expression, we have

$$\text{Range} = \frac{2(\mathbf{v}_i \cos\theta)(\mathbf{v}_i \sin\theta)}{\mathbf{g}} = \frac{\mathbf{v}_i^2 \sin 2\theta}{\mathbf{g}}$$

From this expression, we see that the range is independent of the mass of the projectile and is a maximum when $\sin 2\theta = 1$. This occurs when the launch angle is 45 degrees.

Since vertical motion is independent of horizontal motion, the changes in vertical height are given one-dimensionally as $y = \mathbf{v}_{iy}t - \frac{1}{2}\mathbf{g}t^2$. If you want to know the maximum height achieved, simply use the value for the time required to reach the highest point. To find the trajectory of the projectile, we first substitute for the value of the initial vertical velocity and then substitute for time. The time t is found from the fact that at all times $x = \mathbf{v}_i \cos\theta t$. Solving for t and then substituting, we have the following trajectory:

$$y = (\mathbf{v}_i \sin\theta)\frac{x}{\mathbf{v}_i \cos\theta} - \frac{1}{2}\mathbf{g}\frac{x^2}{\mathbf{v}_i^2 \cos^2\theta}$$

This simplifies to the final equation for the trajectory of the projectile:

$$y = (\tan\theta)x - \frac{\mathbf{g}}{2\mathbf{v}_i^2 \cos^2\theta}x^2$$

Of course, this equation also represents an inverted parabola. This trajectory is seen in Figure 5.

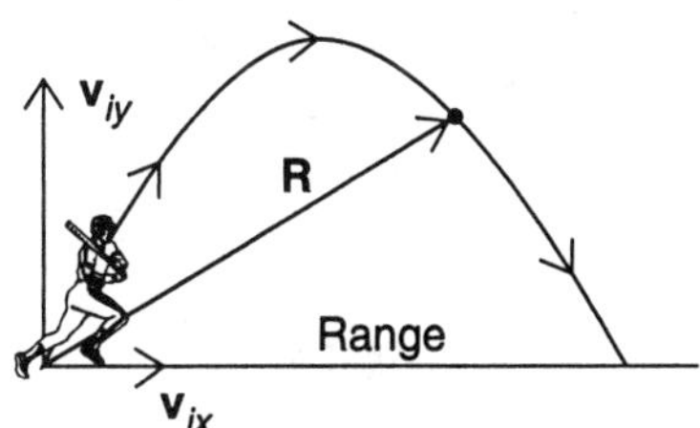

FIGURE 5.

As an example, if the projectile is launched with an initial velocity of 100 m/s at an angle of 30 degrees, the maximum range will be equal to 883.7 m. To find the maximum height, we could first find the time required to reach that height, but a little algebra will give us a formula for the maximum height independent of time. The maximum height can also be expressed as $y_{max} = (\mathbf{v}_i^2 \sin^2\theta)/2\mathbf{g}$. Using the numbers given, the maximum height reached is 127.55 m.

UNIFORM CIRCULAR MOTION

Velocity is a vector. When velocity changes, the magnitude, direction, or both can change. When the direction is the only quantity changing as the result of a centrally directed deflecting force, the result is uniform circular motion. While we will take up the discussion of deflecting forces in the next chapter, consider here an object already undergoing periodic, uniform circular motion. By this we mean that the object maintains a constant speed as it revolves around a circle of radius R in a period of time T. The number of revolutions per second is called the **frequency** f.

Clearly, if the total distance traveled around the circle is its circumference, then the constant average speed is given by $2\pi R/T$. Also, if the circle maintains a constant radius, then the quantity $2\pi/T$ is called the **angular frequency** or **angular velocity** ω. The units of ω involve units related to are length. Before going any further with this topic, we digress a moment to discuss these units.

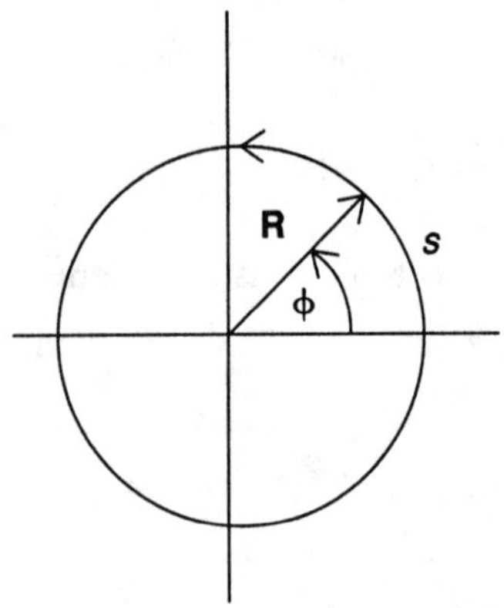

FIGURE 6.

Imagine a circle of radius R as shown in Figure 6. In this circle, a point particle is moving around it counterclockwise, making an angle ϕ with the positive x axis. There is a corresponding arc length s associated with the angle ϕ. From geometry, we know that the length of an arc s is equal to the circumference of the circle multiplied by the ratio of angle ϕ to the whole angle of the circle (360 degrees). That is,

$$S = 2\pi R \frac{\phi}{\text{whole circle angle}}$$

If we define the whole circle angle to be equal to 2π and call the new units radians, then we have the formula $S = R\phi$, which is a common formula from rotational dynamics in physics. Thus, we can define ϕ (rad) = arc length / radius = S / R.

The units of the angle are now such that 360 degrees corresponds to 2π rad. Thus, for a uniformly changing angle in radians, the average angular velocity is defined to be $\Delta\phi / \Delta t$. This is related to the linear velocity along the arc length by the formula

$$\mathbf{v} = \frac{\Delta s}{\Delta t} = R\frac{\Delta\phi}{\Delta t}$$

$$R\omega = \omega R$$

The units of angular velocity (or frequency) are now rad/s. If we define the angular velocity as $\Delta\theta/\Delta t$, then we can also write $\Delta\phi = \boldsymbol{\omega}\,\Delta t$, which looks remarkably similar to the linear formula $\Delta x = \mathbf{v}\,\Delta t$.

Now, let us return to the idea of uniform circular motion. If the object is moving around a circle at a constant rate, its linear (tangential) and angular velocities are related by the angle swept out per second and the radius of the circle. We can see that if we have an object on a rotating platform, or a rotating solid, then the preceding formula informs us that all points have the same angular velocity but that the linear velocity is directly proportional to the radius.

If the velocity is changing (which it is by virtue of its changing direction), then what kind of acceleration does the object have? If we consider the change in velocity $\Delta\mathbf{v}$ to be a change in the quantity $R\boldsymbol{\omega}$, then clearly

$$\frac{\Delta\mathbf{v}}{\Delta t} = R\frac{\Delta\boldsymbol{\omega}}{\Delta t} = R\boldsymbol{\alpha}$$

where α is called the **angular acceleration** in units of rad/s^2. Notice that analogously to linear motion, we can write, for constant angular acceleration, $\Delta\boldsymbol{\omega} = \boldsymbol{\alpha}\,\Delta t$. However, if we have uniform circular motion, the angular frequency is not changing, and so the angular acceleration is zero. Thus, we are still left with the question concerning the nature of the acceleration involved.

This confusion is cleared up easily if we recognize that as the angle ϕ changes, so do the position vector and arc length. In other words, consider Figure 7.

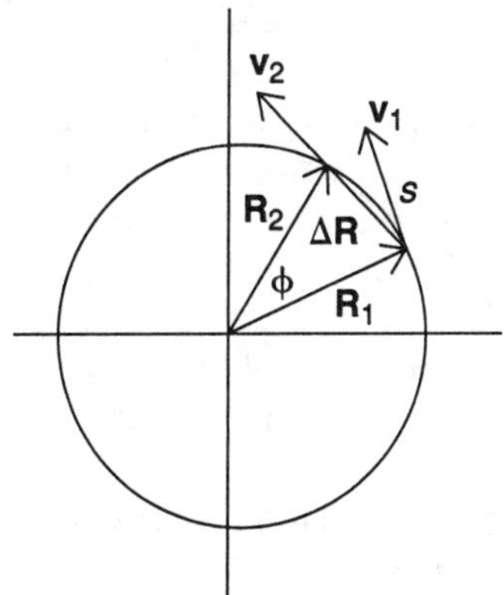

FIGURE 7.

For small changes in angle, the arc length Δs is approximately equal to the displacement $\Delta\mathbf{R}$. Thus we can write, for small approximations, that $\Delta\mathbf{R}/\Delta t = \mathbf{R}\,\Delta\phi/\Delta t$. This means that $\Delta\mathbf{R}/\mathbf{R} = \Delta\phi$. Since velocity is a vector, this means that $\Delta \mathbf{v}$ is swept out in the same time Δt and at the same angle $\Delta\phi$. Thus, $\Delta \mathbf{R}/\mathbf{R} = \Delta \mathbf{v}/\mathbf{v}$, and so we can write

$$\frac{\Delta \mathbf{v}}{\Delta t} = \frac{\mathbf{v}}{\mathbf{R}}\frac{\Delta \mathbf{R}}{\Delta t} = \frac{\mathbf{v}^2}{\mathbf{R}} = \mathbf{a}_c$$

This quantity is known as the **centripetal acceleration**. From the construction it is directed radially inward toward the center of the circle, and the units are standard acceleration units. This is different from the angular acceleration α but is related to the angular velocity in the following way. Since $\omega = \mathbf{v}/R$, we have $\mathbf{a}_c = \mathbf{v}^2/\mathbf{R} = / (\mathbf{v}/\mathbf{R})\mathbf{v} = \omega(\omega\mathbf{R}) = \omega^2\mathbf{R}$. Finally, we can express the centripetal acceleration in terms of the period T. Since we defined $\mathbf{v} = 2\pi\mathbf{R}/T$, it immediately follows that $\mathbf{a}_c = 4\pi^2\mathbf{R}/T^2$.

PROBLEM-SOLVING STRATEGIES

The key to solving projectile motion problems, or any two-dimensional problem, is to remember that the horizontal and vertical motions are independent of each other. Therefore, using the formulas from Chapter 4 on one-dimensional motion can further reduce the complexity of a given problem. The following guidelines should help.

1. Draw a sketch of the situation if none is provided.
2. Choose a suitable frame of reference and coordinate system for the problem.
3. Determine if the components of motion are given in the problem. If not, try to determine the components of velocity and acceleration.
4. Accelerations given in one dimension do not affect the other, and this results in the curved path.
5. Remember that velocity and acceleration are vectors and that both magnitudes and directions should be specified unless otherwise noted.
6. Ask yourself questions about the implications of each change in motion for the total motion of the object.
7. Trajectories are expressed in terms of position variables only. Parametric equations usually relate positions to time.

PRACTICE PROBLEMS FOR CHAPTER 5

Thought Problems

1. Under what conditions can you have two-dimensional motion with one-dimensional acceleration?

2. A ball is dropped vertically from a height h onto a board hinged to the horizontal floor. The board can be raised to some angle of elevation θ relative to the horizontal. How should the angle of elevation be chosen such that when the ball reflects elastically off the incline it will achieve maximum range as a projectile?

3. A car is moving in a straight line with a velocity $\mathbf{v}$. Raindrops are falling vertically downward with a constant terminal velocity $\mathbf{u}$. At what angle does the driver think the drops are hitting the car's windshield? Explain.

Multiple-Choice Problems

1. A projectile is launched at an angle of 45 degrees with a velocity of 250 m/s. Neglecting air resistance, the magnitude of its horizontal velocity at the time it reaches its maximum altitude is equal to _____ m/s.
   (A) 0
   (B) 175
   (C) 200
   (D) 250
   (E) 300

2. A projectile is launched horizontally with a velocity of 25 m/s from the top of a 75-m height. The projectile will take _____ to reach the bottom.
   (A) 15.5
   (B) 9.75
   (C) 6.31
   (D) 4.27
   (E) 3.91

3. An object is launched from the ground with an initial velocity and angle such that the maximum height achieved is equal to the total range of the projectile. The tangent of the launch angle is equal to _____.
   (A) 1
   (B) 2
   (C) 3
   (D) 4
   (E) 5

4. At a launch angle of 45 degrees, the range of a launched projectile is given by _____.
   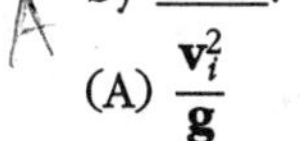
   (A) $\frac{\mathbf{v}_i^2}{\mathbf{g}}$
   (B) $\frac{2\mathbf{v}_i^2}{\mathbf{g}}$
   (C) $\frac{\mathbf{v}_i^2}{2\mathbf{g}}$

(D) $\sqrt{\frac{v_i^2}{2g}}$

(E) $\frac{2v}{g}$

5. A projectile is launched at a certain angle. After 4 s, it is observed that it hits the top of a building 500 m way. The height of the building is 50 m. The projectile was launched at an angle of _____ degrees.
(A) 14
(B) 21
(C) 37
(D) 76
(E) 85

6. A projectile is launched at a velocity of 125 m/s at an angle of 20 degrees. A building is 200 m away. At what height above the ground will the projectile strike the building?
(A) 50.6 m
(B) 90.2 m
(C) 104.3 m
(D) 114.6 m
(E) 125.6 m

7. A boat is crossing a 5-km-wide river that is flowing eastward at 10 m/s. The boat operator wishes to reach the exact point on the opposite shore 15 minutes (min) after starting. With what velocity and in what direction should the boat travel?
(A) 11.2 m/s at 26.6 degrees east of north
(B) 8.66 m/s at 63.4 degrees west of north
(C) 11.2 m/s at 63.4 degrees west of north
(D) 8.66 m/s at 26.6 degrees east of north
(E) 5 m/s due north

8. An object is moving around a circle of radius 1.5 m at a constant velocity of 7 m/s. The frequency of the motion is _____ revolutions/second (rev/s).
(A) 0.24
(B) 0.53
(C) 0.67
(D) 0.74
(E) 0.98

9. A stereo turntable has a frequency of 33 rev/min. A record of radius 15.25 cm is placed on it and begins to rotate. The velocity of a point on the edge of the record is _____m/s.
(A) 2.16
(B) 22.23
(C) 0.53

(D) 7.62
(E) 13.5

10. It is observed that in 3 s, an object moving around a circle sweeps out an angle of 9 rad. If the radius of the circle is 0.5 m, then the centripetal acceleration of the object is _____ m/s^2.
(A) 4.5
(B) 3
(C) 13.5
(D) 15
(E) 18

Free-Response Problems

1. A bomber is diving toward its target at an angle of 45 degrees below the horizontal and at a speed of 320 m/s. When the bomber is 600 ms above the ground, it releases its load, which then hits the target. How long will it take the bomb to reach the target? What horizontal distance will it travel? With what speed will it strike the target?

2. A projectile is launched at an angle of 40 degrees with an initial velocity of 100 m/s. One-hundred meters away is the beginning of a hill that continues to slope upward at an angle of 20 degrees. It is observed that the projectile strikes the hill a distance L up the slope. What is the value of this distance up the slope?

3. Two girls decide to have a swimming race in a river that flows with a velocity **u**. Each girl can swim with the same velocity **v** relative to the water. One girl swims to a point across the river directly opposite her starting point and then swims back. The other girl swims downstream the same distance as the first girl and then swims back upstream.
(a) Which girl wins the race?
(b) What is the ratio of the winning time to the losing time?

4. A football quarterback throws a pass to a receiver at an angle of 25 degrees with the horizontal and an initial velocity of 25 m/s. The receiver is initially at rest 30 m from the quarterback. The instant the ball is thrown, the receiver runs at a constant velocity to catch the pass. In what direction and with what speed should he run?

SOLUTIONS TO PRACTICE PROBLEMS

Thought Problems

1. Two-dimensional motion can result when the acceleration is not along the line of initial motion. For example, a projectile launched horizontally has an initial acceleration that is at right angles to its motion.

2. A sketch of the situation is seen here. If the angle of elevation is 45 degrees, then the ball will be reflected parallel to the ground. If the angle is greater than 45 degrees, then the ball will be projected into the ground,

resulting in a smaller range. Therefore, the angle of elevation must be less than 45 degrees to provide an additional upward velocity.

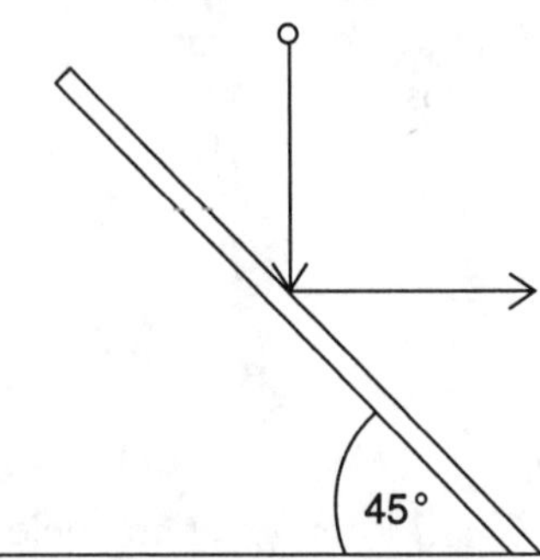

3. In a frame of reference at rest with respect to the ground, the drops appear to be moving in a straight line with constant velocity **u**. However, to a car moving, say to the right, with a velocity **v**, there is an apparent relative velocity (see the accompanying diagram) whose angle is given by tan θ = **u**/**v**.

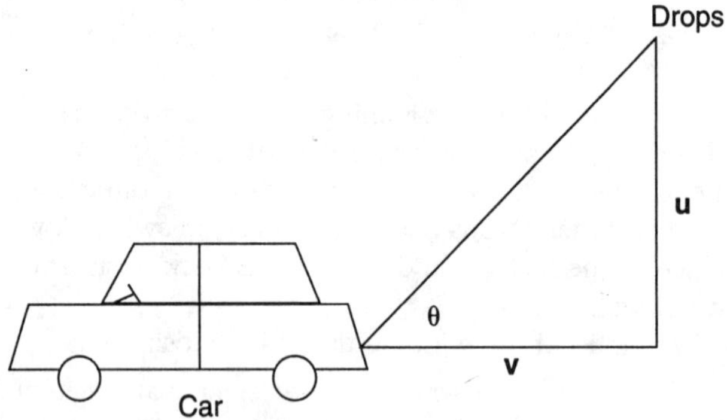

Multiple-Choice Problems

1. **B** The horizontal component of velocity remains constant in the absence of resisitive forces and is equal to $\mathbf{v}_i \cos \theta$. Substituting the numbers, we get $\mathbf{v}_x = (250)(0.7) = 175.0$ m/s.

2. **E** The time required to fall is given by the free fall formula from Chapter 8, namely, $t = \sqrt{2y/\mathbf{g}}$. If we substitute the numbers we get 3.91 s for time.

3. **D** If the maximum height is going to be equal to the range, then set these two equations equal to each other as follows:

$$\frac{\mathbf{v}_i^2 \sin^2 \theta}{2\mathbf{g}} = \frac{\mathbf{v}_i^2 \sin 2\theta}{\mathbf{g}}$$

Recalling that $\sin 2\theta = 2 \sin \theta \cos \theta$, we solve for $\tan \theta$ and find it equals 4.

4. **A** From the formula for range, we see that at 45 degrees, $\sin 2\theta = 1$, and so $R = \mathbf{v}_i^2/\mathbf{g}$.

5. **B** We know that after 4 s the projectile has traveled horizontally 500 m. This means that the horizontal velocity was a constant 125 m/s and is equal to $\mathbf{v}_i \cos \theta$. We also know that the y position of the projectile is 50 m after 4 s. Thus, we can write

$$50 = 4\mathbf{v}_i \sin \theta - (4.9)(4)^2 = 4\mathbf{v}_i \sin \theta - 78.4$$

Therefore, $\mathbf{v}_i \cos \theta = 125$, $\mathbf{v}_i \sin \theta = 32.1$, and $\tan \theta = 0.2568$, and so $\theta = 14.4$ degrees.

6. **A** The x component of velocity is 93.97 m/s, and the y component of velocity is 34.2 m/s. Since the projectile travels 200 m horizontally at 93.97 m/s, the time of flight $t = 2.128$ s. This time is substituted into our equation for the y position, and we have $y = 50.6$ m.

7. **C** The river is flowing at 10 m/s to the right (east), and the resultant desired velocity is 5 m/s upward (north). Therefore, the actual velocity relative to the river is heading west of north. By the Pythagorean theorem, the velocity of the boat must be 11.2 m/s. The angle is given by the tangent function. In the accompanying diagram we see that $\tan \theta = \frac{10}{5} = 2$, and so $\theta = 63.4$ degrees west of north.

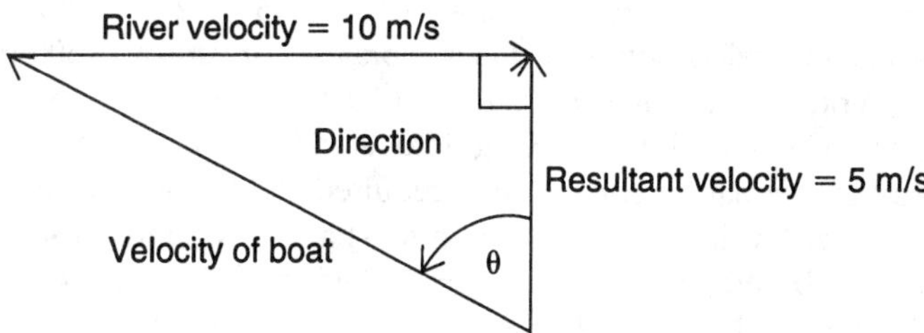

8. **D** The velocity is given by $\mathbf{v} = 2\pi R/T$, where T is the period. But the frequency f is just the reciprocal of the period. Thus, $\mathbf{v} = 2\pi Rf$, and after substituting the numbers $f = 0.74$ rev/s.

9. **C** First convert 33 rev/min into rad/s. We note that there are 2π radians per revolution and 60 s in 1 min. The conversion gives an angular frequency of 3.454 rad/s. Now we convert the radius to meters so that we have SI units. Thus, $R = 0.1525$ m. Now, we know that for uniform circular motion, $\mathbf{v} = R\omega$, and so $\mathbf{v} = (0.1525 \text{ m})(3.454 \text{ rad/s}) = 0.53$ m/s.

10. **A** If the object sweeps out 9 rad in 3 s, then its angualr velocity is 3 rad/s. If the radius is 0.5 m, then the centripetal acceleration is given by $\mathbf{a}_c = \omega^2 R = (9)(0.5) = 4.5 \text{ m/s}^2$.

Free-Response Problems

1. In this problem, we know that the bomber is on a descending flight path and so the angle below the horizontal implies a negative vertical velocity. Since the plane is moving forward, we can treat its forward velocity as positive. Thus, following the problem solving suggestions, we first make a sketch and then determine the components of the velocity.

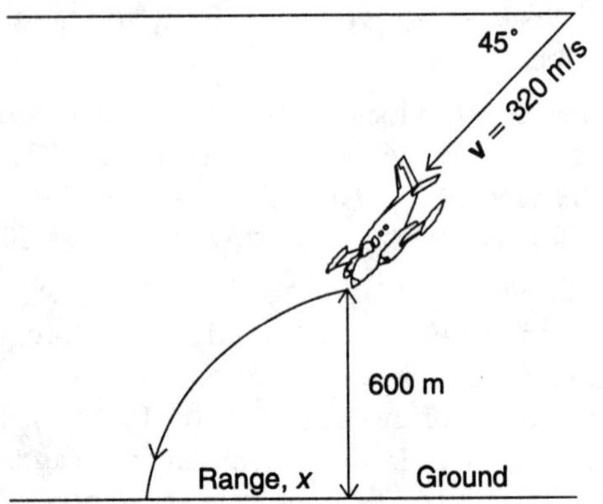

We now determine the two component velocities. We have $\mathbf{v}_x = 320 \cos 45 = 226.27$ m/s and $\mathbf{v}_y = 320 \sin 45 = 226.27$ m/s. The vertical distance traveled downward is algebraically given by -600 m, and so we can write for the time equation $-600 = -226.27t - 4.9t^2$. Dividing all parts by 4.9 and rearranging into standard quadratic form, we get a quadratic equation for the time required to fall the 600 m with an initial downward velocity of 226.27 m/s. That is, we must use the quadratic formula based on $t^2 - 46.18t - 122.45 = 0$. Keeping only the positive solution, we have $t = 2.51$ s.

Knowing the time, we observe that the horizontal velocity of 226.27 m/s is constant, and so the horizontal distance will be $(226.27)(2.51) = 569$ m. For the velocity of impact, we need only determine the change in the vertical velocity. Gravity will accelerate the bomb downward at a rate of 9.8 m/s^2 for 2.51 s. Using the velocity-acceleration formula from Chapter 4, we find that the final vertical velocity will be $(-226.27) - (9.8)(2.51) = -250.87$ m/s. The magnitude of the velocity of impact is determined by the Pythagorean theorem and is equal to 337.84 m/s.

2. This problem is shown in the accompanying diagram. From the diagram we know that $\tan 20 = h/d$. We can also show that $\mathbf{v}_x = 100 \cos 40$ and $\mathbf{v}_y = 100 \sin 40$. This means that $x = 100 + d = (100 \cos 40)t$ or $t = (100$

+ d) / (100 cos 40). The y position at time t is given by h = (100 sin 40)t − 4.9t^2. If we substitute for h in terms of d, and t in terms of d, and apply all the algebra, we are left with a single quadratic equation for the distance d given by $d^2 - 370d - 90{,}598.8 = 0$. Using the quadratic formula and keeping the positive root, we have d = 538.3 m and therefore h = 195.9 m. Using the Pythagorean theorem, we find the hypotenuse ℓ, the distance up the slope, to be 572.8 m.

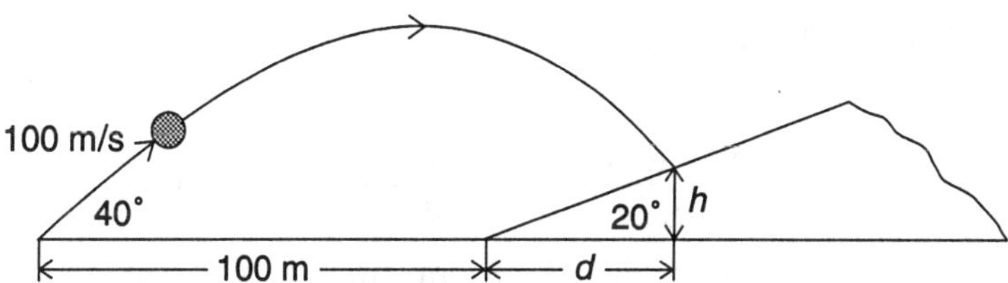

3. The situation is shown in the accompanying diagram. The river flows with velocity **u**, and each girl swims with velocity **v**. The girls travel equal distances ℓ and then back. If the first girl swims to a point directly opposite the starting point, then she must swim in a direction slightly upstream so that her resultant direction ends up straight across. Let's call the two directions perpendicular and parallel. The first swimmer crosses perpendicularly and swims slightly upstream with a relative velocity **v**. The river flows downstream at a velocity **u**. The desired resultant velocity is given by $\sqrt{\mathbf{v}^2 - \mathbf{u}^2}$. The total round-trip time is just twice the distance divided by this resultant velocity, or $t_1 = 2\ell/\sqrt{\mathbf{v}^2 - \mathbf{u}^2}$.

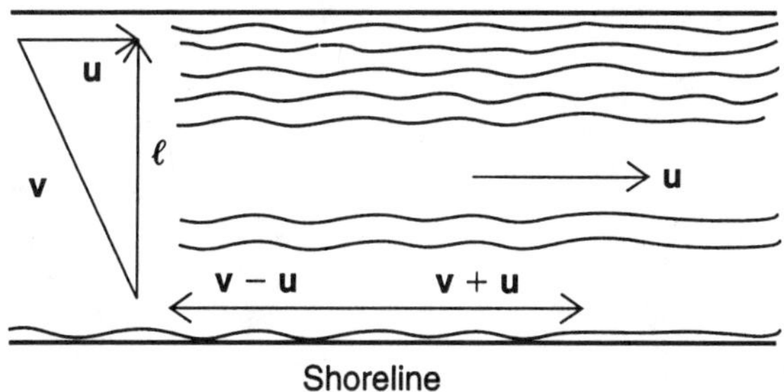

The second swimmer begins swimming downstream with a relative velocity of **v** + **u** and then upstream at a relative velocity of **v** − **u** (keeping the magnitude of the velocity positive). Each distance covered is ℓ, and so the total time is given by

$$t_2 = \frac{\ell}{\mathbf{v} + \mathbf{u}} + \frac{\ell}{\mathbf{v} - \mathbf{u}} = \frac{2\ell\mathbf{v}}{\sqrt{\mathbf{v}^2 - \mathbf{u}^2}}$$

If we take the ratio of these quantities, we have

$$\frac{t_2}{t_1} = \sqrt{1 - \frac{\mathbf{u}^2}{\mathbf{v}^2}}$$

This means that $t_2 = t_1 \sqrt{1 - \mathbf{u}^2/\mathbf{v}^2}$. Since $\mathbf{v}$ is greater than $\mathbf{u}$, the implication is that the second time (parallel) is less than the first time (perpendicular), and hence the second girl wins the race. The ratio factor is a key analogy to understanding the Michelson-Morley experiment we will discuss in a later chapter on special relativity.

4. The first quantity we need is the theoretical range of the football. This turns out to be

$$R = \frac{\mathbf{v}_i^2 \sin 2\theta}{\mathbf{g}} = \frac{(25)^2(\sin 50)}{9.8} = 48.85 \text{ m}$$

The receiver needs to travel 18.85 m away from the quarterback. To determine the speed the receiver must run, we need to know how long it takes the ball to travel 48.85 m. Since the horizontal velocity is assumed to be constant, we can say that the total time will be equal to 48.85 m divided by the horizontal velocity. Thus, $\mathbf{v}_h = 25 \cos 25 = 22.66$ m/s and $t_{total} = 48.85 \text{ m} / 22.66 \text{ m/s} = 2.16$ s. Therefore, to travel 18.85 m in 2.16 s, the receiver must run with a constant speed of 8.73 m/s.

6
FORCES AND NEWTON'S LAWS OF MOTION

INTRODUCTION—FORCES

From our discussion in Chapter 3 about frames of reference, you should be able to convince yourself that if two objects have the same velocity, then the relative velocity between them is zero, and therefore one looks like it is at rest with respect to the other. In fact, it would be impossible to decide whether or not such an *inertial* observer is moving or not. Therefore, accepting this fact, we state that an object that appears to be at rest in the earth frame of reference will simply be stated to be at rest relative to us (the observers).

Observations inform us that an inanimate object will not change its motion freely of its own accord unless an interaction takes place between it and at least one other object. This interaction usually involves contact between the objects, although gravity exerted on the object by the earth or any other body does not involve any direct contact. This type of interaction is sometimes called action at a distance. Electrostatic attraction and repulsion are also examples of this effect (as well as magnetic attraction and repulsion).

When the interaction occurs, we say that a **force** has been created between the objects. If the object was moving (relative to us), then the force may change the direction of the motion or it might change its speed. In other words, there is a change in the velocity, which is a vector quantity. This means that the magnitude and/or direction of the velocity will change. If no change occurs, then we must conclude the presence of another force that resists the changes induced by the one applied. Friction is an example of an opposing force of this type.

Some forces are deflecting in nature, and other forces might be restorative. An example of this is a spring or other elastic material that has been stretched. Once the material has been elongated, it snaps back in an attempt to return to its original status. If all the forces acting on an object produce no net change, then the object is in a state of **equilibrium**. If the object is moving relative to us, we say it is in a state of **dynamic equilibrium**. If the object is at rest relative to us, then we say that it is in a state of **static equilibrium**.

NEWTON'S LAWS OF MOTION

In 1687, at the urging of his friend Edmund Halley, Isaac Newton published his greatest work. It was titled the *Mathematical Principles of Natural*

Philosophy but is more widely known by a shortened version of its original Latin title—*Principia*. In *Principia*, Isaac Newton revolutionized the rational study of mechanics with the introduction of mathematical principles that all of nature obeys. Using his newly developed ideas, Newton set out to explain the observations and analyses of Galileo Galilei and Johannes Kepler.

The ability of an object to resist a change in its state of motion is called **inertia**. This concept is the key to Newton's first law of motion:

> Every body continues in its state of rest, or of uniform motion in a straight line, unless it is compelled to change that state by forces impressed upon it.

In other words, unless acted upon by an external force, an object at rest tends to stay at rest and an object in motion tends to stay in motion. By *rest*, we of course mean the observed state of rest in a particular frame of reference. The concept of *inertia* is taken to mean the tendency of an object to resist a force attempting to change its state of motion. As we will subsequently see, this concept is covered under the new concept of **mass** (a scalar, as opposed to **weight**, a vector force).

If a mass has an unbalanced force incident on it, then the velocity of the mass is observed to change. The magnitude of this velocity change depends inversely on the amount of mass. In other words, a force directed along the direction of motion causes a smaller mass to accelerate more than a larger mass. Newton's second law of motion expresses these observations as follows.

> The change of motion is proportional to the motive force impressed and is made in the direction of that force.

The acceleration produced by a force is in the same direction as the force, however, this does not mean that the object's direction must remain the same. Mathematically, the second law is sometimes expressed as $\mathbf{F} = m\mathbf{a}$. However, to preserve the vector nature of the force and since by *force* we mean *net force*, we can write the second law as

$$\mathbf{F}_{net} = \Sigma\mathbf{F} = m\mathbf{a}$$

The units of force are called **newtons** and are defined to be the force needed to give a 1-kg mass an acceleration of 1 m/s^2. Thus, 1 N = 1 kg-m/s^2. This being the case, the weight of an object is therefore given by $\mathbf{W} = m\mathbf{g}$, where $\mathbf{g}$ is the acceleration of gravity and the units of weight are newtons. In the SI system of units, the kilogram is taken as the standard unit of mass, and it corresponds to 9.8 N of weight or about 2.2 pounds (lb).

Newton's third law of motion is crucial for understanding the laws of conservation we will discuss later on. It stresses the fact that forces are the result of mutual interactions and are thus produced in pairs. The third law is usually stated as follows.

> For every action force, there is an equal and opposite reaction force.

When Newton wrote his second law of motion, his concept of "quantity of motion" was defined to be the product of an object's mass and velocity. In modern times, this concept is known as momentum, and it will be discussed in a separate chapter.

STATIC APPLICATIONS OF NEWTON'S LAWS

If we look more closely at the second law of motion, we will see an interesting implication. If the net force acting on an object is zero, then the acceleration of the object will likewise be zero. Notice, however, that kinematically zero acceleration does not imply zero motion. It simply indicates that the velocity of the object is not changing. If the object is in a state of rest and remains at rest (because of zero net forces), then it is in static equilibrium. Some interesting problems in engineering deal with the static stability of structures. Let's look at a simple example.

Place this book on a table. You will observe that it is not moving relative to you. Its state of rest is provided by the zero net force between the downward force of gravity and the upward reactive force of the table (pushing on the floor, which in turn pushes up on the table, etc.). Figure 1 shows this setup. The upward reactive force of the table is sometimes called the **normal force** and always acts perpendicular to the surface.

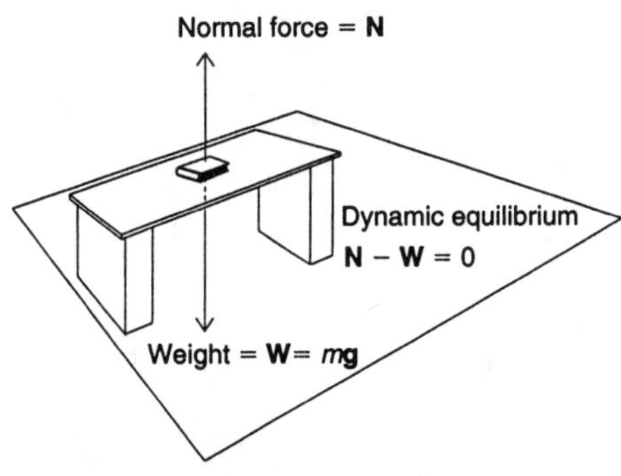

FIGURE 1.

In this example, we have used both the second and third laws. Remember, the second law is a vector equation, and so we must treat the sum of all vector forces in each dimension separately. In this case, we have $\Sigma \mathbf{F}_y = \mathbf{N} - \mathbf{W} = 0$. There are no forces in the x direction to be "analyzed."

A slightly more complex problem concerns a mass suspended by two strings. The tensions in the strings support the mass and are thus vector forces acting at angles. In Figure 2 a 10-kg mass is suspended by two strings making angles of 30 and 60 degrees, respectively, to the horizontal. The question is, What are the tensions in the two strings?

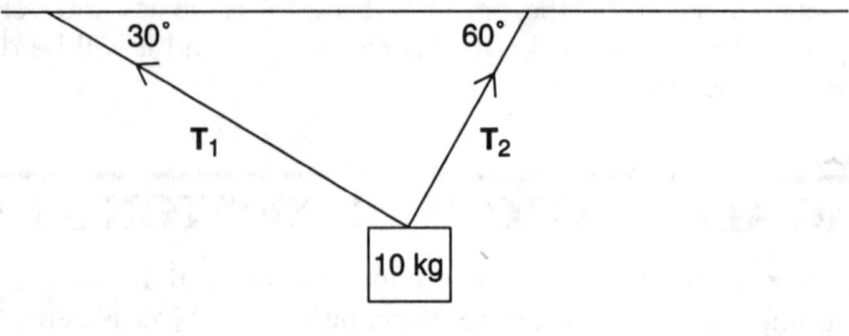

FIGURE 2.

A useful heuristic for solving this type of problem is to construct what is called a *free body diagram*. In such a diagram, you "free" the body of its realistic constraints and redraw it indicating the magnitudes and directions of all applied forces (known or unknown). You then use the diagram to set up static equations for Newton's second law in both the x and y directions. The result is a system of two equations and two unknowns. In our example, a free body diagram for both the x and y directions looks like Figure 3.

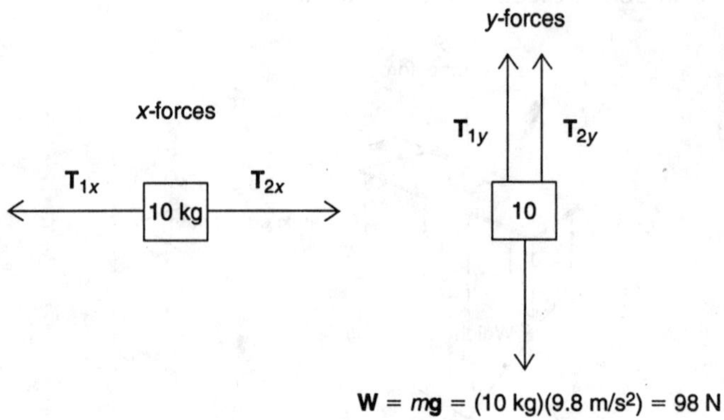

FIGURE 3.

We can now set up our equations for Newton's second law using the techniques of vector analysis reviewed in Chapter 3. We thus determine the x and y components of the tensions, which are given by $\mathbf{T}_{1x} = -\mathbf{T}_1 \cos 30$ and $\mathbf{T}_{2x} = \mathbf{T}_2 \cos 60$. The negative sign is used since $\mathbf{T}_{1x}$ is to the left. Thus, in the x direction we write $\Sigma \mathbf{F}_x = 0 = \mathbf{T}_2 \cos 60 - \mathbf{T}_1 \cos 30 = 0.5\ \mathbf{T}_2 - 0.866\mathbf{T}_1$. Likewise, for the y direction we see that the 10-kg mass weighs 98 N. The downward pull of gravity is compensated for by the upward pull of the two vertical components of the tensions. That is, we can write for the y direction, $\Sigma \mathbf{F}_y = 0 = \mathbf{T}_1 \sin 30 + \mathbf{T}_2 \sin 60 - \mathbf{W} = 0.5\ \mathbf{T}_1 + 0.866\mathbf{T}_2 - 98$. These two simultaneous equations can be solved for $\mathbf{T}_1$ and $\mathbf{T}_2$. Performing the algebra, we find that $\mathbf{T}_1 = 48.97$ N and $\mathbf{T}_2 = 84.87$ N.

Another static situation occurs when a mass is hung from an elastic spring. It is observed that when the mass is attached vertically to a spring that has a certain natural length, the amount of stretching, or elongation, is directly proportional to the applied weight. This relationship is known as **Hooke's law**, and it provides a technique for measuring static forces.

Mathematically, Hooke's law is given as $\mathbf{F} = -k\mathbf{x}$, where k is the spring constant in units of N/m and $\mathbf{x}$ is the elongation beyond the natural length. The negative is used to indicate that the applied force is restorative such that if allowed, the spring would accelerate back in the opposite direction. As a static situation, a given spring can be calibrated for known weights or masses and thus used as a "scale" for indicating weight or other applied forces.

DYNAMIC APPLICATIONS OF NEWTON'S LAWS

If the net force acting on an object is not zero, Newton's second law implies the existence of an acceleration in the direction of the net force. Therefore, if a 10-kg mass is acted on by a 10-N force, the acceleration will be 1 m/s^2. Heuristically, you can construct a free body diagram for the system to analyze all forces acting in all directions and then apply the second law of motion. Be careful to make sure which frame of reference and coordinate system you are using. For example, if the mass is sliding down a frictionless incline, a natural coordinate system to use is one that is rotated such that the x axis is parallel to the incline (see Figure 4).

Therefore, in order to resolve the force of gravity into components parallel and perpendicular to the incline, we must first identify the relevant angle in the geometry. Once this is done, we can see that the magnitude of the downward component of the weight along the incline is given by $-m\mathbf{g} \sin \theta$, where θ is the angle of the incline. The magnitude of the normal force, perpendicular to the incline, is therefore given by $\mathbf{N} = m\mathbf{g} \cos \theta$.

The direction of $m\mathbf{g} \cos \theta$ can be taken as being into the incline's surface and provides the force necessary to keep the block in contact with the surface.

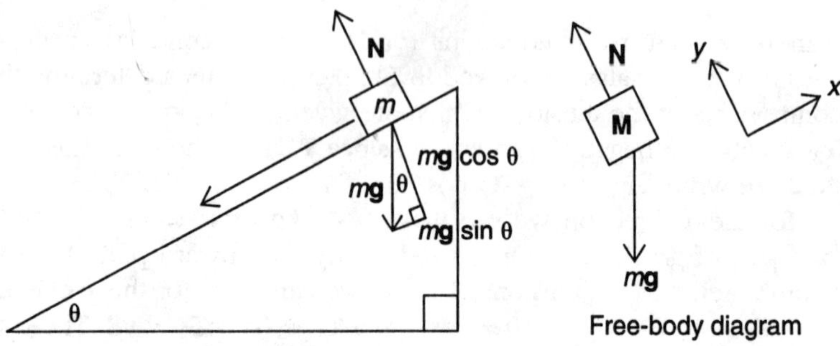

FIGURE 4.

If friction were present, this force would be the main contributor to the frictional force, which would be directed opposite the direction of motion.

For example, if the mass were 10 kg, the weight would be 98 N. If $\theta = 30$ degrees, then we would write for the x forces, $\Sigma \mathbf{F}_x = m\mathbf{a} = -m\mathbf{g} \sin \theta = -98 \sin 30 = -49$ N, and since there is no acceleration in the y direction, we have $\Sigma \mathbf{F}_y = 0 = \text{N} - m\mathbf{g} \cos \theta = \text{N} - 98 \cos 30 = \text{N} - 8.49$. Thus, the normal force is equal to 8.49 N.

CENTRAL FORCES

Consider a point mass moving around a circle, supported by a string making a 45-degree angle to a vertical post (the **conical pendulum**). Let's analyze this situation as seen in Figure 5.

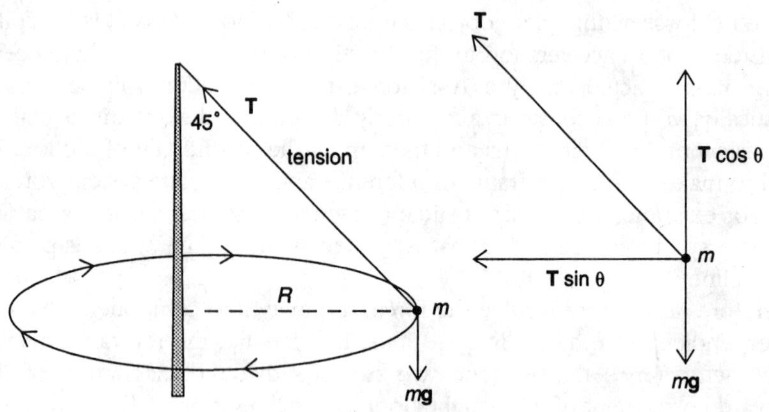

FIGURE 5.

In this case, the magnitude of the weight $m\mathbf{g}$ is balanced by the upward component of the tension in the string, given by $\mathbf{T}\cos\theta$. The inward component of the tension, $\mathbf{T}\sin\theta$, is responsible for providing a **centripetal force**. However, from the frame of reference of the mass, the force experienced is outward and is referred to as **centrifugal force**. This force is often called "fictitious" since it is perceived only in the mass's frame of reference.

If we specify that the radius is 1.5 m and the velocity of the mass is unknown, then we see that vertically, $\mathbf{T}\cos 45 = m\mathbf{g}$. This implies that if $mg = 98$ N, then $T = 138.6$ N. Horizontally, we see that $\mathbf{T}\sin 45 = m\mathbf{v}^2/R$, where the expression for centripetal force is given by $\mathbf{F}_c = m\mathbf{v}^2/R$, using the logic of $\mathbf{F} = m\mathbf{a}$. Using the known information, we solve for velocity and find that $\mathbf{v} = 3.83$ m/s. It is important to remember that the components of forces must be resolved along the principal axes of the chosen coordinate system.

FRICTION

Friction is a contact force between two surfaces that is responsible for opposing sliding motion. Even the smoothest surfaces are microscopically rough with peaks and valleys like a mountain range. When an object is first moved, this friction plus inertia must be overcome. If a spring balance is attached to a mass and then pulled, the reading of the force scale when the mass first begins to move will be a measure of the static friction. Once the mass is moving, if we maintain a steady enough force, it will move with constant velocity. Thus, the acceleration will be zero, indicating that the net force is zero. The reading of the scale will measure the kinetic friction, and this reading will generally be less than the starting reading.

Observations show that frictional force is directly related to the applied load pushing the mass into the surface. From our knowledge of forces, this implies that the normal force is responsible for this action. The proportionality constant linking the normal force with the frictional force is called the **coefficient of friction** μ. There are two coefficients of friction: one for static friction and the other for kinetic friction. For a static case, we would have that $f \leq \mu_s N$. For a kinetic case, we would have that $f = \mu_k N$.

The analysis of a situation involves identifying the normal force. For example, in Figure 6, a mass is being pulled along a horizontal surface by a string making an angle θ.

If we set up our equations of motion, we see that $\Sigma\mathbf{F}_y = 0 = N + \mathbf{F}\sin\theta - m\mathbf{g}$. Thus, the normal force is going to be less than the weight $m\mathbf{g}$ because of the upward component of the applied force. Therefore, $\mathbf{N} = m\mathbf{g} - \mathbf{F}\sin\theta$, and the force of kinetic friction, given a coefficient of kinetic friction μ_k, is written as $f = \mu_k\mathbf{N} = \mu_k(m\mathbf{g} - F\sin\theta)$. Thus, $\Sigma\mathbf{F}_x = \mathbf{F}\cos\theta - \mu_k(m\mathbf{g} - \mathbf{F}\sin\theta) = m\mathbf{a}$.

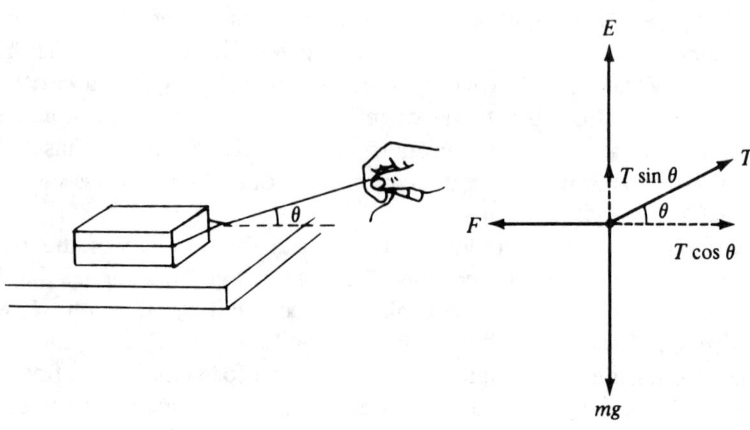

FIGURE 6.

Sample Problem

What is the acceleration of a 5-kg mass being pulled by a string making an angle of 30 degrees with the horizontal when the applied force in the string is 100 N and the coefficient of kinetic friction between the mass and the ground is 0.1?

Solution

If we solve the last equation for the acceleration and substitute in all known values, we have

$$\mathbf{a} = \frac{F \cos \theta - \mu_k(m\mathbf{g} - F \sin \theta)}{m} = 17.34 \ \frac{\text{m}}{\text{s}^2}$$

PROBLEM-SOLVING STRATEGIES

The key to solving force problems, whether static or dynamic, is in constructing the proper free body diagram. Remember, Newton's laws operate on vectors, and so you must be sure to resolve all forces into components once a frame of reference and coordinate system have been chosen. Therefore, you should

1. Choose a coordinate system for your problem.
2. Make a sketch of the situation if one is not provided.
3. Construct a free body diagram for the situation.
4. Resolve all forces into perpendicular components based on the chosen coordinate system.
5. Write Newton's second law as the sum of all forces in a given direction. If it is a static situation, set the summation equal to zero. If the situation

is dynamic, set the summation equal to $m\mathbf{a}$. Be sure to include only applied forces in the diagram.

6. Seek out the normal force, which is always perpendicular to the surface.
7. Remember that the centripetal force is always directed inward toward the circular path and parallel to the plane of the circle. Gravity is always directed vertically downward.
8. Carefully solve your algebraic equations using the techniques for simultaneous equations.

PRACTICE PROBLEMS FOR CHAPTER 6

Thought Problems

1. How does the rotation of the earth affect the apparent weight of a 1-kg mass at the equator?

2. If you overwax a floor, you can actually increase the coefficient of kinetic friction instead of lowering it. Explain why this might happen.

3. If forces occur in *action-reaction* pairs that are equal and opposite, how is it possible for any one force to cause an object to move?

Multiple-Choice Problems

1. In the situation shown, what is the tension in string 1?

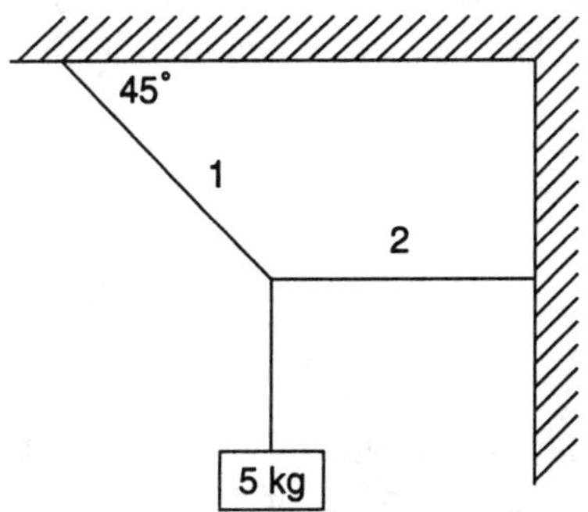

(A) 69.3 N
(B) 98 N
(C) 138.6 N
(D) 147.6 N
(E) 155 N

2. Two masses M and m are hung over a massless, frictionless pulley as shown. If $M > m$, then what is the downward acceleration of mass M?

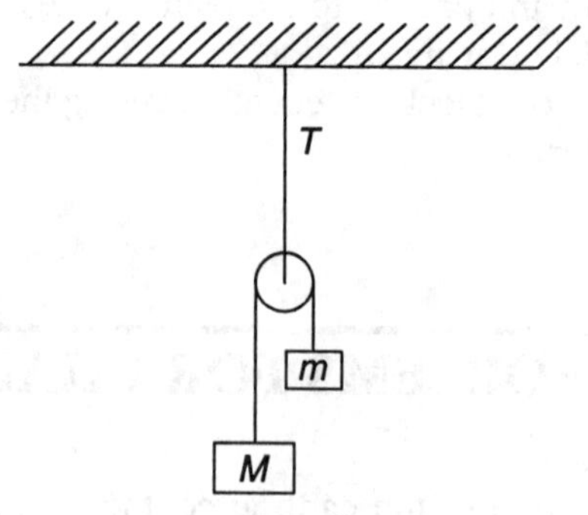

(A) $\mathbf{g}$
(B) $\frac{M - m}{M + m}\mathbf{g}$
(C) $\frac{M}{m}\mathbf{g}$
(D) $\frac{Mm}{M + m}\mathbf{g}$
(E) $Mm\mathbf{g}$

3. A 0.25-kg mass is attached to a string and swung in a vertical circle whose radius is 0.75 m. At the bottom of the circle, the mass is observed to have a speed of 10 m/s. What is the magnitude of the tension in the string at that point?
(A) 2.45 N
(B) 5.78 N
(C) 22.6 N
(D) 35.7 N
(E) 44.7 N

4. A car and driver have a combined mass of 1,000 kg. The car passes over the top of a hill that has a radius of curvature equal to 10 m. The speed of the car at that instant is 5 m/s. What is the force of the hill on the car as it passes over the top?
(A) 7,300 N up
(B) 7,300 N down
(C) 12,300 N up
(D) 12,300 N down
(E) 0 N

5. A hockey puck with a mass of 0.3 kg is sliding along some ice that can be considered frictionless. Its velocity is 20 m/s. The puck then crosses over onto a floor that has a coefficient of kinetic friction equal to 0.35. How far will the puck travel across the floor before it stops?
   (A) 3 m
   (B) 87 m
   (C) 48 m
   (D) 92 m
   (E) 58 m

6. A spring with a stiffness constant of k = 50 N/m has a natural length of 0.45 m. It is attached to the top of a 2.4-m-long incline that makes a 30-degree angle with the horizontal. A mass of 2 kg is attached to the spring, which causes it to be stretched down the incline. How far down the incline does the end of the spring rest?
   (A) 0.196 m
   (B) 0.45 m
   (C) 0.646 m
   (D) 0.835 m
   (E) 1.2 m

7. A 20-N force is pushing two blocks horizontally along a frictionless floor as shown. What is the magnitude of the force that the 8-kg mass exerts on the 2-kg mass?

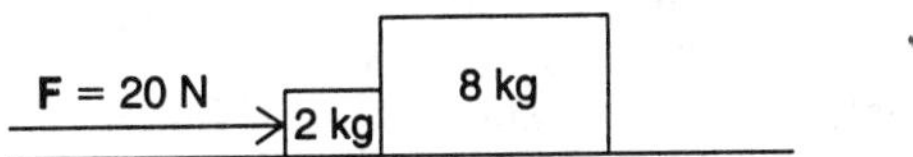

   (A) 4 N
   (B) 8 N
   (C) 16 N
   (D) 20 N
   (E) 24 N

8. A force of 20 N acts horizontally on a mass of 10 kg being pushed on a frictionless incline that makes a 30-degree angle with the horizontal as shown. The magnitude of the acceleration of the mass on the incline is equal to _____ m/s^2.

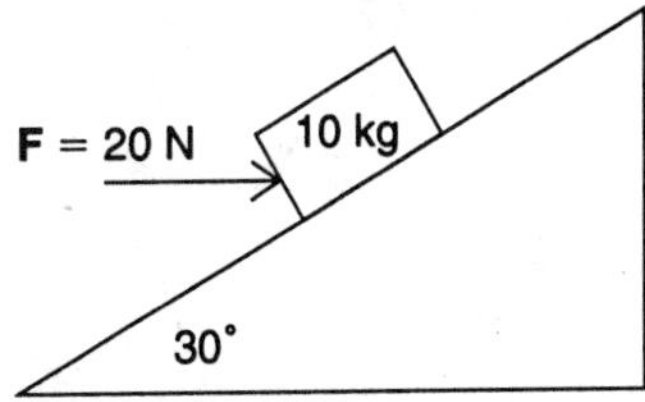

   (A) 1.9
   (B) 2.2

(C) −3.17
(D) −3.87
(E) −4.3

9. Based on the accompanying diagram, what is the tension in the connecting string if the table is frictionless?

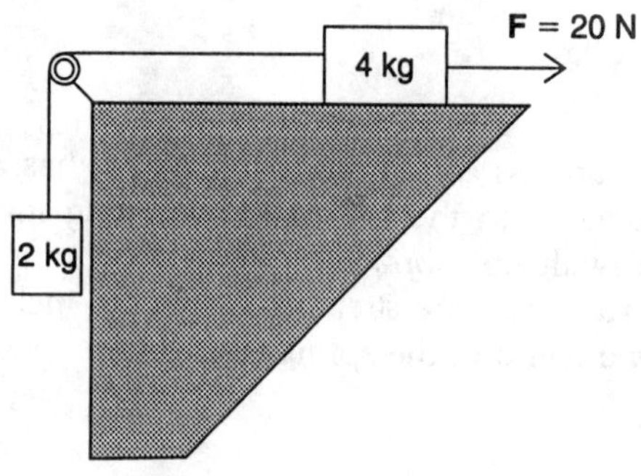

(A) 6.4 N
(B) 13 N
(C) 19.7 N
(D) 25 N
(E) 32 N

10. A mass M is released from rest on an incline that makes a 42-degree angle with the horizontal. In 3 s, the mass is observed to have gone a distance of 3 m. What is the coefficient of kinetic friction between the mass and the surface of the incline?
(A) 0.8
(B) 0.7
(C) 0.6
(D) 0.5
(E) 0.3

Free-Response Problems

1. A 3-kg block is placed on top of a 7-kg block as shown. The coefficient of kinetic friction between the 7-kg block and the surface is 0.35. A horizontal force **F** acts on the 7-kg block.

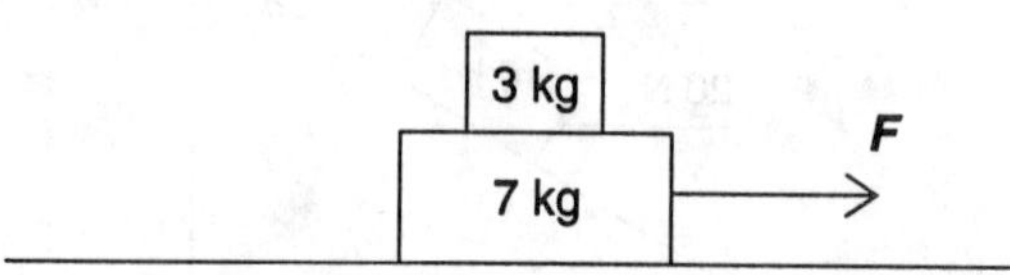

(a) Draw a free body diagram for both blocks.
(b) Calculate the magnitude of the applied force **F** necessary to maintain an acceleration of 5 m/s^2.

(c) Find the minimum coefficient of static friction necessary to prevent the 3-kg block from slipping.

2. A curved road is banked at an angle θ such that friction is not necessary for a car to stay on the road. A 2,500-kg car is traveling at a speed of 25 m/s and the road has a radius of curvature equal to 40 m.
   (a) Draw a free body diagram for the situation described.
   (b) Find the angle θ.
   (c) Calculate the magnitude of the force that the road exerts on the car.
3. The "rotor" is an amusement park ride that can be modeled as a rotating cylinder, with radius *R*. A person inside the rotor is held motionless against the sides of the ride as it rotates with a certain velocity. The coefficient of static friction between a person and the sides is μ.
   (a) Derive a formula for the period of rotation *T* in terms of *R*, *g*, and μ.
   (b) If $R = 5$ m and $\mu = 0.5$, calculate the value of the period *T* in seconds.
   (c) Using the answer to part (b), calculate the angular velocity ω in rad/s.
4. A box of mass *M* rests on a rough surface with coefficient of static friction μ. A force **F** is applied downward to the box at an angle θ.
   (a) Derive an expression for the minimum value of **F** needed to move the box in terms of *M*, **g**, θ, and μ.
   (b) If $M = 50$ kg, $\theta = 15$ degrees, and $\mu = 0.3$, what is the magnitude of this value of **F?**

SOLUTIONS TO PRACTICE PROBLEMS

Thought Problems

1. In the frame of reference of the mass, there is an apparent upward force that tends to reduce the apparent weight of the mass at the equator. This upward force is very small, and only sensitive scales can detect it.
2. Initially the wax fills in the ridges and furrows of a floor on a microscopic level. This reduces the coefficient of kinetic friction. However, each successive layer of wax builds up to the point where the surface condition changes and the wax buildup actually makes the floor sticky by increasing the coefficient of friction.
3. This is a classic question that is very tricky. It is true that for every action there is an equal and opposite reaction. However, these forces act on *different* objects. Thus, the applied force, if it is a net force, can still cause an object to move.

Multiple-Choice Problems

1. **A** We need to apply the second law for static equilibrium. That means we must resolve the tensions into their *x* and *y* components. For $\mathbf{T}_1$, we have $\mathbf{T}_1 \cos 45$ and $\mathbf{T}_1 \sin 45$. Since the system is in equilibrium, we know that the sum of all the forces in the *x* direction must equal zero. This means that $\mathbf{T}_1 \cos 45 = \mathbf{T}_2$ (since the second tension is entirely

horizontal). The y component of $\mathbf{T}_1$ must balance the weight, $\mathbf{W} = m\mathbf{g}$. With a 5-kg mass, $\mathbf{W} = 49$ N. Therefore, we see that $\mathbf{T}_1 \sin 45 = 49$ N and $\mathbf{T}_1 = 69.3$ N.

2. **B** The free body diagrams for both masses look like the accompanying:

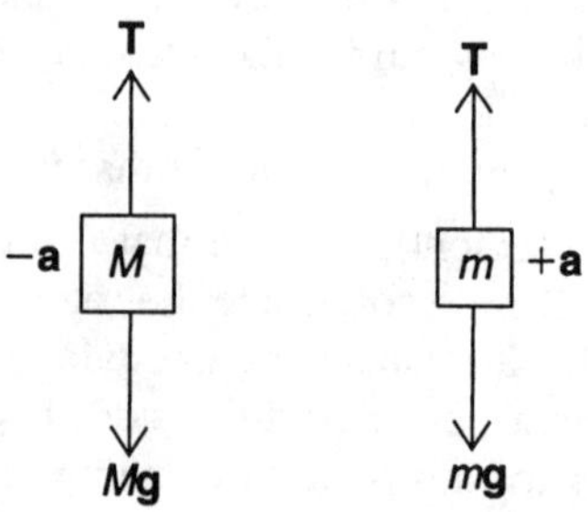

The large mass is accelerating downward, while the small mass is accelerating upward. The tension in the string is directed upward, while gravity, given by the weight, is directed downward. Using the second law for accelerated motion, we must show that $\Sigma\, \mathbf{F}_x = m\mathbf{a}$ and $\Sigma \mathbf{F}_y = m\mathbf{a}$ separately. Thus, we have $\mathbf{T} - M\mathbf{g} = -M\mathbf{a}$ and $\mathbf{T} - m\mathbf{g} = m\mathbf{a}$. Eliminating the tension $\mathbf{T}$ and solving for $\mathbf{a}$ yields $(M - m)\,\mathbf{g} / (M + m)$.

3. **D** The situation is shown in the accompanying diagram. At the lowest point, the downward force of gravity is matched by the upward tension in the string. Thus, we can write $\mathbf{T} - m\mathbf{g} = m\mathbf{v}^2/R$. Substituting the given values and solving for $\mathbf{T}$ yields 35.7 N.

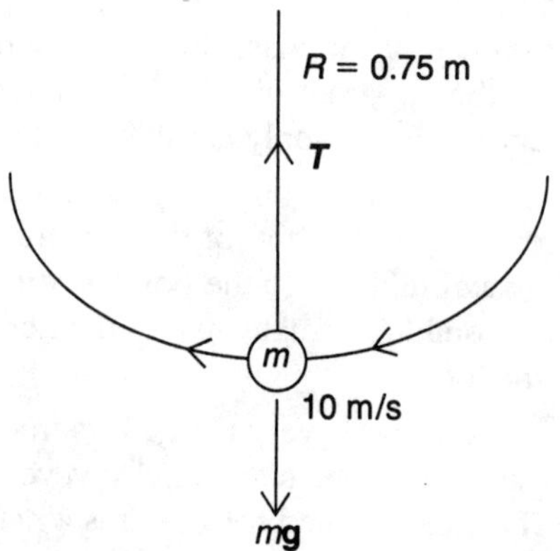

4. **A** As the car goes over the hill, the force that the hill exerts on it is the normal force $\mathbf{N}$. This is modified by the downward force of gravity, and the combination produces (as in Question 6), the centripetal force $\mathbf{F}_c$. Thus, we can write $\mathbf{N} - m\mathbf{g} = -m\mathbf{v}^2/R$. The centripetal force, in this case, is directed downward toward the center of the circular arc as shown and hence is written as a negative. Substituting the given values and solving for $\mathbf{N}$ gives us 7,300 N upward.

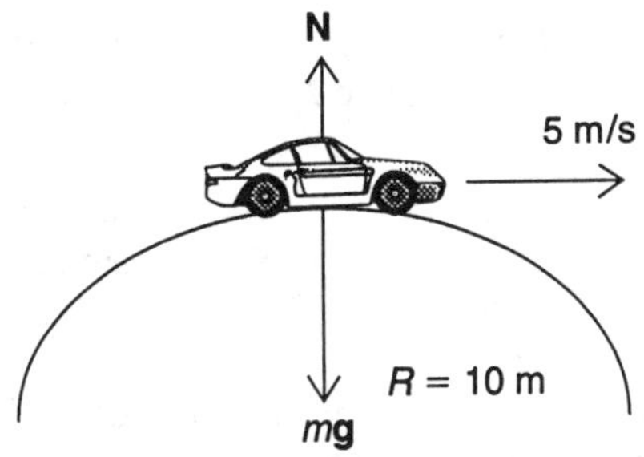

5. **E** With horizontal motion, the normal force is equal to the force of gravity (if there are no other forces with vertical components). When the mass is moving with constant velocity, there are no net forces acting on it. Thus, when friction acts to slow it down, it is the only net force. Therefore, we write $\mathbf{f} = \mu\mathbf{N} = \mu\ m\mathbf{g} = -m\mathbf{a}$ (since it is decelerating). Substituting the given numbers results in a deceleration of $-3.43\ \text{m/s}^2$. Now, the final velocity will be zero, and since the time required to stop is unknown, we use the following kinematic expression to solve for the stopping distance: $-\mathbf{v}_i^2 = 2\mathbf{a}x$. Substituting the given numbers yields approximately 58 m as the distance needed to stop.

6. **C** The situation is shown in the accompanying diagram. The spring constant is $k = 50$ N/m. According to Hooke's law, $\mathbf{F} = k\mathbf{x}$, where $\mathbf{x}$ is the elongation in excess of the spring's natural length (0.45 m in this case). The force, in this case, is provided by the component of weight parallel to the incline which is given by $m\mathbf{g} \sin \theta$. Substituting the given numbers yields an elongation of $\mathbf{x} = 0.196$ m. Since this must be in excess of its natural length, the answer is 0.646 m.

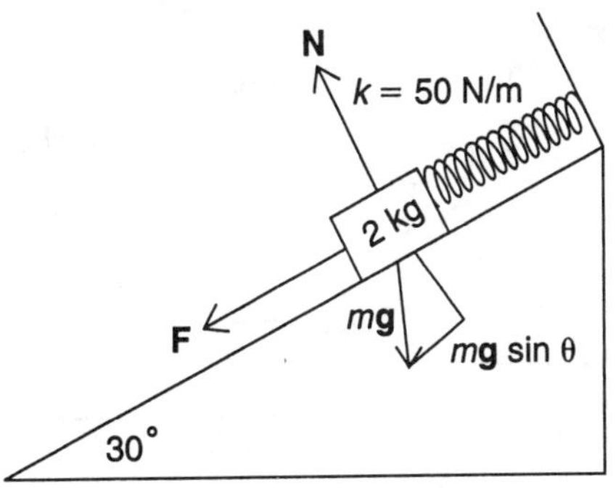

7. **C** The 20-N force is pushing on a total mass of 10 kg. Thus, using $\mathbf{F} = m\mathbf{a}$, the acceleration of both blocks is equal to 2 m/s². If we draw a free body diagram for the 2-kg mass:

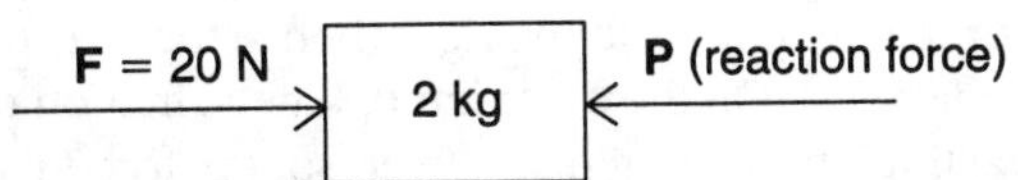

Let **P** represent the force that the 8-kg mass exerts on the 2-kg mass. Writing the second law, we have **F** − **P** = *m***a**. To find **P**, we substitute the given numbers: 20 − **P** = (2)(2) = 4. Thus, **P** = 16 N.

8. **C** From the diagram, we see that the force necessary to move the mass up the incline must be in excess of the component force of gravity trying to push it down the incline. The component of gravity down the incline is always given by *m***g** sin θ. Resolving the given force into a component parallel to the incline (and the other perpendicular to the incline) shows that the force up the incline is (at the same angle) **F** cos θ. Thus, in general we write **F** cos θ − *m***g** sin θ = *m***a**. Substituting our numbers gives **a** = −3.17 m/s^2.
9. **C** From the diagram, we see that the 4-kg mass is accelerating to the right and the 2-kg mass is accelerating upward. Thus, we write that the sum of all forces in each direction equals *m***a**. In the x direction we have (since the tension in the string will try to pull left) −**T** + 20 = 4**a**. In the y direction, the tension pulls up against gravity. We therefore write **T** − 19.6 = 2**a**. Solving for **T** by eliminating **a**, we get **T** = 19.7 N.
10. **A** We know that the mass accelerates from rest uniformly in 3 s and goes a distance of 3 m. Thus, we can say that d = s$\frac{1}{2}$ **a**t^2. Substituting gives us an acceleration of 0.67 m/s^2. On an incline, the normal force is given by *m***g** cos θ, and so friction is given by **f** = μ*m***g** cos θ. Once again, this force opposes the downward force of gravity parallel to the incline given by −*m***g** sin θ. In the downward direction, these two forces are added together and set equal to −*m***a**. Thus, we write for this case, μ *M*(9.8) (cos 42) − *M*(9.8) (sin 42) = − M (0.67). The masses all cancel out, and if we solve for μ, you get 0.8 is the coefficient of kinetic friction.

Free-Response Problems

1. A free body diagram for the two masses is given here.

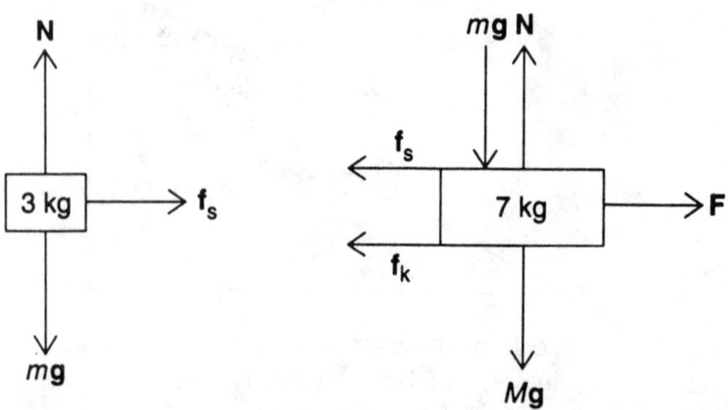

The force of static friction between the two blocks is the force responsible for accelerating the 3-kg block to the right (hence the direction of friction in the free body diagram). On the other hand, kinetic friction opposes the applied force **F** acting on the 7-kg mass. Even so, the force **F** must accelerate both masses combined. Thus, in the horizontal direction we can write for the second law of motion, $\mathbf{F} - \mathbf{f} = M\mathbf{a}$, and $\mathbf{f} = \mu\mathbf{N}$ (since we have horizontal motion, the normal force is equal to the combined weights): $\mathbf{F} - (0.35)(10)(9.8) = (10)(5)$. Solving for **F**, we obtain $\mathbf{F} = 70.15$ N.

Now, for the two masses together, we know that static friction provides the force needed for the 3-kg mass to accelerate at 5 m/s^2. The normal force on the "surface" is just the weight of the 3-kg mass as seen in the free body diagram. Thus, $\mu(3)(9.8) = (3)(5)$. From this we have $\mu = 0.51$.

2. The situation is in the accompanying diagram. In the coordinates chosen, the component of the normal force parallel to the plane of the circular road provides the centripetal acceleration. Thus, we can write, in the absence of friction, $\mathbf{N} \sin \theta = m(\mathbf{v}^2/r)$. We can also see that $\mathbf{N} \cos \theta = m\mathbf{g}$. Thus, eliminating **N** from both equations gives $\tan \theta = \mathbf{v}^2/r\mathbf{g}$. Substituting the given numbers results in $\tan \theta = 1.59$ and $\theta = 57.8$ degrees (rather steep).

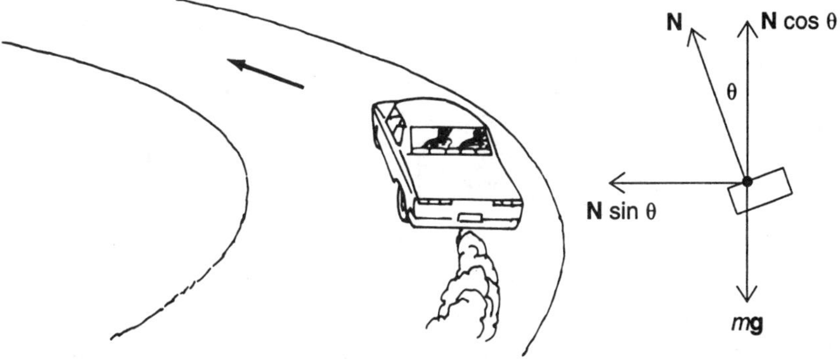

3. The rotor is seen in the accompanying diagram. Here the normal force is perpendicular to the person and supplies the inward centripetal force necessary for rotation. Friction acts along the walls against gravity, $m\mathbf{g}$, tending to slide the person downward. Thus, the key to stability is to be fast enough to maintain equilibrium.

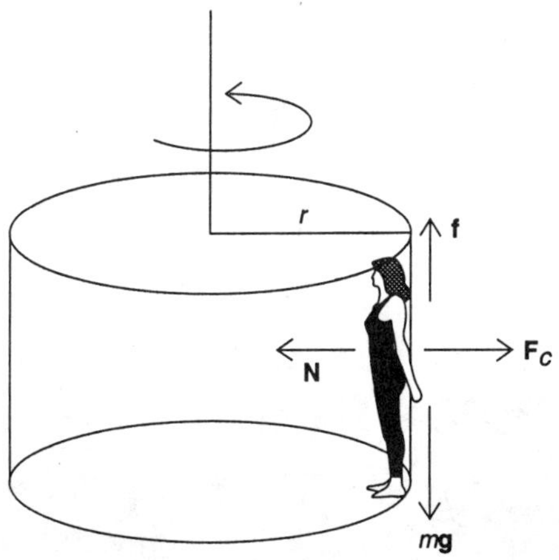

In the frame of reference of the person, there is an outward centrifugal force against the reaction force of the walls. Since $\mathbf{F} = m\mathbf{a}$, $\mathbf{N} = 4M\pi^2 r/T^2$. In the vertical direction $\mathbf{f} = M\mathbf{g}$, and thus $\mathbf{f} = \mu N = \mu M 4\pi^2 R/T^2 = M\mathbf{g}$.

Solving for T, we obtain $T = \sqrt{\mu 4\pi^2 R/\mathbf{g}}$. This makes sense since if the coefficient of friction is high, the rotation rate can be small and thus the period larger. Substituting the given numbers yields $T = 10$ s.

4. (a) The situation is shown in the accompanying diagram. The minimum force $\mathbf{F}$ has a magnitude such that the horizontal component is just equal to the force of static friction: $\mathbf{f} = \mu\mathbf{N}$. From the diagram, we can see that $\mathbf{N} = M\mathbf{g} + \mathbf{F}\sin\theta$ and that $\mathbf{F}\cos\theta = \mu(M\mathbf{g} + \mathbf{F}\sin\theta) = \mu M\mathbf{g} + \mu\mathbf{F}\sin\theta$. Collecting terms, we have $\mathbf{F}(\cos\theta - \mu\sin\theta) = \mu M\mathbf{g}$ and $\mathbf{F} = \mu M\mathbf{g}/(\cos\theta - \mu\sin\theta)$.

   (b) Substituting the given values, we obtain $\mathbf{F} = 165.48$ N.

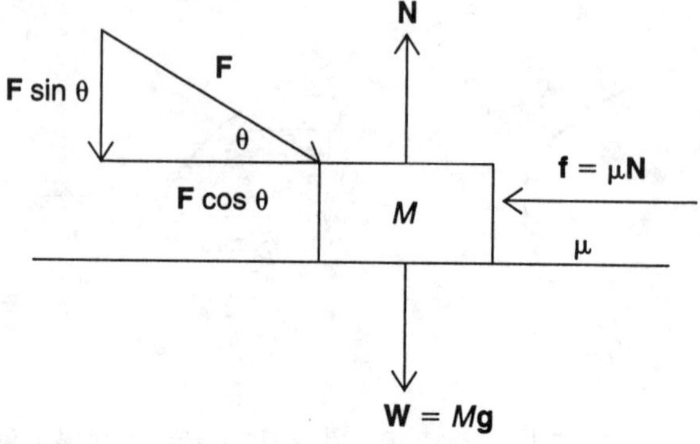

7
WORK AND ENERGY

WORK

Imagine we have two masses m_1 and m_2. Now suppose that two forces act, respectively, on each mass, providing individual accelerations according to Newton's second law of motion. How can we compare the effect of both these forces? One thing we can do is to observe their ratios:

$$\frac{\mathbf{f}_1}{\mathbf{f}_2} = \frac{m_1\mathbf{a}_1}{m_2\mathbf{a}_2}$$

One of the arguments of the Newtonian approach would be to compare the two forces relative to the time each force acts. Let us assume that they act simultaneously. Thus, we can write

$$\frac{\mathbf{f}_1 t}{\mathbf{f}_2 t} = \frac{m_1\mathbf{a}_1 t}{m_2\mathbf{a}_2 t} = \frac{m_1\mathbf{v}_1}{m_2\mathbf{v}_2}$$

where we have used the expression $\mathbf{v} = \mathbf{a}t$, assuming the objects start from rest.

Another way to relate the two forces is to consider the relative displacement through which each force acts. Let us assume they act through the same displacement $\mathbf{d}$. We can therefore write the ratio as

$$\frac{\mathbf{f}_1\mathbf{d}}{\mathbf{f}_2\mathbf{d}} = \frac{m_1\mathbf{a}_1\mathbf{d}}{m_2\mathbf{a}_2\mathbf{d}} = \frac{(1/2)m_1\mathbf{v}_1^2}{(1/2)m_2\mathbf{v}_2^2}$$

where, if we assume that the objects start from rest, we have used $\mathbf{v}^2 = 2\mathbf{ad}$.

Comparing these two representations involves a comparison between vectors and scalars. The first comparison relates a vector quantity expressed as the product of the force and the time (called the **impulse**). This quantity is proportional to the product of the mass and the velocity. However, the product of the force and the displacement is a *scalar* quantity and is proportional to the product of the mass and the square of the velocity. This scalar quantity **fd** is called the **work** done by the force.

Another way to think about this situation is to consider a graph of the force versus the displacement. Since an object's displacement is in the same direction as the net force applied, only the component of force acting in the direction of motion contributes to the work. Imagine that we have a constant force applied to an object. Figure 1 on page 82 is the graph of this relationship. The area under the line is equal to the work done. For variable forces, this method becomes most useful, and one can speak of the average force applied over the interval of distance displaced.

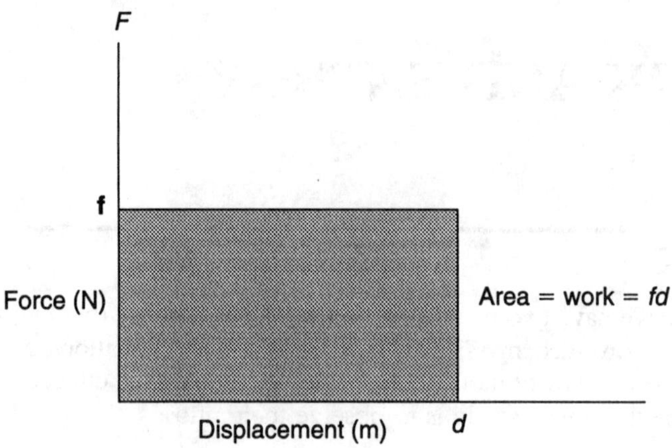

FIGURE 1.

Based on the equation for work, the units of work are N-m such that 1 N-m is equal to the work done by a 1-N force displacing an object 1 m. These units are also called **joules**. Using dimensional analysis, we see that based on the definition of the newton in Chapter 10, 1 J = 1 kg-m^2/s^2.

The notion of work as a physical concept can be seen in the following observation. It can be demonstrated that if a mass m is dropped from a height h onto a nail, it will drive in the nail a distance x. If a mass of $4m$ is dropped from a height of $h/4$ onto a similar nail, the nail will be driven in the same amount. In other words, since the force of gravity provides the acceleration for the mass (i.e., its weight), then this demonstration suggests that work (the effect of driving in nails) is equal to the product of force and displacement. In this case we can write that the work $W = m\mathbf{g}h$. Thus, for a given mass, more work will be done based on the height from which the mass is dropped.

APPLICATIONS OF THE WORK CONCEPT

Suppose a mass of 10 kg is being pulled by a string making a 30-degree angle with the horizontal as shown in Figure 2. The force on the string is 500 N, and the coefficient of friction between the mass and the ground is $\mu = 0.2$. How much work is done in displacing the mass a distance of 10 m?

The horizontal component of the applied force is equal to $\mathbf{F} \cos \theta$, and the vertical component is equal to $\mathbf{F} \sin \theta$. The directions are based on the fact that the string pulls up and to the right. However, $\mathbf{F} \cos \theta$ is not the only horizontal force present. Friction opposes the motion and is proportional to the normal force $\mathbf{N}$. This normal force is not just the reaction force to the weight since the upward component of the string force contributes to a tendency to try to pull up the mass. Thus, the magnitude of $\mathbf{N} = \mathbf{F} \sin \theta - M\mathbf{g}$ and so the magnitude of the

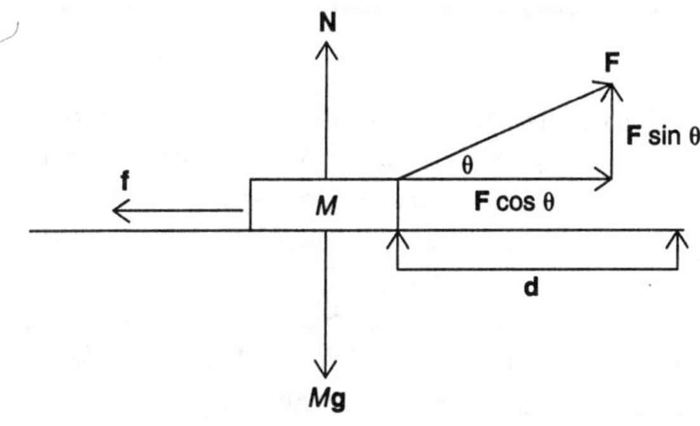

FIGURE 2.

frictional force $\mathbf{f} = \mu\mathbf{N} = \mu(\mathbf{F} \sin \theta - M\mathbf{g})$. The magnitude of the net horizontal force, which does the work, is therefore given by

$$\mathbf{F}_{net} = \mathbf{F} \cos \theta - \mu(\mathbf{F} \sin \theta - M\mathbf{g})$$

It is the product of this net force and the displacement **d** that evaluates the amount of work in joules. Substituting all the given numbers yields $W = 4{,}026$ J.

As another example, consider the work done in stretching a spring. From Hooke's law, we know that the elongation is directly proportional to the applied force: $\mathbf{F} = k\mathbf{x}$, where the stiffness constant k is in N/m. This is not a constant force, and so the graph of force versus elongation looks like Figure 3. The work done in stretching the spring is equal to the area of the triangle formed. Using Hooke's law, this turns out to be equal to $\frac{1}{2}k\mathbf{x}^2$ if we began from zero elongation. If we started at some other elongation, then we would write $W = \frac{1}{2}k\mathbf{x}_f^2 - \frac{1}{2}k\mathbf{x}_i^2$.

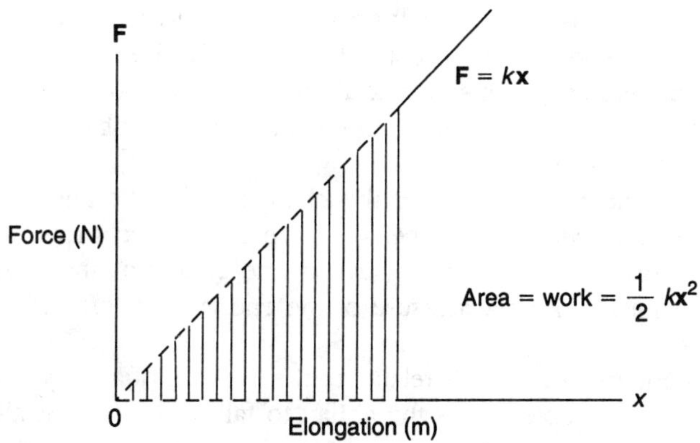

FIGURE 3.

POWER

It is often useful to consider the rate at which work is done. This rate is expressed as the number of joules of work done per second and is called **power**. Algebraically we can write that power = $P = W/t$. The units of power are J/s or **watts**. This relationship also implies that the work done is equal to the product of the power and the time (in seconds). If we have a constant force applied, then we can relate power to the average velocity attained. Since $P = W/t$ and $W = \mathbf{Fd}$, we can write $P = \mathbf{F}(\mathbf{d}/t)$. Recall that the ratio of distance to time is the average velocity. Hence we can say that $P = \mathbf{Fv}$.

ENERGY

While performing experiments on the dynamics of pendulum motion, Galileo discovered that if a nail or a peg is introduced at the vertical position of a swinging pendulum, thereby shortening the original length of string, the pendulum bob will still swing up to its original height. On the way back, the pendulum will swing away from the nail, resume its normal length, and still achieve the same height. While Galileo could not interpret this observation in terms of forces, we can now attribute it to the law of conservation of mechanical energy. To understand this concept better, let's return to our original discussion of work.

Remember from the discussion on page 83 that a comparison between force and displacement leads to the idea that work is proportional to the mass times the square of the velocity. In fact, if we apply a constant force to a mass and write $\mathbf{F} = m\mathbf{a}$, we see something interesting:

$$\mathbf{Fd} = m\mathbf{ad} = \frac{1}{2}m(\mathbf{v}_f^{\,2} - \mathbf{v}_i^2) = \frac{1}{2}m\mathbf{v}_f^2 - \frac{1}{2}m\mathbf{v}_i^2$$

The quantity $\frac{1}{2}m\mathbf{v}^2$ is called the **kinetic energy** (KE) of the object and is defined to be the energy the object has by virtue of its being in motion relative to a frame of reference. This is crucial since we can put ourselves into an inertial frame of reference in which the relative velocity observed is zero and the KE will be zero.

Work is therefore a measure of the change in kinetic energy, and since work implies the action of a force, we have established that forces are the agents of energy transfer. Thus, we can define energy as the capacity to do work and define work as a measure of the energy transferred. The units for KE are joules.

Another concept of energy is related to the position of an object. If we have a mass sitting on a table, it has the ability to fall to the floor if allowed (the table exerts an upward force preventing this from occurring). From our definition of energy, we can state that some type of positional energy exists which has the potential to do work.

Experimentally, we know that if we lift the mass slightly, it will land with a small velocity. However, if we lift the object high, it will land with a large velocity. The displacement of the mass upward implies that work has been done to the object. The ability to fall is due to some kind of positional energy which then converts itself into the kinetic energy of motion. This positional energy is known as **potential energy** (PE) in mechanics and is designated U in some textbooks. The units for PE are again joules.

The gain in potential energy obtained by lifting must be equal to the amount of work done (which we have previously shown to be equal to $m\mathbf{g}h$). This is because the change in kinetic energy is zero. We can therefore state that $\Delta PE = m\mathbf{g}\,\Delta h$ near the earth's surface.

There is a connection between PE and KE when we have vertical displacements in a gravitational field. However, just as in the case of KE, the frame of reference for PE is arbitrary and we can define the zero base level to be anywhere. What's important is only how much energy is changing (remember, work and energy are scalars).

CONSERVATION OF ENERGY

Suppose we lift a mass m a distance h above the ground as in Figure 4. The minimum amount of force needed to overcome gravity is equal to its weight $m\mathbf{g}$. The work done is therefore equal to $m\mathbf{g}h$. This energy goes into a gain in potential energy since if we release the mass, it will fall and gain velocity (i.e., it will gain kinetic energy). If the mass is allowed to fall from the height h, what will its velocity be at the bottom? Kinematically, we know that if a mass is dropped from a height h, then the time required for it to fall is given by $t = \sqrt{2h/\mathbf{g}}$. Given the fact that the initial velocity is zero, the velocity at the bottom, just before impact, will be $\mathbf{v} = -\mathbf{g}t$ or $\mathbf{v} = -\mathbf{g}\sqrt{2h/\mathbf{g}} = -\sqrt{2\mathbf{g}h}$ (where the negative sign indicates the downward direction).

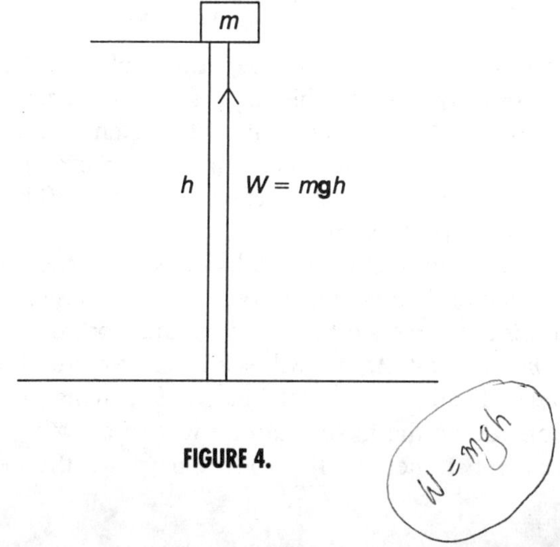

FIGURE 4.

W = mgh

Now, since the work done to raise the mass is equal to the change (gain) in potential energy, the work done by gravity to accelerate the mass downward when it is released must derive from that potential energy. Since the work done through motion is equal to the change in kinetic energy, we can observe that as the mass falls, it loses potential energy and gains a proportional amount of kinetic energy. In the absence of any frictional forces, we can state that $\Delta PE = \Delta KE$.

If we equate the observations that at the bottom it has lost all its potential energy and gained all its kinetic energy, then $m\mathbf{g}h = \frac{1}{2}m\mathbf{v}^2$. This implies, taking into account the direction of motion, that $\mathbf{v} = -\sqrt{2\mathbf{g}h}$ (as before). This discussion leads to the concept of the law of conservation of energy, which briefly states that energy is never created or destroyed. In the the absence of friction (which changes mechanical energy into heat energy), a gain or loss of potential energy is balanced by a loss or gain of kinetic energy.

This conservation of energy concept applies to most mechanical systems. Those involving springs and masses are particularly interesting. Recall that the work done to compress (or stretch) a spring is given by $W = \frac{1}{2}k\mathbf{x}^2$. This is also a measure of the potential energy stored in the spring. If a mass is moving on a frictionless table at a velocity **v** toward an uncompressed spring and collides with it until it is fully compressed (and promptly absorbs all the kinetic energy, causing the mass to stop), then we can calculate the amount of compression as follows.

If we assume that no energy is loss to heat (work by friction), then we can state that the total energy before the interaction is contained in the mass's kinetic energy. After the interaction, the energy is stored in the compressed spring as potential energy. We can therefore write $\frac{1}{2}m\mathbf{v}^2 = \frac{1}{2}k\mathbf{x}^2$. Solving for the compression amount **x**, yields $\mathbf{x} = \mathbf{v}\sqrt{m/k}$.

CONSERVATIVE AND NONCONSERVATIVE FORCES

If the work done by a force is independent of the path taken, the force is called a **conservative force**. This implies that the mechanical energy of the system is conserved. That is, the sum of the potential and kinetic energies of the system remain constant in time. Gravity is an example of a conservative force, and so work done to or by a gravitational force (like lifting or falling) is independent of the path taken.

As an example, consider a mass sliding down a frictionless incline of length L. The magnitude of the component of gravity responsible for the acceleration down the incline is $-m\mathbf{g}\sin\theta$. Therefore, the work done by gravity is given by $W = -(m\mathbf{g}\sin\theta)\,L$. As we will see in Chapter 10, in a right triangle the length L is related to the height of the incline h such that $h = L\sin\theta$. Therefore, the work done down the incline can be written as $W = -m\mathbf{g}h$ which is just the loss of potential energy. The work done down the incline is exactly the same as the work done if the mass falls directly down from the height h. An interesting aspect of this observation is that the accelerating force down the incline is less than the gravitational force (its weight), but the displacement is longer in magnitude. The extent to which we can ideally trade off effort versus displacement is called the **ideal mechanical advantage**.

Friction is an example of a **nonconservative force**. Since friction is derived from the contact between two surfaces, any change in path affects the distance over which the contact takes place and therefore increases the effect of friction. The work done by friction can be expressed as the product of the frictional force and the displacement: $W\mathbf{f} = \mathbf{fd}$. However, an implication of frictional work is the generation of heat, which reveals itself as an apparent nonconservation of mechanical energy. In other words, the work done by any nonconservative force in which the change in total energy ΔE is not equal to zero is given by $W_{nc} = E_f - E_i$.

Sample Problem

A 20-kg girl is sliding down a rough incline 5 m long. She starts at the top of the track, 2.5 m above the ground. Her observed final velocity at the bottom of the track is 5 m/s. (a) How much work is done by friction? (b) What is the magnitude of the average frictional force during the complete slide?

Solution

Since friction is present, we have a nonconservative force acting on the girl. Thus, we may write for part (a):

$$W_{nc} = E_f - E_i = \frac{1}{2} m\mathbf{v}_f^2 - m\mathbf{g}h$$

Substituting the given values yields $W_{nc} = -240$ J of work. (b) Since friction is doing the work, we can write $W_{nc} = \mathbf{fd}$ in magnitude. Thus, $\mathbf{f} = 48$ N.

PROBLEM-SOLVING STRATEGIES

1. Identify the types of energies involved in the situation.
2. Try to determine the initial energy of the system, which is usually the total energy.
3. If no friction is present, write down the equations for the conservation of energy. Remember, if a spring is involved, it has a potential energy different from that of gravity.
4. For work problems, resolve forces into components in the direction of motion. These involve vector analysis techniques. Identify the frame of reference involved.

5. Remember that work and energy are scalar quantities expressed in units of joules.
6. In any dynamics problem, especially one in which energy or work is not expressly stated as a concept, the use of energy considerations to solve the problem may be a useful representation.

PRACTICE PROBLEMS FOR CHAPTER 7

Thought Problems

1. (a) Explain how it is possible for a moving object to simultaneously possess and not possess kinetic energy?
   (b) Based on your answer to part (a) how is it possible for energy to be considered a scalar?

2. When you hold up a 10-kg mass with your arms outstretched, you get tired. However, according to physics, you have not done any work. Explain how this is possible.

3. Does a simple machine reduce the amount of work needed to move an object to a specified location?

Multiple-Choice Problems

1. Which of the following are the units for the spring constant k?
   (A) kg-m^2/s^2
   (B) kg-s^2
   (C) kg-m/s
   (D) kg/s^2
   (E) kg-m^2/s

2. Which of the following is an equivalent expression for mechanical power?
   (A) $\mathbf{F}t/m$
   (B) $\mathbf{F}^2m/\mathbf{a}$
   (C) $\mathbf{F}m^2/t$
   (D) $\mathbf{F}m/t$
   (E) $\mathbf{F}^2t/m$

3. A pendulum consisting of a mass m attached to a light string of length ℓ is displaced from its rest position, making an angle θ with the vertical. It is then released and allowed to swing freely. Which of the following expressions represents the velocity of the mass when it reaches its lowest position?
   (A) $\sqrt{2\mathbf{g}\ell(1 - \cos\theta)}$
   (B) $\sqrt{2\mathbf{g}\ell \tan\theta}$
   (C) $\sqrt{2\mathbf{g}\ell \cos\theta}$

(D) $\sqrt{2\mathbf{g}\ell(1 - \sin\theta)}$
(E) $\sqrt{2\mathbf{g}\ell(\cos\theta - 1)}$

4. An engine maintains constant power on a conveyor belt machine. If the belt's velocity is doubled, the magnitude of its average acceleration will ____.
(A) be doubled
(B) be quartered
(C) be halved
(D) be quadrupled
(E) remain the same

5. A mass *m* is moving horizontally along a nearly frictionless floor with a velocity **v**. Then it encounters part of the floor that has a coefficient of kinetic friction given by μ. The total distance traveled by the mass as it is slowed to a stop by friction is given by ____ .
(A) $2\mathbf{v}^2/\mu\mathbf{g}$
(B) $\mathbf{v}^2/2\mu\mathbf{g}$
(C) $2\mu\mathbf{g}\mathbf{v}^2$
(D) $\mu\mathbf{v}^2/2\mathbf{g}$
(E) $\mu\mathbf{v}\mathbf{g}$

6. Two unequal masses are dropped simultaneously from the same height. The two masses will experience the same change in ____ .
(A) acceleration
(B) kinetic energy
(C) potential energy
(D) velocity
(E) momentum

7. A pendulum consists of a 2-kg mass and swings to a maximum vertical displacement of 17 cm above its rest position. At its lowest point, the kinetic energy of the mass is equal to ____ J.
(A) 0.33
(B) 3.33
(C) 33.3
(D) 333
(E) 3,333

8. A 0.3-kg mass rests on top of spring that has been compressed by 0.04 m. Neglecting any frictional effects and considering the spring to be massless, if the spring has a constant *k* equal to 2,000 N/m, to what height will the mass rise when the system is released?
(A) 1.24 m
(B) 0.75 m
(C) 0.54 m
(D) 1.04 m
(E) 1.34 m

9. A box is pulled along a smooth floor by a force F making an angle θ with the horizontal. As the angle θ increases, the amount of work done to pull the box the same distance d ____ .
   (A) increases
   (B) increases and then decreases
   (C) remains the same
   (D) decreases and then increases
   (E) decreases

10. As the time needed to run up a flight of stairs decreases, the amount of work done against gravity ____ .
    (A) increases
    (B) decreases
    (C) remains the same
    (D) increases and then decreases
    (E) decreases and then increases

Free-Response Problems

1. A 1.3-kg sphere is dropped through a tall column of liquid. When the sphere has fallen a distance of 0.50 m, it is observed to have a velocity of 2.5 m/s.
   (a) How much work was done by the frictional *viscosity* of the liquid?
   (b) What was the average force of friction during that displacement of 0.5 m?

2. A 15-kg mass is attached to a massless spring by a light string that passes over a frictionless pulley as shown. The spring has a force constant of 500 N/m, and the spring is unstretched when the mass is released. What is the velocity of the mass when it has fallen a distance of 0.3 m?

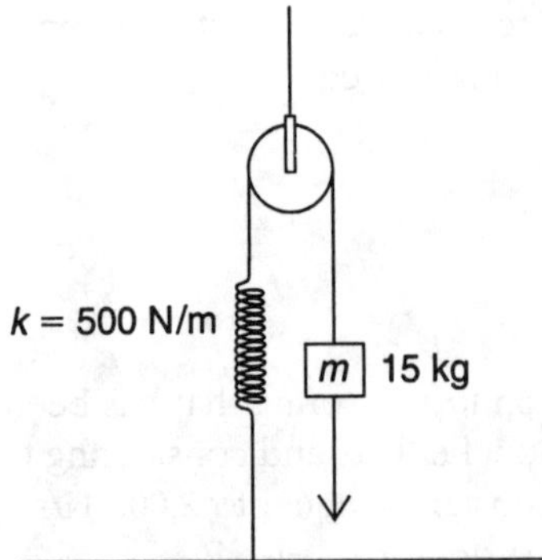

3. A 1.5-kg block is placed on a rough incline. The mass is connected to a massless spring by means of a light string passed over a frictionless pulley as shown. The spring has a force constant k = 100 N/m and is initially unstretched. The block is released from rest and moves down a distance

of 16 cm before coming to rest. What is the coefficient of friction between the block and the surface of the incline?

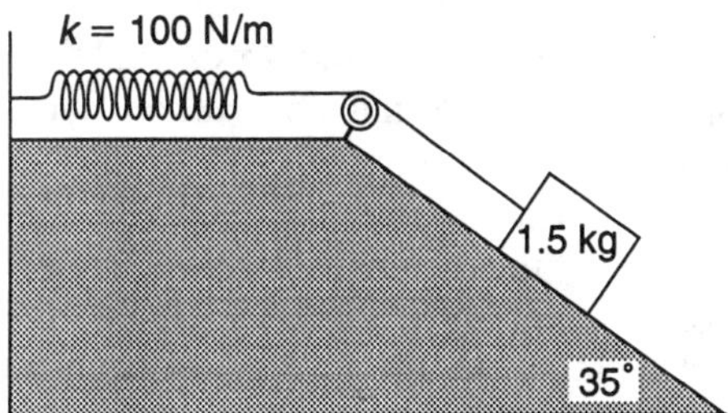

4. A pendulum consists of a mass m attached to a string L. A nail is placed a distance d below the point of suspension and along the vertical direction. What must be the value of d, in terms of L, such that when the pendulum is released from the horizontal position and the string encounters the nail, the mass will swing around the nail in a complete circle. (*Hint*: Let the point of suspension be the zero point for potential energy).

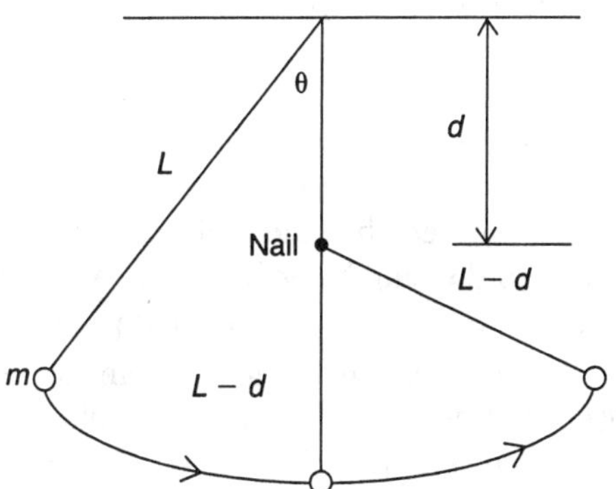

SOLUTIONS TO PRACTICE PROBLEMS

Thought Problems

1. (a) The observed kinetic energy of a moving object is relative to the chosen frame of reference. For example, if you are in an airplane moving with a constant velocity, you are in an inertial frame of reference. Suppose you toss a ball back and forth with a friend while standing in the aisle of the airplane. Your observations of kinetic energy are determined only by the mass of the ball and its relative velocity in the airplane (not very fast). However, to an observer on the ground, the ball is traveling with a velocity that is much faster than you think because of the motion of the plane. Thus, if you hold onto

the ball, it will not have any kinetic energy relative to you. It will, however, have kinetic energy relative to an observer on the ground.

(b) Energy is a scalar, and scalars are supposed to remain the same (invariant) in all inertial frames of reference. However, this is not strictly true. In Chapter 3, you did an example involving the rotation of a coordinate system as a transformation. If you were to apply this transformation to kinetic energy, you would get the same value (kinetic energy is invariant under a rotational transformation of coordinates). However, the example in part (a) indicates that kinetic energy is not invariant under a translational transformation of coordinates. Still, the kinetic energy in a given coordinate system is not a directional quantity and can in a restricted sense be considered a scalar for all practical purposes that we will enounter.

2. When you hold up the weights, your muscles strain to support the weight, and this requires energy from you to maintain the support strength. This occurs even though the weights are not moving and no work is being done.

3. A simple machine is a device that reduces the *effort* required to do work at the expense of the displacement. Ideally, the work done by the machine is the same. If anything, friction increases the amount of work put into a machine. The ratio of work output to work input is called the *efficiency* of the machine.

Multiple-Choice Problems

1. **D** The units for the spring or force constant are provided by Hooke's law, $\mathbf{F} = k\mathbf{x}$. Thus, we see that the units are N/m. Recall, that 1 N = 1 kg-m/s^2, and so dividing by m leaves us with kg/s^2.

2. **E** Power is equal to the work done divided by the time. Therefore, $P = W/t = \mathbf{Fd}/t$. Using some algebra and kinematics we see that $P = \mathbf{Fd}/t = \mathbf{Fv} = \mathbf{F}(\mathbf{a}t) = \mathbf{F}(\mathbf{F}/m)t = \mathbf{F}^2t/m$. You could also obtain the answer by verifying which one has units of J/s.

3. **A** Consider the diagram of the situation shown here.

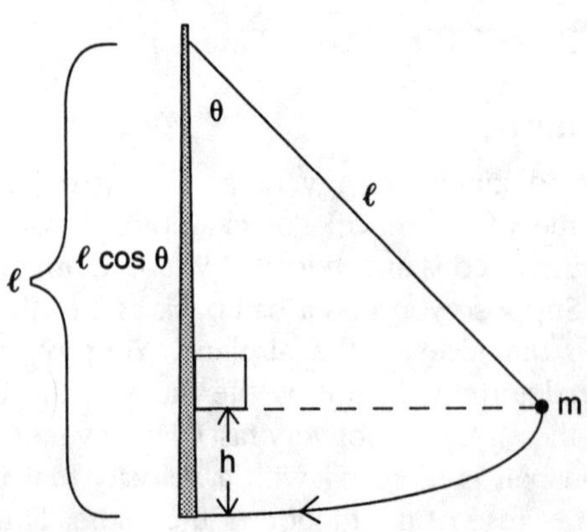

From the geometry of the sketch, notice that $h = \ell - \ell \cos \theta = \ell (1 - \cos \theta)$. Assuming no friction, gravity is the only conservative force acting to do work. Therefore, $\Delta KE = \Delta PE$. This means that at the bottom $\mathbf{v} = \sqrt{2\mathbf{g}h} = \sqrt{2\mathbf{g}\ell(1 - \cos \theta)}$.

4. **C** Power is equal to the product of the average force applied and the velocity. If the velocity is doubled and the power is constant, the average force must be halved. Since $\mathbf{F} = m\mathbf{a}$, the average acceleration of the belt must be halved as well.

5. **B** The only applied force is friction which is doing work to stop the mass. This work is being taken from the initial kinetic energy. For friction, we know that $\mathbf{f} = \mu\mathbf{N}$, and in this case $\mathbf{N} = m\mathbf{g}$. Let $\mathbf{x}$ be the distance traveled before stopping, and then we can write $W\mathbf{f} = \Delta KE$ and $\frac{1}{2}m\mathbf{v}^2 = \mu m\mathbf{g}\mathbf{x}$. Therefore, solving for $\mathbf{x}$, we have $\mathbf{x} = \mathbf{v}^2/2\mu\mathbf{g}$.

6. **D** Objects dropped simultaneously from the same height have the same constant acceleration, which is the change in the velocity. The different masses provide for different energies.

7. **B** Since we have a conservative system, $\Delta KE = \Delta PE = (2)(9.8)(0.17) = 3.33$ J. Remember to change 17 cm to 0.17 m.

8. **C** We are dealing with a conservative system, and so the initial starting energy is just the potential energy of the compressed spring. This energy supplies the work needed to raise the mass a height h, which is a gain in gravitational potential energy. Thus we equate these two expressions and solve for the height: $\frac{1}{2}(2{,}000)(0.4)^2 = (0.3)(9.8)h$. Thus, $h = 0.54$ m.

9. **E** The component of the applied force in the horizontal direction depends on the cosine of the angle. This value decreases with increasing angle. Thus, the work decreases as well.

10. **C** The work done is independent of the time or path taken since gravity is a conservative force. The power generated is affected by time, but the work done in going up the stairs remains the same as long as the same mass is raised to the same height.

Free-Response Problems

1. (a) The change in potential energy is a measure of the initial energy = (0.75)(9.8)(2) = 14.7 J. After falling 2 m, the velocity is 5 m/s, and so the kinetic energy is $\frac{1}{2}(0.75)(5)^2 = 9.375$ J. The work done by friction is due to a nonconservative force which is equal to the difference between these energies: $E_f - E_i = 9.375 \text{ J} - 14.7 \text{ J} = -5.325$ J.

   (b) The average frictional force is equal to the work done divided by the displacement of 2 m. Thus $\mathbf{f} = -2.67$ N, which of course is negative since it opposes the motion.

2. The loss of potential energy is balanced by a gain in elastic potential energy for the spring and kinetic energy for the falling mass if we assume that the starting energy for the system is zero relative to the starting point for the mass. In the absence of friction, the displacement of the mass is equal to the elongation of the spring. Thus, we can equate these energies and write

$$0 = -m\mathbf{g}h + \frac{1}{2}k\mathbf{x}^2 + \frac{1}{2}m\mathbf{v}^2$$

When we substitute the values given, we obtain

$$0 = -(15)(9.8)(0.3) + \frac{1}{2}(500)(0.3)^2 + \frac{1}{2}(15)\mathbf{v}^2$$

Solving for the velocity $\mathbf{v}$, we have $\mathbf{v} = 1.7$ m/s.

3. In this problem, the net work is applied to stretching the spring by an amount equal to the displacement of the mass. Thus we can say $W_g - W_f = W_s$ (where W_g is the work done by gravity, W_f is the work done by friction, and W_s is the work done on the spring). Hence

$$m\mathbf{g}\sin\theta d - \mu m\mathbf{g}\cos\theta d = \frac{1}{2}k\mathbf{x}^2$$

Substituting the numbers given, we have

$$(1.5)(9.8)(0.16)\sin(35) - \mu(1.5)(9.8)(0.16)\cos(35) = \frac{1}{2}(100)(0.16)^2$$

and solving for the coefficient of friction, we have $\mu = 0.036$.

4. Since the pendulum starts off in the horizontal position, $PE_i = 0$. Using the downward direction as negative, we take the potential energy below the point of support as negative. The radius of the desired circular path is $R = L - d$, and the distance from the point of support to the mass at the top of the circle is given by $y = d - (L - d)$.

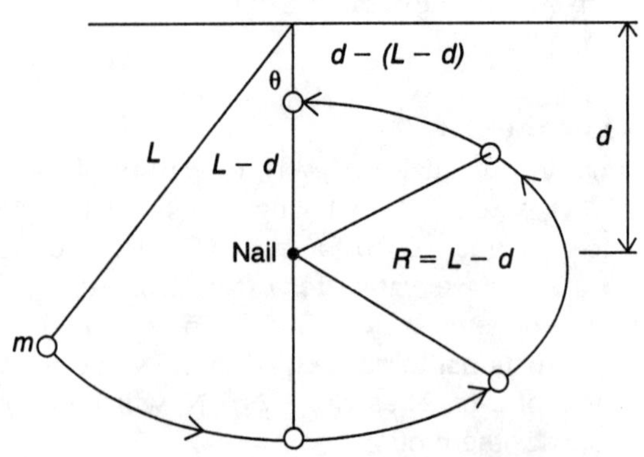

At the top of the path, the mass has a velocity $\mathbf{v}$ and its potential energy is given by $-m\mathbf{g}y$. Thus, using conservation of energy, we can write

$$-m\mathbf{g}(2d - L) + \frac{1}{2}m\mathbf{v}^2 = 0$$
$$\mathbf{v}^2 = 2\mathbf{g}(2d - L) = \mathbf{g}(4d - 2L)$$

The square of the velocity can be found by remembering that if the mass is to complete the circle, it must have a velocity such that the centripetal force is just balanced by gravity:

$$\frac{m\mathbf{v}^2}{R} = m\mathbf{g}$$
$$\mathbf{v}^2 = \mathbf{g}R = \mathbf{g}(L - d)$$

Solving for d, we obtain

$$\mathbf{g}(L - d) = \mathbf{g}(4d - 2L)$$

and therefore, $d = \frac{3}{5}L$.

8
IMPACTS AND LINEAR MOMENTUM

INTRODUCTION—INTERNAL AND EXTERNAL FORCES

In Chapter 6, we discussed Isaac Newton's laws of motion. In reference to the third law of motion, sometimes called the law of action and reaction, we noted that an interaction between two objects in space creates a pair of impact forces that are equal and opposite. Thus, we concluded that forces do not act in isolation. Consider a system of two blocks with masses m and M (with $M > m$). If they were to collide, the forces of impact would be equal and opposite. However, because of the different masses, the response to these forces (i.e., the changes in velocity) would not be equal. In the absence of any outside or external forces acting on the objects (such as friction or gravity), we say that the impact forces are internal.

To Newton, the "quantity of motion" discussed in his *Principia* was the product of the object's mass and velocity. This quantity is called the **linear momentum** or just the **momentum** and is a vector quantity in units of kg-m/s. Algebraically, momentum is designated by the letter **P** such that $\mathbf{P} = m\mathbf{v}$.

To understand Newton's rationale, consider the action of trying to change the motion of a moving object. Do not confuse this with the inertia of the object and note that both the mass and the velocity are important. Consider the following example: A truck moving at a slow 1 m/s can still inflict a large amount of damage because of its mass. A small bullet, weighing maybe a gram or less, does incredible damage because of its high velocity. In each case, (see Figure 1), the damage is the result of a force of impact when the object is intercepted by something else. Let's consider the nature of impact forces.

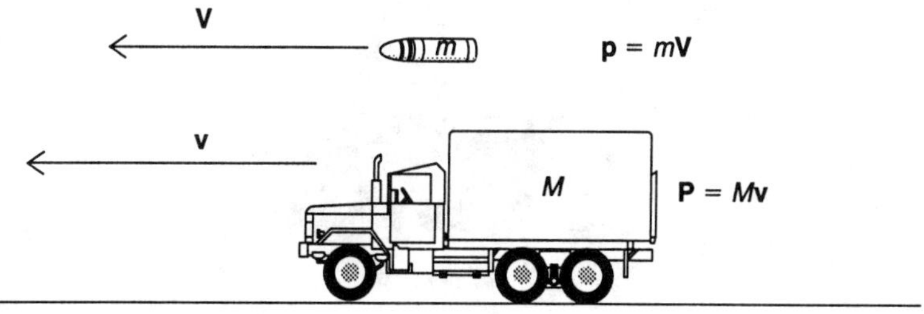

FIGURE 1. *Comparing the momentum of a truck and a bullet.*

IMPACT FORCES AND MOMENTUM CHANGES

Consider a mass m moving with a velocity $\mathbf{v}$ in some frame of reference. If the mass is subjected to external forces, then, by Newton's second law of motion, we can write $\Sigma\mathbf{F} = m\mathbf{a}$. The vector sum of all the forces is referred to as the **net force** and is responsible for changing the velocity of the motion (in magnitude and/or direction).

If we recall the definition of acceleration as the rate of change in velocity, then we can rewrite the second law of motion as $\mathbf{F}_{net} = m\ (\Delta\mathbf{v}/\Delta t)$. This expression is also a vector equation and equivalent to the second law of motion. If we make the assumption that the mass of the object is not changing, then we can again rewrite the second law in the form

$$\mathbf{F}_{net} = \frac{\Delta m\mathbf{v}}{\Delta t} = \frac{\Delta\mathbf{p}}{\Delta t}$$

This equation means that the net external force acting on an object is equal to the rate of change in momentum of the object. This is another alternative form of Newton's second law of motion. The **change in momentum** is a vector quantity in the same direction as the net force applied. Since the time interval is just a scalar quantity, we can multiply both sides by Δt to obtain

$$\mathbf{F}_{net}\ \Delta t = \Delta\mathbf{p} = m\ \Delta\mathbf{v} = m\mathbf{v}_f - m\mathbf{v}_i$$

The quantity $\mathbf{F}_{net}\ \Delta t$ is called the **impulse**. It represents the effect of a force acting on a mass during a time interval Δt and is also a vector quantity. From this expression, it can be stated that the impulse applied to an object is equal to the change in momentum of the object. Another way to consider the impulse is to look at a graph of force versus time for a continuously varying force as shown in Figure 2.

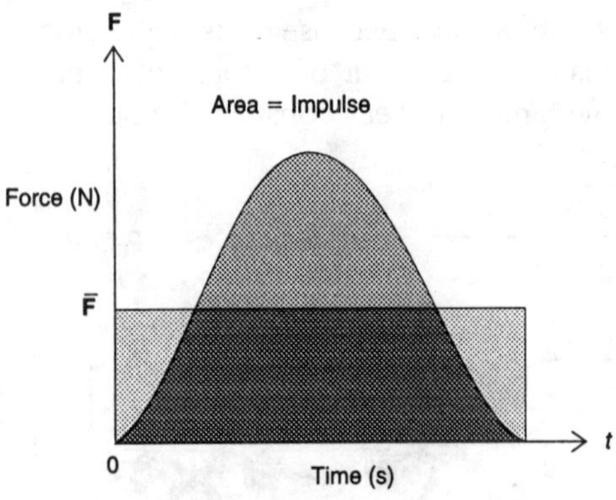

FIGURE 2.

The area under the curve is a measure of the impulse in units of N-s. Another way to view this concept is to identify the average force $\bar{\mathbf{F}}$ such that the area of the rectangle formed by the average force is equal to the area under the entire curve. This is more manageable algebraically, and we can write $\bar{\mathbf{F}}\,\Delta t = \Delta\mathbf{p}$.

As an example, consider the following problem. How long will it take an average braking force of 1,500 N to stop a 2,500-kg car travelling with an initial velocity of 10 m/s? Substitute the numbers into the formula for impulse and momentum change to obtain $-1{,}500\ \Delta t = (2{,}500)(-10)$. The change in the velocity is equal to -10 m/s because the object stops. Solving for the time, we find that it will take about 16.7 s.

THE LAW OF CONSERVATION OF LINEAR MOMENTUM

Newton's third law of motion states that for every action there is an equal and opposite reaction. This reaction force is present whenever we have an interaction between two objects in the universe. Suppose we have two masses m_1 and m_2 that are approaching each other along a horizontal frictionless surface. Let $\mathbf{F}_{12}$ be the force that m_1 exerts on m_2 and let $\mathbf{F}_{21}$ be the force that m_2 exerts on m_1. According to Newton's law, these forces must be equal and opposite; that is, $\mathbf{F}_{12} = -\mathbf{F}_{21}$. Rewriting this expression as $\mathbf{F}_{12} + \mathbf{F}_{21} = 0$ leads to an interesting implication. Since each force is a measure of the rate of change in momentum for that object, we can write $\mathbf{F}_{12} = \Delta\mathbf{p}_1/\Delta t$ and $\mathbf{F}_{21} = \Delta\mathbf{p}_2/\Delta t$. Therefore,

$$\frac{\Delta\mathbf{p}_1}{\Delta t} + \frac{\Delta\mathbf{p}_2}{\Delta t} = 0$$

$$\frac{\Delta(\mathbf{p}_1 + \mathbf{p}_2)}{\Delta t} = 0$$

The change in the sum of the momenta is therefore zero, and this implies that the total momentum for the system $(\mathbf{p}_1 + \mathbf{p}_2)$ is a constant all the time. This conclusion is called the **law of conservation of linear momentum**, and we say simply that the momentum is conserved whenever the net external force is zero.

Another way of writing this law of conservation statement in a general form for any two masses is (after separating all initial and final terms)

$$m_1\mathbf{v}_{1i} + m_2\mathbf{v}_{2i} = m_1\mathbf{v}_{1f} + m_2\mathbf{v}_{2f}$$

Extension of the law of conservation of momentum to two or three dimensions involves the recognition that momentum is a vector quantity. Given two masses moving in a plane relative to a coordinate system, conservation of momentum must hold simultaneously in both the horizontal and vertical directions. These vector components of momentum can be calculated using the standard techniques of vector analysis employed to resolve any vector into components.

ELASTIC AND INELASTIC COLLISIONS

During any collision between two pieces of matter, the momentum is always conserved. This is not, however, necessarily true about the kinetic energy. If two masses stick together after a collision, it is observed that the kinetic energy before is not equal to the kinetic energy after. When the kinetic energy is conserved as well as the momentum, then the collision is described as **elastic**. When the kinetic energy is not conserved after the collision (energy being lost to heat or friction, for example), then the collision is described as **inelastic**.

As an example, suppose it is observed that a mass m has a velocity $\mathbf{v}$, while a mass M is at rest along a horizontal frictionless surface. The two masses collide and stick together. What is the final velocity $\mathbf{u}$ of the system? According to the law of conservation of momentum, the total momentum before the collision $m\mathbf{v}$ must be equal to the total momentum after the collision. Since the two masses are combining, the new mass of the system is $M + m$ and the new momentum is given by $(m + M)\mathbf{u}$. Thus, we find that $\mathbf{u} = m\mathbf{v}/(m + M)$. The initial kinetic energy is $\frac{1}{2}m\mathbf{v}^2$. The final kinetic energy is given by

$$\frac{1}{2}(m + M)\mathbf{u}^2 = \frac{1}{2}\frac{m^2\mathbf{v}^2}{m + M}$$

In an inelastic collision, it is worthwhile to consider the ratio of the final to the initial energy. Using the relationships just derived, we obtain for the ratio of the final to the initial kinetic energy:

$$\frac{\text{KE}_f}{\text{KE}_i} = \frac{m}{(m + M)} < 1.0$$

When we do have an elastic collision, then we write down both conservation laws to obtain two equations involving the velocities of the masses before and after:

$$m_1\mathbf{v}_{1i} + m_2\mathbf{v}_{2i} = m_1\mathbf{v}_{1f} + m_2\mathbf{v}_{2f}$$

$$\frac{1}{2}m_1\mathbf{v}_{1i}^2 + \frac{1}{2}m_2\mathbf{v}_{2i}^2 = \frac{1}{2}m_1\mathbf{v}_{1f}^2 + \frac{1}{2}m_2\mathbf{v}_{2f}^2$$

If we cancel the factor of $\frac{1}{2}$ in the second equation and collect the expressions for each mass on each side, we can rewrite the two expressions as

$$m_1(\mathbf{v}_{1i} - \mathbf{v}_{2i}) = m_2(\mathbf{v}_{2f} - \mathbf{v}_{2i})$$

$$m_1(\mathbf{v}_{1i}^2 - \mathbf{v}_{1f}^2) = m_2(\mathbf{v}_{2f}^2 - \mathbf{v}_{2i}^2)$$

The second equation is factorable, and so it can be simplified. Again rewriting the expressions, we have

$$m_1(\mathbf{v}_{1i} - \mathbf{v}_{1f}) = m_2(\mathbf{v}_{2f} - \mathbf{v}_{2i})$$
$$m_1(\mathbf{v}_{1i} + \mathbf{v}_{1f})(\mathbf{v}_{1i} - \mathbf{v}_{1f}) = m_2(\mathbf{v}_{2f} + \mathbf{v}_{2i})(\mathbf{v}_{2f} - \mathbf{v}_{2i})$$

If we take the ratio of both expressions, we arrive at an interesting result after collecting terms:

$$\mathbf{v}_{1i} - \mathbf{v}_{2i} = \mathbf{v}_{2f} - \mathbf{v}_{1f}$$

This expression states that the relative velocity between the masses before the elastic collision is equal to and opposite the relative velocity between the masses after the elastic collision!

MOTION IN THE CENTER-OF-MASS FRAME OF REFERENCE

When we first introduced the study of motion, we discussed the fact that all motion is relative to some chosen frame of reference. This frame can be "attached" to the moving mass, and if the velocity is constant, the frame is called **inertial**. Another useful frame of reference is one connected with the **center of mass** of a system of masses.

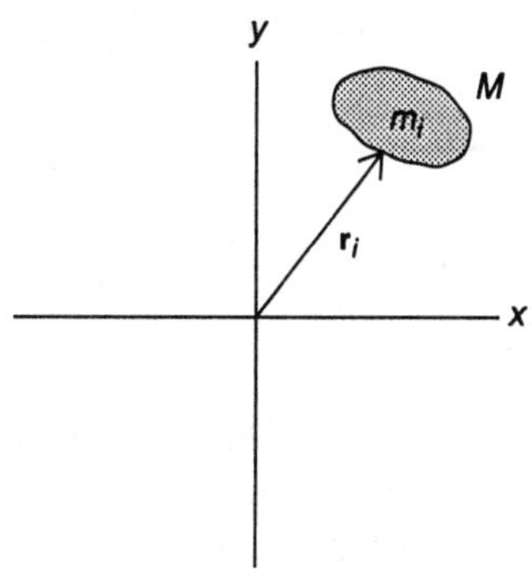

FIGURE 3.

Consider a collection of mass particles as shown in Figure 3. Let us identify an arbitrary mass particle m_i and its position vector $\mathbf{r}_i$ within the chosen coordinate system. If M represents the total mass of the system, such that $M =$

Σm_i, then the weighted distribution of mass relative to the origin is given by the position vector $\mathbf{R}_{cm}$:

$$\mathbf{R}_{cm} = \frac{\Sigma m_i \mathbf{r}_i}{\Sigma m_i}$$

Where the subscript cm stands for center of mass. In other words, the center of mass for the system is the point where all the mass can be considered concentrated and is a "weighted" mean position for the mass distribution.

Another way of writing the expression for center of mass is

$$M\mathbf{R}_{cm} = \Sigma m_i \mathbf{r}_i = m_i \mathbf{r}_i + m_2 \mathbf{r}_2 + \cdots + m_n \mathbf{r}_n$$

Since this is a vector equation, the center-of-mass position vector can be resolved into x and y components for the center of mass:

$$X_{cm} = \frac{m_1 X_1 + m_2 X_2 + \cdots + m_n X_n}{M}$$

$$Y_{cm} = \frac{m_1 Y_1 + m_2 Y_2 + \cdots + m_n Y_n}{M}$$

Dynamically, some interesting effects take place in the center-of-mass frame of reference. First, any coordinate system attached to the center-of-mass frame of reference can provide for zero initial kinetic energy and momentum. Second, if we allow the system to move in a time Δt, then we can write

$$M \frac{\Delta \mathbf{R}_{cm}}{\Delta t} = \Sigma m_i \frac{\Delta \mathbf{r}_i}{\Delta t}$$

This means that $M\mathbf{v}_{cm} = \Sigma\, m_i \mathbf{v}_i$, and we can state that the total momentum for the system can be viewed as the momentum of the center of mass (where the total mass is considered to be concentrated). Additionally, if the system is accelerating, then we can state that

$$M \frac{\Delta \mathbf{v}_{cm}}{\Delta \mathrm{t}} = \Sigma m_i \frac{\Delta \mathbf{v}_i}{\Delta t}$$

$$M\mathbf{a}_{cm} = \Sigma m_i \mathbf{a}_i$$

The implication of this statement is that if an external force is applied to each mass in the system, the responses to the force can be viewed as a single force acting on the center of mass and producing a translational acceleration (as opposed to a rotation; see Chapter 13) since $\mathbf{F} = m\mathbf{a}$. These redescriptions of the dynamics of a collection of masses suggest that the motion or interaction effects may appear simpler in the center-of-mass frame of reference. For example, a projectile in motion has zero momentum when viewed from the center-of-mass frame of reference. The path of this center of mass is a parabola. An explosion will cause the particles to split away while the center of mass continues in its original trajectory.

As an example, consider four masses arranged at the corners of a massless square frame as shown in Figure 4. Where is the center of mass of the system?

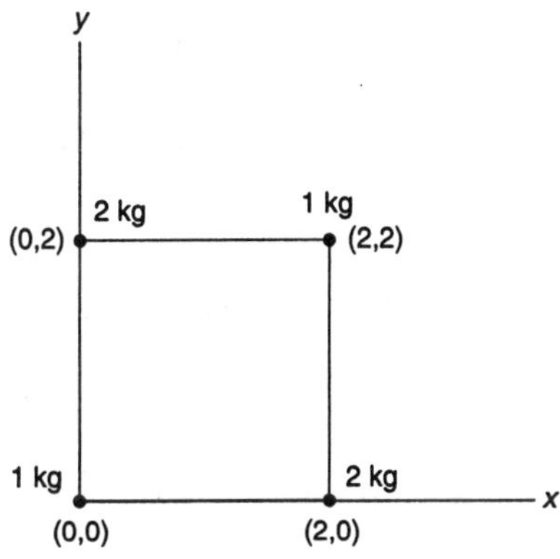

FIGURE 4.

Using our formulas for the x and y components of the center of mass, we have $M = 6$ kg and $X_{cm} = 1$ unit and $Y_{cm} = 1$ unit. Thus $\mathbf{R}_{cm} = (1,1)$, which is in the middle of the square.

CONSERVATION OF MOMENTUM IN VARIABLE-MASS SYSTEMS

Consider a cart of mass M moving to the right with a velocity $\mathbf{v}$. Initially the cart is empty, and then coal pours into it at a uniform rate. As shown in Figure 5, after the cart has passed completely underneath the coal shute, an extra mass Δm has been added and the velocity of the cart, given by the letter $\mathbf{u}$, is now slower. How are these situations related using conservation of momentum?

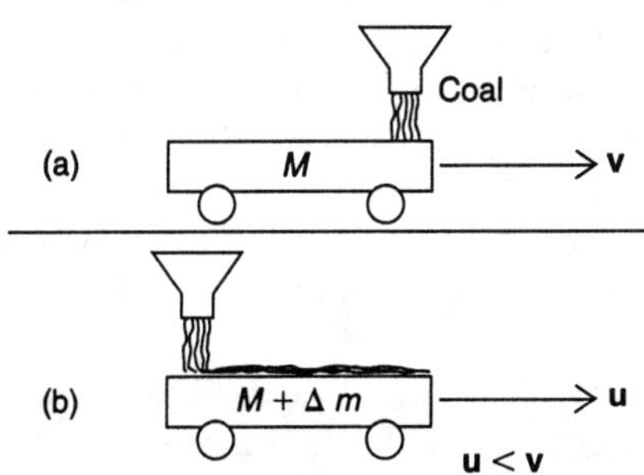

FIGURE 5.

The momentum of the cart before is given by $\mathbf{p}_i = M\mathbf{v}$, and the momentum of the cart after it is filled with coal is $\mathbf{p}_f = (M + \Delta m)\mathbf{u}$. The change in momentum is therefore

$$\mathbf{p}_f - \mathbf{p}_i = \Delta\mathbf{p} = M\mathbf{u} + \Delta m\mathbf{u} - M\mathbf{v} = M(\mathbf{u} - \mathbf{v}) + \Delta m\mathbf{u} = M\Delta\mathbf{v} + \Delta m\mathbf{u}$$

Now, we can divide both sides by the time interval over which the change takes place:

$$\frac{\Delta\mathbf{p}}{\Delta t} = M\frac{\Delta\mathbf{v}}{\Delta t} + \mathbf{u}\frac{\Delta m}{\Delta t}$$

However, since momentum is conserved, $\Delta\mathbf{p}/\Delta t = 0$, and so

$$\mathbf{F} = M\frac{\Delta\mathbf{v}}{\Delta t} = -\mathbf{u}\frac{\Delta m}{\Delta t}$$

Thus, we need to know, or can determine, the rate at which the mass is changing. In a rocket, the change in mass is caused by the burning of fuel, which reduces the inertia of the rocket. The force involved is often called the **thrust**. The velocity $\mathbf{u}$ is known as the **exhaust velocity.**

PROBLEM-SOLVING STRATEGIES

In any closed system, momentum is always conserved. The kinetic energy is not necessarily conserved unless the collision is elastic. In summary then, you should

1. Decide if an impact or collision is involved. If there is such an interaction, observe whether the collision is elastic or inelastic.
2. If the collision is inelastic, masses will usually stick together, so be sure to determine the new combined mass.
3. If the collision is elastic, write down the equations for the conservation of both the momentum and the kinetic energy.
4. Remember that the impulse given to a mass is equal to its change in momentum. The change in momentum is a vector quantity in the same direction as the impulse or net force.
5. Algebraically, be sure to take into account any reversal of directions and remember the sign conventions for left, right, upward, and downward motions.
6. In two dimensions, the center of mass follows a smooth path after an internal explosion since the forces involved are internal and the initial momentum in that frame was zero. For example, the center of mass of a projectile launched at an angle and then exploded still follows the regular parabolic trajectory.

PRACTICE PROBLEMS FOR CHAPTER 8

Thought Problems

1. (a) Can an object have energy without having momentum? Explain.
   (b) Can an object have momentum without having energy? Explain.
2. Explain why there is more danger when you fall and bounce as opposed to falling without bouncing.
3. A cart of mass M is moving with a constant velocity $\mathbf{v}$ to the right. A mass m is dropped vertically onto it, and it is observed that the new velocity is less than the original velocity. Explain what has happened in terms of energy, forces, and conservation of momentum (as viewed from different frames of reference).

Multiple-Choice Problems

1. Which of the following expressions is equivalent to the kinetic energy of a moving particle? (where $\mathbf{p}$ represents the linear momentum of the particle)?
   (A) $m\mathbf{p}^2$
   (B) $m^2/2\mathbf{p}$
   (C) $2\mathbf{p}/m$
   (D) $\mathbf{p}/2m$
   (E) $\mathbf{p}^2/2m$
2. Two carts of mass 1.5 kg and 0.7 kg, respectively, are initially at rest and held together by a compressed massless spring. When released, the 1.5-kg cart moves to the left with a velocity of 7 m/s. What is the velocity of the 0.7-kg cart?
   (A) 15 m/s right
   (B) 15 m/s left
   (C) 7 m/s left
   (D) 7 m/s right
   (E) 0 m/s
3. A water hose is kept horizontal while water flows out at a constant rate of 5 Liters (L)/min. If the velocity of the water is 30 m/s, what is the average force acting on the water as it emerges from the hose?
   (A) 150 N
   (B) 6 N
   (C) 2.5 N
   (D) 360 N
   (E) 12 N
4. A ball with a mass of 0.15 kg has a velocity of 5 m/s. It strikes a wall perpendicularly and bounces off straight back with a velocity of 3 m/s. The ball undergoes a change in momentum equal to ____ kg-m/s.
   (A) 0.30
   (B) 1.20
   (C) 0.15

(D) 5
(E) 7.5

5. What braking force is supplied to a 3,000-kg car that is traveling with a velocity of 35 m/s and is subsequently stopped in 12 s?
(A) 29,400 N
(B) 3,000 N
(C) 8,750 N
(D) 105,000 N
(E) 150 N

6. A 0.1-kg baseball is thrown with a velocity of 35 m/s. It is hit straight back with a velocity of 60 m/s. What is the magnitude of the impulse exerted on the ball by the bat?
(A) 3.5 N-s
(B) 2.5 N-s
(C) 7.5 N-s
(D) 9.5 N-s
(E) 12.2 N-s

7. A 1-kg object is moving to the right with a velocity of 6 m/s. It collides with and sticks to a 2-kg object moving in the same direction with a velocity of 3 m/s. How much kinetic energy is lost in the collision?
(A) 1.5 J
(B) 2 J
(C) 2.5 J
(D) 3 J
(E) 0 J

8. A 2-kg mass moving with a velocity of 7 m/s collides with a 4-kg mass moving in the opposite direction at 4 m/s. The 2-kg mass reverses direction after the collision and has a new velocity of 3 m/s. What is the new velocity of the 4-kg mass?
(A) −1 m/s
(B) 1 m/s
(C) 6 m/s
(D) 4 m/s
(E) 5 m/s

9. A mass m is attached to a massless spring with a force constant k. The mass rests on a horizontal frictionless surface. The system is compressed a distance $\mathbf{x}$ from the spring's initial position and then released. The momentum of the mass when the spring passes its equilibrium position is given by _____.
(A) $\mathbf{x}\sqrt{mk}$
(B) $\mathbf{x}\sqrt{k/m}$
(C) $\mathbf{x}\sqrt{m/k}$
(D) $\mathbf{x}\sqrt{k^2m}$
(E) $\mathbf{x}mk$

10. Three equal masses are arranged at the vertices of a massless triangular frame such that the coordinates of each mass m in the x,y plane are (0,0), (1,2), and (2,0). The coordinates of the center of mass of this system are ____.
(A) (1,1)
(B) $(1,\frac{2}{3})$
(C) $(\frac{1}{2},\frac{1}{2})$
(D) $(\frac{1}{2},\frac{3}{4})$
(E) (0,0)

Free-Response Problems

1. Two blocks with masses 1 kg and 4 kg are moving on a horizontal frictionless surface. The 1-kg mass has a velocity of 12 m/s, and the 4-kg mass is ahead of it and moving at 4 m/s, as shown in the accompanying diagram. The 4-kg mass has a massless spring attached to the end facing the 1-kg mass, and the spring has a force constant k = 1,000 N/m.
(a) What is the maximum compression of the spring after the collision?
(b) What are the final velocities after the collision has taken place?

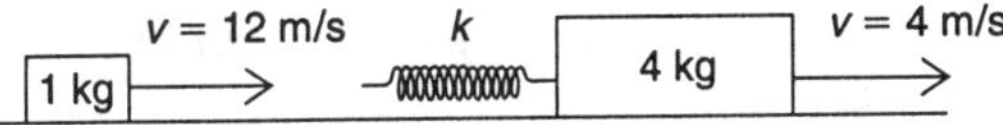

2. A 0.4-kg disk is initially at rest on a frictionless horizontal surface. It is hit by a 0.1-kg disk moving horizontally with a velocity of 4 m/s. After the collision, the 0.1-kg disk has a velocity of 2 m/s at an angle of 43 degrees with the positive x axis.
(a) Determine the magnitude and direction of the velocity of the 0.4-kg disk after the collision.
(b) Determine the amount of KE lost in the collision.

3. A 50-kg girl sits on a platform with wheels on a frictionless horizontal surface as shown. The platform has a total mass of 1,000 kg and is attached to a massless spring with a force constant k = 1,000 N/m. The girl throws a 1-kg object with an initial velocity of 35 m/s at an angle of 30 degrees with the horizontal.
(a) What is the recoil velocity of the platform-and-girl system?
(b) What is the elongation of the spring?

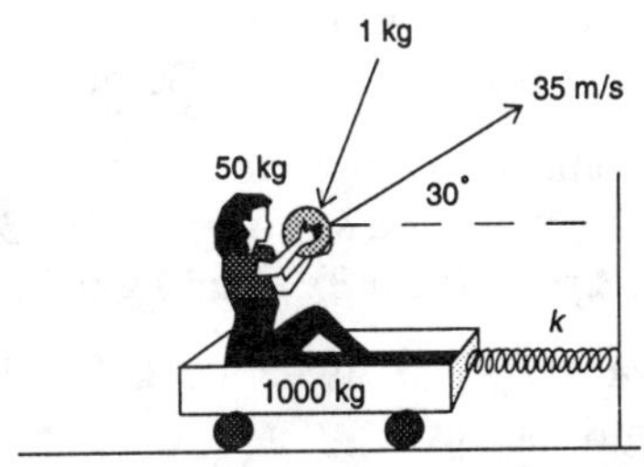

4. A 10-kg block is resting on the edge of a frictionless platform 1.5 m above the ground. A 50-g bullet strikes the block with a velocity **v**. The bullet is embedded in the block and both are projected off the platform, landing a distance of 2.5 m from the base of the platform. With what velocity did the bullet strike the block?

SOLUTIONS TO PRACTICE PROBLEMS

Thought Problems

1. (a) An object can have energy without having momentum if it has potential energy.
   (b) An object cannot have momentum without having energy.
2. When an object bounces, there is an additional upward impulse given to the mass that forces it upward. This impulse provides a relatively large change in momentum that can be potentially dangerous.
3. From the fixed laboratory frame of reference, momentum is conserved, and so the increase in mass must be accompanied by a decrease in velocity. Additionally, we can say that when the mass is placed on the cart, since it had no initial forward motion, because of inertia, it will try to stay at rest, friction will cause the mass to stay on the cart, and the energy needed to bring the mass "up to speed" will be delivered at the expense of the cart's initial kinetic energy. From the cart's frame of reference, the falling mass appears to move diagonally, having a horizontal component in the negative direction. As a result, the momentum of the cart plus its mass is less than the original momentum of the cart, and the cart plus its mass has a velocity less than the original velocity of the cart.

Multiple-Choice Problems

1. **E** If we multiply the formula for kinetic energy by the ratio m/m, we will see that the formula for KE becomes KE $= (\frac{1}{2}m)(m^2)(\mathbf{v}^2) = \mathbf{p}^2/2m$.
2. **A** Momentum is conserved. This means that $(1.5)(7) = 0.7\mathbf{v}$. Thus $\mathbf{v} = 15$ m/s. The direction is to the right since in a recoil the masses go in opposite directions.
3. **C** The average force will be the product of the rate of change in mass (in kg/s) and the velocity. Taking 1 L = 1 kg, we have $\Delta m/\Delta t = 5$ kg/60 s $= 0.083$ kg/s. Thus, $\mathbf{F} = (0.083$ kg/s$)(30$ m/s$) = 2.5$ N.
4. **B** The change in momentum is a vector quantity. The rebound velocity is in the opposite direction, and so $\Delta\mathbf{v} = 5 - (-3) = 8$ m/s. The change in momentum is $\Delta\mathbf{p} = (0.15)(8) = 1.2$ kg-m/s.
5. **C** The formula is $\mathbf{F}\ \Delta t = m\ \Delta\mathbf{v}$. Solving for the force, we get $\mathbf{F} = (3{,}000)(35)/12 = 8{,}750$ N.

6. **D** Impulse is equal to the change in momentum. The change in momentum is $\Delta\mathbf{p} = (0.1)\,[35 - (-60)] = 9.5$ N-s because of the change in the direction of the ball.

7. **D** First, let's find the final velocity of this inelastic collision. Momentum is conserved, and so we can write $(1)(6) + (2)(3) = 3\mathbf{v}'$ since both objects are moving in the same direction. Thus $\mathbf{v}' = 4$ m/s. The initial kinetic energy of the 1-kg object is 18 J, while the initial kinetic energy for the 2-kg mass is 9 J. Thus, the total initial kinetic energy is 27 J. After the collision, the combined 3-kg object has a velocity of 4 m/s and a final kinetic energy of 24 J. Thus, 3 J of kinetic energy has been lost.

8. **B** Momentum is conserved in this collsion, but the directions are opposite and so we must be careful with negative signs. We therefore write $(2)(7) - (4)(4) = -(2)(3) + 4\mathbf{v}'$, and so $\mathbf{v}' = +\,1$ m/s.

9. **A** We set the two energy equations equal to solve for the velocity at the equilibrium position. Thus, $\frac{1}{2}k\mathbf{x}^2 = \frac{1}{2}m\mathbf{y}^2$ since no gravitational potential energy is involved. So we find that the velocity is $\mathbf{v} = \mathbf{x}\sqrt{k/m}$. Now momentum $\mathbf{p} = m\mathbf{v}$, and so we can multiply by m and factor it back under the radical sign where it is squared, and we obtain $\mathbf{p} = \mathbf{x}\sqrt{mk}$.

10. **B** Using our formulas for the x and y positions of the center of mass, we find from our data that in the x direction, $X_{cm} = (m_0 + m_1 + m_2)\,/\,3m = 1$ unit, and in the y direction, $Y_{cm} = (m_0 + m_0 + m_2)\,/\,3m = \frac{2}{3}$ unit.

Free-Response Problems

1. On impact, the spring is compressed, but both masses are still in motion. Therefore, for an instant we have an inelastic collision, and momentum is of course conserved. Thus, we can write for the moment of impact, $m_1\mathbf{v}_{1i} + m_2\mathbf{v}_{2i} = (m_1 + m_2)\mathbf{v}_f$. Solving for the final velocity, we have $\mathbf{v}_f = 5.6$ m/s. Using the initial values for the velocities, we find that the initial kinetic energies are 72 and 32 J, respectively. Thus, the total initial kinetic energy is 104 J. Using the final velocity of 5.6 m/s and the combined mass of 5 kg, we obtain a final kinetic energy of 78.4 J. The difference of 25.6 J is employed to compress the spring in this inelastic collision. Using the formula for the work done against a spring $\frac{1}{2}k\mathbf{x}^2$, we find that $\mathbf{x} = 0.05$ m for the compression.

   After the collision has taken place, the two masses again separate. If we treat the situation as elastic, since we assume that the work done to compress the spring will be used by the spring in rebounding, then we can use the initial velocities as a "before" condition for momentum and kinetic energy and solve for the two final velocities after the rebound has taken place. In order to find both velocities, we need two equations and we use the conservation of kinetic energy to assist us. Thus, we write $m_1\mathbf{v}_{1i} + m_2\mathbf{v}_{2i} = m_1\mathbf{v}_{1f} + m_2\mathbf{v}_{2f}$, and for the kinetic energy, $\frac{1}{2}m_1\mathbf{v}_{1i}^2 + \frac{1}{2}m_2\mathbf{v}_{2i}^2 = \frac{1}{2}m_1\mathbf{v}_{1f}^2 + \frac{1}{2}m_2\mathbf{v}_{2f}^2$. Substituting the numbers given, we get for the momentum $28 = \mathbf{v}_{1f} + 4\,\mathbf{v}_{2f}$, and for the kinetic energy, $208 = \mathbf{v}_{1f}^2 + 4\mathbf{v}_{2f}^2$. If we solve for $\mathbf{v}_{1f}$ from the momentum equation, square it, and substitute into the kinetic energy equation, we obtain a factorable quadratic equation for $\mathbf{v}_{2f}$:

$$\mathbf{v}_{2f}^2 - 11.2\mathbf{v}_{2f} + 28.2 = (\mathbf{v}_{2f} - 7.2)(\mathbf{v}_{2f} - 4) = 0$$

Of the two choices for the second final velocity, only one provides a physically meaningful set of solutions. This is because the first mass must rebound and hence its final velocity must be negative. Our final answers are therefore $\mathbf{v}_{2f} = 7.2$ m/s and $\mathbf{v}_{1f} = -0.8$ m/s.

2. This is a two-dimensional collision, and we consider the conservation of momentum in each direction, before and after. Before, in the x direction we have only the initial momentum of the 0.1-kg disk with a velocity of 4 m/s. After, the 0.1-kg disk has an x component of momentum given by (0.1)(2)(cos 43).

   The 0.4-kg mass initially had zero momentum and now has some unknown velocity at some unknown angle θ to the x axis. Let us assume that the angle is below the x axis, so the x component will be positive and the y component will be negative. If our assumption is correct, we will get a positive answer for the angle θ. If we are incorrect, then a negative answer will let us know. Therefore, the x component of the final momentum for the 0.4-kg mass is given by $0.4\mathbf{v}_{2f}\cos\theta$. The y component of the final momentum is given by $-0.4\mathbf{v}_{2f}\sin\theta$.

   We can therefore write for the x direction,

$$(0.1)(4) + 0 = (0.1)(2)(\cos 43) + 0.4\mathbf{v}_{2f}\cos\theta$$

   and for the y direction,

$$0 = (0.1)(2)\sin 43 - 0.4\mathbf{v}_{2f}\sin\theta$$

   Solving for the angles and velocities in each case, we have the following two equations:

$$\mathbf{v}_{2f}\sin\theta = 0.34$$

$$\mathbf{v}_{2f}\cos\theta = 0.635$$

   Finding the ratio gives $\tan\theta = 0.535$ and $\theta = 28$ degrees. Substituting this angle indicates that the final velocity for the 0.4-kg mass is 0.72 m/s.

   To find the loss of kinetic energy, we just calculate the total initial and final kinetic energies. For the data given, we find that the initial kinetic energy is 0.8 J. Using the final data, we obtain a total final kinetic energy of 0.3 J. Thus, 0.5 J of kinetic energy has been lost.

3. In this problem, the x component of the velocity provides the impulse to elongate the spring using recoil. Thus, the velocity of the ball is 35 cos 30 = 25.98 m/s. Using conservation of momentum, we state that (1)(25.98)

$= 1{,}050\mathbf{v}'$. Thus $\mathbf{v}' = 0.0247$ m/s in the negative x direction (so we can say $\mathbf{v}' = -0.0247$ m/s). This recoil velocity provides kinetic energy that does work to elongate the spring. Thus, we have $\frac{1}{2}(1{,}050)(0.0247^2) = 0.32 = \frac{1}{2}(1{,}000)x^2$. Solving for x gives $x = 0.025$ m for the elongation.

4. A sketch of the situation is shown. Since there are no initial external forces, horizontal momentum must be conserved. The block-bullet system will behave as a horizontally launched projectile with an initial velocity **u**. This velocity is determined by the time and distance of flight.

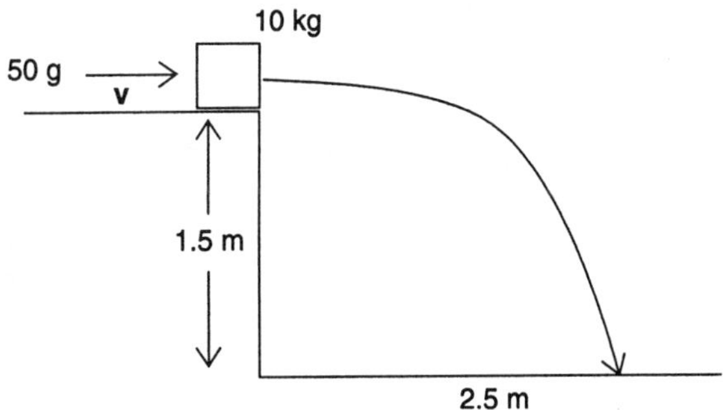

We must therefore have $(0.05\text{ kg})\mathbf{v} = (10.05\text{ kg})\mathbf{u}$. To find the launch velocity **u**, we use the projectile information given. A horizontally launched projectile will be in the air for a time t given by $t = \sqrt{2y/\mathbf{g}} = \sqrt{3/9.8} = 0.55$ s. The projectile then travels 2.5 m in 0.55 s. This means that $\mathbf{u} = 4.5$ m/s. Using conservation of momentum, we finally obtain $\mathbf{v} = 904.5$ m/s.

9
TORQUE AND ANGULAR MOMENTUM

INTRODUCTION—PARALLEL FORCES AND MOMENTS

In Chapter 8, we discussed the fact that if a single force is directed toward the center of mass of an extended object, then the result will be a linear or translational acceleration in the same direction as the force. If however, the force is directed away from the center of mass, then the result will be a rotation about the center of mass (as well as some translational acceleration). In addition, if the object is constrained by a fixed pivot point, then the rotation will take place about that pivot (a hinge, for example). These observations are summarized in Figure 1.

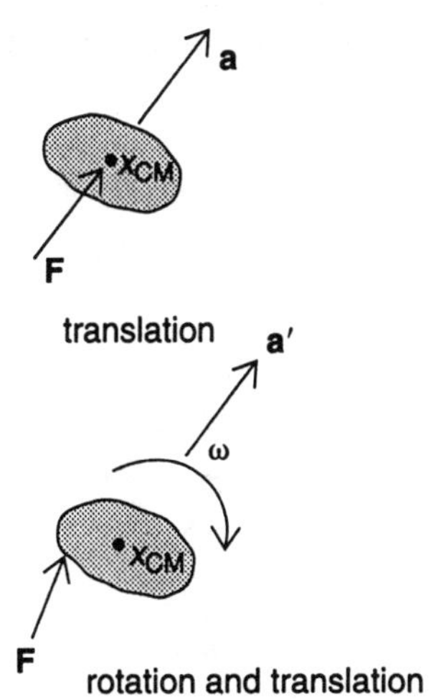

FIGURE 1.

If two forces are used, then it is possible to prevent the rotation if these forces are parallel and of suitable magnitudes. If the object is free to move in space, then translational motion may result (see Figure 2).

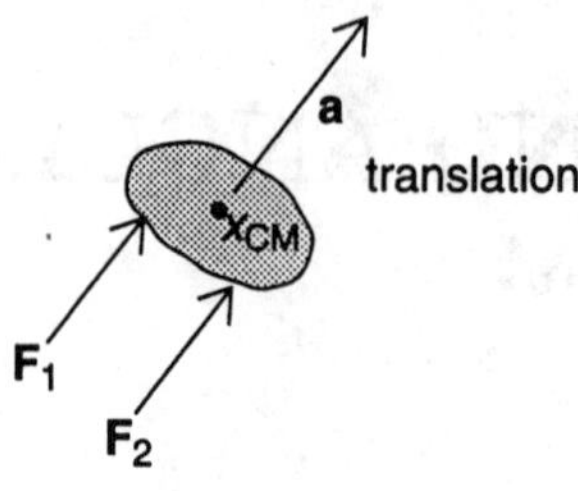

FIGURE 2.

As an example of parallel forces, consider two people on a see-saw as shown in Figure 3. The two people have weights $\mathbf{W}_1$ and $\mathbf{W}_2$, respectively, and are sitting distances d_1 and d_2 from the fixed pivot point (called the fulcrum). From our discussion of forces, we see that the tendency of each force (provided by gravity) is to cause the see-saw to rotate about the fulcrum. Force $\mathbf{W}_1$ tends to cause a counterclockwise rotation, while force $\mathbf{W}_2$ tends to cause a clockwise rotation. Arbitrarily, we state that a clockwise rotation is taken as being a negative one, while a counterclockwise rotation is taken as positive. What factors will influence the ability of the people to remain in balance? That is, what conditions must be met so that the system will remain in rotational equilibrium? (It is already in a state of translational equilibrium and constrained to remain that way.)

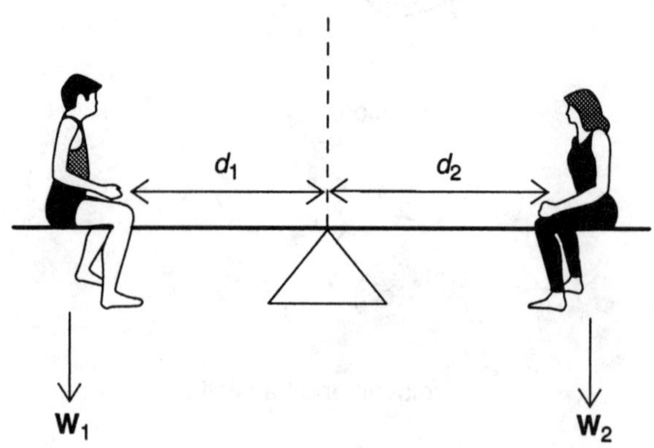

FIGURE 3.

Through experiments, you can show that the distances d_1 and d_2 (called *moment arm distances*) play a crucial role since we take the weights as being constant. If the two weights are equal, it should not be a surprise to find that

$\mathbf{d}_1 = \mathbf{d}_2$. In these examples, the weight of the see-saw is taken to be negligible (not a realistic scenario).

It turns out that if $\mathbf{W}_2$ is greater, the person must sit closer to the fulcrum. If equilibrium is to be maintained, then the following condition must hold:

$$\mathbf{W}_1\mathbf{d}_1 = \mathbf{W}_2\mathbf{d}_2$$

These are force and distance products, but they do not represent work in the translational sense because the two vectors are not parallel or resolvable into parallel components. In rotational motion, the product of a force and a perpendicular moment arm distance (relative to a fulcrum) is called torque. Let us now consider some more examples of torques and equilibria.

TORQUE

When you tighten a bolt with a wrench, you apply a force to create a rotation or twist. This twisting impetus is called **torque** in physics and is represented by the Greek letter τ. Even though its units are N-m, you must not confuse it with translational work. In a static equilibrium, the two conditions met are that the vector sum of all forces acting on the object equal zero and that the vector sum of all torques equal zero:

$$\Sigma \mathbf{F} = 0$$

$$\Sigma \tau = 0$$

When we say that $\tau = \mathbf{F}d$, the distance d is the moment arm distance from the center of mass or the pivot point and $\mathbf{F}$ is the force perpendicular to the vector displacement. If the force is applied at some angle θ to the object as seen in Figure 4, then the component of the force perpendicular to the vector displacement out from the pivot is taken as the force used. Algebraically this is stated as $\tau = \mathbf{F}d \sin \theta$. Remember, torque is a vector quantity while work is not. The direction of the torque is taken as either positive or negative depending on whether the rotation is counterclockwise or clockwise.

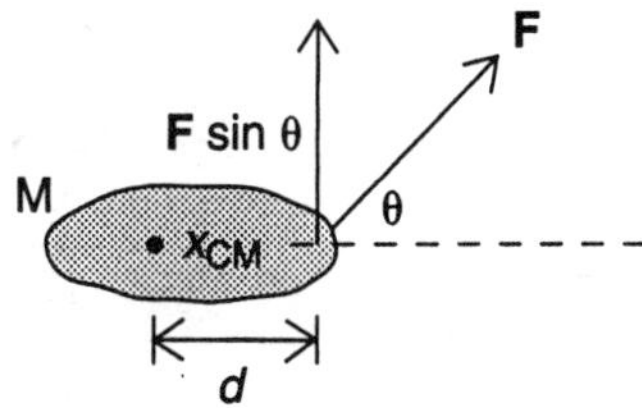

FIGURE 4.

As an example, consider a light string wound around a frictionless, massless wheel as shown in Figure 5. The free end of the string is attached to a 1.2-kg mass that is allowed to fall freely. The wheel has a radius of 0.25 m. What torque is produced?

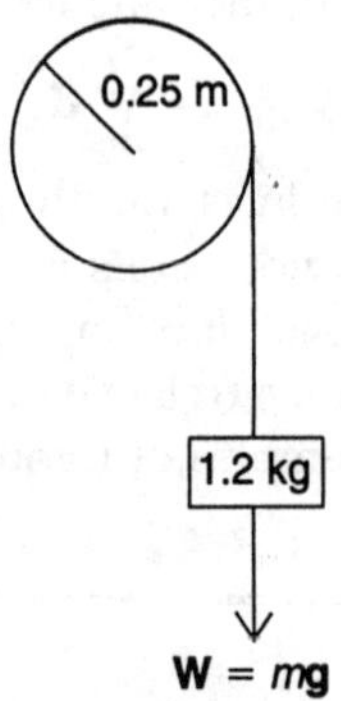

FIGURE 5.

The force acting at right angles to the center of the wheel is the weight of the mass given by $\mathbf{W} = m\mathbf{g} = (1.2)(9.8) = 11.76$ N. The radius serves as the moment arm distance $d = 0.25$ m, and so we can write $\tau = -\mathbf{F}d = (11.76)(0.25) = -2.94$ N-m. The torque is negative since the falling weight induces a clockwise rotation in this example.

MORE STATIC EQUILIBRIUM PROBLEMS USING FORCES AND TORQUES

In the next example (Figure 6), a hinged rod (considered massless) is attached to a wall with a string (and of course the hinge). The string is considered massless and makes an angle θ with the horizontal. A mass M is

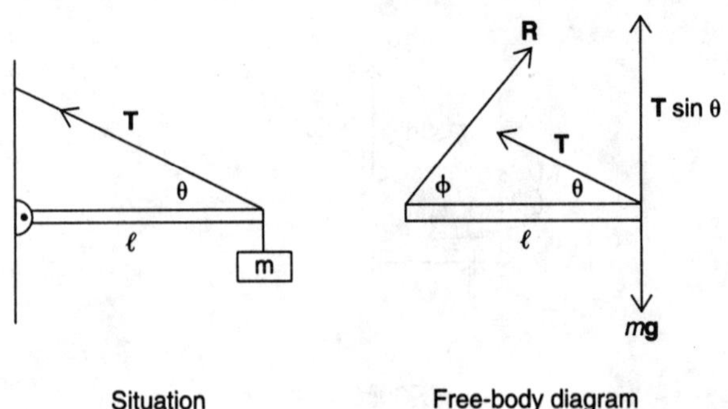

FIGURE 6.

attached to the end of the rod whose length is L. What is the tension **T** in the string?

Since the system is in static equilibrium, we must write that the sum of all forces and torques equals zero. If we choose to focus on the pivot, then all forces acting through that point do not contribute any torques. The reaction force **R** is the response of the wall to the rod and acts at some unknown angle ϕ. Thus, we state

$$\Sigma\boldsymbol{\tau} = 0$$

$$\Sigma\mathbf{F} = 0$$

For the forces, we see from the free body diagram that in the x direction,

$$\underset{\downarrow}{\Sigma\mathbf{F}_x} = 0 = \mathbf{R}\cos\phi - \mathbf{T}\cos\theta$$

and in the y direction,

$$\underset{\downarrow}{\Sigma\mathbf{F}_y} = 0 = \mathbf{R}\sin\phi + \mathbf{T}\sin\theta - m\mathbf{g}$$

Since $\mathbf{T}/\sin\theta = m\mathbf{g}$, it follows that $\mathbf{R}\sin\phi = 0$ by direct substitution. Hence, angle ϕ is also equal to zero and the reaction force is directed horizontally along the rod.

For the torques, we see that the component of tension perpendicular to the rod contributes a counterclockwise torque, while the hanging weight contributes a clockwise torque:

$$\Sigma\boldsymbol{\tau} = 0 = (\mathbf{T}\sin\theta)\,L - m\mathbf{g}L$$

We could specify the length of the rod, but you can see from the torque equation that the length can be eliminated. We can also immediately write $\mathbf{T} = m\mathbf{g}/\sin\theta$. Suppose $m = 10$ kg and $\theta = 30$ degrees. Then $\mathbf{T} = 196$ N. With this information, we see that $\mathbf{R}\sin\phi$ equals zero and that $\mathbf{R}\cos\phi = \mathbf{R}_x = 169.7$ N.

It should be noted that if the rod is considered to have mass, then the torque produced by the rod is its weight taken from the center of mass ($L/2$ in this case and in most cases) and is always clockwise (negative).

ANGULAR MOMENTUM AND ITS CONSERVATION

If we have a force producing a torque perpendicular to a displacement from a pivot point that serves as the radius of an arc, then we can write $\boldsymbol{\tau} = \mathbf{F}r$. If we recall that $\mathbf{F} = m\mathbf{a}$, then we can write $\boldsymbol{\tau} = m\mathbf{a}r$. Now, the acceleration is the rate of change in the velocity; that is, $\mathbf{a} = \Delta\mathbf{v}/\Delta t$, and so we can write (assuming the mass and radius are constant)

$$\boldsymbol{\tau} = \frac{m\Delta\mathbf{v}r}{\Delta t} = \frac{\Delta(m\mathbf{v}r)}{\Delta t} = \frac{\Delta\mathbf{L}}{\Delta t}$$

The quantity mvr is called the **angular momentum** (designated by the letter **L**) and is analogous to the linear momentum discussed in Chapter 8. Also, just as a force is equal to a change in linear momentum over time, a torque is equal to the rate of change in angular momentum. Thus, torque is analogous to force in its ability to produce rotations. Angular momentum is a vector quantity, and its units are kg-m/s.

If the net torque acting on a system is zero, then the rate of change in angular momentum is zero and we say that angular momentum has been conserved. An example of this concept occurs when an ice skater starts to spin and draws her arms inward. Since angular momentum is conserved, the response to a decrease in radius is an increase in angular velocity **ω**, and so she spins faster. Stated as a ratio, for two given times t_1 and t_2,

$$\frac{\mathbf{v}_1}{\mathbf{v}_2} = \frac{r_2}{r_1}$$

From Chapter 9, we know that the angular velocity $\omega = \mathbf{v}/r$, and so we can also write

$$\frac{\omega_1}{\omega_2} = \frac{r_2^2}{r_1^2}$$

Another example of a system in which there is no net torque acting is the solar system. Since the force of gravity is radial, it acts parallel to the radius vector swept out from the sun, thus producing zero torque. Kepler's laws of planetary motion are seen to be a consequence of the law of conservation of angular momentum.

ROTATIONAL KINEMATICS

The use of angular quantities can be traced back to Chapter 5 when we introduced uniform circular motion. If we measure angular displacement in radians, then there are relationships in rotational kinematics that are analogous to these in translational kinematics.

Given an initial angular velocity $\boldsymbol{\omega}_0$ and a constant angular acceleration $\boldsymbol{\alpha}$, then the angular displacement θ at any time t is given by

$$\theta = \omega_0 t + \frac{1}{2}\alpha t^2$$

Additionally, since the angular acceleration is the rate of change in angular velocity, we can state that the angular displacement is equal to the product of the average angular frequency and time:

$$\theta = \frac{\omega_0 + \omega_f}{2}t$$

ROTATIONAL DYNAMICS AND ANGULAR MOMENTUM

Rotational dynamics deals with the application of torques and the resulting angular displacements and rotations. For example, the application of a net force produces a change in the displacement of a moving mass such that the work done is given by $W = \mathbf{F}d$. In rotational dynamics, the application of a torque changes the angular displacement of a rotating object such that the work done is given by $W = \boldsymbol{\tau}\theta$. Rotational power is given by $P = \boldsymbol{\tau\omega}$.

If we consider the angular momentum of a point particle again, we can make the relationship more general. Recall from page 119 that the angular momentum for a point mass is given by $\mathbf{L} = m\mathbf{v}r$. Since the velocity $\mathbf{v}$ is related to the angular velocity through the equation $\mathbf{v} = \boldsymbol{\omega}r$, the angular momentum can be written in the form

$$\mathbf{L} = m\mathbf{v}r = mr(\boldsymbol{\omega}r) = mr^2\boldsymbol{\omega} = I\boldsymbol{\omega}$$

where I is the **rotational inertia** or the **moment of inertia** and is given, for a point mass, by $I = mr^2$. The angular momentum formula, $\mathbf{L} = I\boldsymbol{\omega}$, is analogous to the linear momentum formula, $\mathbf{p} = m\mathbf{v}$. Therefore, for a rotating object, rotational kinetic energy is given by

$$\text{KE}_{\text{rot}} = \frac{1}{2}m\mathbf{v}^2 = \frac{1}{2}m\omega^2r^2 = \frac{1}{2}I\omega^2$$

The application of a torque can change angular momentum. Thus, $\Delta\mathbf{L} = I\Delta\boldsymbol{\omega}$ if we assume no change in the configuration of the object. Now we know that $\boldsymbol{\tau} = \Delta\mathbf{L} / \Delta t$, and this leads to the conclusion that $\boldsymbol{\tau} = I(\Delta\boldsymbol{\omega}/\Delta \text{t}) = I\boldsymbol{\alpha}$. This expression is analogous to Newton's second law for linear motion, $\mathbf{F} = m\mathbf{a}$.

SOLVING ROTATIONAL PROBLEMS USING MOMENTS OF INERTIA

When a system is rotating, it is important to know about which axis it is rotating. The moment of inertia is a measure of the resistance to a change in rotation. If more mass is distributed away from the axis, the moment of inertia is relatively large. Thus, a ring of mass m and radius R has a greater moment of inertia than a solid disk with the same mass and radius.

When angular momentum is conserved, the change in angular frequency is accompanied by a change in the moment of inertia. The ice skater pulling her arms inward decreases her moment of inertia and thus increases her angular velocity. Some different moments of inertia are listed here and should be memorized.

1. A thin ring (hoop) of radius R (axis through the center of mass): $I = MR^2$.
2. A solid disk of radius R (axis through the center of mass): $I = \frac{1}{2}MR^2$.

3. A long, thin rod of length L with axis through the middle (see Figure 7): $I = \frac{1}{12}ML^2$.
4. A long, thin rod of length L with axis attached to one end (see Figure 7): $I = \frac{1}{3}ML^2$.
5. A solid sphere of radius R with the axis through the center: $I = \frac{2}{5}MR^2$
6. A spherical shell of radius R with the axis through the center: $I = \frac{2}{3}MR^2$.

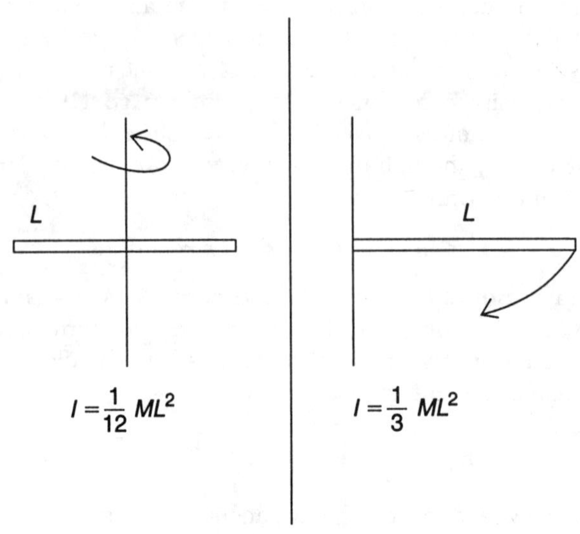

FIGURE 7.

If the moment of inertia about an axis through the center of mass is known, then the moment of inertia about any other axis can be determined using the *parallel axis theorem*: If d is the distance from the center of mass to the new axis and if I is the moment of inertia about an axis through the center of mass (usually given in tables), then

$$I_p = I + Md^2$$

Sample Problem

Suppose we know the moment of inertia of a thin rod of length L about an axis through its center of mass and perpendicular to its orientation in space. This moment of inertia is given by $I = \frac{1}{12}ML^2$. What is the moment of inertia about a parallel axis attached to one end?

Solution

Using the parallel axis theorem, we know that $d = L/2$. Thus,

$$I_p = I + Md^2 = \frac{1}{12}ML^2 + M\left(\frac{L}{2}\right)^2 = \frac{1}{3}ML^2$$

For a rolling object (with no slipping), the instantaneous axis of rotation is located at the point of contact. Thus, the moment of inertia for a rolling object is found by using the parallel axis theorem. For example, a rolling solid sphere (with no slipping) has a moment of inertia equal to

$$I_p = I + Md^2 = \frac{2}{5}MR^2 + MR^2 = \frac{7}{5}MR^2$$

As a general example, consider the problem of a uniformly round object rolling down an incline of height h with no slipping (see Figure 8). The change in potential energy is given by $m\mathbf{g}h$, which is the change in potential energy for the center of mass. The change in kinetic energy is given by the rotational energy and the translational energy of the center of mass.

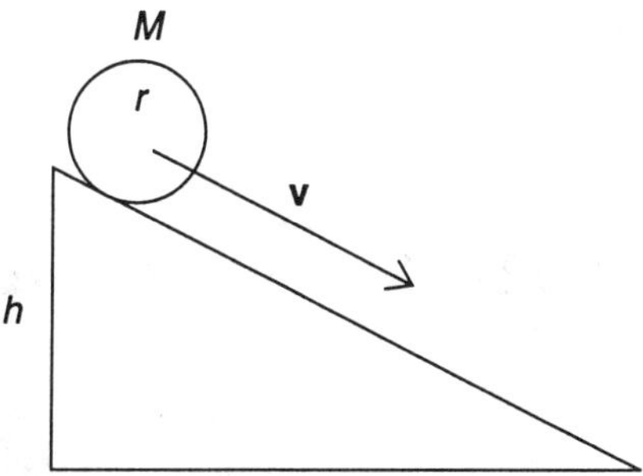

FIGURE 8.

To find the velocity of the object when it reaches the bottom of the incline, we equate the changes in energy (as viewed from the center of mass of the system):

$$M\mathbf{g}h = \frac{1}{2}M\mathbf{v}^2 + \frac{1}{2}I\omega^2$$

$$\omega = \frac{\mathbf{v}}{r}$$

$$M\mathbf{g}h = \frac{1}{2}M\mathbf{v}^2 + \frac{1}{2}I\frac{\mathbf{v}^2}{r^2}$$

$$\mathbf{v} = \sqrt{\frac{2\mathbf{g}h}{1 + (I/Mr^2)}}$$

where I is with respect to the center of mass.

PROBLEM-SOLVING STRATEGIES

Solving torque problems is similar to solving static equilibrium problems with Newton's laws. In fact, Newton's laws provide the first condition for static equilibrium (the vector sum of all forces equals zero). This means that you should once again draw a free body diagram for the situation.

In rotational static equilibrium, the vector sum of all torques must equal zero. Remember that torque is a vector even though its units are N-m. We take clockwise torques as negative and counterclockwise torques as positive. Any force going through the chosen pivot point (or fulcrum) does not contribute any torques. Therefore, it is wise to choose a point that eliminates the greatest number of forces. Also, remember that only the components of forces that are perpendicular to the direction of a radius displacement from the pivot (called the moment arm) are responsible and that this usually involves the sine of the force's orientation angle.

In rotational motion, angular momentum is conserved if no external torques act on an isolated system. This is analogous to the statement that linear momentum is conserved if no external forces act on the system (translationally).

For problems involving rotational dynamics, be sure to identify the axis of rotation. This will determine the moment of inertia for the mass. Use the parallel axis theorem if necessary and be sure to memorize the moments of inertia listed in this chapter.

PRACTICE PROBLEMS FOR CHAPTER 9

Thought Problems

1. A solid disk and thin ring (made of the same material and having the same mass and same radius) are sitting on top of an inclined plane. Both are released from rest simultaneously. Which object reaches the bottom first? Explain.

2. A bicycle wheel is suspended by a pivoting hook located at its central axis. The wheel is oriented vertically in its normal standing-up position, and if released, it will fall. If the wheel is set rotating about its axle, it will not fall. Instead, it will rotate about the fixed pivot in a motion called *precession*. Explain this behavior physically.

3. Explain why a rotating gas tends to flatten out and spin faster as it contracts (this is a simple model for the formation of our solar system).

Multiple-Choice Problems

1. A point mass m is undergoing uniform circular motion with an angular frequency ω in a horizontal circle of radius r. Which of the following is a representation of the angular momentum of the mass?
   (A) $mr^2\omega$
   (B) mr^2/ω

(C) $r\omega^2/m$
(D) $mr\omega$
(E) $m\omega$

2. A skater extends her arms holding a 2-kg mass in each hand. She is rotating about a vertical axis at a given rate. She brings her arms inward toward her body such that the distance of each mass from the axis changes from 1 m to 0.50 m. Neglecting the mass of the skater, her rate of rotation will _____.
(A) be doubled
(B) be halved
(C) be quadrupled
(D) be quartered
(E) be the same

3. A 45-kg girl is sitting on a see-saw 0.6 m from the balance point as shown. How far, on the other side, should a 60-kg boy sit so that the see-saw will remain in balance?

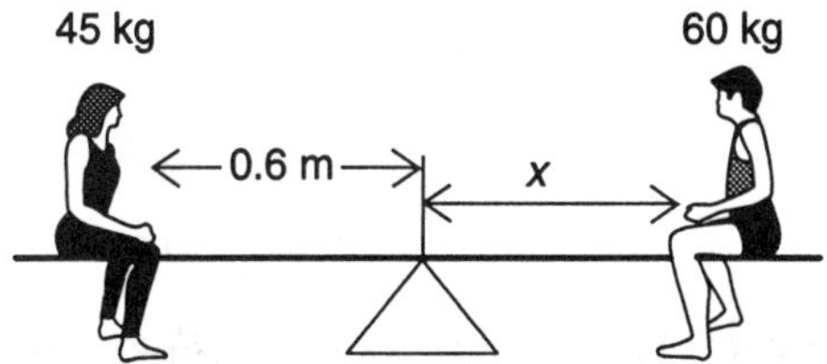

(A) 0.30 m
(B) 0.35 m
(C) 0.40 m
(D) 0.45 m
(E) 0.50 m

4. A balanced meter stick is shown in the accompanying diagram. Each mass has the distance from the fulcrum shown except for the 10-g mass. What is the approximate position of the 10-g mass based on the diagram?

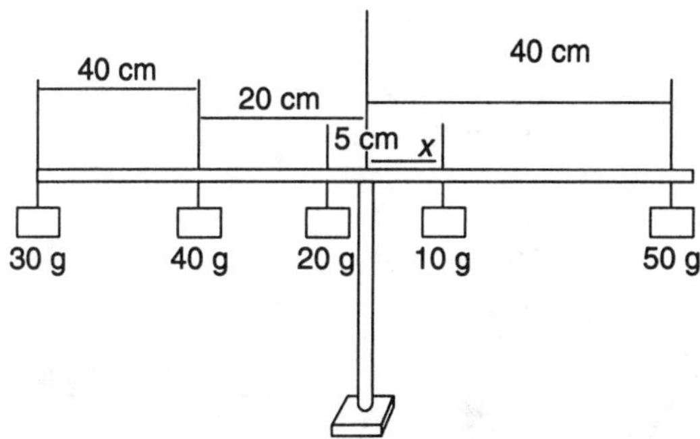

(A) 7 cm
(B) 9 cm

(C) 10 cm
(D) 15 cm
(E) 21 cm

5. A 1-kg mass swings in a vertical circle after being released from a horizontal position with zero initial velocity. The mass is attached to a massless, rigid rod of length 1.5 m. The angular momentum of the mass when it is in its lowest position is approximately _____ kg-m²/s.
(A) 4
(B) 5
(C) 8
(D) 10
(E) 12

6. A rock with a mass of 0.05 kg is swung overhead in a horizontal circle of radius 0.3 m at a constant rate of 5 rev/s. The angular momentum of the rock is _____ kg-m²/s.
(A) 0.14
(B) 0.0056
(C) 0.32
(D) 1.32
(E) 2.45

7. A solid cylinder consisting of an outer radius R_1 and an inner radius R_2 is pivoted on a frictionless axle as shown. A string is wound around the outer radius and pulled to the right with a force $\mathbf{F}_1 = 3$ N. A second string is wound the inner radius and pulled down with a force $\mathbf{F}_2 = 5$ N. If $R_1 = 0.75$ m and $R_2 = 0.35$ m, what is the net torque acting on the cylinder?

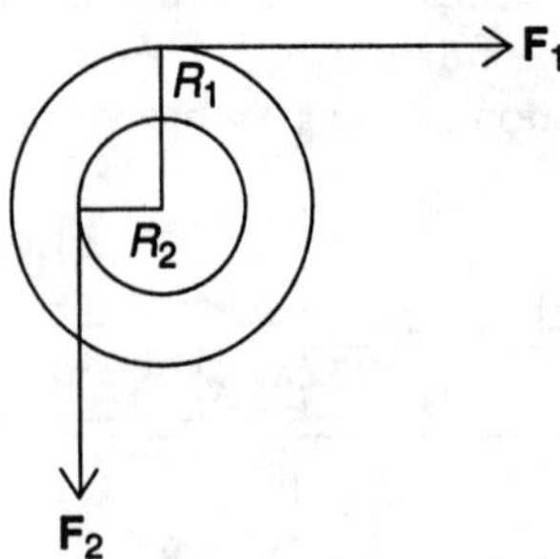

(A) 2.25 N-m
(B) −2.25 N-m
(C) 0.5 N-m
(D) −0.5 N-m
(E) 0 N-m

Answer Problems 11 and 12 based on the accompanying diagram. The rod is considered massless.

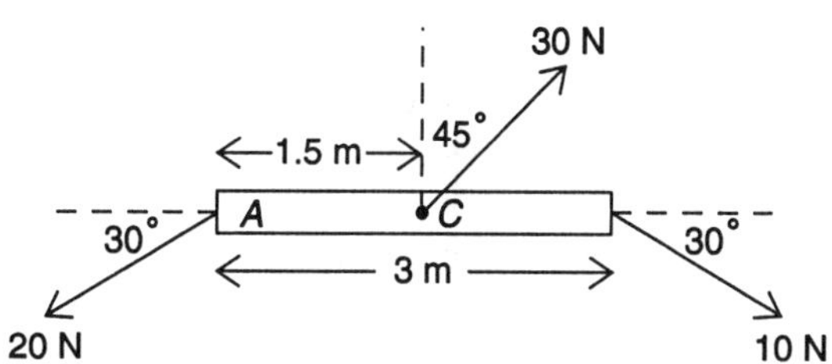

8. What is the net torque about an axis through point *A*?
   (A) 16.8 N-m
   (B) 15.2 N-m
   (C) −5.5 N-m
   (D) −7.8 N-m
   (E) 6 N-m

9. What is the net torque about an axis through point *C*?
   (A) 3.5 N-m
   (B) 7.5 N-m
   (C) −15.2 N-m
   (D) 5.9 N-m
   (E) 7 N-m

10. A small disk of mass 2 kg slides on a frictionless horizontal surface and is constrained to move in a circular path by a light, rigid rod of length 0.5 m. The disk is undergoing uniform circular motion at a velocity of 4 m/s at any instant. A piece of putty with a mass of 0.4 kg is dropped onto the disk. If the radius of the circular path remains constant, what will be the new velocity of the disk-putty system?
    (A) 1.5 m/s
    (B) 2.25 m/s
    (C) 3.3 m/s
    (D) 4.0 m/s
    (E) 5 m/s

Free-Response Problems

1. A conical pendulum consists of a point mass suspended from a light string making a vertical angle θ as shown. The string has length l, and the mass M. If the mass undergoing uniform circular motion with an instantaneous velocity $\mathbf{v}$.
   (a) Draw a free body diagram for this situation.
   (b) Derive an expression for the square of the angular momentum $\mathbf{L}$ in terms of M, l, $\mathbf{g}$, and θ only.

(c) If $\ell = 1$ m, $M = 0.15$ kg, and $\theta = 10$ degrees, determine the magnitude of the angular momentum.

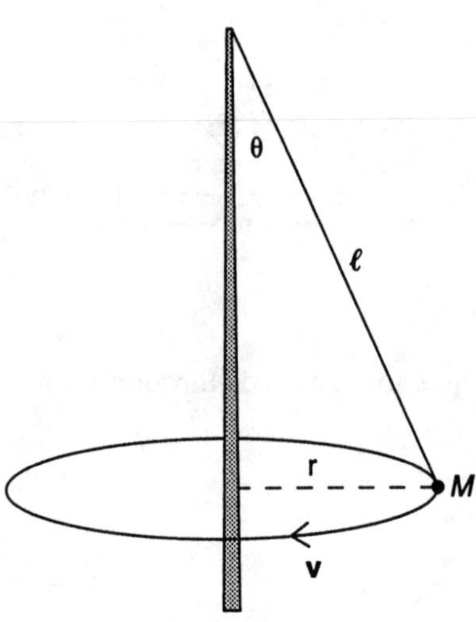

2. A 500-N person stands 2.5 m from a wall to which a horizontal beam is attached. The beam is 6 m long and weighs 200 N (see diagram). A cable attached to the free end of the beam makes an angle of 45 degrees and is attached to the wall.

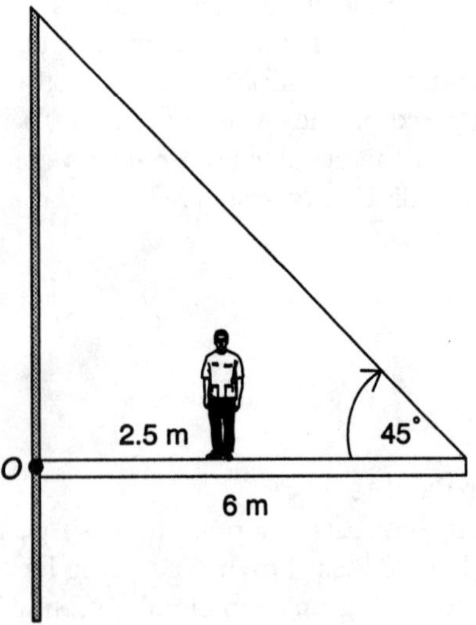

(a) Draw a free body diagram for this situation.
(b) Determine the magnitude of the tension in the cable.
(c) Determine the reaction force that the wall exerts on the beam.

3. A uniform ladder of length L and weight 100 N rests against a smooth vertical wall, and the coefficient of static friction between the bottom of the ladder and the floor is 0.5.
   (a) Draw a free body diagram of this situation.
   (b) Find the minimum angle θ that the ladder can make with the floor such that the ladder will not slip.

4. (a) A solid sphere of radius r is placed in a hemispherical bowl of radius R. If it is released from rest while its center makes an angle θ with the vertical, what will be the angular velocity and linear velocity of its center of mass when it reaches the bottom of the bowl? (Assume no slipping.)
   (b) If $R = 0.2$ m, $r = 0.02$ m, and $\theta = 60$ degrees, what is the magnitude of the linear velocity?

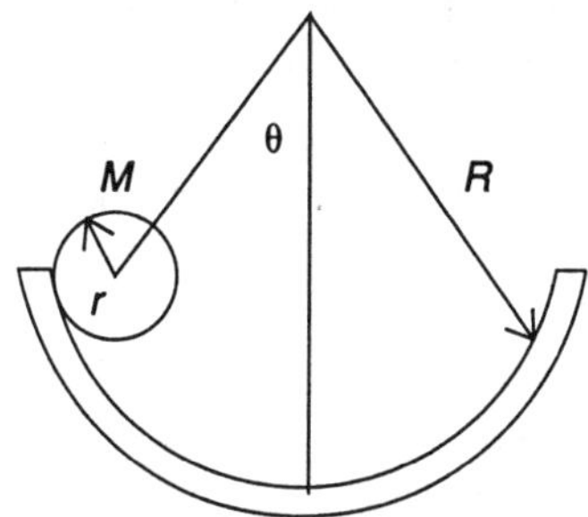

SOLUTIONS TO PRACTICE PROBLEMS

Thought Problems

1. Since the disk has the smaller moment of inertia, it will reach the bottom first because the smaller moment of inertia will provide a greater transfer of energy into translational motion since less goes into rotational energy.

2. The spinning bicycle wheel has angular momentum which resists the action of gravity tending to make it fall. Instead, the force of gravity creates a torque which tends to change the angular momentum. The change in angular momentum is in the same direction as the applied torque. This has the effect of trying to line up the angular momentum with the direction of the torque and the result is a wobbling rotation called precession. A spinning top exhibits precession as it wobbles about its pivot.

3. As the gas contracts (because of gravitational attraction), it rotates faster to conserve angular momentum. It flattens out because centrifugal forces act perpendicular to the rotation axis and resist contraction in this direction, while there is no resistance to contraction in the axial direction.

Multiple-Choice Problems

1. **A** Angular momentum is $m\mathbf{v}r$. Recall that $\mathbf{v} = r\boldsymbol{\omega}$, and so on substitution we have $\mathbf{L} = mr^2\boldsymbol{\omega}$.

2. **C** The ratio of spin rates (angular velocities) is proportional to the inverse ratio of the square of the distance from the axis of rotation. Since the distance decreases by two times, the spin rate must increase by four times.

3. **D** To remain in balance, the two torques must be equal. The force on each is given by $\mathbf{W} = m\mathbf{g}$. The moment arm distances are 0.6 m and x. Since the factor $\mathbf{g}$ appears on both sides of the torque balance equation, we can eliminate it and essentially write $(45)(0.6) = 60x$ and $x = 0.45$ m.

4. **E** In this problem, again, the sum of the torques on the left must equal the sum of the torques on the right. We could convert all masses to kilograms and all distances to meters, but in the balance equation the same factors appear on both sides, and so for simplicity and time efficiency we simply write $(30)(40) + (40)(20) + (20)(5) = 10x + (50)(40)$ and $x = 21$ cm.

5. **C** The formula for angular momentum is $m\mathbf{v}r$, where $m = 1$ kg and $r = 1.5$ m. The velocity at the bottom of the swing can be determined from the law of conservation of energy. The kinetic energy at the bottom, $\frac{1}{2}m\mathbf{v}^2$, equals the loss of potential energy during the swing, $m\mathbf{g}h$, where $h = r = 1.5$ m. Solving for velocity, we have $\mathbf{v} = 5.42$ m/s. Substituting for angular momentum yields 8.1 or approximately 8 kg-m²/s.

6. **A** Angular momentum is $m\mathbf{v}r$. The velocity is given by the circumference $2\pi r$ multiplied by the frequency of 5 rev/s. Making the necessary substitutions gives us an answer of 0.14 units.

7. **D** The net torque is given by the vector sum of all torques. $\mathbf{F}_2$ provides a counterclockwise positive torque, while $\mathbf{F}_1$ provides a clockwise negative torque. Each radius is the necessary moment arm distance. Thus we have $\boldsymbol{\tau}_{net} = (5)(0.35) - (3)(0.75) = -0.5$ N-m.

8. **A** In this problem, the net torque about point A implies that the force passing through this point does not contribute to the net torque. Also, we need the components of the remaining forces perpendicular to the beam. From the diagram, we see that the 30-N force acts counterclockwise (positive), while the 10-N force acts clockwise (negative). Thus, $\boldsymbol{\tau}_{net} = (30)(\cos 45)(1.5) - (10)(\sin 30)(3) = 16.8$ N-m.

9. **B** In this problem, since the pivot is now set at point C, we can eliminate the 30-N force passing through point C as a contributor to the torque. Again, we see that the 20-N force acts in a counterclockwise direction, while the 10-N force acts in a clockwise direction. Each force is 1.5 m from the pivot. We also need the components of each force perpendicular to the beam. Thus, $\boldsymbol{\tau}_{net} = (20)(\sin 30)(1.5) - (10)(\sin 30)(1.5) = 7.5$ N-m.

10. **C** Angular momentum is conserved since the dropping of the putty does not add a net torque to the system. Thus, we can state that $m_1\mathbf{v}_1r_1 = m_2\mathbf{v}_2r_2$. The radius remains the same, and the mass increases by 0.4 kg. Thus, $(2)(4)(0.5) = (2.4)(\mathbf{v}_2)(0.5)$ and $\mathbf{v}_2 = 3.3$ m/s.

Free-Response Problems

1. (a) A free body diagram for a conical pendulum is shown here.

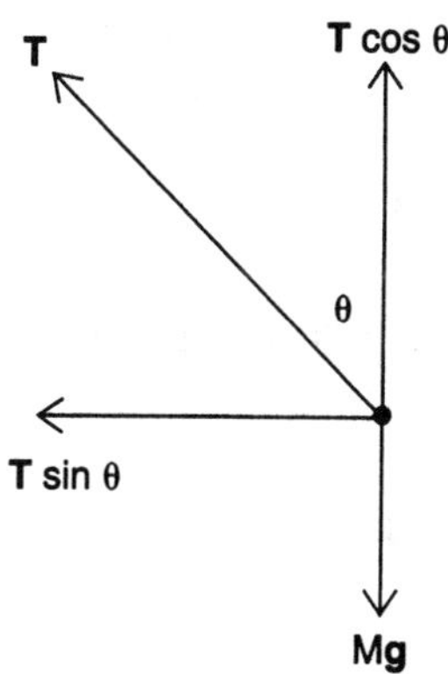

(b) We know that $\mathbf{L} = m\mathbf{v}r$ with a point mass in this case. Thus, $\mathbf{L}^2 = m^2\mathbf{v}^2r^2$. From the geometry we can see that $r = l \sin\theta$, and from Chapter 10 we know that

$$\mathbf{v}^2 = r\mathbf{g}\tan\theta = \frac{(l\sin\theta)(\mathbf{g}\sin\theta)}{\cos\theta} = \frac{l\mathbf{g}\sin^2\theta}{\cos\theta}$$

Making the necessary substitutions, we have $\mathbf{L}^2 = (M^2\mathbf{g}l^3\sin^4\theta)/\cos\theta$

(c) Using the numbers we are given, we find that $\mathbf{L}^2 = 2.077 \times 10^{-05}$, and this implies that $\mathbf{L} = 4.56 \times 10^{-03}$ kg-m²/s.

2. The free body diagram for this situation is shown here.

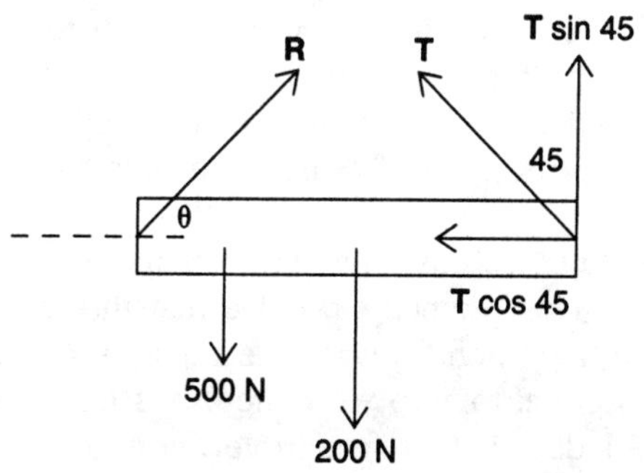

(b) The reaction force acts at some unknown angle θ. We use the two conditions for equilibrium. First, the sum of all forces in the x and y directions must equal zero: $0 = \mathbf{R} \cos \theta - \mathbf{T} \cos 45$ and $0 = \mathbf{R} \sin \theta + \mathbf{T} \sin 45 - 500 \text{ N} - 200 \text{ N}$. Now the sum of all torques through the contact point O must be zero, thus eliminating the reaction force $\mathbf{R}$: $0 = -(500)(2.5) - (200)(3) + (\mathbf{T} \sin 45)(6)$. Both weights produce clockwise torques. Solving for $\mathbf{T}$ in the last equation, we have $\mathbf{T} = 153.21$ N.

(c) Using this result, we can rewrite the first two equations as $\mathbf{R} \cos \theta = 153.21 \cos 45 = 107.25$ N and $\mathbf{R} \sin \theta = 592.75$ N. Taking the ratio of these two equations gives us $\tan \theta = 5.5268$, and therefore $\theta = 79.7$ degrees. Since we know the angle at which the reaction force acts, we can simply write $\mathbf{R} \cos 79.7 = 107.25$, which implies that $\mathbf{R} = 602.37$ N.

3. A sketch of the situation and its free body diagram are shown here.

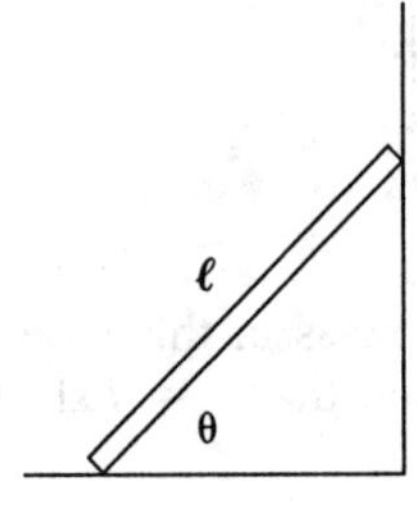

Situation

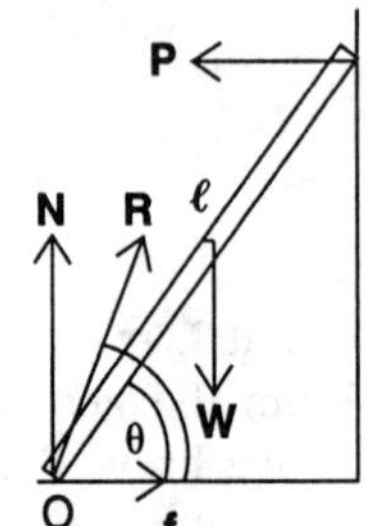

Free-body diagram

The reaction force $\mathbf{R}$ is the vector resultant of the normal force $\mathbf{N}$ (from the floor) and the frictional force $\mathbf{f}$ that opposes slippage. The force $\mathbf{P}$ is the reaction force of the vertical wall, and there is no friction on the vertical wall. The weight $\mathbf{W}$ acts from the center of mass $L/2$ and produces a clockwise torque with respect to point O. Since we have equilibrium at this angle, we can write equations for the conditions of static equilibrium. First, the sum of all x and y forces must be equal to zero. Horizontally, friction is an opposing force, as well as the reaction force $\mathbf{P}$: $0 = \mathbf{f} - \mathbf{P}$. Vertically, we can write $0 = \mathbf{N} - \mathbf{W}$. Since $\mathbf{W} = 100$ N, this means that $\mathbf{N} = 100$ N. Now, for no slippage $\mathbf{f} \leqq \mu\mathbf{N}$, and thus $\mathbf{f} = (0.5)(100) = 50 \text{ N} = \mathbf{P}$.

For the torques to be zero, we need the component of $\mathbf{P}$ perpendicular to the ladder. From the geometry, we see that this is $\mathbf{P} \sin \theta$. Also, the component of weight perpendicular to the ladder is $\mathbf{W} \cos \theta$. Thus, for the sum of all torques equal to zero we write $0 = \mathbf{P}L \sin \theta - \mathbf{W}(L/2) \cos \theta$. The length of the ladder is therefore irrelevant and can be canceled out.

Since we know that $\mathbf{P} = 50$ N and $\mathbf{W} = 100$ N, we find that $\tan\theta = \mathbf{W}/2\mathbf{P} = 1.0$ Therefore, $\theta = 45$ degrees.

4. (a) To solve this problem, we use the law of conservation of energy. The change in the potential energy is equal to the negative of the change in kinetic energy. To find the change in potential energy, we need to know the displacement of the center of mass of the sphere. The accompanying diagram sets up the geometry.

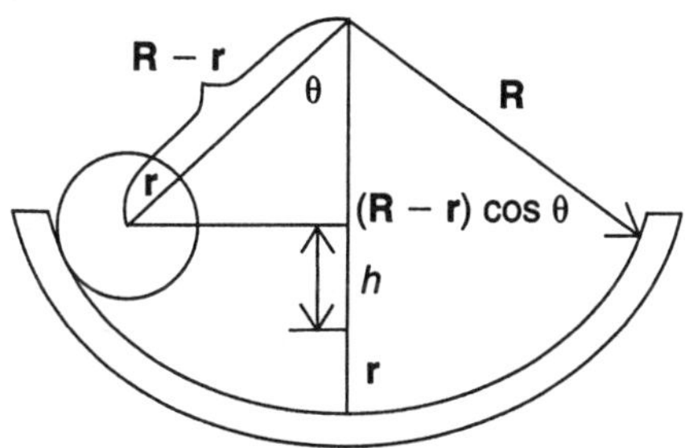

From the geometry, we can see that $h = (\mathbf{R} - \mathbf{r}) - (\mathbf{R} - \mathbf{r})\cos\theta$, and so the change in potential energy can be writen as $\Delta PE = -m\mathbf{g}(\mathbf{R} - \mathbf{r})(1 - \cos\theta)$. At the bottom of the bowl, the sphere has both translational and rotational kinetic energy. Thus, by the law of conservation of energy,

$$m\mathbf{g}(\mathbf{R} - \mathbf{r})(1 - \cos\theta) = \frac{1}{2}m\mathbf{v}^2 + \frac{1}{2}I\omega^2$$

$$m\mathbf{g}(\mathbf{R} - \mathbf{r})(1 - \cos\theta) = \frac{1}{2}m\mathbf{r}^2\omega^2 + \frac{1}{2}\left(\frac{2}{5}m\mathbf{r}^2\right)\omega^2$$

$$\omega = \frac{1}{\mathbf{r}}\sqrt{\frac{10}{7}\mathbf{g}(\mathbf{R} - \mathbf{r})(1 - \cos\theta)}$$

$$\mathbf{v} = \omega\mathbf{r} = \sqrt{\frac{10}{7}\mathbf{g}(\mathbf{R} - \mathbf{r})(1 - \cos\theta)}$$

(b) Substituting all given values yields $\mathbf{v} = 1.12$ m/s.

10
OSCILLATORY MOTION

SIMPLE HARMONIC MOTION I: A MASS ON A SPRING

From Hooke's law in Chapter 6, we know that a spring becomes elongated by an amount directly proportional to the force applied. The force constant k relates the specific amount of force (in newtons) needed to stretch or compress the spring by 1 m:

$$\mathbf{F} = -k\mathbf{x}$$

The negative sign is used to indicate that the force is restorative. One could easily use Hooke's law without the negative sign in the proper context.

Suppose we have a spring with a mass attached to it horizontally such that the mass rests on a flat, frictionless surface as shown in Figure 1.

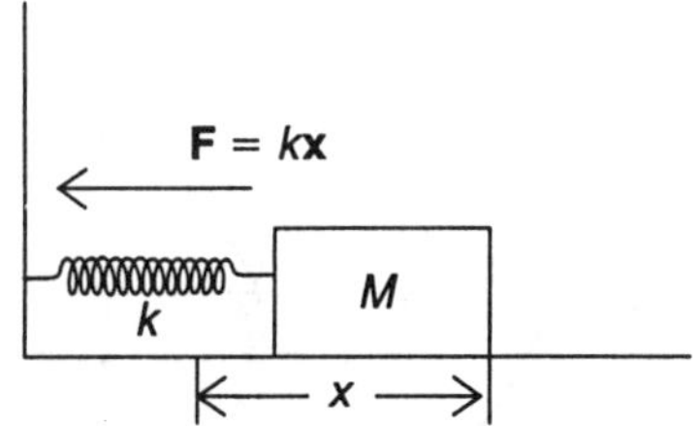

FIGURE 1.

If the mass is pulled a displacement $\mathbf{x}$, then a restoring force of $\mathbf{F} = -k\mathbf{x}$ will act on it when it is released. However, as the mass accelerates past its equilibrium position, its momentum will cause it to keep going, thus compressing the spring. This will slow it down until the same distance $\mathbf{x}$ is reached in compression. The same restoring force will then accelerate the mass back and forth, creating oscillatory motion with a certain period T (in seconds) and frequency f (in cycles/second or hertz).

Now, according to Newton's second law of motion, $\mathbf{F} = m\mathbf{a}$, and so when the mass was originally extended, the restoring force $\mathbf{F} = -k\mathbf{x}$ could also produce an instantaneous acceleration given by $\mathbf{F} = m\mathbf{a}$. In other words, $\mathbf{a} = -(k/m)\mathbf{x}$. The fact that the acceleration is directly proportional to the displacement (but in the opposite direction) is characteristic of a special kind of oscillatory motion called **simple harmonic motion**. From this expression, it can be shown that the acceleration is zero at the equilibrium point (where $\mathbf{x} = 0$).

We can build up a qualitative picture of this type of motion by considering a displacement versus time graph for this mass-spring system. Suppose the system is at rest such that the spring is unstretched. We then pull the mass to the right a distance A (called the amplitude) and release it from rest. This action creates a restoring force that pulls the mass to the left toward the equilibrium point. The velocity gets greater and greater, reaching a maximum as the mass passes through the equilibrium point (at which $\mathbf{a} = 0$). The mass then moves to the left, slowing down as it compresses the spring (since the acceleration is in the opposite direction). The mass momentarily stops when $\mathbf{x} = -A$ (because of the law of conservation of energy) and then accelerates back again to maintain simple harmonic motion (in the absence of friction).

A graph of displacement versus time (see Chapter 8) would look like a graph of the **cosine** function since the mass is beginning at some distance from the origin. We could consider the motion in progress from the point of view of the origin, in which case the graph would be of the **sine** function (some textbooks use this format). A look at Figure 2 shows the characteristics of acceleration and deceleration as the mass oscillates in a period T with a frequency f (for arbitrary units of displacement and time).

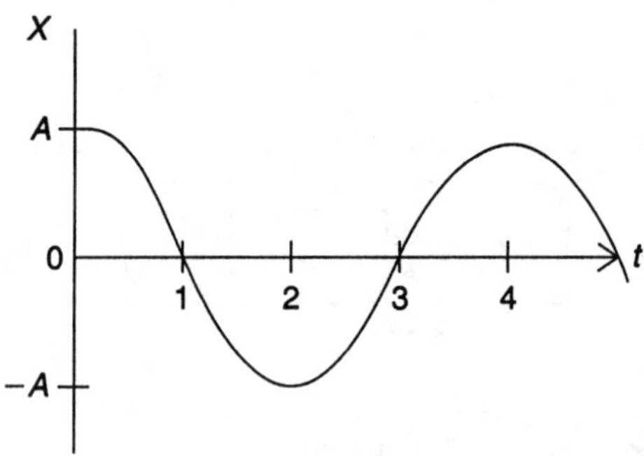

FIGURE 2.

From Chapter 7, we know that the potential energy in a compressed (or stretched) spring is given by $E = \frac{1}{2}kA^2$, for $\mathbf{x} = A$ as in our example. Thus, the constraining points $\mathbf{x} = \pm A$ define the limits of oscillation for the mass.

A graph of velocity versus time can be constructed qualitatively in much the same way. Recall that the slope of the displacement versus time graph represents the instantaneous velocity. From Figure 2, you can see that at $t = 0$, the graph is horizontal, indicating that $\mathbf{v} = 0$. The increasing negative slope shows that the mass is accelerating "backward" until at $t = 1$, the line is momentarily straight, indicating maximum velocity when $\mathbf{x} = 0$. The slope now gradually approaches zero at $t = 2$, indicating that the mass is slowing down as it approaches $\mathbf{x} = -A$. The cycle then repreats itself, producing a graph similar to Figure 3.

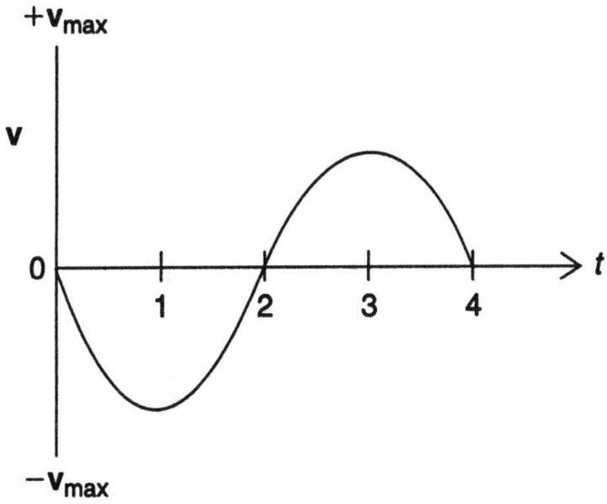

FIGURE 3.

Finally, analyzing the velocity graph with slopes, we produce an acceleration versus time graph. When $t = 0$, the line is momentarily straight with negative slope, indicating a maximum negative acceleration. When the mass crosses the equilibrium point, velocity is maximum but acceleration is momentarily zero. The acceleration (proportional to the displacement), reaches a maximum once again when $\mathbf{x} = -A$. The cycle repeats itself, producing Figure 4. Notice that the acceleration graph is approximately the negative of the displacement graph, as expected.

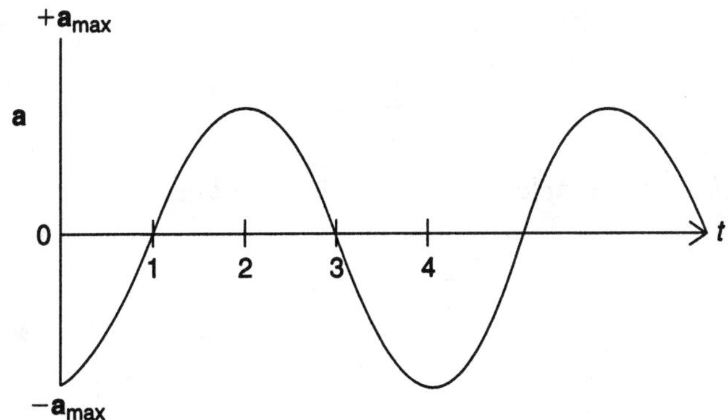

FIGURE 4.

In order to derive the period of oscillation for the mass-spring system, we can consider another form of periodic motion already discussed—uniform circular motion. If a mass is attached to a rotating turntable and then turned on its side, a projected shadow of the rotating mass simulates simple harmonic motion (Figure 5).

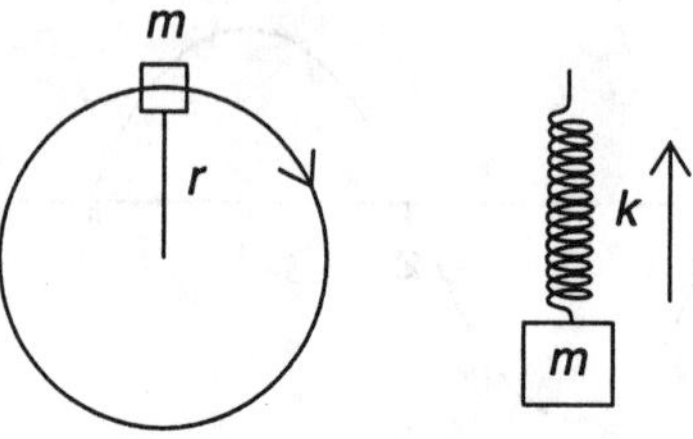

FIGURE 5.

In this simulation, the radius r acts like the amplitude A, and one can adjust the frequency and period of the rotation such that when the shadow appears next to an actual oscillating system, it is difficult to decide which one is actually rotating.

From our understanding of uniform circular motion, we know that the centripetal acceleration is given by

$$\mathbf{a}_c = \frac{4\pi^2 r}{T^2}$$

Since the projected sideways view of this motion appears to approximate simple harmonic motion, we can let the radius r be approximated by the linear displacement $\mathbf{x}$ and write

$$\mathbf{a}_x = \frac{4\pi^2 \mathbf{x}}{T^2}$$

Since $\mathbf{a} = (k/m)\mathbf{x}$ in simple harmonic motion, we can now write

$$\frac{k}{m}\mathbf{x} = \frac{4\pi^2 \mathbf{x}}{T^2}$$

$$T = 2\pi\sqrt{\frac{m}{k}}$$

This is the equation for the period of an oscillating mass-spring system in seconds. To find the frequency of oscillation, recall that $f = 1/T$. Also, recall that the angular frequency (velocity) $\omega = 2\pi/T$, which implies that $\omega = \sqrt{k/m}$.

THE SIMPLE PENDULUM

Imagine a pendulum consisting of a mass M (called a *bob*) and a string of length L that is considered massless (Figure 6). The pendulum is displaced through an angle θ that is much less than 1 rad (about 57 degrees). Under these conditions, the pendulum approximates simple harmonic motion and the period of oscillation is independent of amplitude.

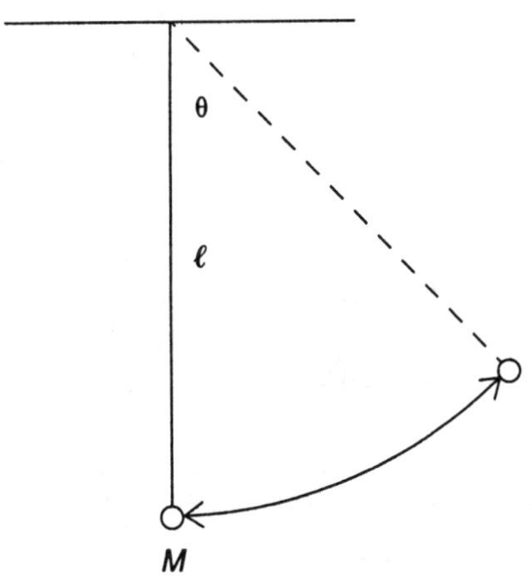

FIGURE 6.

When the pendulum swings through an arc of length s, it appears to be following a straight path for a suitably chosen small period of time or small section of the arc. In this approximation, we can imagine the mass as being accelerated down an incline. The oscillations occur because gravity accelerates the mass back to its lowest position. Its momentum maintains the motion through that point, and then the constraining action of the string causes the mass to swing in an upward arc. Conservation of energy brings the mass to the same vertical displacement (or causes it to swing through the same arc length), and then it stops momentarily until gravity begins to pull it down again.

If we imagine that an incline of set angle θ is causing the acceleration, then from our discussion of kinematics we know that $\mathbf{a} = -\mathbf{g} \sin \theta$ (where the angle is measured in radians). Now if θ is sufficiently less than 1 rad, we can write $\sin \theta \approx \theta$, and so

$$\mathbf{a} = -\mathbf{g} \sin \theta \approx -\mathbf{g}\theta \approx \frac{\mathbf{g}s}{L}$$

In the preceding equation we have used the known relationship from trigonometry that in radian measure, if s is the arc length and L corresponds to the effective "radius" of swing, then $s = L\theta$.

Remember that these are only approximations, but for angles of about 10 or 20 degrees, the approximations are fairly accurate. Since the pendulum now approximates simple harmonic motion, we know that we can find a suitable rotational motion that when viewed in projection (like the pendulum swing viewed in projection) simulates simple harmonic motion. Therefore, as in Simple Harmonic Motion I on page 133, we can write

$$\mathbf{a} = \frac{4\pi^2 s}{T^2} = \frac{\mathbf{g}s}{L}$$

where the arc length s is the amplitude of swing.

If we solve for the period T, we finally get:

$$T = 2\pi\sqrt{\frac{L}{\mathbf{g}}}$$

Notice that the period is independent of the mass and is very sensitive to the local acceleration of gravity. This is an excellent way to independently measure the value of **g** in various locations.

THE KINEMATICS OF SIMPLE HARMONIC MOTION

From the previous discussion, we know that if a mass is attached to a horizontal spring, pulled to one side, and then released, it undergoes simple harmonic motion. A graph of displacement versus time for this motion can be expressed in terms of a cosine function as was shown. Algebraically, note that for $\mathbf{x} = A$ at $t = 0$, $\phi = 0$ and $\mathbf{v} = 0$. Thus, we can write

$$\mathbf{x} = A\cos(\omega t + \phi)$$

Recall from Chapter 5 that the angular frequency ω is in rad/s, and so the product ωt is an angle in radians. The angle ϕ is called the **phase angle** and can be used to adjust the initial condition for the starting point at $t = 0$. When this angle is suitably chosen, we can convert the formula for displacement to $\mathbf{x} = A\sin\omega t$ (which some textbooks present as the displacement).

Let us use the first formula for displacement and choose $\phi = 0$ so that $\mathbf{x} = A\cos\omega t$. Since $\omega = 2\pi/T$, we can write $\mathbf{x} = A\cos(2\pi/T)t$. Thus, when $t = 0$, $\mathbf{x} = A$, and when $t = T/2$, $\mathbf{x} = -A$ (as expected since $T/2$ is one half the period of oscillation). The velocity of the mass is a maximum when the mass passes through the point $\mathbf{x} = 0$, and this occurs at one quarter of a period, $T/4$. Recall that in uniform circular motion, $\mathbf{v} - r\omega$. In this case, we associate the amplitude A with the equivalent circular radius r such that $\mathbf{v}_{max} = A\omega$. In a similar fashion, we can show that the acceleration $\mathbf{a} = -\omega^2 A$.

From Chapter 7, we know that the work done to compress a spring is also equal to the potential energy stored in it. This value is expressed as $\frac{1}{2}k\mathbf{x}^2$. If the system is set into oscillation by stretching the spring an amount $\mathbf{x} = A$, then the maximum energy in the oscillating system is given by $U = \frac{1}{2}kA^2$. As the mass oscillates, it reaches maximum velocity when it passes through $\mathbf{x} = 0$. At that point, the potential energy of the spring is zero, and since we are treating cases without friction, the loss of potential energy is balanced by this gain in kinetic energy:

$$\text{KE}_{\text{max}} = \frac{1}{2}m\mathbf{v}_{\text{max}}^2 = \frac{1}{2}mA^2\omega^2$$

Thus, we can now write, for conservation of energy,

$$\frac{1}{2}mA^2\omega^2 = \frac{1}{2}kA^2$$

which implies that $\omega = \sqrt{k/m}$ as expected.

If we want to treat cases in which the mass is between the two extremes of $\mathbf{x} = 0$ and $\mathbf{x} = \pm A$, we note that the spring still possesses some elastic potential energy (no gravitational potential energy is involved since the mass is oscillating horizontally). This implies

$$\frac{1}{2}kA^2 = \frac{1}{2}m\mathbf{v}^2 + \frac{1}{2}k\mathbf{x}^2$$

If we solve for the velocity, we obtain

$$\mathbf{v} = \pm\sqrt{\frac{k}{m}(A^2 - \mathbf{x}^2)}$$

Sample Problem

A body executes simple harmonic motion with a displacement that varies in time according to the relationship $\mathbf{x} = 0.02 \cos \pi t$.

(a) Find the amplitude, angular frequency, frequency, and period for the body.
(b) Find the maximum velocity attained by the body.
(c) If the body is a mass of 0.3 kg attached to a spring, find the force constant k.
(d) Find the velocity when $t = 0.2$ s.

Solution

(a) From discussions in this chapter we know that $\mathbf{x} = A \cos \omega t$, and so $A = 0.02$ m and $\omega = \pi$ rad/s. Now, $\omega = 2\pi f$, and so solving for frequency we get $f = 0.5$ cycles/s. Period $T = 1 / f$, and so $T = 2$ s.

(b) $\mathbf{v}_{\text{max}} = A\omega = (0.02)(3.14) = 0.0628$ m/s.

(c) Recall that in simple harmonic motion $\omega = \sqrt{k/m}$. Using $\omega = 3.14$ rad/s and $m = 0.3$ kg, we find that $k = 2.96$ N/m.

(d) When $t = 0.2$ s, $\mathbf{x} = 0.02 \cos (\pi 0.2)$. The argument of the cosine function is in radians, and so we find that $\mathbf{x} = 0.016$ m. Now the velocity at that position (i.e., time) is given by $\mathbf{v} = \pm\sqrt{k/m(A^2 - \mathbf{x}^2)}$. Substituting the given numbers yields $\mathbf{v} = \pm 0.037$ m/s.

PHASE SPACE ANALYSIS OF SIMPLE OSCILLATING SYSTEMS

The One-Dimensional Harmonic Oscillator

Consider a mass attached to a spring with a force constant k. The spring is considered to be massless and frictionless. The mass rests on a horizontal frictionless surface and is displaced horizontally and released such that it executes simple harmonic motion.

The total energy for this system can written as

$$E = \frac{1}{2} m\mathbf{v}^2 + \frac{1}{2} k\mathbf{x}^2$$

$$E = \frac{\mathbf{p}^2}{2m} + \frac{k\mathbf{x}^2}{2}$$

where $\mathbf{p}$ is the momentum at any position $\mathbf{x}$.

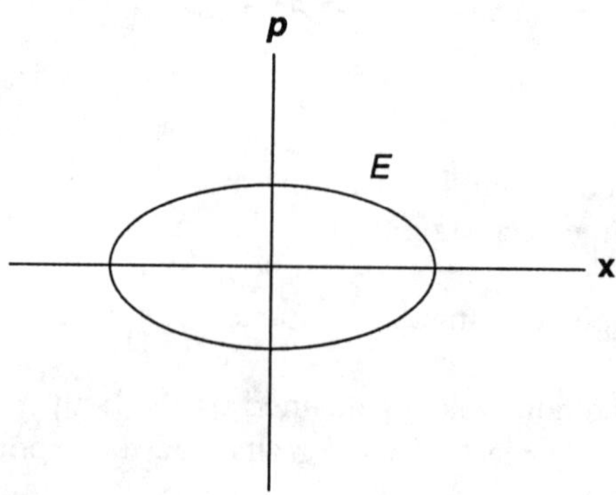

FIGURE 7. *Phase space ellipse.*

If we divide both sides by the energy E, then we can rewrite this equation in the form of an ellipse in *phase space* (Figure 7), where the momentum and position of the mass are treated as coordinates:

$$1 = \frac{\mathbf{p}^2}{2mE} + \frac{\mathbf{x}^2}{2E/k}$$

This ellipse has a semimajor axis $a = \sqrt{2E/k}$ and a semiminor axis $b = \sqrt{2mE}$.

The area of this ellipse is in units of energy times time, which is called *action*. When we consider the area of this ellipse, we look at the changes in momentum and position over one complete cycle or phase. Hence, the time factor involved is the period of oscillation. Thus, the action in this case is equal to the product of the total energy and the period:

$$\text{Area} = \text{action} = \pi ab = \pi\sqrt{2mE\frac{2E}{k}} = E2\pi\sqrt{\frac{m}{k}}$$

$$T = 2\pi\sqrt{\frac{m}{k}}$$

The Simple Pendulum

Consider a simple pendulum consisting of a point mass m attached to a light string of length l. The pendulum is pulled back an angle θ and then released. If $\theta << 1$ (in radians), then the movement of the pendulum approximates simple harmonic motion.

For the swinging mass, the moment of inertia for the pendulum is given as I. The total energy for the system can be written as

$$E = \frac{1}{2}I\omega^2 + m\mathbf{g}l(1 - \cos\theta)$$

The phase space coordinates for the pendulum are the angular momentum $\mathbf{L}$ and the angular displacement θ. Since $\mathbf{L} = I\omega$, we can multiply through by the moment of inertia to obtain the square of the angular momentum. Additionally, since $\theta << 1$, we can use the series expansion for cosine and write $\cos\theta = 1 - (\theta^2/2)$. Therefore, we rewrite the total energy equation as

$$2EI = I^2\omega^2 + m\mathbf{g}lI\ \theta^2$$

$$2EI = L^2 + m\mathbf{g}lI\ \theta^2$$

$$1 = \frac{\mathbf{L}^2}{2EI} + \frac{\theta^2}{2E/m\mathbf{g}l}$$

The area of this ellipse is also the action for the system and again is equal to the product of the total energy and the period:

$$\text{Area} = \pi ab = \sqrt{2EI\frac{2E}{m\mathbf{g}l}} = E2\pi\sqrt{\frac{I}{m\mathbf{g}l}}$$

$$I = ml^2$$

$$T = 2\pi\sqrt{\frac{l}{\mathbf{g}}}$$

where we have substituted for the moment of inertia for a point mass swinging on a string and arrived at the expected period for a simple pendulum.

The Torsion Pendulum

A simple torsion pendulum consists of a circular disk of radius R attached to a stiff wire at its center (see Figure 8). The disk is given a deflecting twist through an angle θ and then released.

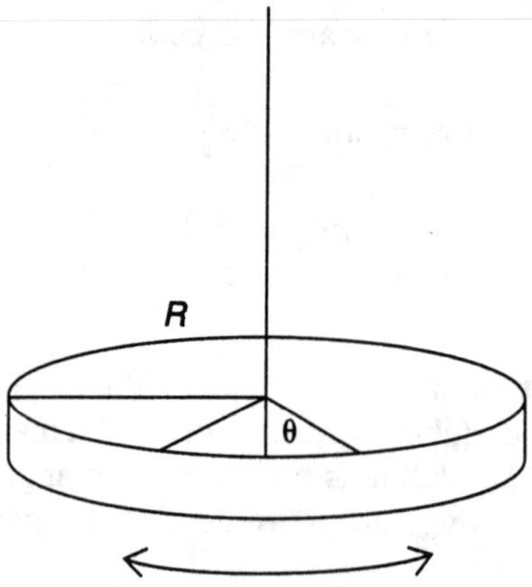

FIGURE 8. *Torsion pendulum*

The restoring torque on the disk is given by $\tau = -\kappa\theta$, where κ is a torsion constant for the system. The moment of inertia for the disk is designated I. The total energy for this system is given by

$$E = \frac{1}{2} I\omega^2 + \frac{1}{2} \kappa\theta^2$$

The phase space coordinates for this system are the angular momentum L and the angular displacement θ. Following the procedure outlines for the simple pendulum, the total energy equation can be reduced to

$$1 = \frac{\mathbf{L}^2}{2EI} + \frac{\theta^2}{2E/\kappa}$$

If we again take the area of the phase space ellipse to represent the action,

$$A = \pi ab = \pi\sqrt{2EI\frac{2E}{\kappa}} = E2\pi\sqrt{\frac{I}{\kappa}}$$

$$T = 2\pi\sqrt{\frac{I}{\kappa}}$$

A Rod Attached to a Spring

As a final example of this methodology, consider a slender rod of mass M attached at one end to a frictionless pivot. The length of the rod is l, and at some distance d from the hinge a massless spring is vertically attached (see Figure 9).

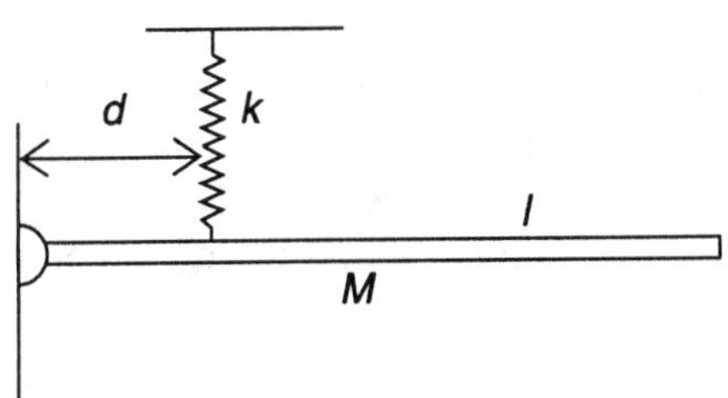

FIGURE 9.

When the rod is pulled down through a small angle θ, the system oscillates and approximates simple harmonic motion. What is the period of oscillation?

The moment of inertia for this system is given by $I = \frac{1}{3}Ml^2$. For $\theta << 1$, the vertical displacement of the spring is given by $\mathbf{x} = d\theta$. Therefore, the total energy equation for this system is given by

$$E = \frac{1}{2} I\omega^2 + \frac{1}{2} kd^2\theta^2$$

The phase space coordinates are given by the angular momentum $\mathbf{L}$ and the angular displacement θ. After multiplying through by the moment of inertia I, the equation becomes

$$2EI = \mathbf{L}^2 + kId^2\theta^2$$

$$1 = \frac{L^2}{2EI} + \frac{\theta^2}{2E/kd^2}$$

Following our standard procedure, we find the area of the phase space ellipse and identify it as the product of the total energy and the period:

$$\text{Area} = \pi ab = \pi\sqrt{2EI\frac{2E}{kd^2}} = E2\pi\sqrt{\frac{I}{kd^2}}$$

$$I = \frac{1}{3} Ml^2$$

$$\mathbf{T} = 2\pi\frac{l}{d}\sqrt{\frac{M}{3k}}$$

PROBLEM-SOLVING STRATEGIES

1. Remember that simple harmonic motion is related to circular motion and is characterized by the fact that the acceleration varies directly with the displacement (but in the opposite direction).

2. A simple pendulum approximates simple harmonic motion only if its displacement angle is much less than 1 rad (about 57 degrees). The period is nearly independent of amplitude under these circumstances and is completely independent of the mass.
3. It is sometimes easier to use energy considerations since the equations involve scalars and no free body diagrams have to be drawn. Another alternative is to use phase space analysis.
4. For pendulum or pendulum-like problems, if $\theta << 1$ rad, then $\sin \theta \approx \tan \theta \approx \theta$.

PRACTICE PROBLEMS FOR CHAPTER 10

Thought Problems

1. In terms of moment of inertia, explain why the period of a pendulum increases as its length increases.
2. Explain why an oscillating pendulum only approximately represents simple harmonic motion.
3. Explain how you could experimentally determine the moment of inertia of an irregular object.

Multiple-Choice Problems

1. What is the length of a pendulum whose period at the earth's equator is 1 s?
(A) 0.15 m
(B) 0.25 m
(C) 0.30 m
(D) 0.45 m
(E) 1.0 m

2. On the planet Xorph, an astronaut determines the acceleration of gravity by means of a pendulum. She observes that a 1-m-long pendulum has a period of 1.5 s. The acceleration of gravity on the planet Xorph is _____ m/s^2.
(A) 7.5
(B) 15.2
(C) 10.2
(D) 26.3
(E) 17.5

3. When a 0.05-kg mass is attached to a vertical spring, it is observed that the spring stretches 0.03 m. The system is then placed horizontally on a frictionless surface and set into simple harmonic motion. What is the period of the oscillations?
(A) 0.75 s
(B) 0.12 s

(C) 0.35 s
(D) 1.3 s
(E) 2.3 s

4. A mass-spring system is observed to have a period of oscillation equal to 1.45 s. If the amplitude of the oscillations is 0.12 m, then the maximum velocity attained by the mass is _____ m/s.
(A) 0.17
(B) 0.32
(C) 12.08
(D) 0.03
(E) 0.52

5. Which of the following is an equivalent expression for the maximum velocity of an oscillating mass-spring system?
(A) $\sqrt{Ak/m}$
(B) $\sqrt{A^2mk}$
(C) $\sqrt{A^2m/k}$
(D) $\sqrt{A^2k/m}$
(E) Akm

6. A mass of 0.5 kg is connected to a massless spring with a force constant k = 50 N/m. The system is oscillating on a frictionless horizontal surface. If the amplitude of the oscillations is 2 cm, then the total energy of the system is _____ J.
(A) 0.01
(B) 0.1
(C) 0.5
(D) 0.3
(E) 0.2

7. A mass of 0.3 kg is connected to a massless spring with a force constant k = 20 N/m. The system oscillates horizontally on a frictionless surface with an amplitude of 4 cm. What is the velocity of the mass when it is 2 cm from its equilibrium position?
(A) 0.28 m/s
(B) 0.08 m/s
(C) 0.52 m/s
(D) 0.15 m/s
(E) 0.34 m/s

8. If the length of a simple pendulum is doubled, then its period will _____.
(A) decrease by 2
(B) increase by 2
(C) decrease by $\sqrt{2}$
(D) increase by $\sqrt{2}$
(E) remain the same

9. The pendulums of two grandfather clocks have equal lengths, and Clock A runs faster than Clock B. Which of the following statements is true?
(A) Pendulum A is more massive.
(B) Pendulum B is more massive.
(C) Pendulum A swings through a smaller arc.
(D) Pendulum B swings through a smaller arc.
(E) None of these statements are correct.

10. A 2-kg mass is oscillating horizontally on a frictionless surface when attached to a spring. The total energy of the system is observed to be 10 J. If the mass is replaced by a 4-kg mass but the amplitude of oscillations (and the spring) remain the same, then the total energy of the system will be ____ J.
(A) 10
(C) 5
(C) 20
(D) 3.3
(E) 15.5

Free-Response Problems

1. A mass M is attached to two springs with force constants k_1 and k_2, respectively. The mass can slide horizontally over a frictionless surface. For each of the two arrangements for the mass and springs shown here, determine the period of oscillation.

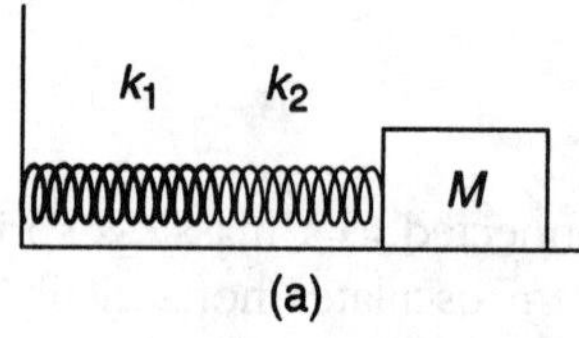

(a)

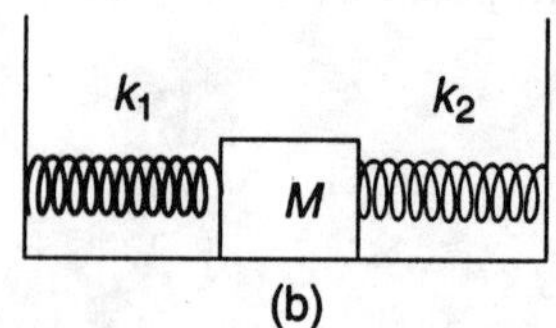

(b)

2. A mass M is attached to two light elastic strings both of length L and made of the same material. The mass is displaced vertically upward by a small displacement $\Delta \mathbf{y}$ such that equal tensions $\mathbf{T}$ exist in the strings. The mass is released and begins to oscillate up and down. Assume that the displacement is small enough so that the tensions do not change appreciably.

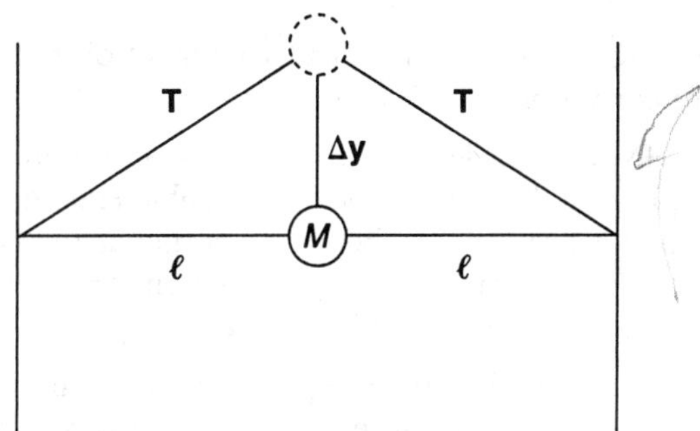

(a) What was the net restoring force acting on the mass when it was initially displaced and held at rest?
(b) What is the frequency of oscillation?

3. A block of mass 0.5 kg rests on top of a flat, rectangular plate. The system is undergoing simple harmonic motion as shown, with a period of 2.7 cycles per second. The plate rests on a horizontal frictionless surface, and the coefficient of static friction between the block and the plate is 0.4.
   (a) Draw a free body diagram for the block.
   (b) What is the maximum amplitude of oscillation such that the block will not slip?
   (c) If the spring has a force constant $k = 1{,}000$ N/m, what is the maximum energy possessed by the system (without causing the block to slip)?

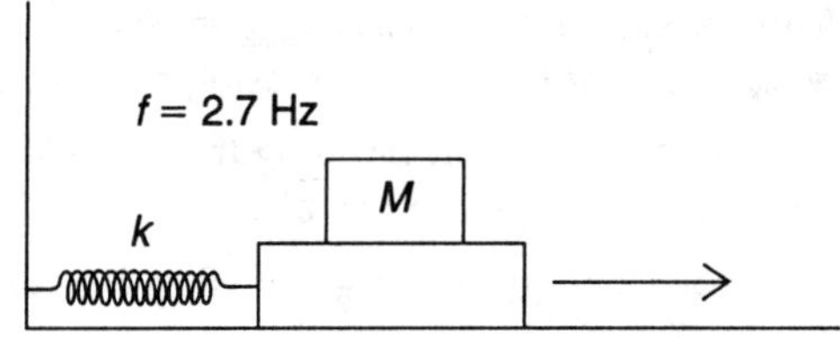

4. Using the information from Problem 4 in Chapter 9, determine the period of oscillation of a solid sphere of radius r at rest in a hemispherical bowl of radius R making an angle θ with the vertical. Use the phase space analysis method outlined in this chapter and also make use of the approximation $\cos\theta = 1 - (\theta^2/2)$. Assume mechanical energy is conserved.

SOLUTIONS TO PRACTICE PROBLEMS

Thought Problems

1. A pendulum can be considered a point mass rotating through a circular arc of radius L. Therefore, if the length is changed, so is the moment of inertia for the point mass. A shorter pendulum moves more quickly

through the arc because of the smaller moment of inertia, and the period is therefore corresponding shorter.

2. Mathematically, the acceleration of the mass is proportional to the sine of the angle. The ability of a body to exhibit simple harmonic motion depends on the acceleration being proportional to the negative of the displacement. This is true as an approximation for a pendulum only if the angle is very small (as measured in radians).

3. The period of a pendulum is dependent on its moment of inertia as outlined on page 142. Therefore, if an irregular object is suspended from a support, it will oscillate with simple harmonic motion for small displacement angles. If the period is measured, then one can establish a relationship to calculate the moment of inertia.

Multiple-Choice Problems

1. **B** If we use the formula for the period of a pendulum, $T = 2\pi\sqrt{l/\mathbf{g}}$, and square both sides, we can solve for the length: $l = T^2\mathbf{g}/4\pi^2$. Using the values given in the problem, $l = 0.25$ m.

2. **E** Again using the formula for the period of a pendulum, this time we solve for the acceleration of gravity: $g = 4\pi^2 l/T^2$. For the values given in the problem, $g = 17.5$ m/s^2.

3. **C** First find the spring constant. From $\mathbf{F} = k\mathbf{x}$, we see that $\mathbf{F} = mg = (0.05)(9.8) = 0.49$ N Thus, $k = \mathbf{F}/\mathbf{x} = 0.49/0.03 = 16.3$ N/m. Now the period $T = 2\pi\sqrt{m/k}$. For the values given in the problem, $T = 0.35$ s.

4. **E** The formula for maximum velocity is $\mathbf{v}_{max} = A\omega$. Also, recall that $\omega = 2\pi/T$. Thus $\mathbf{v}_{max} = 2\pi\ A/T = 0.52$ m/s using the values given.

5. **D** Since $\mathbf{v}_{max} = A\omega$ and $\omega = \sqrt{k/m}$, bringing the A back under the radical sign makes the equation equal to $\sqrt{A^2k/m}$.

6. **A** The formula for total energy is $E = \frac{1}{2}\ kA^2$. For the values given in the problem, we have $E = 0.01$ J.

7. **A** The formula for any intermediary velocity is $\mathbf{v} = \sqrt{(k/m)(A^2 - \mathbf{x}^2)}$. Before substituting values, we state that 4 cm = 0.04 m and 2 cm = 0.02 m. Now, using these values for amplitude and position, we substitute and find that $\mathbf{v} = 0.28$ m/s.

8. **D** Since the period of a pendulum varies directly as the square root of the length, if the length is doubled, the period will increase by $\sqrt{2}$.

9. **C** If the clock runs faster, this means that its period is quicker and therefore that it swings through a smaller arc to maintain the faster period (since the lengths and shapes are the same).

10. **A** The maximum energy in the mass-spring system can be expressed independently of the mass. (It depends on the force constant and the square of the amplitude.) Thus, if the spring and the amplitude remain the same, so will the maximum energy of the system.

Free-Response Problems

1. (a) When the mass is stretched a distance $\mathbf{x}$, spring k_1 is stretched $\mathbf{x}_1$ and spring k_2 is stretched a distance $\mathbf{x}_2$. Newton's third law states that the forces must be equal and opposite at the point of connection. Thus, we can write

$$\mathbf{F}_1 = k_1\mathbf{x}_1 = k_2\mathbf{x}_2 = \mathbf{F}_2 = \mathbf{F} = k_{eq}\mathbf{x}$$

where $\mathbf{x} = \mathbf{x}_1 + \mathbf{x}_2$. Now

$$\frac{\mathbf{F}}{k_{eq}} = \frac{\mathbf{F}_1}{k_1} + \frac{\mathbf{F}_2}{k_2}$$

and since $\mathbf{F} = \mathbf{F}_1 = \mathbf{F}_2$,

$$\frac{1}{k_{eq}} = \frac{1}{k_1} + \frac{1}{k_2}$$

or

$$k_{eq} = \frac{k_1 k_2}{k_1 + k_2}$$

Thus,

$$T = 2\pi\sqrt{m/k_{eq}} = 2\pi\sqrt{\frac{m(k_1 + k_2)}{k_1 k_2}}$$

This is the final expression for part (a).

(b) In this case, each spring is displaced the same distance as the mass since they are on either side of it. Thus, we can write $\mathbf{F} = -(k_1 + k_2)\mathbf{x}$ and set $k' = (k_1 + k_2.)$ Therefore, the period $T = 2\pi\sqrt{m/(k_1 + k_2)}$.

2. (a) In the diagram, we can see that since $\Delta\mathbf{y}$ is small, $\theta = \Delta\mathbf{y}/L$. Now, since θ is small (in radians), $\theta \approx \sin\theta \approx \Delta\mathbf{y}/L$. From the geometry, we see that the net downward force is given by $\Sigma\mathbf{F} = -2T\sin\theta = -2T\,\Delta\mathbf{y}/L$.

(b) Since this approximates simple harmonic motion, we can write

$$\mathbf{F} = -ky$$

$$-2T\frac{\Delta\mathbf{y}}{L} = -k\,\Delta\mathbf{y}$$

$$k = \frac{2T}{L}$$

and

$$\omega = 2\pi f = \sqrt{\frac{k}{m}} = \sqrt{\frac{2T}{Lm}}$$

$$f = \frac{1}{2}\pi\sqrt{\frac{2T}{Lm}}$$

3. (a) A free body diagram for the upper block is shown here.

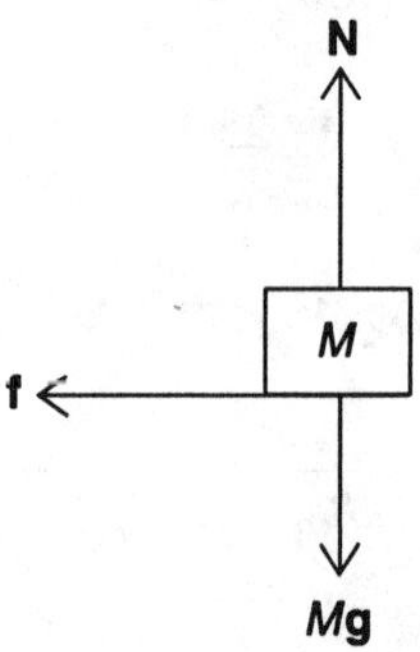

(b) The formula for maximum acceleration is $\mathbf{a} = A\omega^2$. Now the force that accelerates the block on top is the force of static friction. Thus, $\mathbf{f} = m\mathbf{a} = \mu_s N = \mu_s m\mathbf{g}$ and $\mu m\mathbf{g} = A\omega^2$. These equations are for the maximum amplitude (acceleration; force) and while technically they should be expressed in the form of "less than or equal to," we are interested in the maximum values here. Thus, since $\omega = 2\pi f$ (where f is the frequency given as 2.7 Hz), we can solve for the maximum amplitude $A = 0.0136$ m.

(c) The total energy of the system at maximum amplitude is given by $E = \frac{1}{2}kA^2$. Substituting the values given yields $A = 0.0929$ J.

4. From Problem 4, in Chapter 9, we know that the potential energy of the sphere is given by

$$\text{PE} = m\mathbf{g}(\mathbf{R} - \mathbf{r})(1 - \cos\theta)$$

The total energy at the bottom of the bowl is therefore

$$\begin{aligned} E &= m\mathbf{g}(\mathbf{R} - \mathbf{r})(1 - \cos\theta) + \frac{1}{2}m\mathbf{v}^2 + \frac{1}{2}I_{cm}\omega^2 \\ &= m\mathbf{g}(\mathbf{R} - \mathbf{r})\frac{\theta^2}{2} + \frac{1}{2}m\mathbf{r}^2\omega^2 + \frac{1}{2}I_{cm}\omega^2 \\ &= \frac{m\mathbf{g}(\mathbf{R} - \mathbf{r})}{2}\theta^2 + \frac{1}{2}(m\mathbf{r}^2 + I_{cm})\omega^2 \\ &= \frac{m\mathbf{g}(\mathbf{R} - \mathbf{r})}{2}\theta^2 + \frac{1}{2}I\omega^2 \end{aligned}$$

where we have used the approximation for cos θ, the fact that $\mathbf{v} = \mathbf{r}\boldsymbol{\omega}$, and the parallel axis theorem to state that $I = m\mathbf{r}^2 + I_{cm}$.

To form the phase space ellipse, we need to form the square of the angular momentum by multiplying both sides by I.

$$2EI = m\mathbf{g}I(\mathbf{R} - \mathbf{r})\theta^2 + \mathbf{L}^2$$

$$1 = \frac{\theta^2}{2E/m\mathbf{g}(\mathbf{R} - \mathbf{r})} + \frac{\mathbf{L}^2}{2EI}$$

This is the phase space ellipse. The area of this ellipse is equal to the sum of the total energy (which is a constant) and the period of oscillation. Following the procedure outlined in the text,

$$\text{Area} = \pi\sqrt{\frac{2E}{m\mathbf{g}(\mathbf{R} - \mathbf{r})}2EI} = E2\pi\sqrt{\frac{I}{m\mathbf{g}(\mathbf{R} - \mathbf{r})}}$$

$$T = 2\pi\sqrt{\frac{m\mathbf{r}^2 + I_{cm}}{m\mathbf{g}(\mathbf{R} - \mathbf{r})}}$$

where we have used the approximation [illegible] the [illegible] and the [illegible] to solve [illegible]

To find the phase space ellipse [illegible] the [illegible] multiply both sides by [illegible]

[illegible]

This is the [illegible] space ellipse. The area of this ellipse [illegible] of the [illegible] which is a constant [illegible] the [illegible] remains [illegible]

[illegible]

11 GRAVITATION

KEPLER'S LAWS OF PLANETARY MOTION

In the early seventeenth century, Johannes Kepler developed three laws of planetary motion in an effort to support the belief of Copernicus that the sun is the center of the universe. Briefly, these laws state that

1. All the planets orbit the sun in elliptical paths (with the sun at one focus).
2. A planet sweeps out an equal area of space in an equal amount of time in its orbit.
3. For any planet, the ratio of the cube of its mean distance from the sun to the square of its period is a constant.

We can understand these three laws by looking at Figure 1. An **ellipse** consists of two fixed points called **foci** (a single point is called a *focus*) and is defined to be the set of all points such that sum of the distances to each of the fixed points is a constant. Thus, if point P is a distance P_{f1} from the first focal point and a distance P_{f2} from the second one, then $P_{f1} + P_{f2}$ is always a constant. In the solar system, the sun is located at one of the foci. The extent to which an ellipse differs from a circle is referred to as its **eccentricity**.

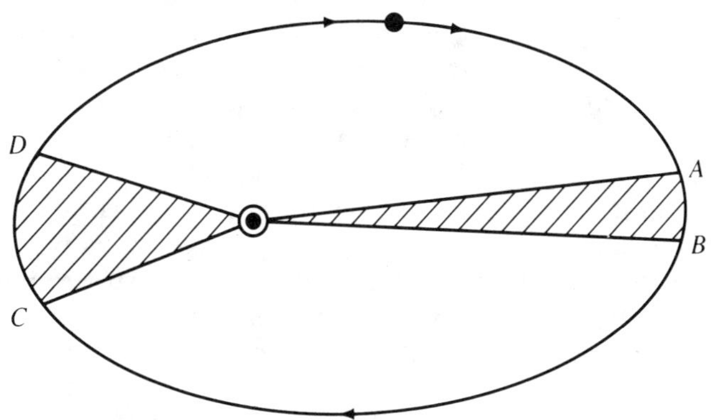

FIGURE 1. *Kepler's laws.*

The second law can be understood from the law of conservation of angular momentum. In reference to Figure 1, if it takes 1 month to go from point A to point B and 1 month to go from point C to point D, then Kepler's second law asserts that the areas swept out (as outlined) are equal. From our knowledge of forces, we suspect that the force acting on the planet becomes stronger as the planet gets closer since the planet begins to accelerate in its orbit (just as the centripetal acceleration directed toward the center does).

The eccentricity of planetary orbits is so small that they can almost be considered circles. Therefore, since the force acting on the planets is radial, we already know that if the force is parallel to the radius vector from the pivot or action point (in this case, the sun), then no net torque acts on the system.

If no net torque acts on the system, then there is no change in the angular momentum of the system (it is conserved). Therefore, as a planet gets closer to the sun it must orbit faster in order to maintain the same amount of angular momentum. In the next section, we shall look at Newton's law of universal gravitation and explain in more detail the meaning behind Kepler's second law.

Kepler's third law of motion can written mathematically in the form

$$\frac{R^3}{T^2} = K_s = 3.35 \times 10^{18}\ m^3/s^3$$

This constant is the same for all planets in the solar system. The universality of Kepler's third law lies in the fact that for any system (such as the earth-moon system), a suitable Kepler's constant can be found. The moon also obeys Kepler's laws, as do the satellites of Jupiter and stars in distant galaxies.

NEWTON'S LAW OF UNIVERSAL GRAVITATION

We already know that all objects falling near the surface of the earth have equal accelerations given by $\mathbf{g} = 9.8\ \text{m/s}^2$. The value of this constant can be determined from the independence of the period of a pendulum from the mass of the bob. This *empirical* verification is independent of any theory of gravity.

The weight of an object on the earth is given by $\mathbf{W} = m\mathbf{g}$, and this represents the magnitude of the force of gravity acting on the object due to the earth. From our discussion in Chapter 9, we know that gravity causes a projectile to assume a parabolic path. Isaac Newton, in his work *Principia* extended the idea of projectile motion to an imaginary situation in which the velocity of the projectile was so great that it would fall and fall but that the curvature of the earth would bend away and leave the projectile in orbit. Newton conjectured that this might be the reason why the moon orbits the earth. Is the moon falling toward the earth like the proverbial apple?

To answer this question, Newton first had to determine the centripetal acceleration of the moon based on observations from astronomy. He knew the relationship between centripetal acceleration and period, which can be written as

$$\mathbf{a}_c = \frac{4\pi^2 R}{T^2}$$

where R is the distance from the moon in meters and T is the orbital period in seconds. From astronomy we know these values to be $R = 3.8 \times 10^8$ m, and since the moon orbits the earth in 27.3 days, we have $T = 2.3 \times 10^6$ s. Using these values, we find that $\mathbf{a}_c = 2.8 \times 10^{-3}$ m/s^2.

Using the formula for centripetal acceleration, Newton was able to observe that since $\mathbf{F} = m\mathbf{a}$ by his second law, the formula for centripetal force is given by

$$\mathbf{F}_c = \frac{M4\pi^2R}{T^2}$$

Using Kepler's third law of planetary motion and setting M equal to the mass of the moon, Newton found he could express the force of gravity in the form

$$\mathbf{F}_g = \frac{M_m 4\pi^2 K_e}{R^2}$$

Kepler's constant would be evaluated for the earth-moon system using Kepler's third law.

Using his third law of motion, Newton realized that the force that the earth exerts on the moon should be exactly equal to, but opposite, the force that the moon exerts on the earth (observed as tides). This implied that the constant K should be dependent on the mass of the earth and that this mutual interaction implied that the force of gravity should be proportional to the product of both masses. In other words, Newton's law of gravity can be expressed as

$$\mathbf{F}_g = \frac{GM_1M_2}{R^2}$$

This is a vector equation in which the force of gravity is directed inward toward the center of mass for the system (in this case, near the center of the earth). Since the earth is many times more massive than the moon, it is the moon that orbits the earth and not vice versa. Actually, they both orbit about the common center of mass which is located near the center of the earth.

The value of G, called the **universal gravitational constant**, was experimentally determined by Henry Cavendish in 1795. In modern units it is equal to 6.67×10^{-11} N-m^2/kg^2. If we recognize that $\mathbf{F} = m\mathbf{a}$, then if we let M_1 equal the mass of an object of mass m and let M_2 equal the mass of the earth M_e, then the acceleration of a mass m near the surface of the earth is given by

$$\mathbf{a} = \frac{GM_e}{R_e^{\,2}}$$

where R_e is the radius of the earth in meters. Using known values for these quantities, we discover that $\mathbf{a} = 9.8$ m/s^2 $= \mathbf{g}$. Thus, we have a theory that accounts for the value of the known acceleration due to gravity. In fact, if we replace the mass of the earth by the mass of any planet, and the radius by the corresponding radius of the planet, the previous formula allows one to determine the value of **g** on any planet or astronomical object in the universe. For example, the value of **g** on the moon is approximately **g** = 1.6 m/s^2 or about one sixth of the value on the earth. Thus, objects on the moon weigh one sixth as much as they do on earth.

Now if you recall Newton's prediction for the acceleration of the moon toward the earth, his theory claims that the value of **g** decreases as the square of the distance from the earth. Since Greek times, it has been known that the mean distance to the moon is approximately equal to 60 times the radius of the earth. Thus, the acceleration of the moon should be 1/3,600 the acceleration of an object near the earth's surface. Thus **a** = 9.8/3,600 = 0.0028. Newton's theory of gravity was confirmed in one simple triumphant demonstration. Further proof came when his friend, Edmund Halley, used Newton's law to predict that a certain comet (now known as Halley's comet) would return every 76 years, next returning in 1758.

NEWTON'S INTERPRETATION OF KEPLER'S LAWS

We are now in a position to discuss Kepler's third law of planetary motion in light of Newton's law of gravitation. First, observe that Newton's theory merely describes the relationship between two masses and their displacement from each other. It is a vector law, and so the familiar rules for vector mechanics apply. However, as Newton himself wrote, nowhere is the "cause" of gravity discussed. In his seminal work, *Principia*, Newton himself boldly stated that he "frames no hypotheses" regarding the origin of the force of gravity.

Nevertheless, the theory allows us to understand the meaning behind the laws of Kepler. For example, we already noted that Kepler's second law can be thought of as a consequence of the law of conservation of angular momentum. The reason that the orbits of the planets are ellipses (actually they are from a set of figures called *conic sections*) was determined by Newton and involves the total amount of energy available to the planet when it formed. We will not go into any more detail about this question. The third law of Kepler relates the distance and velocity of a planet's orbit around the sun:

$$\frac{R^3}{T^2} = K_s$$

To begin with, let us write the equation for the force of gravity between a planet and the sun:

$$\mathbf{F} = \frac{GM_pM_s}{R^2}$$

Recall now that if the eccentricities are small, then the ellipses can be treated as though they are approximately circles. In this case, the mean radius of the orbit can be thought of as the radius of a circular path in which a centripetal force (provided by gravity) acts on the planet:

$$\frac{M_p 4\pi^2 R}{T^2} = \frac{GM_p M_s}{R^2}$$

The mass of the planet can be eliminated from both sides, and we are simply left with an equation that depends on the mass of the sun and the distance to the sun (which is Kepler's law):

$$\frac{R^3}{T^2} = \frac{GM_s}{4\pi^2} = K_s$$

If you want to determine the value of the constant for the earth system, simply use the mass of the earth instead of the sun's mass in the preceding equation. The mass of the sun is approximately equal to 2×10^{30} kg, and the mass of the earth is approximately equal to 6×10^{24} kg.

Kepler's third law implies that for any system the ratio of distance cubed to time squared is a constant for any two intervals of time:

$$\frac{R_1^3}{T_1^2} = \frac{R_2^3}{T_2^2}$$

We can use this relationship to find any corresponding period for any given distance from the center of the earth (or sun) if we know any *standard.* For example, if we take the earth system and use the moon's motion as our standard, then we can compute the orbital period of an artificial satellite or spacecraft (recognizing that R is actually the sum of the height above the surface of the earth and earth's radius).

This leads directly to the concept of *geostationary* or *geosynchronous* orbits. At a certain altitude (about 22,000 mi), the orbital period of an orbiting spacecraft is exactly equal to the rotational period of the earth (about 24 hours). The spacecraft (or satellite) appears to remain stationary above a specified point on the earth. These satellites are used for communication relays and weather forcasting since a series of them, strategically placed, can always monitor a given region of the earth's surface.

GRAVITATIONAL ENERGY

Recall that the amount of work done (by or against gravity) when vertically displacing a mass is given by the change in the gravitational potential energy:

$$\Delta\text{PE} = \Delta m\mathbf{g}h$$

We now know that the value of **g** is variable and varies inversely with the square of the distance from the center of the earth. Also, since we want the potential energy to become "weaker" the closer we get to the earth, we can use the results from page 156 to rewrite the potential energy formula as

$$\text{PE} = \frac{-GM_0M_e}{R}$$

where $R = R_e + h$ and h is the height above the earth's surface. If h is very small compared with the radius of the earth, it can effectively be eliminated from the equation.

The *escape velocity* from the gravitational force of the earth can be determined by considering the situation where the object just reaches infinity with zero final velocity given some initial velocity at any direction away from the surface of the earth. We designate that velocity as $\mathbf{v}_{(esc)}$ and state that when the final velocity is zero (at infinity), the total energy must be zero, which implies

$$\frac{1}{2}M_0V_{esc}^2 = \frac{GM_0M_e}{R_e}$$

The mass of the object can be eliminated from the relationship, leaving

$$\mathbf{v}_{esc} = \sqrt{\frac{2GM_e}{R_e}}$$

One can leave the surface of the earth with any velocity. There is, however, one minimum starting velocity at which the spacecraft would not fall back to the earth as a result of gravity.

The orbital velocity can be determined by assuming that we have an approximately circular orbit. In this case, we can set the centripetal force equal to the gravitational force:

$$\frac{GM_0M_e}{R_e^2} = \frac{M_0\mathbf{v}_0^2}{R_e}$$

Eliminating the mass of the object from the equation leaves us with

$$\mathbf{v}_{orbit} = \sqrt{\frac{GM_e}{R_e}}$$

We can determine the period of a circular orbit near the surface of the earth by equating the centripetal force to the gravitational force:

$$\frac{4m\pi^2R_e}{T^2} = \frac{GmM_e}{R_e^2}$$

$$T = \sqrt{\frac{4\pi^2R_e^3}{GM_e}}$$

The total energy for the orbiting satellite can be found from

$$E = \frac{1}{2} m\mathbf{v}^2 - \frac{GmM_e}{R_e} = \frac{GmM_e}{2R_e} - \frac{GmM_e}{R_e} = -\frac{GmM_e}{2R_e}$$

KEPLER'S SECOND LAW AND ANGULAR MOMENTUM

Kepler's second law states that a planet sweeps out equal areas in equal time. We can now demonstrate that this is a consequence of the law of conservation of angular momentum. In Figure 2, we see an elliptical orbit in which the angle $\Delta\theta$ is taken to be small enough that the arc AB is approximately equal to the line segment AB. Thus, instead of a wedge-shaped area, we have a triangle AOB whose area is approximately equal to $\frac{1}{2}\, r\,(r\,\Delta\theta)$:

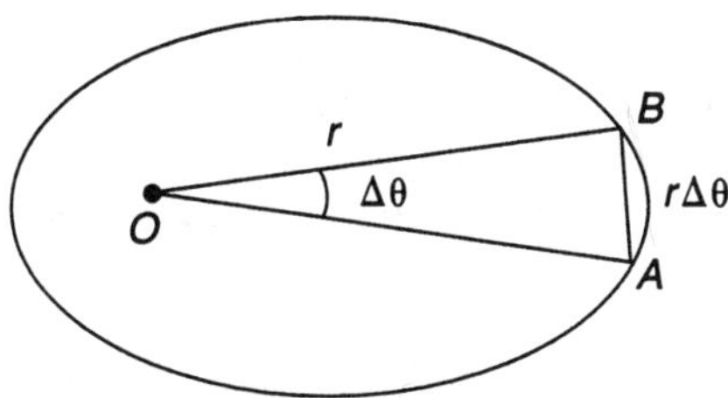

FIGURE 2.

Now we can see how angular momentum enters into the discussion:

$$\Delta A \approx \frac{1}{2} r(r\ \Delta\theta) = \frac{1}{2} r^2\ \Delta\theta$$

$$\frac{\Delta A}{\Delta t} \approx \frac{1}{2} r^2 \frac{\Delta\theta}{\Delta t}$$

If we let Δt approach zero, then we can state

$$\frac{\Delta A}{\Delta t} = \frac{1}{2} r^2\omega = \frac{mr^2\omega}{2m} = \frac{\mathbf{L}}{2m}$$

where we have introduced the angular momentum $\mathbf{L}$ of the planet. Since we know that no external torques act on the planet, angular momentum is conserved and $\mathbf{L}$ is a constant. Thus, the rate of change of area (if the time interval approaches zero) is a constant. This is Kepler's second law.

Sample Problem

(a) What is the magnitude of the acceleration of gravity at an altitude of 400 km above the surface of the earth?

(b) What percentage loss in the weight of an object results?

Solutions

(a) The formula for the acceleration of gravity above the earth's surface is

$$\mathbf{g} = \frac{GM_e}{(R_e + h)^2}$$

We have $M_e = 5.98 \times 10^{24}$ kg $R_e = 6.38 \times 10^6$ m, and h = 400 km = 0.4×10^6 m. Substituting these values, as well as the known value for G given in this chapter, we have $\mathbf{g} = 8.67$ m/s^2.

(b) The fractional change is found by comparing the value of **g** at 400 km to its value at the earth's surface. That is, 8.67/9.8 = 0.885. Thus, there is an 11.5% loss of weight at that height.

Sample Problem

Three uniform spheres of masses 1, 2, and 4 kg are fixed at the corners of a right triangle as shown. The positions relative to the coordinate system indicated are also given. What is the magnitude of the resultant gravitational force on the 4-kg mass?

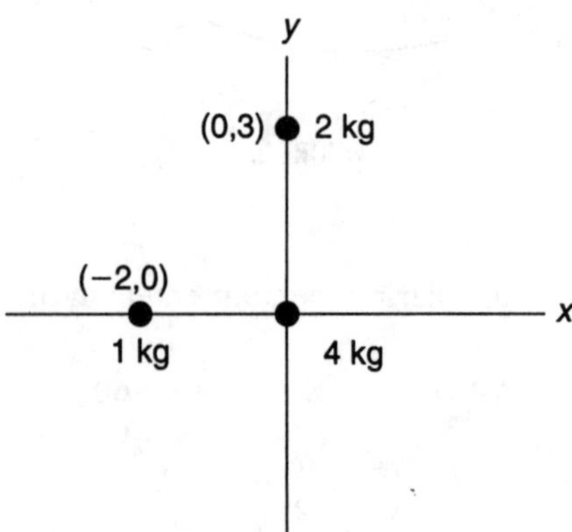

Solution

In this problem, we have to determine the force of gravitational attraction between each mass and the 4-kg mass separately. Each force is calculated separately. Let's begin with the 1-kg mass and the 4-kg mass:

$$\mathbf{F}_{14} = \frac{(6.67 \times 10^{-11})(1)(4)}{(-2)^2} = 6.67 \times 10^{-11} \text{ N}$$

The direction is to the right and is subsequently negative since we want the force on the 4-kg mass.

For the 2-kg mass and the 4-kg mass, the force on the 4-kg mass is directed upward (positive):

$$\mathbf{F}_{24} = \frac{(6.67 \times 10^{-11})(2)(4)}{(3)^2} = -5.93 \times 10^{-11}\ \text{N}$$

The magnitude of the resultant force is given by the Pythagorean theorem:

$$\mathbf{F} = \sqrt{(6.67 \times 10^{-11})^2 + (5.93 \times 10^{-11})^2} = 8.92 \times 10^{-11}\ \text{N}$$

PROBLEM-SOLVING STRATEGIES

There are not many special considerations for solving problems involving Kepler's laws or Newton's law of gravitation. You should remember the connection between Kepler's second law and the law of conservation of angular momentum as well as the connection between Kepler's third law and newton's law of gravity. Also, remember that each system has its own Kepler's constant that orbiting objects obey.

Remember that gravity is a force and therefore a vector quantity. The distances are measured in meters from the center of masses of each object, and so if an object is above the surface of the earth, you must add the height to the radius of the earth. If you remember that the force of gravity is an inverse square law relationship, you might be able to deduce an answer from logic rather than algebraic calculations (which take time).

PRACTICE PROBLEMS FOR CHAPTER 11

Thought Problems

1. Show that the units of $\mathbf{L}/2m$ are the same as the rate of change in area in Kepler's second law.
2. If the force of gravity acting on a mass near the earth's surface is proportional to the mass, why doesn't a heavier mass accelerate faster than a smaller mass (with no air resistance)?
3. If astronauts orbiting the earth are considered "weightless," explain one method for possibly determining their mass in space.

Multiple-Choice Problems

1. What is the value of $\mathbf{g}$ at a position above the earth's surface equal to the radius of the earth?
   (A) 9.8 N/kg
   (B) 4.9 N/kg

(C) 6.93 N/kg
(D) 2.45 N/kg
(E) 1.6 N/kg

2. A planet has half the mass of the earth and half the radius. Compared to the acceleration of gravity near the surface of the earth, the acceleration of gravity near the surface of this planet is _____.
(A) twice as much
(B) one fourth as much
(C) half as much
(D) the same
(E) zero

3. Which of the following is an expression for the acceleration of gravity with uniform density ρ and radius *R*?
(A) $G(4\pi\rho/3R^2)$
(B) $G(4\pi\rho R^2/3)$
(C) $G(4\pi\rho/3R)$
(D) $G(4\pi\rho/3)$
(E) none of these

4. Given that the mean radius of the moon's orbit is 3.84×10^8 m and its period is 2.36×10^6 s, what is the altitude above the surface of the earth at which a geostationary satellite orbits? (The radius of the earth is 6.38×10^6 m.)
(A) 4.23×10^7 m
(B) 3.59×10^7 m
(C) 6.2×10^8 m
(D) 2.2×10^7 m
(E) 5.8×10^7 m

5. What is the orbital velocity for a satellite at a height of 300 km above the surface of the earth? (The mass of the earth is approximately $6 \times^{24}$ kg, and its radius is 6.4×10^6 m)
(A) 5.42×10^1 m/s
(B) 1.15×10^6 m/s
(C) 7.7×10^3 m/s
(D) 6×10^6 m/s
(E) 3×10^8 m/s

6. What is the escape velocity from the moon given that the mass of the moon is 7.2×10^{22} kg and the radius of the moon is 1.778×10^6 m?
(A) 1.64×10^3 m/s
(B) 2.32×10^3 m/s
(C) 2.69×10^6 m/s
(D) 5.38×10^6 m/s
(E) 3×10^8 m/s

7. A black hole theoretically has an escape velocity that is greater than or equal to the velocity of light (3×10^8 m/s). If the effective mass of the black hole is equal to the mass of the sun (2×10^{30} kg), what is the effective "radius" (called the *Schwarzchild radius*) of the black hole?
   (A) 3×10^3 m
   (B) 1.5×10^3 m
   (C) 8.9×10^6 m
   (D) 4.45×10^6 m
   (E) 0 m

8. Kepler's second law of planetary motion is a consequence of the law of conservation of _____.
   (A) linear momentum
   (B) charge
   (C) energy
   (D) mass
   (E) angular momentum

9. What is the gravitational force of attraction between two trucks, each of mass 20,000 kg, separated by a distance of 2 m?
   (A) 0.057 N
   (B) 0.013 N
   (C) 0.0067 N
   (D) 1.20 N
   (E) 0 N

10. The gravitational force between two masses is 36 N. If the distance between them is tripled, then the force of gravity will be _____.
   (A) the same
   (B) 18 N
   (C) 9 N
   (D) 4 N
   (E) 27 N

Free-Response Problems

1. Four masses are fixed at the corners of a rectangle as shown.

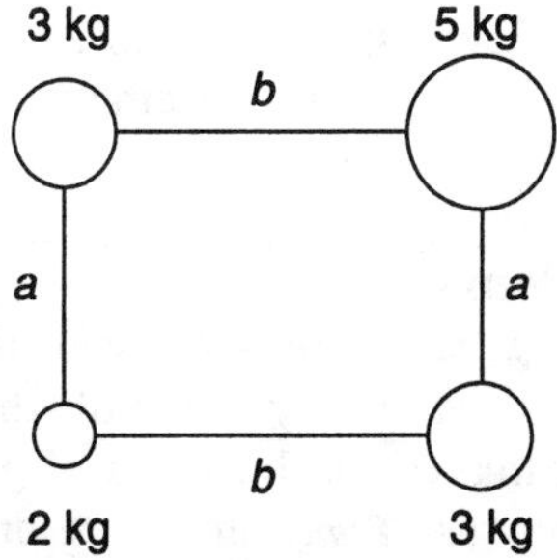

(a) Derive an expression for the x and y components of the resultant force acting on the 2-kg mass.

(b) If $a = 0.1$ m and $b = 0.2$ m, what are the magnitude and direction of the resultant gravitational force?

2. A satellite orbits around a planet of uniform density ρ, mass M, and radius R at a height h above the surface of the planet. The orbital period of the satellite is T. Derive an expression for the density of the planet in terms of h, R, G, and T only. What is the approximate density if $h << R$?

3. Find the magnitude of the gravitational field strength **g** at a point P along the perpendicular bisector between two equal masses M separated by a distance $2b$ as shown.

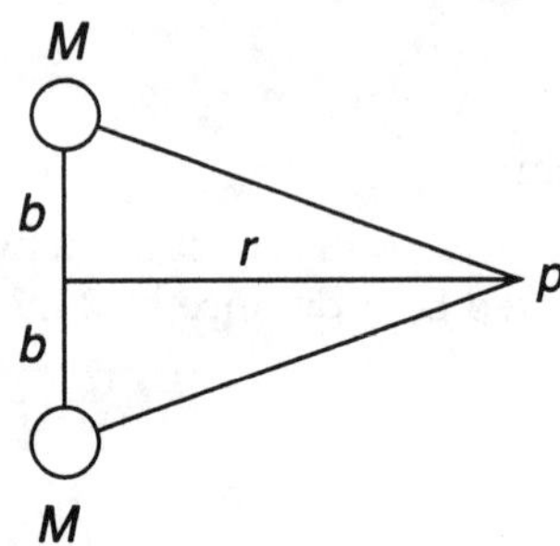

SOLUTIONS TO PRACTICE PROBLEMS

Thought Problems

1. The units of angular momentum are kg-m²/s. The expression $\mathbf{L}/2m$ is angular momentum divided by mass, which is in units of m^2/s. This is equivalent to area over time.

2. The force of gravity increases proportionally to the mass. However, $\mathbf{a} = \mathbf{F}/m$, and so as the force increases, so does the inertia (in the same proportion). Therefore, the ratio $\mathbf{F}/m$ is a constant (called **g** near the surface of the earth).

3. Scientists are concerned about astronaut mass loss after many weeks in space. Since a conventional scale cannot be used in a free fall condition, one method is to employ a spring system and determine the astronaut's *inertial mass* by using an oscillating system that is independent of the force of gravity.

Multiple-Choice Problems

1. **D** Since the value of **g** varies inversely with the square of the distance from the center of the earth, if you double the distance from the center (where you are in this case), the value of **g** decreases by one fourth: $\frac{1}{4}(9.8) = 2.45$. Since $g = \mathbf{F}/m$, alternative units are N/kg.

2. **A** Using the formula for **g**, if you take half the mass, the value will decrease by one half. If you decrease the radius by half, the value will increase by four times. Combining both effects results in an overall increase of two times as much in comparison.

3. **D** The formula for **g** is $\mathbf{g} = GM/R^2$. The planet is essentially a sphere of mass M and density ρ. The radius is R, and the planet has a volume $V = \frac{4}{3}\pi R^3$. Thus, $M = V\rho$. Making the substitutions yields $\mathbf{g} = G4\pi\rho R/3$.

4. **B** From Kepler's third law, we can determine the value of the constant K for the earth system using the data for the moon: $(3.84 \times 10^8)^3/(2.36 \times 10^6)^2 = 1.01664 \times 10^{13}\ \text{m}^3/\text{s}^2$. Now, use Kepler's third law again, this time using a period of 86,400 s (which equals the 24-h period of a geostationary orbit), solving for R, and using the above constant for K:

$$\frac{R^3}{T^2} = K = 1.01664 \times 10^{13} = \frac{R^3}{(86{,}400)^2}$$

Solving for R (don't forget to take the cube root!) yields $R = 4.23 \times 10^7$ m as the distance from the center of the earth. To obtain the answer, you must subtract the radius of the earth (which is given in the question) and you are left with approximately $h = 3.59 \times 10^7$ m.

5. **C** The formula for orbital velocity is $\mathbf{v}_{\text{orbit}} = \sqrt{GM/R}$, where R is the distance from the center of the earth. In this case we must add 300 km = 300,000 m to the radius of the earth. Thus, $R = 6.7 \times 10^6$ m. Substituting the given values yields $\mathbf{v}_o = 7{,}728$ m/s.

6. **B** The formula for escape velocity is $\mathbf{v}_{\text{esc}} = \sqrt{2GM_m/R_m}$. Substituting the values given yields $\mathbf{v}_{(\text{esc})} = 2{,}324$ m/s.

7. **A** From Problem 6 we know that the escape velocity is given by $\mathbf{v}_{\text{esc}} = \sqrt{2GM/R}$. To find R, we need to square both sides and then solve for the radius. This results in $R = 2GM/\mathbf{v}^2$. Substituting the values given yields $R = 3{,}000$ m, which is only a theoretical size for the black hole. As an interesting example, try calculating the value of **g** on such an object!

8. **E** Kepler's second law can be explained in terms of the conservation of angular momentum. Since the force of gravity is radial, the net torque acting on a planet in the solar system is zero. This means that the change in angular momentum is zero and that it is therefore conserved.

9. **C** Using the formula for gravitational force, $\mathbf{F} = GM_1M_2/R^2$, and substituting the values given (don't forget to square the distance) yields $\mathbf{F} = 0.0067$ N of force.

10. **D** The force of gravity is an inverse square law relationship. This means that as the distance is tripled, the force is decreased by one ninth. One ninth of 36 N is 4 N.

Free-Response Problems

1. (a) To find the component forces, we can see from the diagram that the force along b is only horizontal and the force along a is only vertical. The force connecting the 2-kg and 5-kg masses is directed along the diagonal of the rectangle, and this length is given by the Pythagorean theorem.

   If θ is the acute angle from the positive x axis, then $\tan\theta = a/b$. Thus, we see that the x component of the diagonal force is related to the vector component involving $\cos\theta$, while the y component involves $\sin\theta$. We can therefore combine these ideas to write for the magnitudes:

$$\mathbf{F}_x = \frac{G(2)(3)}{b^2} + \frac{G(2)(5)}{a^2 + b^2}\cos\theta$$

$$\mathbf{F}_y = \frac{G(2)(3)}{a^2} + \frac{G(2)(5)}{a^2 + b^2}\sin\theta$$

   (b) To find the value of the magnitude and direction of the net resultant force, we first need the value of θ. We know that $\tan\theta = a/b$. Given that $a = 0.1$ m and $b = 0.2$ m, we find that $\theta = 26.57$ degrees. Therefore, $\cos\theta = 0.8944$ and $\sin\theta = 0.4473$. Using the values given, we find that $\mathbf{F}_x = 2.19 \times 10^{-8}$ N and $\mathbf{F}_y = 4.6 \times 10^{-8}$ N. The magnitude of the resultant force is given by the Pythagorean theorem: $\mathbf{F} = 5.09 \times 10^{-8}$ N.

   To find the direction of the resultant, we note that since we consider the 2-kg mass to be fixed and gravity is attractive, the resultant force is directed toward the lower left part of the page. Specifically, we recall from our work on vectors that if ϕ is the direction of the resultant, then $\tan\phi = \mathbf{F}_y/\mathbf{F}_x$. Now both $\mathbf{F}_x$ and $\mathbf{F}_y$ are directed in what we have, by convention, termed negative directions. However, from trigonometry recall that the tangent function is positive (since we have a ratio of two negative numbers) in the *third quadrant.* Thus, taking the ratio, we find that $\tan\phi = 2.1$ and thus $\phi = 64.5$ degrees below the negative x axis (244.5 degrees).

2. The density $\rho = M/V$. Since the planet is a sphere, we can write $\rho = 3M/4\pi R^3$. From Kepler's third law, we also know that

$$T^2 = \frac{4\pi^2}{GM}r^3 = \frac{4\pi^2}{GM}(R + h)^3$$

   where we have used $r = R + h$. Now, factoring out an R leaves us with

$$T^2 = \frac{4\pi^2}{GM}R^3\left(1 + \frac{h}{R}\right)^3$$

Solving for the cube of the radius, which appears in the density formula, we have

$$R^3 = \frac{GMT^2}{4\pi^2\left(1 + \frac{h}{R}\right)^3}$$

We must now substitute this expression for the cube of the radius into the first density formula. Canceling out all like terms leaves us with the desired expression (independent of mass):

$$\rho = \frac{3\pi}{GT^2}(1 + \frac{h}{R})^3$$

If $h << R$, then h/R is $<< 1$, and so the term can be eliminated as an approximation. Thus, for small heights above the surface of the planet (relative to the radius of the planet) we can write $\rho \approx 3\pi/GT^2$.

3. Let's call the field strength at P caused by the top mass $\mathbf{g}_1$. Likewise, let's call the field strength at point P due to the bottom mass $\mathbf{g}_2$. Both of these field strengths are accelerations and therefore they are vectors. The distance from point P to the line connecting the masses is r, and to the midpoint distance connecting the masses, b. Therefore, the distance from point P to each mass is given by the Pythagorean theorem and is equal to $\sqrt{r^2 + b^2}$.

   The direction of each acceleration $\mathbf{g}$ is directed toward each mass from point P. Since each mass is identical and the distance to each mass is the same, the angles formed by the vectors to the x axis are the same. Let's call each angle θ such that $\tan \theta = b/r$ in magnitude.

   From the preceding analysis we can conclude that

$$\mathbf{g}_1 = \mathbf{g}_2 = \frac{GM}{r^2 + b^2}$$

   The vector components of each field strength result in a symmetric cancellation of the y components. This is true because the direction of $\mathbf{g}_1$ is toward the upper left, and thus its x component is directed to the left and its y component is directed upward. The field strength $\mathbf{g}_2$ has an x component that is also directed toward the left and equal in magnitude to the x component of $\mathbf{g}_1$. The y component of $\mathbf{g}_2$ is directed downward and is also equal in magnitude to the y component of $\mathbf{g}_1$. Since these two vectors are equal and opposite, they sum to zero and do not contribute to the net resultant field (which is just directed horizontally toward the left). What remains to be done is to determine the expression for the x component of $\mathbf{g}_1$ or $\mathbf{g}_2$ and then multiply by 2. Since $\mathbf{g}_{1x} = \mathbf{g}_1 \cos \theta$, we can see from the geometry that $\cos \theta = r/\sqrt{r^2 + b^2}$. Combining both results, we finally obtain $\mathbf{g}_{(net)} = 2\ GMr/(r^2 + b^2)^{3/2}$.

12
MECHANICAL PROPERTIES OF MATTER

DENSITY AND SPECIFIC GRAVITY

Most matter can be characterized by its **density** ρ, which is defined to be a measure of the mass per unit volume (in kg/m^3 or g/cm^3) such that $\rho = M/V$. A sample list of densities is given in Table 1.

TABLE 1. DENSITY OF VARIOUS SUBSTANCES

Substance	Density, ρ (kg/m^3)
Air	1.3
Alcohol, ethyl	790
Aluminum	2,700
Carbon dioxide	2.0
Copper	8,900
Gold	19,000
Helium	0.18
Hydrogen	0.09
Ice	920
Iron	7,800
Lead	11,000
Oxygen	1.4
Water, pure	1,000
Water, sea	1,030

The **specific gravity** of a substance is the ratio between the density of the substance and the density of water. For example, the specific gravity of iron is 7.8. The specific gravity of mercury is 13.6, which means its density is 13,600 kg/m^3.

ELASTICITY

We have seen that Hooke's law expresses the relationship between an applied force and the elongation of a spring: $\mathbf{F} = -k\mathbf{x}$. When solids are impacted by forces, their shape can become distorted in a number of ways (see Figure 1). For example, the forces can act to stretch the solid and place

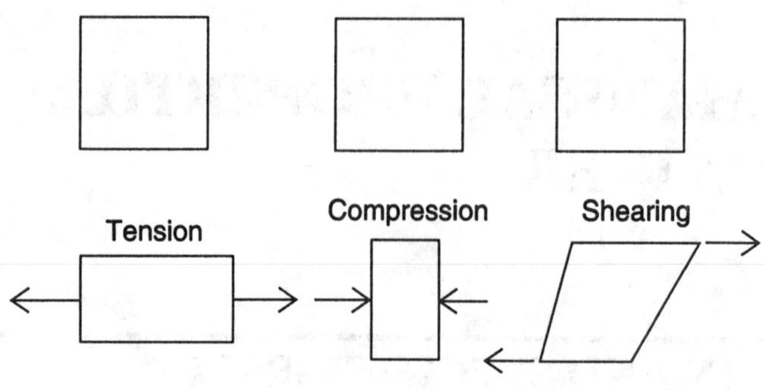

FIGURE 1.

it under a state of **tension**, the forces can squeeze the solid inward and place it under a state of **compression**, or the forces can cause layers of solid to be displaced differently by sliding over one another (called **shearing**).

In physics, the **stress** applied is a measure of the force per unit cross-sectional area. The response to the stress is called a **strain**. Hooke's law can then be expressed as

$$\frac{\text{Stress}}{\text{strain}} = \text{constant (k)} = \text{modulus of elasticity}$$

In a wire, the strain is a measure of the fractional change in length due to a stress:

$$\text{Strain} = \frac{\Delta L}{L_o}$$

where L_o is the original length of the solid. We define the **modulus of elasticity** to be the ratio of the stress to the strain. If we apply our definition of stress, then in this case the modulus of elasticity is called **Young's modulus** and is given by the relationship

$$Y = \frac{\mathbf{F}/A}{\Delta L/L_o}$$

The fractional change in length is therefore

$$\frac{\Delta L}{L_o} = \frac{1}{Y}\frac{\mathbf{F}}{A}$$

Young's modulus for different samples is presented in Table 2.

The elasticity of any material is strictly limited to a range of applied stresses. Hooke's law, a special case of the application of Young's modulus to a spring, is valid as long as the elastic limit is not exceeded. For example, the elastic limit of aluminum is 1.8×10^8 N/m^2, the elastic limit of copper is 1.5×10^8 N/m^2, and the elastic limit of steel is 2.5×10^8 N/m^2.

TABLE 2. YOUNG'S MODULUS FOR VARIOUS SUBSTANCES

Substance	Young's Modulus ($\times 10^{10}$ N/m^2)
Aluminum	7.0
Concrete	2.0
Copper	11
Glass	5.5
Iron	19
Lead	1.6
Steel	20

SHEAR

When a book's covers are shifted out of alignment, a shearing force has been applied. This stress changes the shape of the book but not its volume. We denote the angle of displacement (or angle of shear) as a measure of the displacement over the thickness (see Figure 2). We can therefore write

$$\text{Shear stress} = \frac{\mathbf{F}}{A}$$

$$\text{Shear strain} = \phi = \frac{s}{d} \text{ radians}$$

and, after defining the shear modulus S to be equal to stress/strain,

$$S = \frac{\mathbf{F}/A}{\phi} = \frac{\mathbf{F}/A}{s/d}$$

$$\phi = \frac{1}{S}\frac{\mathbf{F}}{A}$$

Table 3 presents a listing of the shear modulus for various substances.

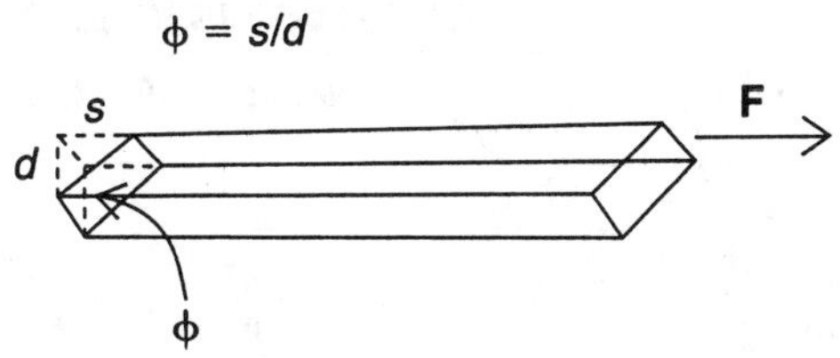

FIGURE 2. *Shear*

TABLE 3. SHEAR MODULUS FOR VARIOUS SUBSTANCES

Substance	Shear Modulus ($\times 10^{10}$ N/m^2)
Aluminum	3.0
Copper	4.2
Glass	2.3
Iron	7.0
Lead	0.56
Steel	8.4

BULK MODULUS

If forces act perpendicularly on all sides to squeeze a solid together uniformly, we obtain a change in volume that is characterized by the substance's **bulk modulus**. The compressive force per unit surface area is also known as **pressure** and can be expressed in units of **pascals** instead of N/m^2 (these units are equivalent). At sea level, standard air pressure can be expressed in units of **atmospheres** (atm) such that 1 atm = 101,000 Pa.

We can therefore write (using a negative sign for uniform compression)

$$\text{Volume stress} = \frac{\mathbf{F}}{A}$$

$$\text{Volume strain} = \frac{\Delta V}{V_o} = \frac{\text{change in volume}}{\text{original volume}}$$

$$\text{Bulk modulus } B = \frac{-\text{stress}}{\text{strain}} = -\frac{\mathbf{F}/A}{\Delta V/V_o}$$

Finally, if we solve for the fractional change in volume,

$$\frac{\Delta V}{V_o} = -\frac{1}{B}\frac{\mathbf{F}}{A}$$

In Tables 4 and 5 we present the bulk modulus for various solids and liquids.

TABLE 4. BULK MODULUS FOR VARIOUS SOLIDS

Substance	Bulk Modulus ($\times 10^{10}$ N/m^2)
Aluminum	7.0
Copper	14
Glass	3.7
Iron	10
Lead	0.77
Steel	16

TABLE 5. BULK MODULUS FOR VARIOUS LIQUIDS AT ROOM TEMPERATURE

Substance	Bulk Modulus ($\times 10^9$ N/m^2)
Alcohol, ethyl	0.90
Benzene	1.05
Kerosene	1.3
Mercury	26
Water	2.3

The relatively high bulk moduli of liquids makes them fairly incompressible. Gases, however, are highly compressible. Both liquids and solids are classified as *fluids* and are distinguished by their compressibility. We will treat fluids in more detail in the next chapter.

Sample Problem

How much pressure is required to compress water by 5% of its original volume at room temperature?

Solution

The bulk modulus of water is 2.3×10^9 N/m^2. To decrease its volume by 5% means that $\Delta V/V_o = -0.05$ and hence the pressure $\mathbf{P} = \mathbf{F}/A = (-2.3 \times 10^9 \text{ Pa})(-0.05) = 1.15 \times 10^8$ Pa.

Sample Problem

A copper wire of length 1.7 m has a weight of 100 N hung from it. If the diameter of the wire is 1.5 mm, by what percentage will the wire stretch?

Solution

The first thing we need to know is the cross-sectional area of the wire. Assume that it is circular. Thus, the radius is $r = 0.75$ mm $= 0.00075$ m. The area is given by

$$A = \pi r^2 = (3.14)(0.00075)^2 = 1.766 \times 10^{-6} \text{ m}^2$$

Thus, the fractional change in length is (using Young's modulus for copper)

$$\frac{\Delta L}{L_o} = \frac{1}{Y}\frac{F}{A} = \frac{1}{11 \times 10^{10}}\frac{100}{1.766 \times 10^{-6}} = 0.00051$$

which corresponds to a change of 0.05%.

SURFACE TENSION AND CAPILLARITY

Fill a glass with water until it reaches the brim. Add a little more, and the water will rise above the brim, forming a convex shape. The force holding the water up above the brim is called **surface tension**. The water is behaving like skin or membrane being stretched. Because of surface tension, small insects can stand on the surface of water, and steel needles can float.

Surface tension arises because of intermolecular forces. Under the water, the forces between molecules act equally in all directions. At the surface, however, there is a net downward force because there are no molecules above the surface (Figure 3).

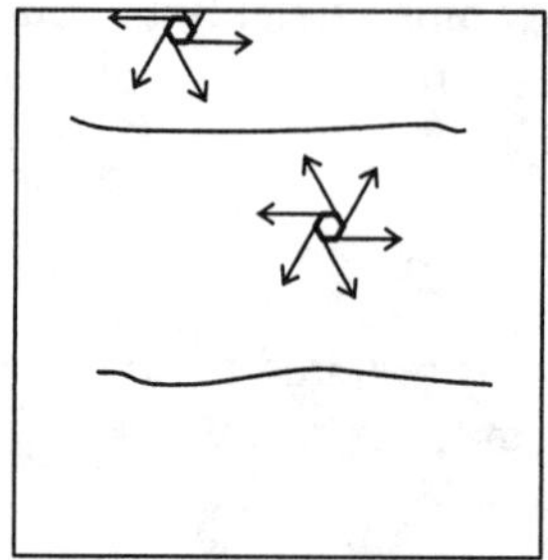

FIGURE 3. *Surface tension forces.*

The degree of surface tension γ can be measured using a wire frame dipped into the water. The amount of force per unit length necessary to pull the frame out of the water is defined to be a measure of the surface tension (in N/m):

$$\gamma = \frac{\mathbf{F}}{L}$$

Another way of looking at surface tension is to consider it a measure of the excess energy contained above the brim per unit area (in the same units). Water at 20°C has a surface tension of about 0.073 N/m which decreases with an increase in temperature. Mercury has a surface tension of about 0.44 N/m (with water having one of the highest values of γ).

When a thin tube is placed in the water, the level of water in the tube rises. This phenomenon is called **capillarity**. The two forces involved in capillarity action are cohesion and adhesion. *Cohesion* is a measure of the intermolecular forces in the liquid. *Adhesion* is the ability of the liquid to stick to a surface ("wetness"). If adhesion is greater than cohesion, the liquid sticks to or wets the tube. A concave shape is formed, and we define a characteristic angle of contact called θ. For water and clean glass, $\theta \approx 0$. When cohesion is greater than adhesion, which occurs with mercury and glass, the level of liquid in the tube is lower than the adjacent liquid in the reservoir (Figure 4).

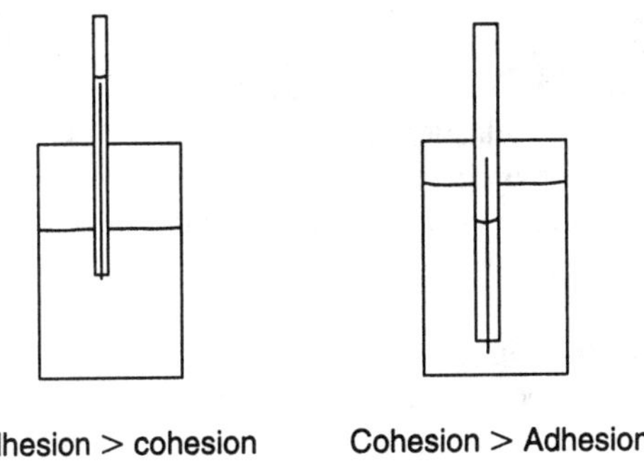

FIGURE 4.

The level h to which the liquid rises or falls can be calculated from

$$h = \frac{2\gamma \cos \theta}{r\rho \mathbf{g}}$$

where r is the inner radius of the tube, $\mathbf{g}$ is the acceleration of gravity, ρ is the density of the liquid, γ is the surface tension factor, and θ is the angle of contact between the tube and the liquid.

PROBLEM-SOLVING STRATEGIES

When solving problems involving the mechanical properties of matter, it is very important to keep track of the units. Often, small measurements appear in millimeters, but the calculations should be done in meters. Additionally, a problem may not explicitly state which modulus is required or if it is indeed a problem involving surface tension or capillarity. Look for key words that can provide clues to the exact nature of the problem's representation. As always, try to draw a sketch of the situation if one is not provided. Finally, make sure you understand the relationship between the variables in a formula (inverse, direct, etc.).

PRACTICE PROBLEMS FOR CHAPTER 12

Thought Problems

1. Explain the dynamics of jiggling Jell-O (a brand of gelatin).
2. When a glass rod is inserted in a dish of water and then removed, it looks "wet." If a similar glass rod is inserted in a dish of mercury and then removed, it still looks "clean." Explain this difference? (*Caution*: Never work with exposed mercury unless you are strictly supervised by your professor or instructor!)

3. A rubber band is not strictly elastic. If you hang weights from a rubber band and record the elongations, you will observe that the rubber band is easily stretched (and elastic) for small weights, but it becomes harder to stretch with heavier weights. Sketch a graph that might result from this observation and give a plausible explanation of what is happening.

Multiple-Choice Problems

1. A wire cable is stretched a distance d under a load **W**. If it is replaced by a cable made of the same material but with twice the length and twice the diameter, the cable will stretch a distance _____.
   (A) $2d$
   (B) d
   (C) $d/2$
   (D) $4d$
   (E) $d/4$

2. What force is needed to punch a hole, 8 mm in diameter, in a sheet of steel 4.0 mm thick such that it can withstand a maximum shear stress of 3.5×10^8 N/m^2?
   (A) 1.76×10^4 N
   (B) 3.5×10^4 N/m^2
   (C) 1.2×10^3 N/m^2
   (D) 7.3×10^5 N/m^2
   (E) 6.2×10^4 N/m^2

3. A sample of lubricating oil compresses by 0.3% when a pressure of 5 MPa is applied to it. What is the bulk modulus of the oil in GPa?
   (A) 0.9
   (B) 0.6
   (C) 1.3
   (D) 1.2
   (E) 1.7

4. A wire is supporting a load **W** that puts it under stress. The stress is independent of _____.
   (A) the wire's length
   (B) the wire's density
   (C) the weight of the load
   (D) the diameter of the wire
   (E) the acceleration of gravity

5. A copper wire is used to support a 3-kg mass. If the elastic limit of copper is 1.5×10^8 Pa, what is the minimum diameter of wire that can be used to support this load?
   (A) 0.4 mm
   (B) 0.5 mm
   (C) 0.6 mm

(D) 0.7 mm
(E) 0.8 mm

6. A steel wire has a Young's modulus of 2×10^{11} Pa and is 1.5 m long. What will be the percentage change in length when a 200-N weight is hung from it?
(A) 0.015%
(B) 0.002%
(C) 0.013%
(D) 0.2%
(E) 0.02%

7. A solid lead sphere is dropped into the ocean where it rests on the bottom. If the pressure at the bottom is 400 atm, by what percentage will the sphere be compressed? (The bulk modulus for lead is $B = 0.77 \times 10^{10}$ Pa.)
(A) 0.5%
(B) 0.4%
(C) 0.3%
(D) 0.2%
(E) 0.1%

8. When a solid has applied forces that act in opposite directions but not along the same line of action, the nature of the stress is called _____.
(A) compression
(B) tension
(C) shearing
(D) torsion
(E) expansion

9. A thin glass tube of diameter d is inserted in a beaker of water. The water rises to a level h as a result of capillarity. If a similar tube with half the diameter is inserted in the water, the level will become _____.
(A) h
(B) $2h$
(C) $h/2$
(D) $4h$
(E) $h/4$

10. A clean glass rod with radius r is placed in a dish of water. Which of the following is an expression for the potential energy of the mass of liquid whose center of gravity is at a distance h relative to the remaining liquid in the dish?
(A) $2\gamma/V r \mathbf{g}$
(B) $2r\mathbf{g}\gamma$
(C) $2\gamma/r\mathbf{g}$
(D) $2V\gamma/r$
(E) $2\gamma/\mathbf{g}$

Free-Response Problems

1. (a) Hooke's law for an elastic spring of constant diameter d is a special case using Young's modulus of elasticity. Show that the expression $\mathbf{F} = k\mathbf{x}$ can be derived from the expression for Young's modulus Y.
   (b) Using the results from part (a), express the potential energy for a spring (stretched from its original length) in terms of Young's modulus.
2. A 25-kg hammer strikes a blow to a cylindrical steel spike that has a diameter of 2 cm. The speed at impact is 20 m/s, while the rebound speed is observed to be 12 m/s. The impact lasts for 0.07 s. What was the average strain on the spike during the time of impact?
3. A steel wire is stretched between two cylindrical turning posts 1.5 m apart, and there is negligible tension in the wire at the time. The diameter of the wire is 0.5 mm, while the diameter of each post is 4 mm. Through what angle must one of the posts be turned so that a tension of 200 N is exerted on the wire?

SOLUTIONS TO PRACTICE PROBLEMS

Thought Problems

1. The dynamics of jiggling Jell-O involve oscillation of the gelatin as a result of shearing forces. As you shake the gelatin, shearing forces and inertia move the top layers more easily than the bottom layers. A restoring force is set up which enables the gelatin to jiggle until the oscillations are damped out. It's a delicious way to test physics!
2. Water wets a glass rod because the force of adhesion is greater than the force of cohesion. Mercury does not wet a glass rod because the cohesive forces in mercury are greater than the adhesive forces. Thus, you can push down a layer of mercury and in a capillary tube, the level of mercury in the tube will be below the remaining layer of liquid.
3. A rubber band is stretched to its elastic limit fairly quickly. This means that a plot of force versus elongation will be a diagonal line for a short interval, and then, as the elastic limit is approached, the curve will begin to flatten out (see the accompanying graph).

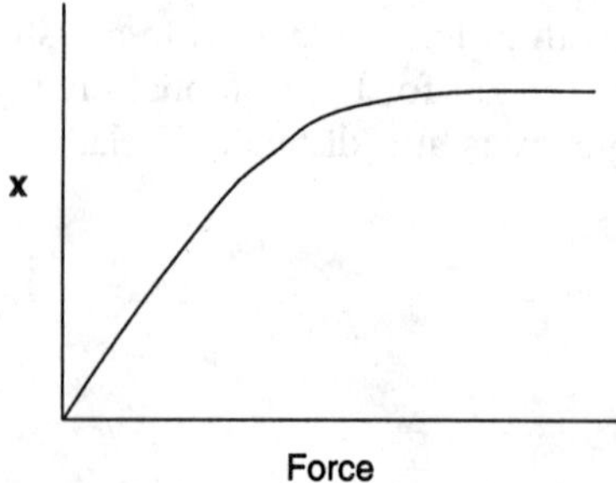

Multiple-Choice Problems

1. **C** The formula for stretching a cable is $\Delta L = L_o\mathbf{F}/YA$, where Y is Young's modulus. In this problem, $\Delta L = d$ and the original length and diameter are being doubled. This has the effect of reducing ΔL to $d/2$.

2. **B** The stress is equal to the ratio of the applied force to the cross-sectional area. We are given the maximum shear stress of 3.5×10^8 Pa. The surface area of the punched circular hole is given by $A = \pi hd = (3.14)(0.004 \text{ m})(0.008 \text{ m}) = 1.0048 \times 10^{-4} \text{ m}^2$. The force is equal to the product of these two numbers and is 35,168 N.

3. **E** To find the bulk modulus of the oil, we first note that a compression of 0.3% is equal to a fractional change in volume of 0.003. From the formula for bulk modulus, if we divide 5×10^6 Pa by this fractional change, we will obtain a bulk modulus for oil equal to 1.7×10^9 Pa or 1.7 GPa.

4. **A** The stress applied to a wire is independent of the wire's length.

5. **B** The weight of a 3-kg mass is $\mathbf{W} = 29.4$ N. Using the given information, we find that the cross-sectional area is equal to the ratio of the applied load to the elastic limit. Thus, $A = (29.4 \text{ N})/(1.5 \times 10^8 \text{ Pa}) = 1.96 \times 10^{-7} \text{ m}^2$. Since a cross section of the wire is circular, we can easily find the radius (and hence the diameter):

$$d = 2\sqrt{\frac{A}{\pi}} = 5 \times 10^{-4} \text{ m} = 0.5 \text{ mm}$$

6. **E** The first thing we need to do is to find the radius of the wire. Since $d = 2.5$ mm, $r = 1.25$ mm $= 0.00125$ m. The area of a cross section is therefore equal to $A = \pi r^2 = 4.9 \times 10^{-6} \text{ m}^2$. The percentage change in length is equal to the fractional change in length multiplied by 100:

$$\frac{\Delta L}{L_o} = \frac{1}{Y}\frac{\mathbf{F}}{A} = 0.000203$$

The percentage change is therefore 0.02%. Notice that the original length of the wire is not relevant to this problem.

7. **A** The pressure of 400 atm is equivalent to 4.04×10^7 Pa. Thus, for the bulk modulus of lead, the fractional change in volume is the ratio of the applied pressure to the bulk modulus: $\Delta V/V_o = 0.0052$. This corresponds to a percentage change of 0.5%.

8. **C** The description of the action of the forces is applicable to shearing.

9. **B** The capillary height the water rises is inversely proportional to the radius of the tube. If the diameter is reduced by half, so is the radius. The level therefore increases by a factor of 2.

10. **D** The change in PE relative to the surrounding liquid is given by $\Delta PE = mgh$. If we recall that $m = \rho V$, and that for water and a clean glass rod the angle of contact is equal to zero and so $\cos\theta = 1.0$, then we have

$$mgh = \rho Vg \frac{2\gamma}{r\rho g} = \frac{2V\gamma}{r}$$

Free-Response Problems

1. (a) The formula for the fractional change in length of an elastic "wire" (in this case a spring) is

$$\frac{\Delta L}{L_o} = \frac{1}{Y}\frac{\mathbf{F}}{A}$$

$$\mathbf{F} = \frac{YA}{L_o}\Delta L$$

If we let $k = -YA/L_o$ and recognize ΔL as the elongation $\mathbf{x}$, then we can see that

$$\mathbf{F} = -k\mathbf{x}$$

(b) The work done in stretching a spring from its natural length L_o by some amount $\mathbf{x} = \Delta L$ is equal to the potential energy in the spring and is given by

$$W = \frac{1}{2}k\mathbf{x}^2 = \frac{YA}{2L_o}(\Delta L)^2$$

2. The ratio of the stress to the strain in a solid is equal to a constant. In this case, the elastic modulus of rigidity is Young's modulus. Thus, the average strain is equal to the ratio of the average stress applied to Young's modulus for steel (2×10^{11} N/m^2).

The average stress applied is equal to the ratio of the average force applied to the cross-sectional area. In this problem, the magnitude of the average force applied is equal to the total change in momentum divided by the time interval of 0.07 s. Thus,

$$\mathbf{F} = \frac{\Delta p}{\Delta t} = \frac{m\,\Delta\mathbf{v}}{\Delta t} = \frac{(25\text{ kg})(32\text{ m/s})}{0.07\text{ s}} = 11{,}429\text{ N}$$

The change in the velocity is 32 m/s.

Now the radius of the spike is $r = 2$ cm $= 0.02$ m, and so if we assume a circular area, we have $A = \pi r^2 = (3.14)(0.02)^2 = 0.001256$ m^2. The average stress is therefore equal to

$$\text{Stress} = \frac{\mathbf{F}}{A} = 9.1 \times 10^6\text{ N/m}^2$$

and the average strain is therefore equal to (in dimensionless units)

$$\text{Strain} = \frac{\text{stress}}{Y} = 4.5 \times 10^{-5}$$

3. From the formula for elastic stretching, we know that

$$\Delta L = \frac{L_0}{Y}\frac{\mathbf{F}}{A} = \frac{1.5}{(2 \times 10^{11})}\frac{200}{(3.14)(0.00025)^2} = 0.00764 \text{ m}$$

where we have used the fact that the radius is $r = 0.25$ mm $= 0.00025$ m and $L_o = 1.5$ m.

This length is wrapped around a cylindrical post whose circumference is equal to

$$C = \pi d = (3.14)(0.004 \text{ m}) = 0.01256 \text{ m}$$

To find the angle of rotation, we need to know what fraction the change in length corresponds to compared with the total circumference. Taking this ratio, we find that the ratio = 0.00764/0.01256 = 0.608, which means the angle must be equal to $\theta = (360)(0.608) = 219$ degrees.

13
FLUIDS

STATIC FLUIDS

Fluids represent states of matter that take the shape of their containers. We can, in a limited sense, classify a fluid by its compressibility, which is the reciprocal of the bulk modulus discussed in Chapter 12. In this sense, liquids are generally *incompressible* fluids, while gases are *compressible fluids.* In a Newtonian sense, liquids do work by being displaced, while gases do work by compressing or expanding. As we shall see later, the compressibility of gases leads to other effects described by the subject of thermodynamics.

The application of a stress to a fluid is a measure of the external pressure as defined in the last chapter. Additionally, fluids themselves can exert pressure by virtue of their weight or force of motion. We have already defined the unit of pressure to be the pascal, which is equivalent to 1 N of force per square meter of surface area. An additional unit used in physics is the **bar**, where 1 bar = 100,000 Pa. Atmospheric pressure is sometimes measured in *millibars.* In this section, we look at fluids that are considered static with respect to themselves and the earth frame of reference.

PASCAL'S PRINCIPLE

In a fluid, static pressure is exerted on the walls of the container. Within the fluid these forces act perpendicular to the walls. If an external pressure is applied to the fluid, this pressure will be transmitted uniformly to all parts of the fluid. The last sentence is also known as **Pascal's principle** since it was developed by the French physicist Blaise Pascal.

Pascal's principle refers only to an external pressure. Within the fluid, the pressure at the bottom of the fluid is greater than at the top. We can also state that the pressure exerted on a small object in the fluid is the same regardless of the orientation of the object.

As an example of Pascal's principle consider the hydraulic press shown in Figure 1. The small-area piston A_1 has an external force $\mathbf{F}_1$ applied to it. At the other end, the large-area piston A_2 has some unknown force $\mathbf{F}_2$ acting on. How do these forces compare? According to Pascal's principle, the force per unit area represents an external pressure which will be transmitted uniformly through the fluid. Thus, we can write

$$\frac{\mathbf{F}_1}{A_1} = \frac{\mathbf{F}_2}{A_2}$$

Sample Problem

Referring to Figure 1, suppose a force of 10 N is applied to the small piston of area 0.05 m^2. If the large piston has an area of 0.15 m^2, what is the maximum weight the large piston can lift?

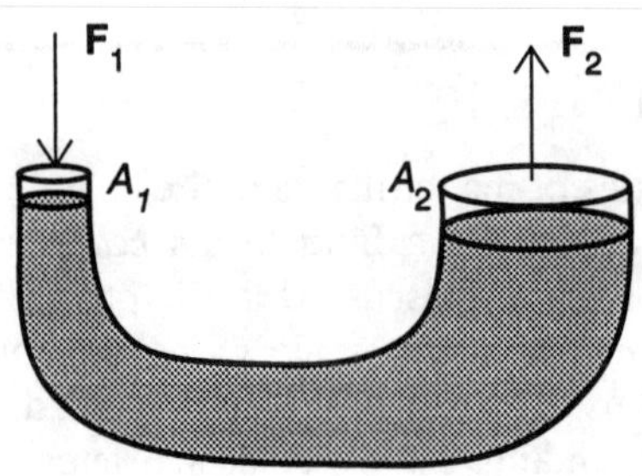

FIGURE 1.

Solution

Since the secondary force is proportional to the ratio of the areas, $\mathbf{F}_2 = 30$ N.

STATIC PRESSURE AND DEPTH

Figure 2 shows a tall column of liquid in a sealed container. What is the pressure exerted on the bottom of the container? To answer this question, we first consider the weight of the column of liquid of height h. Since $\mathbf{W} = m\mathbf{g}$ and $m = \rho V$, the weight is $\rho \mathbf{g} V$. This is the force applied to the bottom of the container.

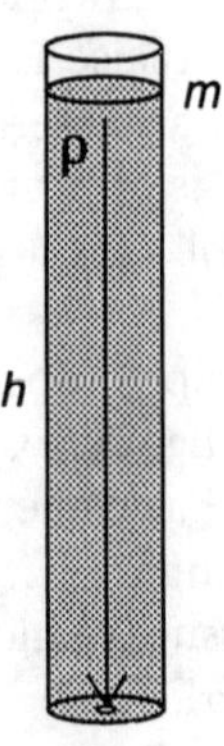

FIGURE 2.

Now, in a container with a regular shape, $V = AH$, where A is the cross-sectional area (in this case, we have a cylinder whose cross sections are uniform circles). Thus, $\mathbf{F} = \rho \mathbf{g} V = \rho \mathbf{g} A h$. Using the definition of pressure, we obtain

$$p = \frac{\mathbf{F}}{A} = \rho \mathbf{g} h$$

If the container is open at the top, then air pressure adds to the pressure of the column of liquid. The total pressure can therefore be written as $\mathbf{p} = \mathbf{p}_{\text{ext}} + \rho \mathbf{g} h$.

Sample Problem

A column of mercury is held up at 1 atm of pressure in an open-tube barometer (see the accompanying diagram). To what height does it rise? The density of mercury is 13.6 times the density of water.

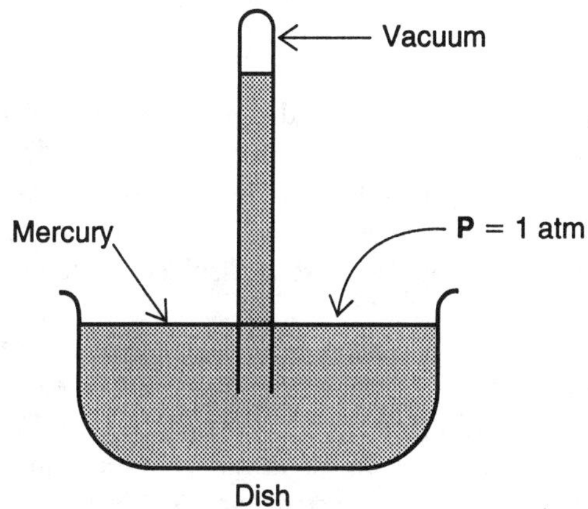

Solution

At 1 atm the pressure is 101 kPa. Thus, we can write

$$1.01 \times 10^5 \text{ N/m}^2 = (13.6 \times 10^3 \text{ kg/m}^3)(9.8 \text{ m/s}^2)h$$

$$h = 0.76 \text{ m} = 76 \text{ cm}$$

BUOYANCY AND ARCHIMEDES' PRINCIPLE

When an object is immersed in water, it feels lighter. In a cylinder filled with water, the action of inserting a mass in the liquid causes it to displace upward. The volume of the water displaced is equal to the volume of the object (even if it is irregularly shaped), as illustrated in Figure 3.

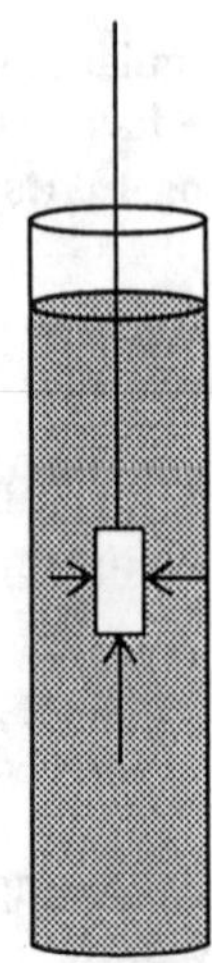

FIGURE 3. *Measuring volume by water displacement.*

Archimedes' principle states that the upward force of the water (called the **buoyant force**) is equal to the weight of the water displaced. Normally, one might think that an object floats if its density is less than water. This statement is only partially correct. A steel needle floats because of surface tension, and a steel ship floats because it displaces a volume of water equal to its weight.

The weight of the water displaced can be found mathematically. The fluid displaced has a weight $\mathbf{W} = m\mathbf{g}$. Now, the mass can be expressed in terms of the density of the liquid and its volume, $m = \rho V$. Hence, $\mathbf{W} = \rho V\mathbf{g}$.

The volume of the object can be determined in terms of the apparent loss of weight in water. Suppose an object weighs 5 N in the air and 4.5 N when submerged in water. The difference of 0.5 N is the weight of the water displaced. The volume is therefore given by

$$V_f = \frac{\Delta m}{\rho_{\text{fluid}}} = \frac{\Delta \mathbf{W}}{\mathbf{g}\rho}$$

Using this relationship, we have $V_f = (0.5\ \text{N})/(9.8\ \text{N/kg})(1 \times 10^3\ \text{kg/m}^3) = 5.1 \times 10^{-5}\ \text{m}^3$.

FLUIDS IN MOTION

The situation regarding static pressures in fluids changes when they are in motion. Microscopically, we could try to account for the motion of all molecular particles that make up the fluid, but this would not be very practical. Instead, we treat the fluid as a whole and consider what happens as the fluid passes through a given cross-sectional area each second. Sometimes the word *flux* is used to describe the volume of fluid passing through a given area each second.

Consider the fluid shown in Figure 4 moving uniformly with a velocity **v** in a time *t* through a segment of a cylindrical pipe. The distance traveled is given by the product **v***t*. Since the motion is ideally smooth, there is no resistance offered by the fluid as different layers move relative to one another. This resistance is known as **viscosity**, and the type of fluid motion we are considering here is called **laminar flow**.

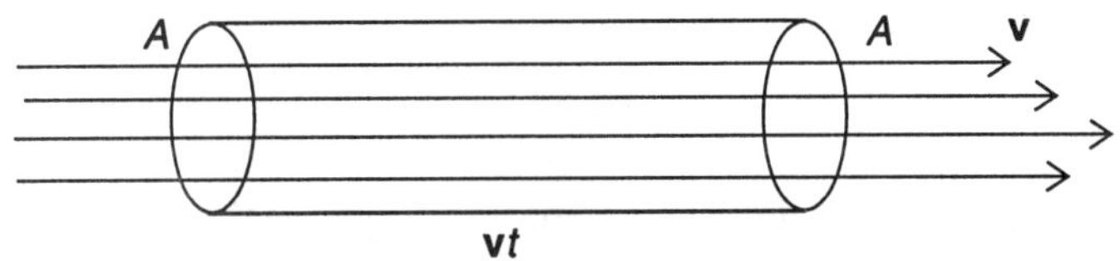

FIGURE 4.

The rate of flow *R* is defined to be the volume of fluid flowing out of the pipe each second (in m^3/s):

$$R = \frac{\mathbf{v}tA}{t} = \mathbf{v}A$$

If the flow is laminar, then the **equation of continuity** states that the rate of flow *R* will remain constant. Therefore, as the cross-sectional area decreases, the velocity must increase:

$$\mathbf{v}_1A_1 = \mathbf{v}_2A_2$$

The power generated by the fluid is given by the product of the force exerted per unit area and the rate of flow (which is equal to the pressure times the rate of flow):

$$\text{Power} = \mathbf{P}R$$

BERNOULLI'S EQUATION

Consider a fluid moving through an irregularly shaped tube at two different levels given by h_1 and h_2 as shown in Figure 5. At the lower level, the fluid exerts a pressure $\mathbf{P}_1$ while moving through an area A_1 with a velocity $\mathbf{v}_1$. At the top, the fluid exerts a pressure $\mathbf{P}_2$ while moving through an area A_2 with a velocity $\mathbf{v}_2$. Bernoulli's equation is related to changes in pressure as a function of velocity.

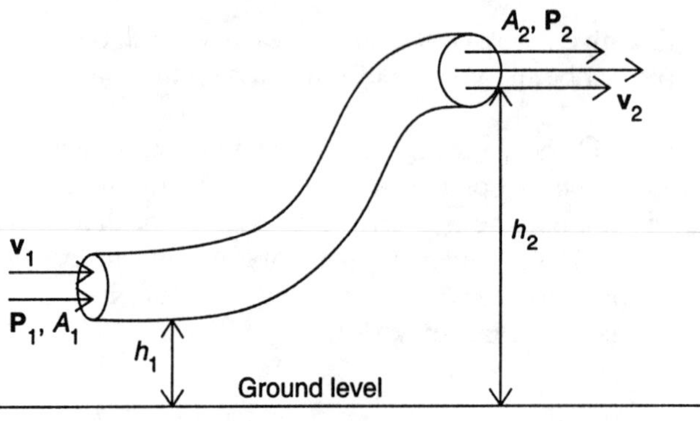

FIGURE 5.

Let us begin by considering the work done in moving the fluid from position 1 to position 2. The power generated is equal to the product of the pressure and the rate of flow (**P***R*), and so the work, which is equal to the product of power and time, can be written as

$$\Delta W = P_1 A_1 \mathbf{v}_1 t - P_2 A_2 \mathbf{v}_2 t = \frac{P_1 m}{\rho} - \frac{P_2 m}{\rho}$$

The change in the potential energy is given by

$$\Delta \mathrm{PE} = m\mathbf{g}h_2 - m\mathbf{g}h_1$$

The change in kinetic energy is given by

$$\Delta \mathrm{KE} = \frac{1}{2} m\mathbf{v}_2^2 - \frac{1}{2} m\mathbf{v}_1^2$$

Adding up both changes in energy and equating it with the work done, we obtain

$$\frac{\mathbf{P}_1 m}{\rho} - \frac{\mathbf{P}_2 m}{\rho} = m\mathbf{g}h_2 - m\mathbf{g}h_1 - \frac{1}{2} m\mathbf{v}_2^2 - \frac{1}{2} m\mathbf{v}_1^2$$

$$\boxed{\mathbf{P}_1 + \rho\mathbf{g}h_1 + \frac{1}{2}\rho\mathbf{v}_1^2 = \mathbf{P}_2 + \rho\mathbf{g}h_2 + \frac{1}{2}\rho\mathbf{v}_2^2}$$

The last equation is known as **Bernoulli's equation**. Now let us consider some applications of this equation.

A Fluid at Rest

In Figure 6, we see a static fluid. The two layers at heights h_1 and h_2 have static pressures $\mathbf{P}_1$ and $\mathbf{P}_2$. Since the fluid is at rest, Bernoulli's equation reduces to

$$\Delta\mathbf{P} = \mathbf{P}_2 - \mathbf{P}_1 = \rho\mathbf{g}\,\Delta h = \rho\mathbf{g}(h_2 - h_1)$$

The difference in pressure is just proportional to the difference in levels.

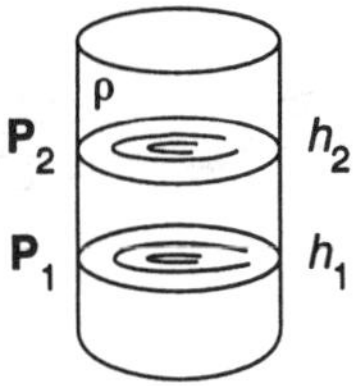

FIGURE 6.

A Fluid Escaping Through a Small Orifice

Earlier in this chapter, we saw that the pressure difference is proportional to the difference in height. Suppose a small hole of circular area A is punched into the container below h_1 (Figure 7). This pressure difference will force the fluid out of the hole at a rate of flow $R = \mathbf{v}A$. What is the velocity of the fluid as it escapes? And what is the rate of flow?

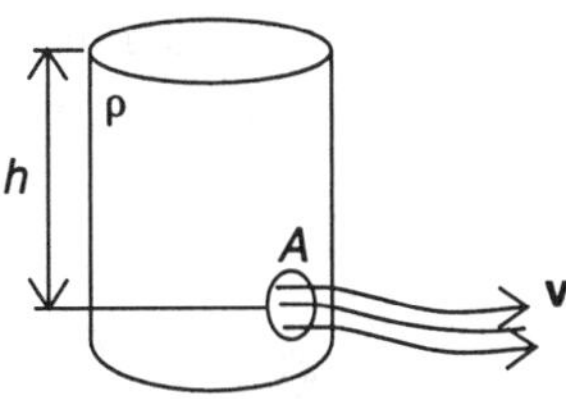

FIGURE 7.

To answer these questions, we consider Bernoulli's equation as an analog of the conservation of energy equation. Let's choose the potential energy to be zero at h_1. The top fluid at h_2 is essentially at rest. Thus, using Bernoulli's equation, we can write

$$\frac{1}{2}\rho\mathbf{v}^2 = \rho\mathbf{g}\,\Delta h$$

$$\mathbf{v} = \sqrt{2\mathbf{g}\,\Delta h}$$

$$R = \mathbf{v}A = A\sqrt{2\mathbf{g}\,\Delta h}$$

A Fluid Moving Horizontally

Consider a fluid moving horizontally through a tube that narrows in area. Bernoulli's equation states that as the velocity of a moving fluid increases, its static pressure decreases. We can analyze this in Figure 8.

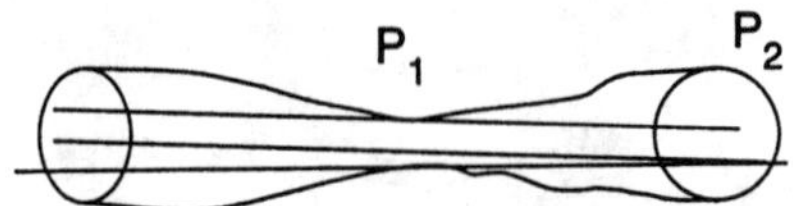

FIGURE 8. *Venturi tube*

Since the level is horizontal, there is no change in pressure resulting from a change in level. We can therefore write

$$\mathbf{P}_1 + \frac{1}{2}\rho\mathbf{v}_1^2 = \mathbf{P}_2 + \frac{1}{2}\rho\mathbf{v}_2^2$$

$$\Delta\mathbf{P} = \mathbf{P}_1 - \mathbf{P}_2 = \frac{1}{2}\rho(\mathbf{v}_2^2 - \mathbf{v}_1^2)$$

In aerodynamics, a wing moving in level flight has a lifting force acting on it exactly equal to its load. This force is caused by the pressure difference between the upper and lower surfaces of the wing. The wing is curved or *cambered* on top (see Figure 9), which has the effect of lowering the pressure on the top of wing (caused by faster-moving air molecules). It is this shape that causes the pressure difference and lift.

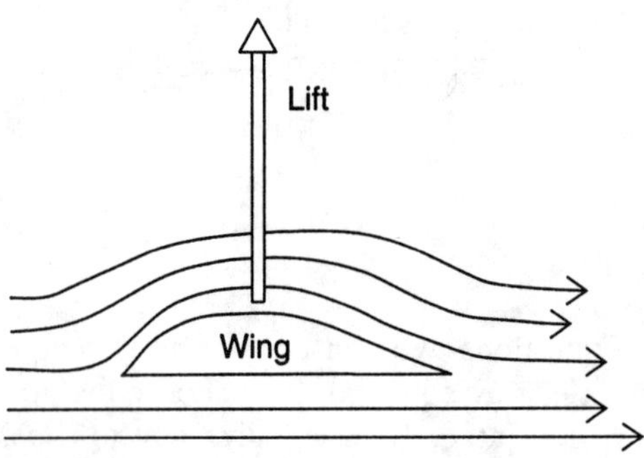

FIGURE 9. *Air flow pattern over a wing section.*

PROBLEM-SOLVING STRATEGIES

Solving fluid problems is similar to solving particle problems. We treat the fluid as a whole unit (i.e., macroscopically) as opposed to microscopically. The units of pressure must be either pascals or N/m^2 in order to use the formulas derived.

The concepts of Bernoulli's principle and Archimedes' principle should be thoroughly understood, as well as their applications and implications. Buoyancy is an important physical phenomenon and an important part of your overall physics education.

As always, drawing a sketch helps. Often, conceptual knowledge will be enhanced if you understand how the variables in a formula are related. Consider questions that involve changing one variable and observing the effect on others.

PRACTICE PROBLEMS FOR CHAPTER 13

Thought Problems

1. Which has more pressure on the bottom? A large tank of water 30 cm deep or a cup of water 35 cm deep? Explain your answer.
2. Two paper cups are suspended by strings and hung near each other (separated by about 10 cm). When you blow air between them, the cups are attracted to one another. Explain why this occurs.
3. A cup of water is filled to the brim and held above the ground. Two holes on either side of the cup and horizontally level with each other are punched into the cup, causing water to run out. When the cup is released and allowed to fall, it is observed that no water escapes during the fall. Explain why.

Multiple-Choice Problems

1. The rate of flow of liquid from a hole in a container depends on all of the following except _____.
   (A) the density of the liquid
   (B) the height of the liquid above the hole
   (C) the area of the hole
   (D) the acceleration of gravity
   (E) the diameter of the hole
2. A person is standing on a railroad station platform when a high-speed train passes by. The person will tend to be _____.
   (A) pushed away from the train
   (B) pulled in toward the train
   (C) pushed upward into the air
   (D) pulled down into the ground
   (E) unaffected by the passage of the train
3. Bernoulli's equation is based on which law of physics?
   (A) conservation of linear momentum
   (B) conservation of angular momentum
   (C) Newton's first law of motion
   (D) Newton's law of gravity
   (E) conservation of energy

4. Which of the following expressions represents the power generated by a liquid flowing out of a hole of area A with a velocity $\mathbf{v}$?
(A) (pressure) × (rate of flow)
(B) (pressure)/(rate of flow)
(C) (rate of flow)/(pressure)
(D) (pressure) × (velocity)/(area)
(E) (rate of flow)/(area)

5. Which of the following statements is an expression of the *equation of continuity?*
(A) Rate of flow equals the product of velocity and cross-sectional area.
(B) Rate of flow depends on the height of the fluid above the hole.
(C) Pressure in a static fluid is transmitted uniformly throughout.
(D) Fluid flows faster through a narrower pipe.
(E) None of these statements.

6. A moving fluid has an average pressure of 600 Pa as it exits a circular hole with a radius of 2 cm at a velocity of 60 m/s. What is the approximate power generated by the fluid?
(A) 32 W
(B) 62 W
(C) 1,200 W
(D) 45 W
(E) 12 W

7. Alcohol has a specific gravity of 0.79. If a barometer consisting of an open-ended tube placed in a dish of alcohol is used at sea level, to what height in the tube will the alcohol rise?
(A) 8.1 m
(B) 7.9 m
(C) 15.2 m
(D) 13.1 m
(E) 0.79 m

8. An ice cube is dropped into a mixed drink containing alcohol and water. The ice cube sinks to the bottom. From this, you can conclude _____.
(A) that the drink is mostly alcohol
(B) that the drink is mostly water
(C) that the drink is equally mixed with water and alcohol
(D) nothing unless you how much liquid is present
(E) nothing unless you know the temperature of the drink

9. A 2-N force is used to push a small piston 10 cm downward in a simple hydraulic machine. If the opposite large piston rises by 0.5 cm, what is the maximum weight the large piston can lift?
(A) 2 N
(B) 40 N

(C) 20 N
(D) 4 N
(E) 1 N

10. Balsa wood with an average density of 130 kg/m_3 is floating in pure water. What percentage of the wood is submerged?
(A) 87%
(B) 13%
(C) 50%
(D) 25%
(E) 5%

Free-Response Problems

1. A U-tube open at both ends is partially filled with water. Benzene ($\rho = 0.897 \times 10^3$ kg/m^3) is poured into one arm, forming a column 4 cm high. What is the difference in height between the two surfaces?

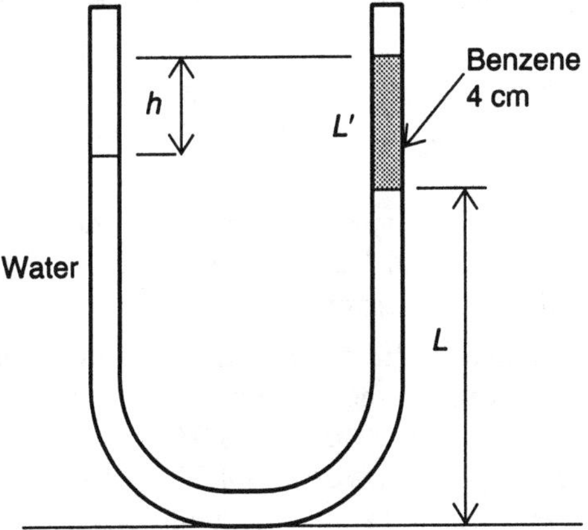

2. A venturi tube has a pressure difference of 15,000 Pa. The entrance radius is 3 cm, while the exit radius is 1 cm. What are the entrance velocity, exit velocity, and flow rate if the fluid is gasoline ($\rho = 700$ kg/m^3)?
3. A cylindrical tank of water (height H) is punctured at a height h above the bottom. How far from the base of the tank will the water stream land (in terms of h and H)? What must the value of h be such that the distance at which the stream lands will be equal to H?

SOLUTIONS TO PRACTICE PROBLEMS

Thought Problems

1. The pressure at the bottom of a container of water is dependent on the depth of the water, not on its overall volume. Hence the 35-cm cup has more pressure at the bottom.

2. Bernoulli's principle states that as the velocity of a moving fluid increases, the pressure it exerts decreases. Thus, by blowing air between two cups, you reduce the air pressure between them relative to the other sides of the cups. This pressure difference gives rise to a force directed inward.
3. The water streams out the holes because of the relative difference in pressure. However, in free fall, the water is falling along with the cup, and so no water streams out.

Multiple-Choice Problems

1. **A** The flow rate of a liquid from a hole does not depend on the density of the liquid.
2. **B** Because of the Bernoulli effect, the speeding train reduces the air pressure between the person and the train. This pressure difference creates a force tending to pull the person into the train. Be very careful when standing on a train platform!
3. **E** Bernoulli's principle was developed as an application of conservation of energy.
4. **A** Power is expressed in J/s. The product of pressure (force divided by area) and flow rate (m^3/s) leads to newtons times meters over seconds (J/s).
5. **D** The equation of continuity states that the product of velocity and area is a constant for a given fluid. A consequence of this is that a fluid moves faster through a narrower pipe.
6. **D** The power is equal to the product of the pressure and the flow rate. The flow rate is equal to the product of the velocity and the cross-sectional area (which is a circle of radius 0.02 m). The area is given by $\pi r^2 = 1.256 \times 10^{-3}$ m^2. When we multiply this area by the velocity and the pressure, we get 45.2 W as a measure of the generated power.
7. **D** The pressure exerted by the air is balanced by the column of liquid alcohol in equilibrium:

$$h = \frac{\mathbf{P}}{\rho \mathbf{g}} = \frac{1.01 \times 10^5 \text{ N/m}^2}{(790 \text{ kg/m}^3)(9.8 \text{ m/s}^2)} = 13.1 \text{ m}$$

8. **A** Since the density of alcohol is less than that of water, ice floats "lower" in alcohol than in water. From the given information, the drink appears to be mostly alcohol.
9. **B** We need to use the conservation of work-energy in this problem. The work done on the small piston must equal the work done by the large piston. Since the ratio of displacements is 20:1, the large piston will be able to support a maximum load of 40 N (since 2 N × 20 = 40 N).
10. **B** The percentage submerged is given by the ratio of its density to that of pure water (1,000 kg/m^3). Thus, 130/1,000 = 0.13 = 13%.

Free-Response Problems

1. Let L' be the level of benzene that will float on top of the water, let L be the level of water, and let h be the difference in levels. Since the tube is open, the pressures are equalized at both ends. Thus, we can write

$$(L + L' - h)\mathbf{g}\rho_w = L'\mathbf{g}\rho_b + L\mathbf{g}\rho_w$$

$$h = L'\left(1 - \frac{\rho_b}{\rho_w}\right)$$

Since the ratio of the densities is 0.879 and $L' = 4$ cm, $h = 0.484$ cm.

2. Using Bernoulli's equation and the equation of continuity, we can write

$$\Delta\mathbf{P} = \frac{\rho}{2}(\mathbf{v}_2^2 - \mathbf{v}_1^2)$$

$$\mathbf{v}_1 A_1 = \mathbf{v}_2 A_2$$

$$\mathbf{v}_2 = \mathbf{v}_1 \frac{A_1}{A_2}$$

$$\Delta\mathbf{P} = \frac{\rho}{2}\mathbf{v}_1^2\left(1 - \frac{A_1^2}{A_2^2}\right)$$

From the given information, we know that the ratio of areas $A_1/A_2 = 9$. Thus, substituting all given values into the last equation yields $\mathbf{v}_1 = 0.732$ m/s, $R = 0.0021$ m^3/s, and $\mathbf{v}_2 = 6.56$ m/s.

3. The change in potential energy must be equal to the change in horizontal kinetic energy:

$$m\mathbf{g}(H - h) = 1/2m\mathbf{v}_x^2$$

$$\mathbf{v}_x = \sqrt{2\mathbf{g}(H - h)}$$

Now if we assume, as in projectile motion, that the horizontal velocity remains the same and the only acceleration of the stream is vertically downward because of gravity, we can write

$$\mathbf{x} = \mathbf{v}_x t$$

$$y = h = \frac{1}{2}gt^2$$

$$t = \sqrt{\frac{2h}{g}}$$

$$\mathbf{x} = 2\sqrt{h(H - h)}$$

For $\mathbf{x} = H$, we must have $h = H/2$, as can be easily verified using the preceding range formula.

14
TEMPERATURE AND HEAT

INTRODUCTION

We have already learned that friction is a nonconservative force. When work is done against friction, the kinetic energy of the system is not conserved. Where does the energy go? Observations made by "touch" indicate that even though the mass and shape may stay the same, the mass "feels" different. This sensation is attributable to a form of energy we call **heat**.

TEMPERATURE AND ITS MEASUREMENT

Temperature can be defined in terms of relative measure of heat flow. Even more simply, we can state it is a measure of how *hot* or *cold* a body is (relative to some standard reference like your body or another arbitrary mark). Human touch is unreliable in measuring temperature since what feels hot to one person may feel cold to another.

During the seventeenth century, Galileo invented a method for measuring temperature using water and its ability to react to changes in heat content by expanding or contracting. Water is not the best substance to use for a variety of reasons. First, it has a relatively high freezing and low boiling point and, second, it has a peculiar variation in density as a function of temperature because of its chemical structure.

Even so, the phenomenon of **thermal expansion** is important not only for this topic but also for engineering buildings, roads, and bridges. We will discuss this aspect in more detail later in the chapter.

Most modern thermometers use mercury or alcohol in a small capillary tube. In the metric system, the standard sea level points of reference are the freezing point (0°C) and boiling point (100°C) of water. This temperature scale used to be called the *centrigrade* scale but is now called the *Celsius* scale, after Anders Celsius. The English *Farenheit* system is not used in physics.

One difficulty with the Celsius scale is that the zero mark is not a true zero since some substances freeze below this point. Some gases (like oxygen and hydrogen) do not even condense until their temperatures reach far below 0°C. The discovery of a theoretical "absolute" zero temperature has lead to the absolute or Kelvin temperature scale. On this scale, a temperature of −273°C is redefined to be **absolute zero** on the Kelvin scale since there is no theoretical lower temperature.

An easy way to remember how to convert from Celsius to Kelvin is to add the number 273 to your Celsius temperature. This is just adding a scale factor that preserves the Celsius metric scaling. In other words, there is still a 100°C temperature difference between melting ice and boiling water (273 K and 373 K, respectively).

THERMAL EXPANSION OF SOLIDS, LIQUIDS, AND GASES

Solids

You know from your previous studies in science that matter exists in three states: solid, liquid, and gas. These states react differently when heat is applied because of their different structures. A metal rod when heated expands primarily linearly because it is longer than it is wide. And when a sidewalk is constructed, spacing is left between panels to allow for expansion. Similarly, on a bridge you sometimes see metal spacers in the roadway to allow for roadway expansion. Experiments can be performed to indicate that a direct relationship exists between the fractional amount of expansion and the change in temperature. If we let L_o be the original length of the solid and L be the new length of the solid after expansion, then $\Delta L/L_o$ represents the fractional change in length. It is this quantity that is directly proportional to the change in temperature:

$$\frac{\Delta L}{L_o} = \alpha \Delta T$$

where α is called the **coefficient of linear expansion** in units of (1/°C). Notice that this relationship is *dimensionless* since the ratio of length to length cancels the units out. Some selected coefficients of linear expansion are given in Table 1.

TABLE 1. COEFFICIENT OF LINEAR EXPANSION FOR VARIOUS SUBSTANCES

Substance	Coefficient ($\times 10^{-6}$ °C^{-1})
Aluminum	23
Brass	19
Concrete	12
Copper	17
Glass, Pyrex	3.2
Steel	12
Lead	30
Silver	51

The expression for the fractional change in length can also be written in the form

$$L = L_o(1 + \alpha\Delta T)$$

As an example, consider the following problem. A copper rod is 0.5 m in length at 20°C What is its length at 100°C?

To solve this problem, we note that the coefficient of linear expansion for copper is $\alpha = 17 \times 10^{-6}\,°C^{-1}$. Using the formula for linear expansion, we find that for an 80°C change in temperature, $\Delta L = 0.00068$ m and thus $L = 0.50068$ m. There would be no observable change to three decimal places.

There is a change in area as well. Since area can be considered as "length squared," we can write the area change for uniform expansion as

$$A = L^2 = L_o^2(1 + \alpha\ \Delta T)^2 = A_o(1 + 2\alpha\ \Delta T + \alpha^2\ \Delta T^2)$$

If $\alpha << 1$, then to a first approximation we can write

$$A = A_o(1 + 2\alpha\ \Delta T)$$

We can infer from this expression that the coefficient of area expansion for a solid is approximately twice the value of the coefficient of linear expansion. In a similar way, we can observe that if any volume changes occur in the solid, then following the logic that volume is "length cubed," we can write (again assuming that $\alpha << 1$)

$$V = V_o(1 + 3\alpha\ \Delta T)$$

The coefficient of volume expansion for solids is approximately three times the value of the linear expansion coefficient.

Liquids and Gases

The expansion of an unconstrained gas is uniform in volume. In this way, we define the volume relationship to be similar to the linear relationship for solids:

$$\frac{\Delta V}{V_o} = \beta\ \Delta T$$

A list of some values for volume expansion is given in Table 2.

TABLE 2. COEFFICIENT OF VOLUME EXPANSION FOR VARIOUS SUBSTANCES

Substance	Coefficient ($\times 10^{-6}\,°C^{-1}$)
Acetone	150
Air	3670
Ethanol	1,100
Mercury	180
Water	210

Consider the following example of volume expansion. By what percentage will a given volume of air expand if its temperature is changed by 10°C?

To solve this problem, we note that the coefficient of volume expansion for air is $3670 \times 10^{-6}\ °C^{-1}$. Using the formula for volume expansion, we can write

$$\frac{\Delta V}{V} = \beta\ \Delta T = (3.67 \times 10^{-4}\ °C^{-1})(10°C) = 0.00367$$

Thus, the percentage change in volume is 0.367%.

HEAT CAPACITY

You may have noticed that during the hottest days of summer the water in a pool, lake, or ocean is usually much cooler than the air or land. If equal masses of water and sand are placed under sun lamps, it can be observed that the temperature of the sand rises faster than the temperature of the water. Additional experiments reveal that when heated to the same temperature and then placed on a plate of wax, equal masses (same shapes and sizes) of metals melt the wax to different depths.

These observations suggest that different substances maintain different heat capacities when equal masses and equal changes in temperature are compared. In physics, a quantity of heat is designated by the letter Q, any change in heat ΔQ, usually accompanied by a change in temperature ΔT, and these two quantities are directly proportional to each other.

Historically, the unit of heat content was the **calorie** (cal). One calorie was defined to be the amount of heat needed to raise the temperature of 1 g of water 1°C. Based on this definition, the specific heat capacity of any substance was defined to be equal to the amount of heat needed to raise the temperature of 1 g of the substance 1°C, with water having a specific heat equal to 1 cal/(g-°C). The heat content of a substance also depends on how much of the substance is involved. Thus, algebraically, we can write

$$\Delta Q = mc\ \Delta T$$

where c is the *specific heat capacity* of the substance in cal/(g-°C).

Experiments in the nineteenth century by the physicist James Joule revealed that if 4.19 J of mechanical work is done, 1 cal of heat is added. In standard SI units of physics (followed in this book), the unit for heat is the same as all energy units: the joule. The fact that 4.19 J is equivalent to 1 cal is referred to as the *mechanical equivalent of heat.* Thus, to convert from calories to joules, we simply multiply by 4.19. Be careful with units since in SI units we should be using kilograms instead of grams. Through a factor analysis, we can convert to kilojoules when kilograms are used. Thus, the unit of specific heat can be expressed as kJ/(kg-°C), and this will ensure that the units for heat are in kilojoules (provided we use Celsius temperatures and kilograms for mass). Some specific heats are presented in Table 3.

TABLE 3. SPECIFIC HEAT OF VARIOUS SUBSTANCES

Substance	Specific Heat [kJ/(kg-°C)]
Alcohol, ethyl	2.43 (liquid)
Aluminum	0.90 (solid)
Copper	0.39 (solid)
Iron	0.45 (solid)
Lead	0.13 (solid)
Mercury	0.14 (solid)
Silver	0.24 (solid)
Tungsten	0.13 (solid)
Water	
Ice	2.05 (solid)
Water	4.19 (liquid)
Steam	2.01 (gas)
Zinc	0.39 (solid)

As an example, consider the following problem. How much heat was added to 250 g of copper that was heated from 20°C to 100°C? To answer this question, we look at Table 3 and find that the specific heat for copper is 0.39 kJ/(kg-°C). Since 250 g = 0.25 kg and $\Delta\ T$ = 80 degrees, we have $\Delta\ Q$ = (0.25)(0.39)(80) = 7.8 kJ.

CONSERVATION OF HEAT

When heat is transferred from one substance to another, all the energy transferred is conserved. Now it is quite possible that some of the heat is leaked to the environment, but even so, all the heat is accounted for. In an ideal situation, an isolated system allows for the direct transfer of heat without the additional gain or loss of heat to the environment being considered. In such a situation we can say that the heat gained by one substance is equal to the heat lost by the other. That is, we can write $\Delta\ Q_{(gain)} = \Delta\ Q_{(loss)}$.

Consider the following situation as an example. A 0.05-kg sample of lead at a temperature of 100°C is placed in 2 kg of water at a temperature of 20°C. What is the final temperature of the mixture when equilibrium has been achieved? From Table 3, we can obtain the specific heats of lead and water. Since we are making a comparison, we must be sure that all units are consistent. We are given masses in kilogram units, and so everything is correct.

We can now state that the final temperature reached by the mixture will involve equal changes in heat content. Thus,

$$M_w c_w \ \Delta T_w = M_l c_l \ \Delta T_l$$
$$(2)(4.19)(T_f - 20) = (0.05)(0.13)(100 - T_f)$$

The different forms of the expressions for ΔT are used to keep the change in temperature a positive value. Solving for the temperature of the final mixture, we obtain $T_f = 20.06°C$ (not much of a change in temperature!).

LATENT HEAT AND CHANGES OF STATE

In the last section, we found that a change in temperature implies a change in heat content (and vice versa). However, you may have observed that when water boils, its temperature does not change even though more heat is being added. This apparent contradiction can be explained by recognizing that when water boils, it changes into steam. Therefore, instead of changing the temperature, the heat being added is doing work to change the state of the water from a liquid to a gas (vaporization). Similar observations demonstrate that when water freezes, its temperature does not go below 0°C (assuming it is relatively pure).

The amount of heat needed to change the state of a substance is referred to as *latent* (or hidden) heat. Specifically, in going from a liquid to a gas (or vice versa), we use the term **latent heat of vaporization**. When changing from a liquid to a solid (or vice versa), we use the term **latent heat of fusion**. Both these quantities are measured in J/g or kJ/kg, and a list of some latent heats is given in Table 4. For any change in state, if L represents the appropriate latent heat quantity, then we can write $\Delta Q = mL$ (where mass is expressed in kilograms). Notice also that the latent heat of vaporization is larger than the latent heat of fusion for all substances.

As an example, consider a 20-g piece of ice at −15°C at sea level. Heat is added until it becomes 20 g of steam at 120°C. How much heat was added (assuming there are no losses)?

TABLE 4. HEATING CONSTANTS

Substance	Heat of Fusion (kJ/kg)	Heat of Vaporization (kJ/kg)
Aluminum	396	10,500
Copper	205	4,790
Iron	267	6,290
Lead	25	866
Mercury	11	295
Silver	105	2,370
Water	334	2,260
Zinc	113	1,770

To answer this question, we first note that this is a five-step problem. First, the ice is heated up to its melting point. Second, the ice melts. Third, the water is heated up to its boiling point. Fourth, the water vaporizes into steam. Fifth, the steam is heated up to 120°C.

Each of these five steps can be expressed algebraically using the results of this and the previous sections.

1. $\Delta Q = mc\,\Delta T = (0.02)(2.05)(15) = 0.615$ kJ
2. $\Delta Q = mL = (0.02)(335) = 6.7$ kJ
3. $\Delta Q = mc\,\Delta T = (0.02)(4.19)(100) = 8.38$ kJ
4. $\Delta Q = mL = (0.02)(2{,}260) = 45.2$ kJ
5. $\Delta Q = mc\,\Delta T = (0.02)(2.01)(20) = 0.804$ kJ

The total amount of heat added is therefore $\Delta Q = 61.699$ kJ.

If we were to plot the change in heat versus the change in temperature, we would see these five steps as five distinct regions. In the regions corresponding to the three states of matter (solid, liquid, and gas), there would be three diagonal lines corresponding to the direct relationship between change in temperature and change in heat content. In the other two regions, where the state was changing, there would be horizontal lines indicating no change in temperature. The length of each line would be proportional (not equal) to the two latent heats. A typical graph of this type is shown in Figure 1.

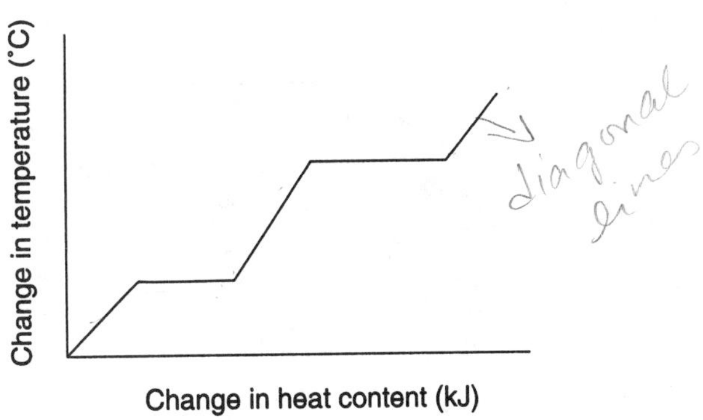

FIGURE 1.

To find the value of the latent heat quantity desired, simply measure off the horizontal scale and divide by the mass. (Be sure the horizontal axis is in units of joules. Occasionally, the horizontal axis is in units of time. Examples of these changes are found in the problems at the end of the chapter.) The slopes of the diagonal lines are related to the specific heats (but reciprocally). From the formula for heat transfer within a substance, we can show that $c = \Delta Q/m\,\Delta T$. This is actually the reciprocal of the slope divided by the mass. Thus, a substance has a high specific heat if it has a line with a small slope (indicating that it takes a large amount of heat to change its temperature by a small amount).

HEAT FLOW AND THERMAL CONDUCTIVITY

Heat can flow through a substance by the process of conduction. Metals are good conductors of heat because they contain a large number of free electrons (this feature also makes metals good conductors of electricity). If we consider a small, rectangular slab of material, then the rate at which heat flows from one side to the other through a thickness Δx can be shown to be related to the variation in temperature from one side to the other (ΔT):

$$\frac{\Delta Q}{\Delta t} = kA\frac{\Delta T}{\Delta x}$$

where k is the **coefficient of thermal conductivity** and is in W/(m-°C). The quantity A is the cross-sectional area of the slab of thickness Δx (see Figure 2). Additionally, the quantity $\Delta T/\Delta x$ is sometimes referred to as the **temperature gradient**. A list of some coefficients of thermal conductivity (rated at 25°C) is given in Table 5.

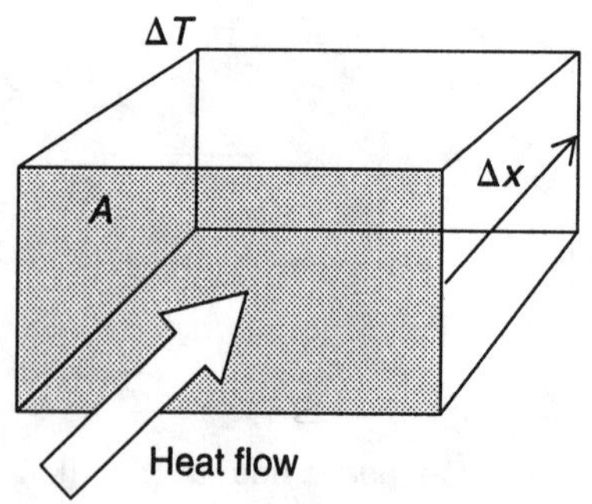

FIGURE 2.

Another useful way to look at the expression for heat flow is to divide by the cross-sectional area A. If we define the left-hand side of the equation to be H, which is the rate of heat flow per unit area (in W/m²), then H can be referred to as a **heat flux** since it indicates the power generated per unit area. This aspect of heat flow is very important in engineering construction. Thus, we can write $H = k/\Delta T/\Delta x$ as the expression for heat flux.

TABLE 5. COEFFICIENTS OF THERMAL CONDUCTIVITY

Substance	Coefficient (W/(m-°C)
Aluminum	238
Copper	397
Gold	314
Silver	427
Lead	34.7
Iron	79.5
Glass (Pyrex)	0.8
Concrete	0.8
Wood	0.08
Ice	2.0
Air	0.0234

PROBLEM-SOLVING STRATEGIES

There are no special considerations to be concerned with when solving problems involving heat transfer or heat flow. Make sure you are aware of the proper units based on the tables. For example, suppose you are given a problem in which a bullet is shot into a block of wood. The kinetic energy of the bullet is transferred to heat as it enters the wood (assuming that the wood does not move when struck by the bullet). In order to compare the two energies, you must make sure that the units match. The kinetic energy (using kilograms for mass) is in joules using the standard formula from mechanics. However, using the same kilogram mass unit, the thermal heat energy would be in units of kilojoules, and that could lead to an inconsistent value for changes in temperature, specific heat, or thermal conductivity.

PRACTICE PROBLEMS FOR CHAPTER 14

Thought Problems

1. Explain how it is possible for water to moderate the seasonal effects of winter and summer for shore areas.
2. Explain why steam at 100°C feels hotter than boiling water at the same temperature.
3. Under what conditions can a substance absorb heat without changing temperature?

Multiple-Choice Problems

Solve Problems 1 through 7 based on the accompanying graph representing the response of 2.0 kg of a gas at 140°C. The gas is releasing heat at a rate of 6 kJ/min.

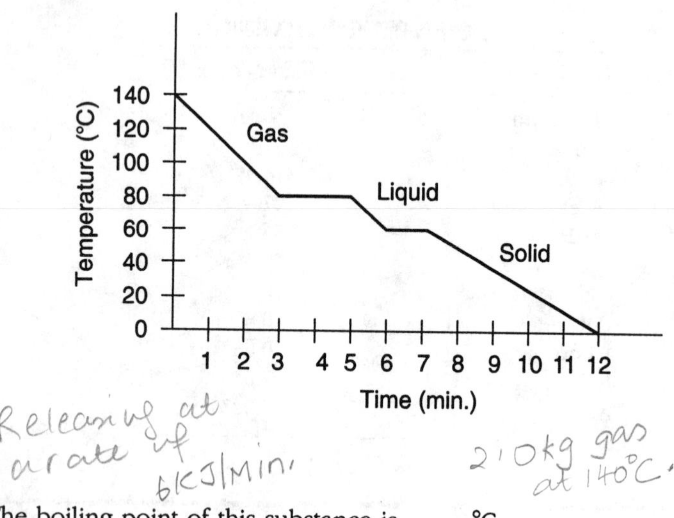

1. The boiling point of this substance is _____ °C.
   (A) 0
   (B) 60
   (C) 80
   (D) 140
   (E) not enough information given
2. The melting point of this substance is _____ °C.
   (A) 0
   (B) 60
   (C) 80
   (D) 140
   (E) not enough information given
3. The heat of vaporization for this substance is _____ kJ/kg.
   (A) 80
   (B) 12
   (C) 6
   (D) 2
   (E) 24
4. The heat of fusion for this substance is _____ kJ/kg.
   (A) 3
   (B) 6
   (C) 12
   (D) 20
   (E) 24
5. Compared to the specific heat of the substance in the gas state, its specific heat as a solid is _____.
   (A) more
   (B) less
   (C) the same

(D) sometimes more, sometimes less
(E) not enough information to determine.

6. The specific heat of the substance in the liquid state is _____ kJ(kg-°C).
   (A) 0.02
   (B) 0.15
   (C) 0.45
   (D) 0.62
   (E) 0.98
7. How much heat is released by the substance from the time the gas starts to condense until it completely solidifies?
   (A) 4 kJ
   (B) 8 kJ
   (C) 12 kJ
   (D) 24 kJ
   (E) 34 kJ
8. A change in temperature of 50°C is equal to a change of _____ kelvins.
   (A) 323
   (B) 283
   (C) −223
   (D) 100
   (E) 50
9. An experiment is performed in which 0.02 kg of a substance is heated in a bath of boiling water until its temperature is approximately 100°C. The substance is then placed in an insulated container holding 0.10 kg of water at 25°C. The final equilibrium temperature of the mixture is 32°C. The specific heat of the substance is _____ kJ/(kg-°C).
   (A) 2.16
   (B) 2.095
   (C) 2.57
   (D) 4.19
   (E) 1.0
10. A steel rod is heated from 20°C to 200°C. If the rod was originally 0.45 m long, by what percentage did it expand?
    (A) 0.09%
    (B) 0.216%
    (C) 0.24%
    (D) 0.15%
    (E) 0.45%

Free-Response Problems

1. Two square slabs of thicknesses d_1 and d_2 with cross-sectional areas A_1 and A_2, have outward-facing temperatures of T_1 and T_2 ($T_2 > T_1$), respectively. The two slabs are in contact with each other (face to face) and have thermal conductivities of k_1 and k_2, respectively.

(a) Derive an expression for the temperature at the boundary where the two slabs are touching.

(b) If each slab has sides of length 0.25 m, $d_1 = 0.01$ m, $d_2 = 0.03$ m, $T_1 = 50°C$, $T_2 = 100°C$, $k_1 = 9 \times 10^{-6}$ W/(m-°C), and $k_2 = 15 \times 10^{-6}$ W/(m-°C), calculate the temperature at the boundary between the two slabs.

2. Concrete is poured into sections of a sidewalk at a temperature of 15°C. Each slab has a length of 5 m. If the concrete is to reach a temperature of 60°C, what minimum spacing must be allowed such that no buckling in the sidewalk occurs?

3. A Pyrex flask is calibrated at 25°C. It is filled to the 150-mL mark with liquid acetone at 40°C. Assume that the Pyrex flask is in equilibrium with the acetone initially.
   (a) What is the volume of the acetone when it is cooled to 25°C?
   (b) What is the percentage change in the volume of the flask itself compared to the change in the acetone?

SOLUTIONS TO PRACTICE PROBLEMS

Thought Problems

1. The high specific heat of water means that it is difficult to change its temperature. Thus, in the winter the water is often at a temperature higher than the air (except in extreme northern areas), which has the effect of keeping beach areas warmer. In the summer, the water is often cooler than the air, and so this has a cooling effect at beach areas.
2. Steam is hotter than boiling water (at the same temperature) because of the release of energy called the latent heat of vaporization.
3. When a substance is changing states, the addition of more energy does not change its temperature.

Multiple-Choice Problems

1. **C** Since the substance starts out as a gas (when $t = 0$ min), the temperature at which it changes from a gas to a liquid is the boiling point. According to the graph, this change occurs at 80°C.
2. **B** Since the substance becomes a liquid at 80°C, the next transition occurs at a temperature of 60°C, where it changes from a liquid to a solid.
3. **C** During the transition from gas to liquid ($t = 3$ min to $t = 5$ min), 12 kJ of heat is released. This corresponds to the heat of vaporization if we divide by the mass. Thus, 12 k J/2 kg = 6 kJ/kg.
4. **A** During the transition from liquid to solid ($t = 6$ min to $t = 7$ min), 6 kJ of heat is released. Thus, the heat of fusion is 6 kJ/2 kg = 3 kJ/kg.
5. **A** The change in temperature for both complete states is 60°C. The gas

begins at $t = 0$ and starts changing phase at $t = 3$ min. The solid begins at $t = 7$ min and ends at $t = 12$ min. Since the slope of the gas phase is greater, the corresponding specific heat is smaller because a shorter time is required to change the temperature of the gas than is required to change the temperature of the solid.

6. **B** The liquid slope begins at $t = 5$ min and ends at $t = 6$ min. Thus, 6 kJ of heat is released. The corresponding change in temperature is 20°C. The mass is 2 kg, and using our formula for specific heat, $c = 6$ kJ/(20°C)(2 kg) = 0.15 kJ/(kg-°C).

7. **D** The gas starts to condense at $t = 3$ min and completely solidifies at $t = 7$ min. Thus, 4 min have elapsed at 6 kJ/min which implies that 24 kJ was released during that time period.

8. **E** A *change* of 50°C is equal to a *change* of 50 kelvins. You do not have to convert 50°C to kelvins since what we are interested in is the *change*.

9. **A** Since heat is conserved, we can state that the heat gained equals the heat lost. The hot substance loses heat, and the water gains an equal amount. We therefore can write $(0.02)c(68) = (0.1)(4.19)(7)$. Solving for the specific heat yields $c = 2.16$ kJ/(kg-°C).

10. **B** The coefficient for the linear expansion of steel is $\alpha = 12 \times 10^{-6}$. Now, the percentage change in expansion is related to the fractional change in expansion multiplied by 100. Thus, using our formula, we need to multiply: $\alpha\,\Delta T = (12 \times 10^{-6})(180) = 0.00216$. Multiplying the answer by 100 gives us a percentage change of 0.216%.

Free-Response Problems

1. (a) At equilibrium, the rate of heat transfer between the two slabs must be the same. Thus, we can write

$$\frac{k_1 A(T - T_1)}{d_1} = \frac{k_2 A(T_2 - T)}{d_2}$$

where T is the equilibrium temperature at the boundary interface. Solving for T yields the final result for part (a):

$$T = \frac{k_1 d_2 T_1 + k_2 d_1 T_2}{k_1 d_2 + k_2 d_1}$$

which is independent of area.

(b) Substituting all the given numbers, we find a numerical value of $T =$ 67.86°C.

2. This is actually a simple problem once you recognize that the minimum expansion distance for the concrete slabs is just the change in length $\Delta L = L\alpha\,\Delta T = (5)(12 \times 10^{-6})(45) = 0.0027$ m.

3. (a) The new volume of the acetone can be determined from a modified version of the volume expansion formula:

$$V = V_o(1 + \beta\ \Delta T) = (150)[1 + (150 \times 10^{-6})(-15)] = 149.66 \text{ mL}$$

(b) We know that $\Delta V_{(acetone)} = (\beta\ \Delta T)_{(acetone)}$ and $\Delta V_{(flask)} = (\beta\ \Delta T_{(flask)}$. Now, for the same initial volume and change in temperature, we can state that

$$\frac{\Delta V_a}{\Delta V_f} = \frac{\beta_{acetone}}{\beta_{Pyrex}} = \frac{150 \times 10^{-6}}{3(3.2 \times 10^{-6})} = \frac{1}{6.4 \times 10^{-2}}$$

which implies a 6.4% comparison.

15
THERMODYNAMICS AND THE KINETIC THEORY OF GASES

GAS PRESSURE

Pressure is defined to be the measure of force per unit area applied to matter. In Chapter 13, we studied the behavior of fluids under stress. Gases represent compressible fluids, and when confined, they exert a force per unit area on the walls of their containers. This effect is called the **gas pressure** and is related to the collisions of gas molecules. In this chapter, we will investigate the nature of the effect of these collisions on the temperature, pressure, and volume of a confined gas.

MOLAR QUANTITIES

Under normal conditions, most gases behave according to a simple relationship involving their pressure, volume, and absolute temperature called the **ideal gas law**. This **equation of state** applies to what is referred as an **ideal gas**. The condition known as **STP** which stands for *standard temperature and pressure*, is usually applied to ideal gases and means that the gas is at a temperature of 0° C and 1 atm of pressure. Real gases are very complex, and a large number of variables are required to understand all the intricate movements taking place within a gas.

To help us understand ideal gases, we introduce a quantity called the **molar mass**. In thermodynamics (and chemistry) the molar mass or **gram molecular mass** is defined to be the mass of a substance that contains 6.02×10^{23} molecules. This number is called **Avogadro's number** N_a, and 1 mol of an ideal gas occupies a volume of 22.4 L at 0° and 1 atm of pressure or STP. For example, the gram molecular mass of carbon is 12 g, while the gram molecular mass of oxygen is 32 g. If m represents the actual mass, then the number of moles of gas is given by the relationship $n = m/M$. For example, 64 g of oxygen equals 2 mol, and 36 g of carbon equals 3 mol.

THE IDEAL GAS LAW EQUATION OF STATE

Imagine that we have an ideal gas in a sealed, insulated chamber such that no heat can escape or enter. In Figure 1 we see that an external force is placed

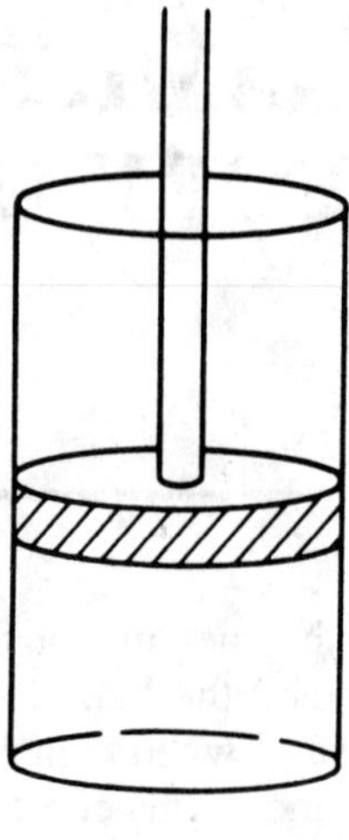

FIGURE 1.

on the piston at the top of the chamber so that the pressure change causes the piston to move downward. The chamber is a cylinder with cross-sectional area A and height h. If a constant force $\mathbf{F}$ is applied, work is done to move the piston (frictionless) a distance Δh. If the temperature of the gas remains constant, then the relationship between pressure and volume is known as **Boyle's law** and can be expressed as

$$\mathbf{P}V = \text{constant}$$

or

$$\mathbf{P}_1V_1 = \mathbf{P}_2V_2$$

A plot of Boyle's law would be a hyperbola, and this fact will be important when we consider the subject of heat engines later in this chapter.

Experiments demonstrate that if a gas is enclosed in a chamber of constant volume, then there is a direct relationship between the pressure in the chamber and the Kelvin temperature of the gas. This relationship states that the ratio of pressure to absolute temperature for an ideal gas always remains constant (indicating a diagonal straight-line plot at constant volume). We can write this relationship algebraically as

$$\frac{\mathbf{P}}{T} = \text{constant}$$

or

$$\frac{\mathbf{P}_1}{T_1} = \frac{\mathbf{P}_2}{T_2}$$

where the temperatures must be in kelvins.

The French scientist Jacques Charles (and independently Joseph Gay-Lussac) developed a relationship between the volume and absolute temperature of an ideal gas at constant pressure. This law is commonly known as **Charles' law** and states that the ratio of volume to absolute temperature for an ideal gas remains constant. That is,

$$\frac{V_1}{T_1} = \frac{V_2}{T_2}$$

These three gas laws can be summarized by one **ideal gas law** which has four forms.

The first form is a direct consequence of all three gas laws:

$$\frac{\mathbf{P}_1 V_1}{T_1} = \frac{\mathbf{P}_2 V_2}{T_2}$$

where the temperatures must be in kelvins. The second form states that the product of pressure and volume divided by the absolute temperature remains a constant:

$$\frac{\mathbf{P}V}{T} = \text{constant}$$

The final and more general form of the ideal gas law (and the one found in most advanced textbooks) involves the number of moles of gas present and a universal gas constant R:

$$\mathbf{P}V = nRT$$

where $R = 8.31$ J/mol K. Finally, if $n = N/N_a$, where N is the actual number of molecules and N_a is Avogadro's number, then a constant $k = R/N_a$ (called **Boltzmann's constant**) can be introduced such that the ideal gas law takes the form

$$\mathbf{P}V = NkT$$

Again, the temperatures must always be in kelvins. Additionally, if we express the volume in liters, the pressure in atmospheres, and the temperature in kelvins, then the number of moles n is given by

$$n = 12.2\,\frac{\mathbf{P}V}{T}$$

THE FIRST LAW OF THERMODYNAMICS

If work is being done on an ideal gas at constant temperature, then Boyle's law states that the work done is equal to the product of the pressure and the volume (this product remaining constant). A plot of Boyle's law is shown in Figure 2 and is a hyperbola. Pressure and volume changes that include

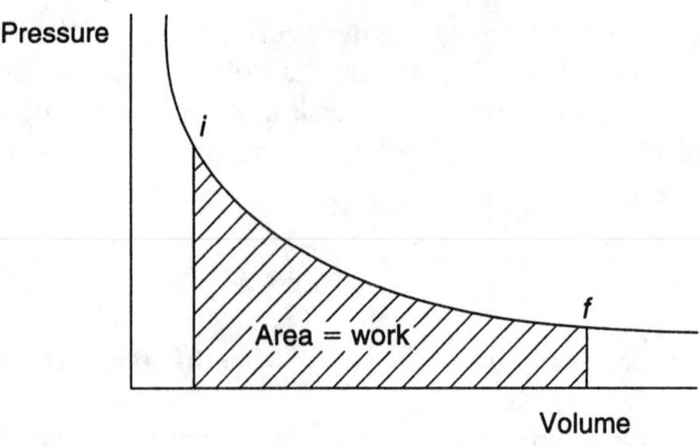

FIGURE 2.

temperature changes and are adiabatic obey a different kind of inverse relationship in which the volume is raised to a higher exponential power (making the hyperbola steeper). We will encounter these kinds of curves later in this chapter. The **P***V* diagram for Boyle's law shows that the area enclosed by the curve from some initial state to a final state is equal to the work done.

The **first law of thermodynamics** is basically the law of conservation of energy. Let ΔQ represent the change in heat energy supplied to a system, let ΔU represent the change in internal molecular energy, and let ΔW represent the amount of net work done by the system. The first law of thermodynamics states that

$$\Delta Q = \Delta U + \Delta W$$

In thermodynamics, there are certain terms that are often used and are worth defining:

1. *Isothermal*—meaning constant temperature
2. *Isobaric*—meaning constant pressure
3. *Isochoric*—meaning constant volume
4. *Adiabatic*—meaning no gain or loss of heat to the system ($\Delta Q = 0$).

Examples of each term are shown in Figure 3 in a **P***V* diagram.

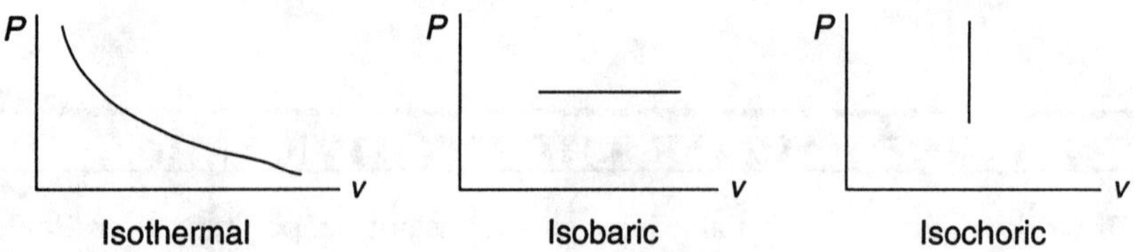

FIGURE 3.

THE SECOND LAW OF THERMODYNAMICS AND HEAT ENGINES

No system is ideal. When a machine is manufactured and operated, heat is always lost because of friction. This friction creates heat that can be accounted for by the law of conservation of energy. Recall that heat flows from a hotter body to a cooler body. Thus, if there is a reservoir of heat at high temperature, the system will draw energy from that reservoir (hence its name). Only a certain amount of energy is needed to do work, and so the remainder (if any), is released as exhaust at a lower temperature. This phenomenon was investigated by a French physicist named Sadi Carnot who was interested in the cyclical changes that take place in a *heat engine* that converts thermal energy into mechanical energy.

In the nineteenth century, the German physicist Rudolf Clausius coined the term *entropy* to describe the increase in randomness or disorder of a system. His studies can be summarized as follows:

1. Heat never flows from a cooler to a hotter body of its own accord.
2. Heat can never be taken from a reservoir and have nothing else happen in the system.
3. The entropy of the universe is always increasing.

These three statements embody the second law of thermodynamics, namely, that it is impossible to construct a cyclical machine that produces no other effect than to transfer heat continuously from one body to another at a higher temperature.

Connected with the second law of thermodynamics is the concept of a reversible process. Heat flows spontaneously from hot to cold, and the flow is irreversible (unless acted on by an outside agent). A system may be reversible if it passes from the initial to the final state by way of intermediary equilibrium states. This reversible process was studied extensively by Carnot and can be represented on a $\mathbf{P}V$ diagram.

Consider the simple Carnot cycle in Figure 4.

In order to change the system from $\mathbf{P}_1$ to $\mathbf{P}_2$ isochorically, heat must be supplied and the temperature rises. Because of the pressure change we allow the gas to expand isobarically from $(\mathbf{P}_2, V_1)$ to $(\mathbf{P}_2, V_2)$. The gas is doing work to expand, and so we supply energy to replace whatever is "lost." Now suppose we allow the system to cool by placing it in contact with a reservoir at a lower temperature. If this is to occur isochorically from $\mathbf{P}_2$ back to $\mathbf{P}_1$ at the same volume, then additional energy must be again added to the system from the outside. Finally, we allow the gas to contract from V_2 to V_1 isobarically. The temperature decreases according to Charles' law, and so additional energy must be supplied. In each isochoric change, $\Delta W = 0$ since there were no changes in volume. During isobaric changes, $\Delta W_2 = \mathbf{P}_2(V_2 - V_1)$ and $\Delta W_1 = \mathbf{P}_1(V_1 - V_2)$. The net work is equal to $(\mathbf{P}_2 - \mathbf{P}_1)(V_2 - V_1)$ and is equal to the area of the rectangle.

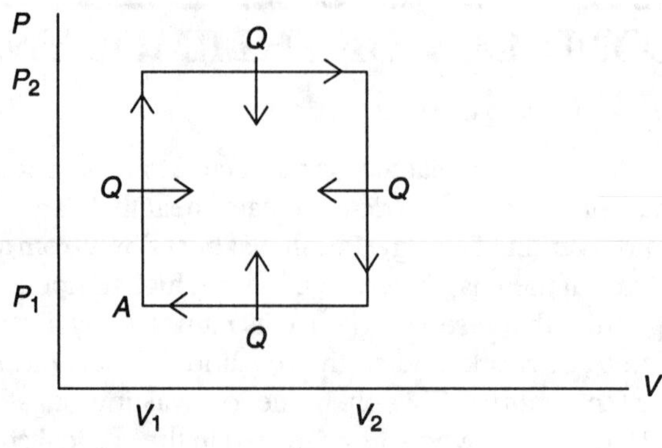

FIGURE 4.

A Carnot cycle in which adiabatic temperature changes occur is shown in Figure 5.

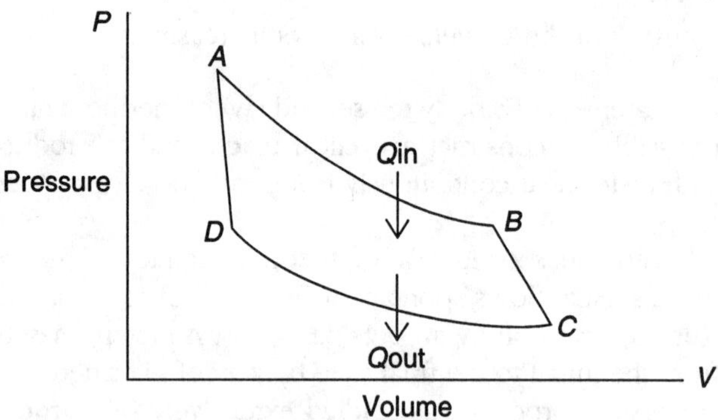

FIGURE 5.

The isothermal changes are from *A* to *B* and from *C* to *D*. The adiabatic changes are segments *BC* and *AD* (in which no heat is gained or lost). Therefore, heat must be added in segment *AB* and released in segment *CD*. In other words, there is an isothermal expansion in *AB* at a temperature T_1. Heat is absorbed from the reservoir, and work is done. *BC* is an adiabatic expansion in which an insulator has been placed around the container of the gas. The temperature falls, and work is done. In *CD*, we have an isothermal compression when the container is connected to a reservoir of temperature T_2 that is lower than T_1. Heat is released in this process, and work is done. Finally, in *DA*, the gas is compressed adiabatically and work is done. The total work done is equal to the area of the figure enclosed within the curves.

The efficiency of this heat engine is given by the fractional ratio of the total work done to the original amount of heat added:

$$\text{Efficiency} = \frac{W}{Q_1} = 1 - \frac{Q_2}{Q_1}$$

The reason for this is that the net work done in one cycle is equal to the net heat transferred ($Q_1 - Q_2$) since the change in internal energy is zero.

THE KINETIC THEORY OF GASES

There are six assumptions made by the kinetic theory of gases:

1. All gases are made up of very large numbers of molecules, and their separations are large compared with their sizes.
2. The molecules obey Newton's laws of motion, but individually they move randomly.
3. The molecules undergo elastic collisions with each other.
4. The forces between molecules are negligible except during a collision.
5. The gas involved at any time is a pure gas.
6. The gas is in thermal equilibrium with the walls of the container.

Accepting these assumptions, imagine that we have such a gas in a cubical container of length L (see Figure 6). Suppose there are N molecules moving about randomly inside the container. Under the assumption of large numbers, we can expect that approximately one third of them are moving along one of the principal coordinate axes x, y, or z at any given time.

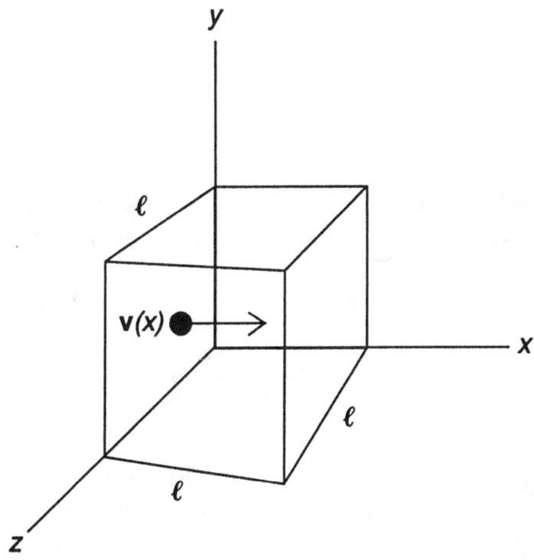

FIGURE 6.

Imagine one molecule of mass M moving from one side of the container to the other with a constant horizontal velocity $\mathbf{v}$. The molecule has a momentum $M\mathbf{v}$ in the x direction. If an elastic collision takes place with one of the walls, the magnitude of the change in momentum will be $2M\mathbf{v}$.

In a time t, the distance traveled by the molecule will be equal to $\mathbf{v}t$. Compared with the round trip distance $2L$, the total number of round trips is given by $\mathbf{v}t/2L$. Thus, in a time t, the total change in momentum for this molecule will be given by $M\mathbf{v}^2t/L$. The change in momentum is equal to the impulse applied to the molecule. This means that when we divide the change in momentum by the time, the result is equal to the average force applied. We can therefore state that $\mathbf{F} = M\mathbf{v}^2/L$. The total force on the wall is the sum of all such terms, and the total pressure is this force divided by the area of the wall. We have $N/3$ molecules in the x direction all moving with some average velocity $\overline{\mathbf{v}_x}$. When we divide the force by the area, we end up with the cube of the length of a side of the cube, which equals the volume V.

The final expression is the pressure exerted on the wall in the x direction. Now, before writing that expression, we know that for all three directions we have equally probable average velocities. Thus, the average of the squares of the velocities in each direction is just one third of the average of the square of the actual velocity. Therefore, the final expression for the pressure is

$$\mathbf{P} = \frac{N}{3}\frac{M}{V}\overline{\mathbf{v}^2}$$

where $\mathbf{v}$ is the actual velocity. Rearranging this expression, we see that the pressure is proportional to twice the average kinetic energy of the molecules (at constant volume):

$$\mathbf{P}V = \frac{2}{3}N(KE_{\text{avg}})$$

where $\text{KE}_{\text{avg}} = \frac{1}{2}M\overline{\mathbf{v}^2}$. This is the kinetic theory equivalent of the ideal gas law derived previously.

If we recall that $\mathbf{P}V = NkT$ (where the temperature T is in kelvins), then we can write that the absolute temperature is proportional to the average kinetic energy of the molecules:

$$T = \frac{2}{3k}\frac{1}{2M\overline{\mathbf{v}^2}}$$

The average kinetic energy can also be expressed as $\text{KE}_{\text{avg}} = \frac{3}{2}kT$, and the square root of $\overline{\mathbf{v}^2}$ is called the **root mean square velocity** such that

$$\mathbf{v}_{\text{rms}} = \sqrt{\frac{3kT}{M}}$$

If we let m represent the molecular mass, usually expressed in g/mol, then the root mean square velocity can be written as

$$\mathbf{v}_{\text{rms}} = \sqrt{\frac{3RT}{m}}$$

Finally, the change in internal energy ΔU is given by the expression

$$\Delta U = \frac{3}{2}\,nRT$$

where $R = 8.31$ J/mol.

HEAT CAPACITIES OF GASES

The molar heat capacity of a monotomic gas is based on the fact that the only contribution to the total internal energy of the gas is the translational kinetic energy of the molecules. If a quantity of heat ΔQ is supplied to the gas, either the internal energy will change or work will be done by the system. If we specify that no work is to be done, then $\Delta Q = \Delta U = \frac{3}{2}\,R\,\Delta T$ per mole. Thus, we define the molar heat capacity of the gas (at constant volume) to be

$$c_v = \frac{\Delta Q}{\Delta T} = \frac{3}{2}\,R$$

in cal/(mol-K). For all monatomic gases, the molar heat capacities are the same, while the specific heat capacities (heat capacity per gram) differ.

If we allow work to be done on the monatomic gas, then we can define the molar specific heat at constant pressure. First, if work is done at constant pressure, then $\Delta Q = \Delta U + \mathrm{P}\,\Delta v$. From the ideal gas law, if P is constant, then $\mathbf{P}\,\Delta v = R\,\Delta T$. Thus, $\Delta Q = \frac{3}{2}R\,\Delta T + R\,\Delta T = \frac{5}{2}R\,\Delta T$. Thus,

$$c_p = \frac{5}{2}\,R$$

and $\Delta Q = nc_p\,\Delta T$.

The ratio of heat capacities is defined to be

$$\gamma = \frac{c_p}{c_v}$$

For monatomic gases, $\gamma = 1.67$. Finally, we note that

$$c_p - c_v = R$$

Sample Problem

A 20-m^3 sample of an ideal gas is at 50°C and a pressure of 50 kPa. If the temperature is raised to 200°C and the volume is compressed by half, what will be the new pressure of the gas?

Solution

We must first remember to change all temperatures to kelvins. Therefore,

$$50°\text{C} = 323 \text{ K} \quad \text{and} \quad 200°\text{C} = 473 \text{ K}$$

We now use the ideal gas law for two sets of conditions:

$$\frac{\mathbf{P}_1 V_1}{T_1} = \frac{\mathbf{P}_2 V_2}{T_2}$$

Substitute the values given, and we have

$$\frac{(50)(20)}{323} = \frac{\mathbf{P}_2(10)}{473}$$

Solving for the pressure, we find that $\mathbf{P}_2 = 146.4$ kPa.

PROBLEM-SOLVING STRATEGIES

It is important to remember that all temperatures must be converted to kelvins before doing any calculations. Also, since the units for the gas constant *R* and Boltzmann's constant *k* are in SI units, all masses and molecular masses must involve kilograms. For example, the molecular mass of nitrogen is usually given as 28 g/mol. However, in an actual calculation, this value must be converted to 0.028 kg/mol.

You also should remember that unless a comparison relationship is used (like Boyle's law or as in the preceding sample problem), all pressures must be in pascals (not kilopascals, atmospheres, or centimeters of mercury) and all volumes must be in cubic meters. However, in a comparison relationship, the choice of units is irrelevant (except for temperature which is always in kelvins regardless), and you can use atmospheres for pressure or liters (1,000 cm^3) for volume. The molar specific heat capacities depend on the number of moles—not the number of grams (specific heat capacity).

PRACTICE PROBLEMS FOR CHAPTER 15

Thought Problems

1. What happens to the portion of heat put into an engine that does not do any work?
2. Explain the evaporation of water in terms of the kinetic molecular theory.
3. When you rub alcohol on your body, you feel cooler. Explain why.

Multiple-Choice Problems

1. At constant temperature, an ideal gas is at a pressure of 30 cm of mercury and a volume of 5 L. If the pressure is increased to 65 cm of mercury, the new volume will be ____L.
   (A) 10.8
   (B) 2.3
   (C) 0.43
   (D) 1.7
   (E) 5.6

2. At constant volume, an ideal gas is heated from 75°C to 150°Celsius. If the original pressure was 1.5 atm, then the new pressure will be ____.
   (A) doubled
   (B) halved
   (C) the same
   (D) less than doubled
   (E) quadrupled

3. At constant pressure, 6 m^3 of an ideal gas at 75°Celsius is cooled until its volume is halved. The new temperature of the gas will be ____°C.
   (A) 174
   (B) 447
   (C) −99
   (D) 37.5
   (E) 100

4. Water is used in an open-tube barometer. If the density of water is 1,000 kg/m^3, what will be the level of the column of water at sea level (assuming 1.0 atm of air pressure)?
   (A) 10.3 m
   (B) 12.5 m
   (C) 13.6 m
   (D) 11.2 m
   (E) 9.7 m

5. As the temperature of an ideal gas increases, the average kinetic energy of its molecules____.
   (A) increases and then decreases
   (B) decreases
   (C) remains the same
   (D) decreases and then increases
   (E) increases

6. The product of pressure and volume is in units of ____.
   (A) pascals
   (B) k/N
   (C) watts

(D) joules
(E) newtons

7. Which of the following is equivalent to 1 Pa of gas pressure?
(A) 1 kg/s^2
(B) 1 kg-m/s
(C) 1 $kg\text{-}m^2/s^2$
(D) 1 $kg/m\text{-}s^2$
(E) 1 kg/m

8. What is the efficiency of a heat engine that performs 700 J of useful work from each 2,700 J of heat absorbed from its hot reservoir?
(A) 74%
(B) 26%
(C) 35%
(D) 65%
(E) 50%

9. Hydrogen has a molecular mass of 2g/mol, while oxygen has a molecular mass of 32 g/mol. At the same temperature, compared to the molecular velocity of hydrogen, the molecular velocity of oxygen is ____.
(A) one fourth
(B) one sixteenth
(C) 4 times greater
(D) 16 times greater
(E) the same

10. Given that the molecular mass of hydrogen is 2 g/mol, what is its molecular velocity at a temperature of 27°C?
(A) 1,933 m/s
(B) 61.2 m/s
(C) 3,745 m/s
(D) 580 m/s
(E) 360 m/s

Free-Response Problems

1. (a) How many molecules of helium are required to fill up a balloon with a diameter of 50 cm at a temperature of 27°C?
(b) What is the average kinetic energy of each molecule of helium?
(c) What is the average velocity of each molecule of helium?

2. An engine absorbs 2,000 J of heat from a hot reservoir and expels 750 J to a cold reservoir during each operating cycle.
(a) What is the efficiency of the engine?
(b) How much work is done during each cycle?
(c) What is the power output of the engine if each cycle lasts for 0.5 s?

3. A monotomic ideal gas undergoes a reversible process as shown in the accompanying **P***V* diagram.

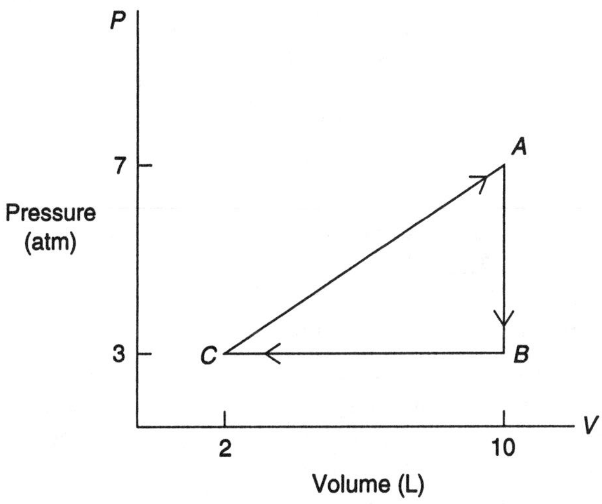

(a) If the temperature of the gas is 500 K at point *A*, how many moles of gas are present?

(b) For each part of the cycle (*AB*, *BC*, and *CA*), determine the values of ΔW, ΔU, and ΔQ.

SOLUTIONS TO PRACTICE PROBLEMS

Thought Problems

1. Heat that is not used to do work is released as output exhaust.
2. The kinetic molecular theory states that the molecules of a liquid are in a constant state of motion. Near the surface of the liquid, adhesive forces create surface tension forces. However, with the addition of heat, molecular velocities increase to the point where molecules can escape from the surface of the liquid. This occurs dramatically at the boiling point. During evaporation, molecules occasionally escape over time based on the absorption of heat from the liquid's container or the surface it is adhering to (like water on your skin; see Problem 3).
3. Alcohol absorbs a large amount of heat from your skin and evaporates quickly. This has a pronounced cooling effect, and it is easily felt within a matter of seconds.

Multiple-Choice Problems

1. **B** At constant temperature, an ideal gas obey's Boyle's law. Since we are making a two-position comparison, we can use the units given for

pressure and volume. Substituting these values into Boyle's law, we obtain (30)(5) = (65) V_2, which implies that V_2 = 2.3 L.

2. **D** At constant volume, the changes in pressure are directly proportional to the changes in Kelvin temperature. Therefore, even though the Celsius temperature is being doubled, the Kelvin temperature is not. Thus, the new pressure must be less than doubled. Alternately, you can calculate the new pressure since 75°C = 348 K and 150 C = 423 K. Thus, 1.5/348 = $\mathbf{P}_2$/423 and $\mathbf{P}_2$ = 1.8 atm. Again, since a comparison relationship is involved, we can keep the pressure in atmospheres for convenience.

3. **C** This is a problem involving Charles's law, and the temperatures must be in kelvins. Thus, 75°C = 348 K, and we can write 6/348 = $3/T_2$. However, recall that the volume change is directly proportional to the Kelvin temperature change. Thus, T_2 = 174 K = − 99°C.

4. **A** The formula for pressure is $\mathbf{P} = \rho \mathbf{g} h$. The density of water is 1,000 kg/m^3 and $\mathbf{g}$ = 9.8 m/s^2. We also know that at sea level $\mathbf{P}$ = 101 kPa, which must be converted back to pascals or N/m^2. Thus, $h = \mathbf{P}/\rho g$, and making the necessary substitutions yields h = 10.3 m (a very large barometer over 35 ft tall!).

5. **E** The absolute temperature of an ideal gas is directly proportional to the average kinetic energy of its molecules. Hence, the average kinetic energy increases with temperature.

6. **D** The units of $\mathbf{P}V$ are joules, which are the units of energy.

7. **D** One pascal is equivalent to 1 N/m^2. But, one newton is equal to 1 kg-m/s^2. Thus, the final unit is 1 kg/ms^2.

8. **B** The molecular velocity varies inversely with the square root of the molecular mass. Thus, heavier molecules travel more slowly than lighter molecules at the same temperature.

9. **C** The ratio of molecular masses is $\frac{32}{2}$ = 16. However, the velocities vary inversely as the square root of the molecular mass. Thus, the velocity of hydrogen is four times the velocity of oxygen at the same temperature.

10. **A** The formula for molecular velocity (root mean square velocity) given the molecular mass is $\mathbf{v}_{rms} = \sqrt{3RT/m}$. The temperature T must be in kelvins, and therefore 27°C = 300 K. Also, even though the molecular mass of hydrogen is 2 g/mol, the equation demands that the units of mass be kilograms. Thus, m = 0.002 kg/mol and R = 8.31 J/mol. Substituting the given values, we have $\mathbf{v}_{(rms)}$ = 1,933 m/s.

Free-Response Problems

1. (a) From the ideal gas formula, we know that $\mathbf{P}V = NkT$, where N is the number of molecules and k is Boltzmann's constant. For a 50-cm-diameter balloon, the radius is 0.25 m, and 1 atm is equal to 101,000 Pa

of pressure. The volume of a sphere is given by $V = \frac{4}{3}\pi R^3$, which equals 0.065 m^3 on substitution. Now, we know that $N = PV/kT$, where T is equal to 300 K in this problem. Substituting all known values,

$$N = \frac{(101{,}000)(0.065)}{(1.38 \times 10^{-23})(300)} = 1.59 \times 10^{24} \text{ molecules}$$

(b) The average kinetic energy is equal to $\overline{KE} = \frac{3}{2}kT = (1.5)(1.38 \times 10^{-23})(300) = 6.21 \times 10^{-21}$ J.

(c) The molecular velocity is given by $\mathbf{v}_{rms} = \sqrt{3RT/m} = \sqrt{(3)(8.31)(300)/(0.004)} = 1367$ m/s.

2. (a) The efficiency of a heat engine is given by the formula

$$e = \frac{\Delta Q}{Q_{hot}} = \frac{2{,}000 - 750}{2{,}000} = 0.625 \text{ or } 62.5\%$$

(b) The work done during each cycle is equal to $\Delta Q = 2{,}000 - 750 = 1{,}250$ J.

(c) The power output equals the work done during each cycle divided by the time required for the cycle: $P = 1{,}250/0.5 = 2{,}500$ W.

3. (a) Since we know the pressure, volume, and absolute temperature of the ideal (monotomic) gas at point A, we can use the ideal gas law, $\mathbf{P}V = nRT$, to find the number of moles n. First, we must use standard units: 7 atm = 707,000 Pa, and since 1 L = 0.001 m^3, 10 L = 0.01 m^3. Thus, $n = (707{,}000)(0.01)/(8.31)(500) = 1.7$ mol

(b) (i) The cycle AB is isochoric, and so $\Delta W = 0$. Thus, $\Delta U = \Delta Q$. Now, as the pressure decreases from 7 atm to 3 atm, the absolute temperature decreases proportionally. Thus, we can write $7/500 = 3/T_B$, which implies that the absolute temperature at point B is 214.29 K. To find the change in the internal energy, recall that $\Delta U = \frac{3}{2}nR\,\Delta T = (1.5)(1.7)(8.31)(-285.7) = -6{,}054$ J $= \Delta Q$ also!

(ii) The cycle BC is isobaric, and the decrease in volume from 10 L to 2 L is accompanied by a corresponding decrease in absolute temperature. We know from step (i) that the temperature at B is 214.29 K, and so we can use Charles's law to find the temperature at point C. We can therefore write $10/214.29 = 2/T_C$, which implies that the absolute temperature at point C is 42.86 K. This means $\Delta T = -171.43$ K. Also, note that the work done is negative since we are expelling energy to reduce pressure and volume.

Now, in this process, work is done to compress the gas and $\Delta W = \mathbf{P}(V_C - V_B)$. The pressure is constant at $\mathbf{P} = 3$ atm = 303,000 Pa, and $\Delta V = -8$ L $= -0.008$ m^3. Thus, $\Delta W = (303{,}000)(-0.008) = -2424$ J. To find ΔU, note that the temperature change for the 1.7 mol was -171.43 K, and so $\Delta U = (1.5)(1.7)(8.31)(-171.43) = -3{,}632.69$ J. To find ΔQ, we use the first law of thermodynamics: $\Delta Q = \Delta U + \Delta W = -3{,}632.69 - 2{,}424 = -6{,}057$ J.

(iii) The last part of the reversible cycle puts back all the work and energy that was removed in the first two steps. Thus, in the first two steps a total of 12,111 J of heat was removed [the sum of ΔQ in steps (i) and (ii)]. So, for *CA*, $\Delta Q = +\ 12{,}111$ J. In a similar way, the total change in internal energy equaled $-\ 9{,}686.69$ J. Thus, in cycle *CA*, $\Delta U = +\ 9{,}686.69$ J. Finally, $\Delta W = \Delta Q - \Delta U = 12{,}111 - 9{,}686.69 = 2{,}424.31$ J. Notice that since no work was done in step (i), this is equal to the work done in step (ii).

16
ELECTROSTATICS

THE NATURE OF ELECTRIC CHARGES

The ancient Greeks used to rub pieces of amber on wool or fur. The amber was then able to pick up small objects that were not made of metal. The Greek word for amber is ελεκτρον (*elektron*), from which the word *electric* is derived. These pieces of amber retained their attractive properties for some time, and so the effect appeared to be **static**. This was different from the behavior of magnetic ores (lodestones), which were naturally occurring rocks that attracted only metallic objects).

In modern times, hard rubber (such as ebonite) is used with cloth or fur to dramatically demonstrate the properties of electrostatic force. If you rub an ebonite rod with cloth and then bring it near a small cork sphere painted silver (called a *pith ball*) and suspended on a thin thread, the pith ball will be attracted to the rod. When they touch each other, the pith ball is repelled. If a glass rod that has been rubbed with silk is then brought near the pith ball, it is attracted to it. When you touch the pith ball with your hand, it returns to its normal state.

To explain these effects, chemical experiments in the nineteenth century demonstrated the presence of molecules called **ions** which possessed similar *affinities* for certain objects (like carbon or metals) placed in a solution with them. These objects are called **electrodes**, and these experiments confirmed the existence of two types of ions called **positive** and **negative**. This is similar to the two types of effects produced when ebonite and glass are rubbed. Even though both attract small objects, these objects become **charged** oppositely when rubbed, as indicated by the behavior of the pith ball. Further, chemical experiments coupled with an atomic theory demonstrated that in solids it was the negative charges that were transferred. Additional experiments by Michael Faraday in England during the first half of the nineteenth century suggested the existence of a single, fundamental carrier of **electric charge** which was later called the **electron**. The corresponding carrier of positive charge was later called the "**proton**."

The essentially neutral condition of all matter is due to the balance of charges in atoms containing equal numbers of protons and electrons. Ions consist of atoms or molecules that contain more or fewer electrons. When an acid is placed in a solution, it dissociates into positive and negative ions. When ebonite is rubbed with cloth, electrons are transferred to the ebonite, giving it a net negative charge and leaving the cloth positive.

THE DETECTION AND MEASUREMENT OF ELECTRIC CHARGES

When an ebonite rod is rubbed with cloth, only the part of the rod in contact with the cloth becomes charged, and the charge remains localized for some time (hence the term *static*). For this reason (among others), rubber is called an **insulator** (along with plastic and glass). A metal rod held in your hand cannot be charged statically for two reasons. First, metals are **conductors**, meaning that they allow electric charges to flow through them. Second, your body is a conductor, and any charges placed on the metal rod are conducted out through you (and into the earth). This effect is called **grounding**. The silver-coated pith balls can become statically charged since they are suspended by a thread (which is an insulator). They can be used to detect the presence and sign of electric charges, but they are not very practical for obtaining a qualitative measurement of the magnitude of charge they possess.

An instrument that is often used for qualitative measurement is the **electroscope**. One form of an electroscope consists of two "leaves" made of gold foil (Figure 1a). The leaves are vertical when the electroscope is uncharged, but as a negatively charged rod is brought near, the leaves diverge. Recalling the hypothesis that only negative charges move in solids, we see that the electrons on the knob of the electroscope are repelled downward toward the leaves through the conducting stem. The knob is positively charged (which can be verified with a charged pith ball) as long as the rod is near but not touching (Figure 1b). On contact, electrons are directly transferred to the knob, stem, and leaves, and the whole electroscope then becomes negatively charged (Figure 1c). The extent to which the leaves are spread apart is a relative indication of how much charge is present (but only qualitatively). If you touch the electroscope, you will ground it and the leaves will collapse together.

The electroscope can also be charged by **induction**. If you touch the electroscope shown in Figure 1b with your finger (see Figure 2a), the repelled

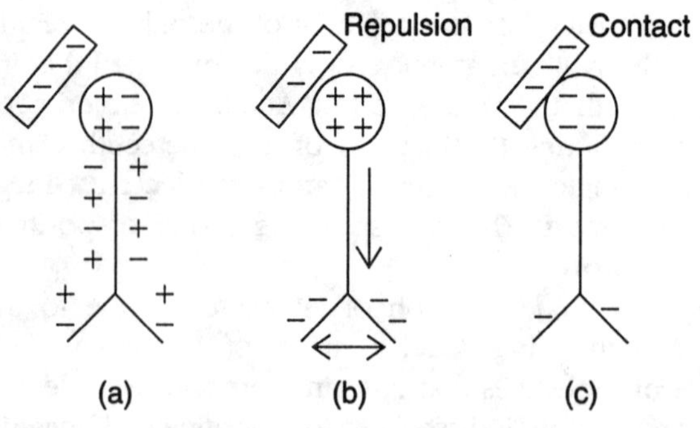

FIGURE 1.

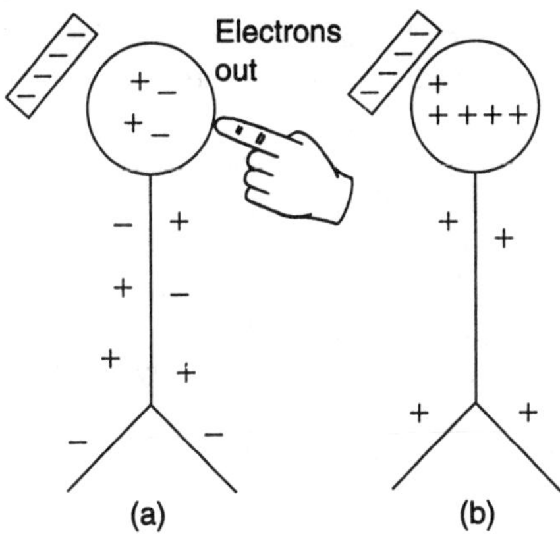

FIGURE 2.

electrons will be forced out into your body. If you remove your finger, keeping the rod near, the electroscope will be left with an overall positive charge by induction (Figure 2b).

Finally, we can state that electric charges in any distribution obey a law of conservation. When we transfer charges, we always maintain a balanced accounting. Suppose we have two charged metal spheres. Sphere A has +5 elementary charges, and sphere B has a +1 elementary charge (thus both are positively charged). When two spheres are brought into contact, which way will the charges flow? Since negative charges always flow from a higher concentration to a lower one, sphere B will have an overall greater negative tendency (since it is less positive). On contact, it will give up negative charges until equilibrium is achieved. Notice that, out of the countless number of total charges in the sphere, it is only the excess charges that flow out. A little arithmetic shows that sphere B will give up two negative charges, leaving it with an overall charge of +3. Sphere A will gain the two negative charges, reducing its charge to +3. If the two spheres are separated, they will each have a charge of +3 elementary charges.

COULOMB'S LAW

From the previous two sections, we can conclude that like charges repel each other while unlike charges attract. The electrostatic force between two charged objects can act through space and even in a vacuum. This ability makes it similar to the force of gravity. However, there are two key differences (other than the fact that charge and mass are two different physical quantities). First, gravity always attracts. Electrostatic force can either repel or attract depending on the signs of the charges. Second, charge is relativistically invariant. This means that, unlike mass, which can actually vary as an object's velocity approaches the speed of light, an electric charge remains observationally the same in any frame of reference regardless of what the frame is doing (accelerating or moving at a constant velocity).

In the SI system of units, charge is measured in **coulombs**, named for the French scientist Charles Coulomb, who studied the nature of electrostatic force in the late eighteenth century. He discovered that two point charges (designated q_1 and q_2) separated by a distance r experience a mutual force along a line connecting the charges that varies directly as the product of the charges and inversely as the square of the distance between them. This is known as **Coulomb's law** and, as in the case of gravity, it involves an inverse square law acting on matter at a distance. Mathematically, Coulomb's law can be written as

$$\mathbf{F} = \frac{kq_1q_2}{r^2}$$

The constant k has a value $k = 8.9875 \times 10^9 \approx 9 \times 10^9$ N-m^2/C^2. An alternative form of Coulomb's law is written as

$$\mathbf{F} = \frac{1}{4\pi\epsilon_0}\frac{q_1q_2}{r^2}$$

The new constant, ϵ_0, is called the **permittivity of free space** or the **permittivity of the vacuum** and has a value of $\epsilon_0 = 8.8542 \times 10^{-12}$ C^2/N-m^2. In the SI system, one elementary charge is designated as e and has a magnitude of 1.6×10^{-19} C.

Coulomb's law is a vector equation just like Newton's law of universal gravitation. The direction of the force is along a radial vector connecting the two point sources. Note that Coulomb's law applies only to point sources (or sources that can be treated as point sources such as charged spheres). If we have a distribution of point charges, then the net force on one charge will be the vector sum of all the other electrostatic forces. This aspect of force addition is sometimes known as **superposition**.

THE ELECTRIC FIELD

Another way to consider the force between two point charges is to recall that the force can act through free space. If a sphere has a charge of $+Q$ and a small test charge $+q$ is then brought near it, the test charge will be repelled according to Coulomb's law. The test charge will be repelled everywhere along a radial vector out from the charge $+Q$. We can state that even if the charge $+Q$ is too small to be visible, the influence of the electrostatic force can be observed and measured (since charges have mass and $\mathbf{F} = m\mathbf{a}$ as usual).

In this way, the charge $+Q$ is said to set up an electric field which pervades the space surrounding the charge and produces a force on any other charge that enters the field (just as a gravitational field does). The strength of the electric field **E** is defined to be the measure of the force per unit charge experienced at a particular location. In other words, $\mathbf{E} = \mathbf{F} / q$, and the units are N/C. The electric field strength **E** is a vector quantity since it is derived from a force (a vector) and a charge (a scalar).

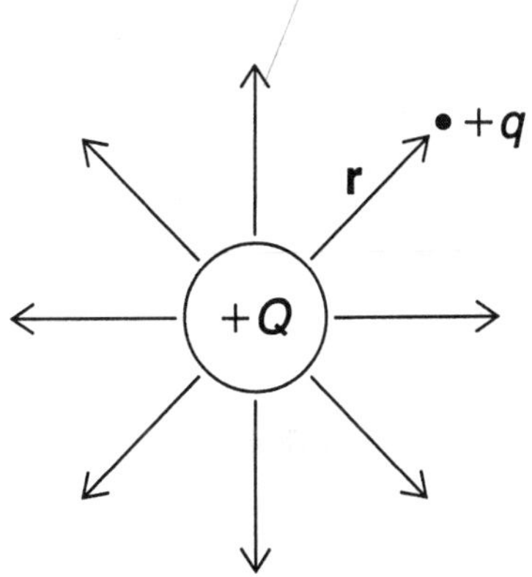

FIGURE 3.

An illustration of this behavior is shown in Figure 3. The charge $+Q$ is represented by a charged sphere drawn as a circle. A small positive test charge, $+q$, is brought near and is repelled along a radial vector drawn outward from the charge $+Q$. In fact, anywhere in the vicinity of the charge $+Q$, the test charge will be repelled along an outward radial vector, and we therefore draw these **force field lines** as radial vectors coming out from the charge $+Q$. By convention, we always consider the test charge to be a positive one. Since the force varies according to Coulomb's law, the electric field strength **E** also varies depending on the location of the test charge (assuming it and the main charge remain constant in magnitude). Thus, we can write

$$\mathbf{E} = \frac{\mathbf{F}}{q} = \frac{kQ}{r^2} = \frac{1}{4\pi\epsilon_0}\frac{Q}{r^2}$$

We can also interpret field strength by observing the density of field lines per square meter. With point charges, the radial nature of their construction causes them to converge near the surface of the charge, and this indicates a relative increase in field strength. In a later section, we shall encounter a configuration in which the field strength remains constant throughout.

There are several other configurations of electric field lines that we can consider. In Figure 4, the field between two point charges with different sign arrangements is illustrated. Notice that the field lines are always perpendicular to the "surface" of the source and that they never cross each other. The arrows on the field lines always indicate the direction in which a positive test charge will move.

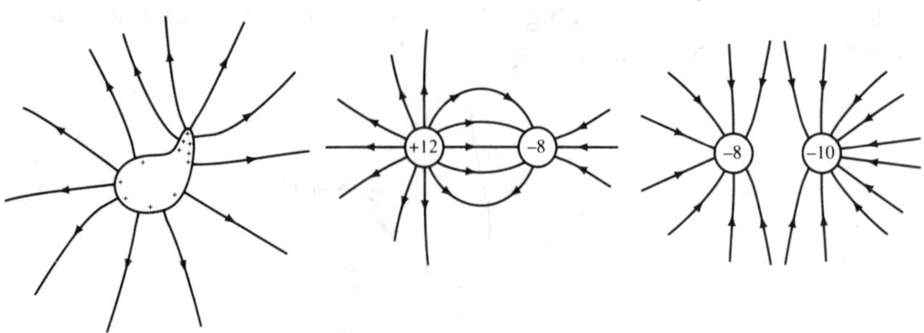

FIGURE 4.

ELECTRIC POTENTIAL

We have reviewed the fact that gravity is a conservative force. Any work done by or against gravity is independent of the path taken. Work done was measured as a change in the gravitational potential energy. Recall that work is equal to the magnitude of force times displacement: $W = \mathbf{F}\,\Delta\mathbf{x}$. Suppose that the force is acting on a charge q in an electric field $\mathbf{E}$. The work done by or against the electric field, which is also a conservative field, is given by $W = \mathbf{E}q\,\Delta r$ (in one dimension, radially). Since electrostatic force is conservative, the work done is equal to the change in electric potential energy between two points A and B.

The two points A and B cannot be just anywhere in the field. Consider the original electric field diagram in Figure 3. If the test charge $+q$ is a distance r from the source charge $+Q$, then it will experience a certain radial force whose magnitude is given by Coulomb's law. At any position around the source at a distance r, we can observe that the test charge will experience the same force. The localized potential energy will be the same as well. The set of all such positions defines a circle of radius r (actually a sphere in three dimensions) which is called an **equipotential surface**.

In Figure 5, we observe a series of equipotential curves of varying radii. Since the electric field is conservative, work is done by or against it only when a charge is moved from one equipotential surface to another. To better understand this effect, we define a quantity called the **electric potential** which is a measure of the magnitude of electric potential energy per unit charge at a particular location in the field. Thus, the work done per unit charge in moving from equipotential surface A to equipotential surface B is a measure of what is called the **potential difference** or **voltage**. Thus, we define V to be the potential difference ($V_A - V_B$) and to be equal to W/q, and so the units

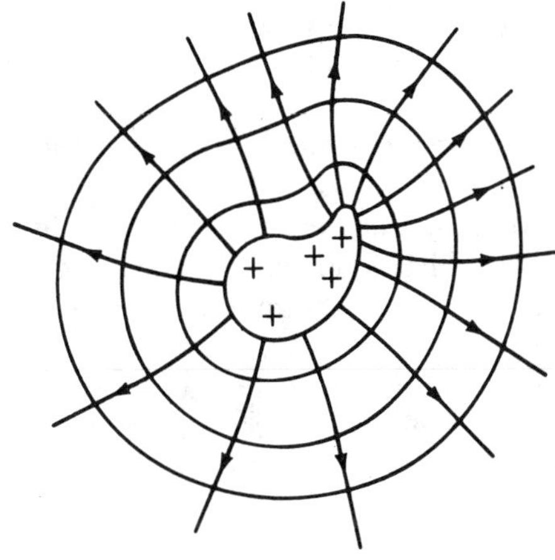

FIGURE 5.

of potential difference are J/C or **volts**. Since $W = q\mathbf{E}\,\Delta r$, calculating the work can be difficult if the electric field can vary (as in the case of a point source). However, if we can find a configuration in which the electric field is uniform, **E** becomes a constant. Since potential difference is the work done per charge, we see that $V = \mathbf{E}\,\Delta r$. The units of electric field strength can now be defined in terms of the potential difference between two points. Usually, one of these points is infinity, and the potential is defined to be zero there. The localized electric potential is then given by $V = kQ/r$, using Coulomb's law. In a uniform electric field, we can now define the electric field strength, $\mathbf{E} = V/d$, and the units are V/m.

The desired configuration is one in which we have two charged parallel plates separated by a distance d. When each side is oppositely charged (see Figure 6), the electric field (and hence the electric force on a test charge within the field) is uniform. If a charge q and mass m enter the electric field with a velocity $\mathbf{v}$, it will be subject to an electric force $\mathbf{F}$ such that $\mathbf{F} = m\mathbf{a}$. In this

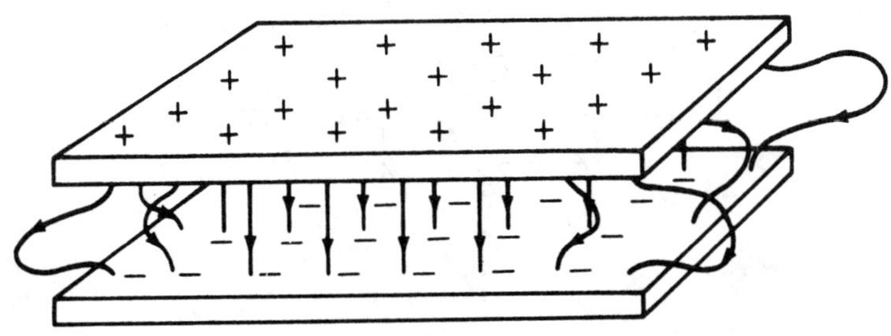

FIGURE 6.

case, $\mathbf{F} = \mathbf{E}q$, and the kinematics of charged particles in uniform electric fields can be developed analogously to the kinematics of particles in a uniform gravitational field. The applications of this kinematics extend into the field of electronics and atomic theory.

There are some sign conventions that have been developed for dealing with electric potential. By agreement, we state that a positive test charge in a uniform electric field loses potential energy as it moves in the direction of the field. In other words, if the change in electric potential energy is given by ΔU, then $\Delta U = -q\mathbf{E}d$. Consequently, because of conservation of energy, the charge gains kinetic energy as it moves in the direction of the field.

Another convenient unit of energy is the **electronvolt** (eV). By definition, if one elementary charge experiences a potential difference of 1 V, its energy (or the work done to transfer the charge) is equal to 1 eV. Consequently, we can state that 1 eV = 1.6×10^{-19} J.

Charges flow when a difference in potential exists (just as water flows when there is a difference in pressure). To maintain a constant flow of charge (current), a continuous potential difference must be maintained. Chemically, this is achieved through the use of a **battery** and attached conducting wires which can create a potential difference between two points. In the case of two parallel plates (see Figure 7), the potential difference sets up an electric field between the plates. If a conductor is placed between the plates, charge can flow. If no conductor is placed between the plates, charge is simply stored on the plates. If an insulating material such as wax or paper (called a **dielectric**) is placed between the plates, static charges can be stored to a much higher concentration. This situation defines a simple **parallel plate capacitor**.

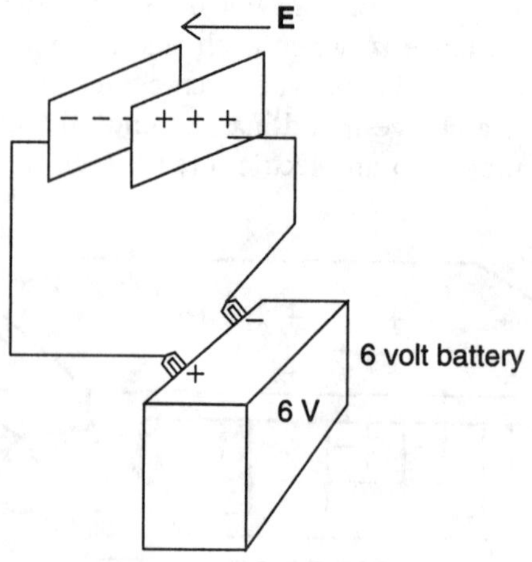

FIGURE 7.

CAPACITANCE

When we have two charged parallel metal plates, there is a proportional relationship between the potential difference and the total charge. This configuration is called a *capacitor.* The capacitance C of a capacitor is defined to be the ratio of the magnitude of the charge on either conductor to the potential difference between them:

$$C = \frac{Q}{V}$$

This is a positive scalar quantity that is essentially a measure of the capacitor's ability to store charge. The reason for this is that for a given capacitor, the ratio Q/V is a constant since potential difference increases as the charge on either capacitor increases.

The SI unit of capacitance is the **farad**, which is equal to 1 C/V. The different types of capacitors used in electronic devices have capacitances that range from microfarads to picofarads (see Appendix B for a definition of these prefixes if you are not familiar with them).

In this section we deal only with simple parallel plate capacitors. However, in general, if we have a spherical distribution of charge, the electric field at some point greater than r_0 (the radius of the charge distribution) is given by

$$\mathbf{E} = \frac{1}{4\pi\epsilon_0}\frac{Q}{r^2} = \frac{Q}{\epsilon_0}\frac{1}{4\pi r^2} = \frac{Q}{\epsilon_0 A}$$

Now, with two parallel plates, the electric field is uniform and is given by $\mathbf{E} = Vd$, and A is the area of the rectangular plates. The potential difference is then given by

$$V = \mathbf{E}d = \frac{Qd}{\epsilon_0 A}$$

Finally, given the definition of capacitance, we can write

$$C = \frac{Q}{V} = \frac{Q}{Qd/\epsilon_0 A} = \frac{\epsilon_0 A}{d}$$

The implication of the last formula is that the capacitance of the parallel plate capacitor is directly proportional to the area of the plates and inversely proportional to their separation.

The energy stored in a capacitor can be studied by recognizing that the definition of capacitance can be written as $Q = CV$. This direct relationship between the plates' charge and voltage would produce a diagonal straight line if you plotted charge versus voltage (see Figure 8). Notice that the units of QV are joules (the units of work). The area under this line is equal to the amount of energy stored in the capacitor, which is given by $E = \frac{1}{2}CV^2$. The behavior of capacitors in electric circuits will be reviewed in the next chapter.

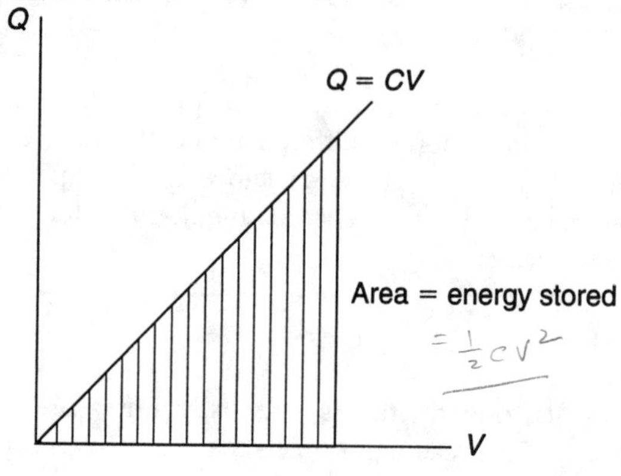

FIGURE 8.

Sample Problem

A charge of 5 μC is located at the origin of a Cartesian coordinate system. A second charge of −2 μC is placed horizontally 0.35 m to the right of the first charge. What is the magnitude of the electric field strength at the location (0,0.7 m)?

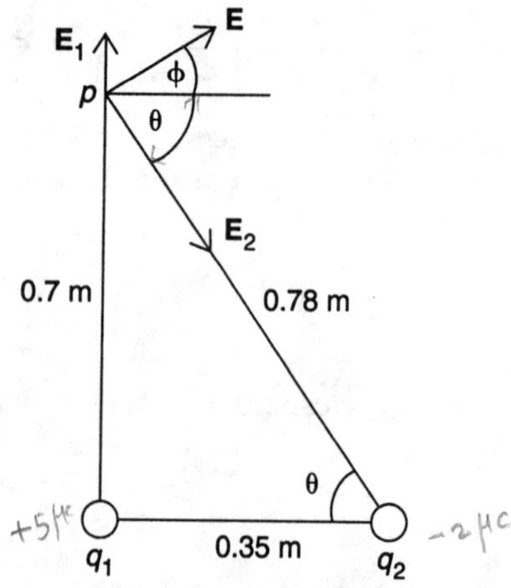

Solution

First draw a sketch of the situation as shown. The electric field is a vector at a particular point p, and the direction of this vector is the direction a positive test charge would take. Since q_1 is positive, the direction is upward away from q_1. However, charge q_2 is negative, and so the electric field vector is directed along a line from point p to charge q_2. Using the given information and the Pythagorean theorem, we find that the distance is 0.78 m. The law of superposition states that the net electric field at point p will be equal to the vector sum of the two fields $\mathbf{E}_1$ and $\mathbf{E}_2$ (as shown). Using the tangent function, vector $\mathbf{E}_2$ makes an angle of $\theta = -63.4$ degrees.

The magnitude of $\mathbf{E}_1$ is given by

$$\mathbf{E}_1 = \frac{k|q_1|}{r_1^2} = \frac{(9 \times 10^9)(5 \times 10^{-6})}{0.49} = 9.18 \times 10^4 \text{ N/C}$$

The magnitude of $\mathbf{E}_2$ is given by

$$\mathbf{E}_2 = \frac{k|q_2|}{r_2^2} = \frac{(9 \times 10^9)(2 \times 10^{-6})}{0.6125} = 2.9 \times 10^4 \text{ N/C}$$

Now, the resultant field $\mathbf{E}$ is the vector sum of these two. Recall that if we resolve vectors into components, we can simply add the components and then recombine them using the Pythagorean theorem. $\mathbf{E}_1$ is already in the positive y direction, and so it has no other components. The direction of $\mathbf{E}_2$ is -63.4 degrees, and it has components given by $\mathbf{E}_2 \cos\theta$ and $\mathbf{E}_2 \sin\theta$. Recall that the cosine function is positive in the fourth quadrant, and so

$$\mathbf{E}_{2(x)} = \mathbf{E}_2 \cos\theta = 2.9 \times 10^4 \cos(-63.4) = 12{,}985 \text{ N/C}$$

and

$$\mathbf{E}_{2(y)} = \mathbf{E}_2 \sin\theta = 2.9 \times 10^4 \sin(-63.4) = 25{,}930.47 \text{ N/C}$$

The magnitude of the components of the net electric field are found by simply adding like components: $\mathbf{E}_y = 65{,}870$ N/C and $\mathbf{E}_x = 12{,}985$ N/C. Using the Pythagorean theorem gives us the magnitude of the net electric field strength at point P, which is $\mathbf{E} = 67{,}137$ N/C.

PROBLEM-SOLVING STRATEGIES

The preceding sample problem reminds you that we are dealing with vector quantities when the electrostatic force and field are involved. Drawing a sketch of the situation and using techniques from vector construction and algebra were most effective in solving the problem. Also, be sure to keep track of units. If a problem involves potential or capacitance, remember to maintain the standard SI system of units.

Remember that electric field lines are drawn as though they follow a positive test charge and that Coulomb's law applies only to point charges. The law of superposition allows you to combine the effects of many forces (or fields) using vector addition. Additionally, symmetry may allow you work with a reduced situation that can be simply doubled or quadrupled. In this case, a careful look at the geometry of the situation is called for.

For charges in motion, if the electric field is uniform, the charges will experience a uniform acceleration that will allow you to use Newtonian kinematics to analyze the motion. Electrons in oscilloscopes are an example of this kind of application.

Potential and potential difference (voltage) are scalar quantities. In a practical sense, it is the change in potential that is electrically significant and not the potential itself.

PRACTICE PROBLEMS FOR CHAPTER 16

Thought Problems

1. A positively charged rod attracts a suspended object. What conclusions can we make about the electrostatic charge on the suspended object?
2. Why is it safe to be inside a car during a lightning storm?
3. (a) Are charges required to be at a location where electric potential energy is detected?
   (b) Are charges required to be at a location where an electric potential is detected?
   (c) Under what conditions can a location have a low electric potential energy but a relative large electric potential?

Multiple-Choice Problems

1. An isolated metal sphere, A, is charged to a value of $+Q$ elementary charges. It is then touched to and separated from an identical but neutral metal sphere, B. The second sphere is then touched to a third identical, neutral metal sphere, C. Finally, spheres A and C are touched together and then separated. Which of the following represents the distribution of charge on each sphere, respectively, after the above process has been completed?
   (A) $Q/3$, $Q/3$, $Q/3$
   (B) $Q/4$, $Q/2$, $Q/4$
   (C) $3Q/8$, $Q/4$, $3Q/8$
   (D) $3Q/8$, $Q/2$, $Q/4$
   (E) none of these
2. Electric appliances are usually grounded in order to _____.
   (A) maintain a balanced charge distribution
   (B) prevent the buildup of heat on them
   (C) run properly using household electricity
   (D) prevent the buildup of an overloaded circuit
   (E) prevent the buildup of static charges on them
3. Two point charges experience a force of repulsion equal to 3×10^{-4} N when separated by 0.4 m. If one of the charges is 5×10^{-4} C, what is the magnitude of the other charge?
   (A) 1.07×10^{-11} C
   (B) 2.7×10^{-11} C
   (C) 3.2×10^{-9} C

(D) 5×10^{-4} C
(E) 0 C

4. A parallel plate capacitor has a capacitance of C farads. If the area of the plates is doubled while the separation between the plates is halved, the new capacitance will be _____F.
(A) $2C$
(B) $4C$
(C) $C/2$
(D) C
(E) $C/4$

5. What is the capacitance of a parallel plate capacitor made of two aluminum plates 4 cm in side length and separated by 5 mm?
(A) 2.832×10^{-11} F
(B) 2.832×10^{-10} F
(C) 2.832×10^{-12} F
(D) 2.832×10^{-9} F
(E) 1 F

6. If it takes 10 J of work to move 2 C of charge in a uniform electric field, the potential difference present is equal to _____V.
(A) 20
(B) 12
(C) 8
(D) 5
(E) 10

7. Which of the accompanying diagrams represents the equipotential curves in the region between a positive point charge and a negatively charged parallel plate?

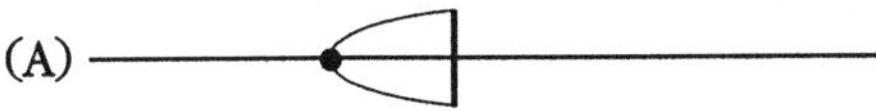

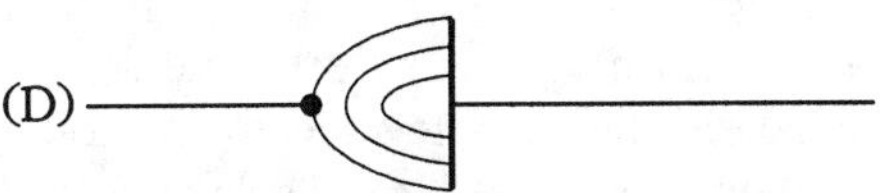

(E) none of these diagrams

8. An electron is placed between two charged parallel plates as shown. Which of the following statements is true?

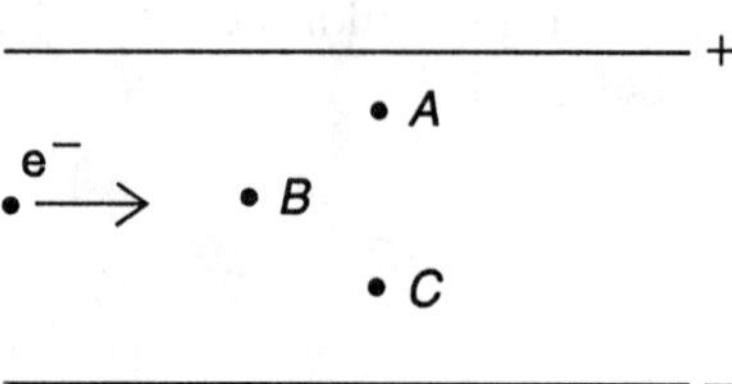

I. The electrostatic force at *A* is greater than at *B*.
II. The work done in moving from *A* to *B* to *C* is the same as the work done in moving from *A* to *C*.
III. The electrostatic force is the same at points *A* and *C*.
IV. The electric field strength decreases as the electron is repelled upward.
(A) I and II
(B) I and III
(C) II and III
(D) II and IV
(E) I and IV

9. How much kinetic energy is given to a doubly ionized helium atom as the result of a potential difference of 1,000 V?
(A) 3.2×10^{-16} eV
(B) 1,000 eV
(C) 2,000 eV
(D) 3.2×10^{-19} eV
(E) not enough information given

10. Which of the following is equivalent to 1 F of capacitance?
(A) $C^2s^2/kg\text{-}m^2$
(B) $kg\text{-}m^2/C^2s^2$
(C) C/kg-s
(D) $Cm/kg^2\text{-}s$
(E) C m/s

Free-Response Problems

1. An electron enters a uniform electric field between two vertically separated parallel plates with $\mathbf{E} = 400$ N / C. The electron enters with an initial velocity of 4×10^7 m/s, and the length of each plate is 0.15 m.
   (a) What is the acceleration of the electron?
   (b) How long does the electron travel through the electric field?
   (c) Assuming that the electron enters the field level with the top plate (which is negatively charged), what is the vertical displacement of the electron during that time?

(d) What is the magnitude of the velocity of the electron as it exits the field?

2. A 1-g cork sphere is coated with silver paint. It is in equilibrium, making a 10-degree angle in a uniform electric field of 100 N/C as shown. What is the charge on the sphere?

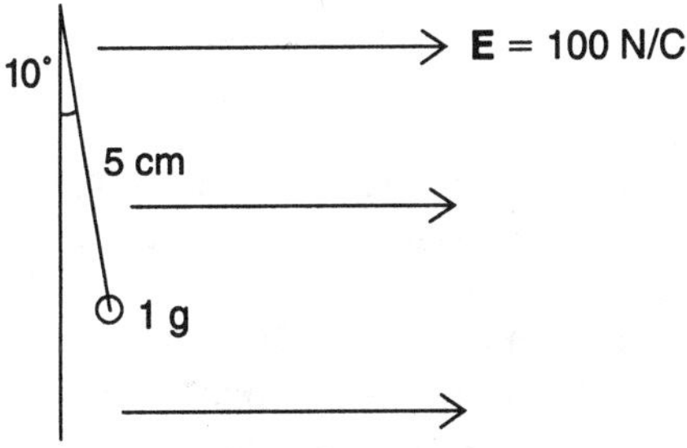

3. Four identical point charges ($Q = +3\ \mu C$) are arranged in the form of a rectangle as shown. What are the magnitude and direction of the net electric force on the charge in the lower left corner attributable to the other three charges?

+Q +Q

5 cm

+Q 7 cm +Q

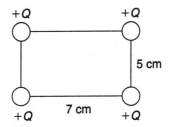

SOLUTIONS TO PRACTICE PROBLEMS

Thought Problems

1. If the object is attracted to the positively charged rod, we can only conclude that it is either negatively charged or neutral (since a neutral object will be attracted because of charge polarization).
2. If your car is hit by lightning, the charges will be distributed around the outside and then dissipated since the car is grounded. This situation is similar to Faraday's "ice pail" experiment in which it can be demonstrated that the electrostatic field inside a hollow conductor is zero.
3. (a) In order for electric potential energy to be detected, there must be electric charges at the observed location. This is because the electric potential energy is determined by the work done on an actual charge owing to the presence of an electric field.

(b) Electric charges do not have to be present at a location in which an electric potential exists. This quantity is a measure of the work done per unit charge and is determined solely by the characteristics of the field at the specified location.

(c) If a small amount of charge is present at a location, it is possible for the electric potential energy to be low (because of the small charge), but the work per unit charge can be large (since algebraically one would be dividing by a small magnitude).

Multiple-Choice Problems

1. **C** We use conservation of charge to determine the answer. Sphere A has a $+Q$ charge and sphere B has zero charge. After they are touched and separated, each will have one half of the total charge (which is $+Q$). Thus, both now have $+Q/2$. When B is touched with C (which also has zero charge), we again distribute the charge evenly by averaging. Thus, B and C now have $+Q/4$, while A still has $+Q/2$. Finally, when A and C are touched together, we take the average of $+Q/4$ and $+Q/2$, which is $+3Q/8$. The final distribution is therefore (C).

2. **E** Electric appliances are grounded to prevent the buildup of static charges on their surfaces. Additionally, grounding can act as a safety in case an internal live wire shorts to a conducting part of the appliance.

3. **A** We use Coulomb's law to calculate the answer. Everything is given except one charge. The working equation therefore looks like

$$3 \times 10^{-4} = \frac{(9 \times 10^{9})(5 \times 10^{-4})q}{(0.4)^2}$$

$$q = 1.07 \times 10^{-11}\ \text{C}$$

4. **B** The formula for the capacitance of a parallel plate capacitor is $C = \epsilon_0 A/d$. Therefore, if the area is doubled and the separation distance halved, the capacitance will increase by four times.

5. **C** We can use the formula from Problem 4, but first all measurements must be in meters and square meters. Converting the given lengths, 4 cm = 0.04 m and 5 mm = 0.005 m. $A = (0.04)(0.04) = 0.0016\ \text{m}^2$, and $d = 0.005$ m. Using the formula and the value for the permittivity of free space given in this chapter, we have $C = 2.832 \times 10^{-12}$ F.

6. **D** Potential difference is the work done per unit charge. Thus $V = W/q = \frac{10}{2} = 5$ V.

7. **C** The equipotential curves around a point charge are concentric circles. The equipotential curves between two parallel plates are parallel lines (since the electric field is uniform). The combination of these two makes (C) the solution.

8. **C** The electric field is the same at all points between two parallel plates since it is uniform. Thus, the force on an electron is the same every-

where. Also, we stated in this chapter that the electric field is a conservative field just like gravity. This means that work done to or by the field is independent of the path taken. Thus, both statements II and III are true and (C) is the answer.

9. **C** One electronvolt is defined to be the energy given to one elementary charge through 1 V of potential difference. A doubly ionized helium atom has a charge of +2 elementary charges. Since the potential difference is 1,000 V, the kinetic energy is 2,000 eV. When using electronvolt units, we eliminate the need for the small numbers associated with actual charges of particles.

10. **A** One farad is defined to be 1 C/V. One volt is 1 J/C. One joule is 1 kg-m^2/s^2. The combination results in (A) as the solution.

Free-Response Problems

1. (a) Since the electron is in a uniform electric field, the kinematics of the electron follow the standard kinematics of mechanics. Since $\mathbf{F} = m\mathbf{a}$ and $\mathbf{F} = -e\mathbf{E}$, $\mathbf{a} = -e\mathbf{E}/m$ (the negative sign is needed since an electron is negatively charged). Using the information given in the problem and the values for the charge and mass of the electron from Appendix A we find that $\mathbf{a} = -(1.6 \times 10^{-19})(400)/(9.1 \times 10^{-31}) = -7.03 \times 10^{13}$ m/s^2.

   (b) The electron is subject to a uniform acceleration in the downward direction. The path will therefore be a parabola, and the electron will be a projectile in the field. The initial horizontal velocity remains constant, and therefore the time $t = L/\mathbf{v}_0 = 0.15/4 \times 10^7 = 3.75 \times 10^{-9}$ s.

   (c) If we take the initial entry level as "zero" in the vertical direction, the electron initially has zero velocity in that direction. Thus, the vertical displacement is given by $y = \frac{1}{2}\mathbf{a}t^2$. Using the known information from parts (a) and (b), we have $y = -\frac{1}{2}(7.03 \times 10^{13})(3.75 \times 10^{-9})^2 = -4.94 \times 10^{-4}$ m.

   (d) The magnitude of the velocity as the electron exits the field, after time t has elapsed, is given by the vector nature of two-dimensional motion. The horizontal velocity is the same throughout the encounter and equal to 4×10^7 m/s. The vertical velocity is given by $\mathbf{v}_y = -\mathbf{a}t$ since there is no initial vertical velocity. Using the known information from the previous parts, we find that $\mathbf{v}_y = -(7.03 \times 10^{13})(3.75 \times 10^{-9}) = -2.6 \times 10^5$ m/s. Now the magnitude of the velocity is given by the Pythagorean theorem since the velocity is the vector sum of these two component velocities:

$$\mathbf{v} = \sqrt{(4 \times 10^7)^2 + (2.6 \times 10^5)^2} = 40{,}000{,}845 \text{ m/s} \approx 4 \times 10^7 \text{ m/s}$$

   The result is not a significant change in the magnitude of the initial velocity. There is of course a significant change in direction. The angle of this change can be determined by the tangent function if desired.

2. If the pith ball is in equilibrium, then the vector sum of all forces acting on it must equal zero. There are several forces involved. Mechanically, gravity acts to create a tension in the string. This tension has an upward component that directly balances the force of gravity (*m***g**) acting on the pith ball. A second horizontal component acts to the left and counters the horizontal electric force established by the field and the charge on the pith ball.

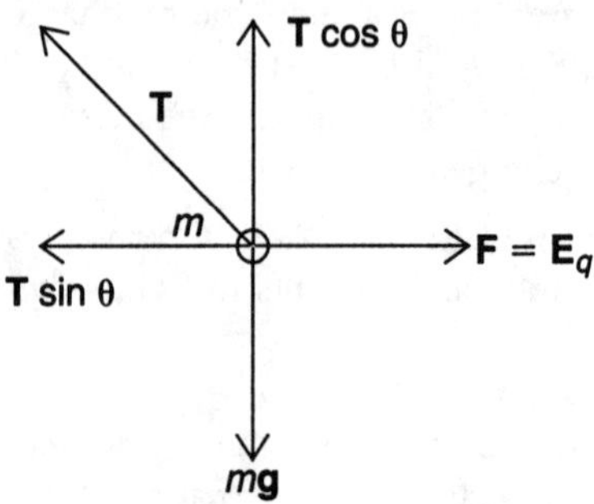

From the accompanying free body diagram, we see that **T** cos θ = *m***g**, where θ = 10 degrees and m = 1 g = 0.001 kg. Thus, **T** = 0.00995 N using the known value for the acceleration of gravity **g**. Now, in the horizontal direction, **T** sin θ = **E**q and **E** = 100 N/C. Using all known values given and derived, we arrive at $q = 1.75 \times 10^{-5}$ C.

3. Since all the charges are positive, the lower left corner charge will experience mutual repulsions from the other three as shown in the diagram. The magnitude of each force is given by Coulomb's law. The diagonal distance was determined from the Pythagorean theorem, and each force magnitude is therefore calculated to be

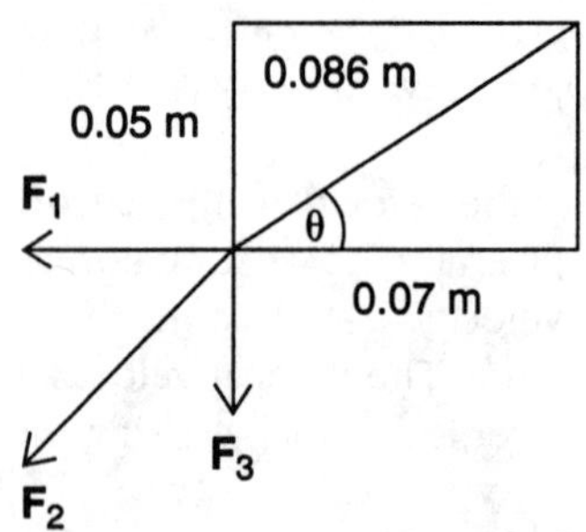

$$\mathbf{F}_1 = \frac{(9 \times 10^9)(3 \times 10^{-6})(3 \times 10^{-6})}{(0.07)^2} = 16.53 \text{ N}$$

$$\mathbf{F}_2 = \frac{(9 \times 10^9)(3 \times 10^{-6})(3 \times 10^{-6})}{(0.086)^2} = 10.95 \text{ N}$$

$$\mathbf{F}_3 = \frac{(9 \times 10^9)(3 \times 10^{-6})(3 \times 10^{-6})}{(0.05)^2} = 32.4 \text{ N}$$

Now, $\mathbf{F}_1$ acts to the left and $\mathbf{F}_3$ acts downward, and so no further vector reductions are necessary. However, $\mathbf{F}_2$ acts at some angle toward the lower left given by θ. To find the angle, recall that $\tan \theta = \frac{5}{7}$, which implies that $\theta = 35.53$ degrees. Both components of $\mathbf{F}_2$ will be negative by the sign convention in Chapter 7.

We can now write $\mathbf{F}_{2(x)} = -10.95 \cos 35.53 = -8.91$ N and $\mathbf{F}_{2(y)} = -10.95 \sin 35.53 = -6.36$ N. The resultant or net force acting on the lower left corner charge is the vector sum of these charges. We can find the components of this force by simply adding up the forces in the same directions:

$$\mathbf{F}_x = -16.53 - 8.91 = -25.44 \text{ N}$$

$$\mathbf{F}_y = -32.40 - 6.36 = -38.76 \text{ N}$$

The magnitude of the net force $\mathbf{F}$ is given by the Pythagorean theorem and is equal to 46.36 N. The direction of this force is toward the lower left and is given by $\tan \phi = \mathbf{F}_y/\mathbf{F}_x = 38.76/25.44 = 1.523$. This implies that $\phi = 56.72$ degrees relative to the negative x axis.

17
ELECTRIC CIRCUITS

CURRENT AND ELECTRICITY

In the last chapter, we observed that if two points have a potential difference between them and are connected with a conductor, then negative charges will flow from a higher concentration to a lower one. In this sense charge flow is very similar to the flow of water in a pipe due to a pressure difference.

Moving electric charges are referred to as electric **current**, which measures the amount of charge passing a given point every second. The units of measurement are C/s. Algebraically, we designate current by the letter *I* and state:

$$I = \frac{\Delta Q}{\Delta t}$$

In electricity, it is the battery that supplies the potential difference needed to maintain a continuous flow of charge. In the nineteenth century, physicists thought that this potential difference was an electric force that pushed an electric fluid through a conductor and called it **electromotive force** (EMF). Today, we know that EMF is not a force but a potential difference measured in volts. Do not be confused by the designation EMF in the course of reviewing or solving problems.

From chemistry, recall that a battery uses the action of acids and bases on different metals to free electrons and maintain a potential difference. In the process, two terminals are created and designated positive and negative. When a conducting wire is attached and looped around to the other end, a complete circle of wire (circuit) is produced, allowing for the continuous flow of charge. The battery acts like a "pump," forcing electrons off the positive side onto the negative side using chemical reactions (see Figure 1). The moving electrons

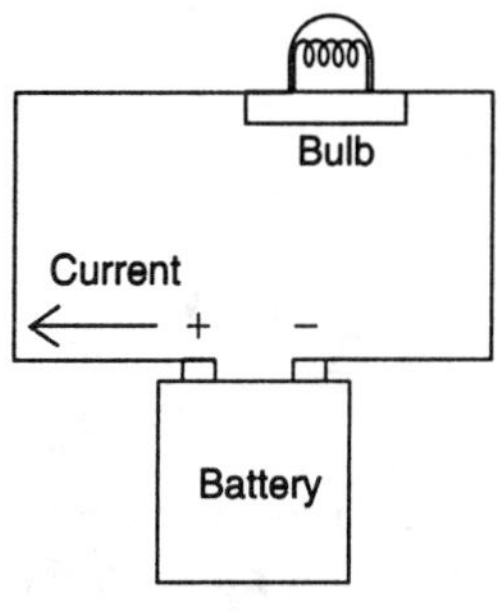

FIGURE 1.

can then do work since they carry energy. This work is the electricity we have become so familiar with in today's modern world.

Figure 1 shows a simple electric circuit. The direction of the conventional current is from the positive terminal, as in most college textbooks. In order to maintain a universal acceptance of concepts and ideas (recall our earlier discussions of concepts and labels), schematic representations for electric devices have been developed and accepted by physicists and electricians worldwide. When drawing or diagramming an electric circuit, these schematics are used, and it is important that you be able to interpret and draw them to fully understand this material. In Figure 2, schematics for some of the most frequently encountered electric devices are presented.

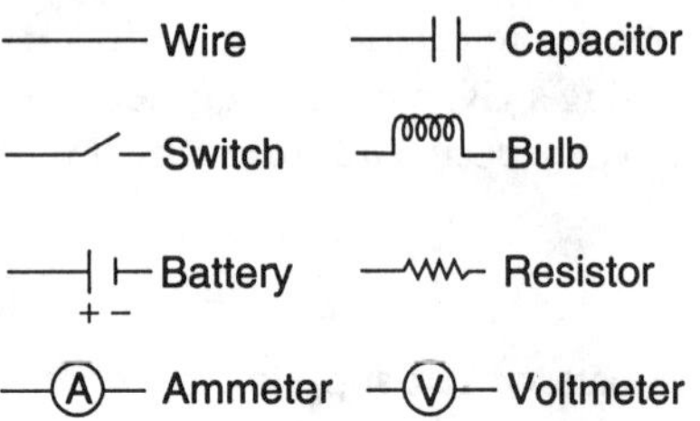

FIGURE 2. *Electrical schematic diagrams.*

The simple circuit shown in Figure 1 can now be diagrammed schematically as in Figure 3. A switch has been added to the schematic, and of course charge cannot flow unless the switch is closed.

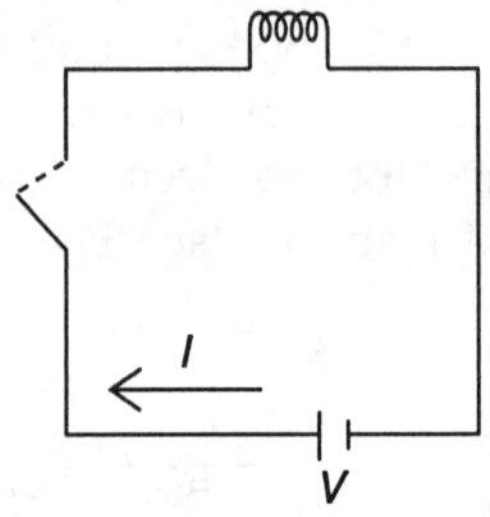

FIGURE 3.

Figure 2 shows three schematics that we have not yet discussed. A **resistor** is a device whose function is to reduce the current. We will investigate resistors in more detail in the next section. An **ammeter** is a device that measures the current. If you live in a house, you can locate your water meter (or the apartment building's water meter if you do not live in a house) and note that it is placed within the flow line. The reason for this is that the meter must measure the flow of water per second through a given point. An ammeter is placed within an electric circuit in much the same way. This is referred to as a *series connection*, and it maintains the singular nature of the circuit. In practical terms, you can imagine cutting a wire in Figure 3 and hooking up the bare leads to the two terminals of the ammeter.

A **voltmeter** is a device that measures the potential difference or *voltage* between two points. Unlike an ammeter, a voltmeter cannot be placed within the circuit since it would effectively be connected to only one point. A voltmeter is therefore attached in a *parallel connection*, creating a second circuit. (But only a small amount of current flows through the voltmeter to operate it.) In Figure 4, the simple circuit is redrawn with the ammeter and voltmeter in place.

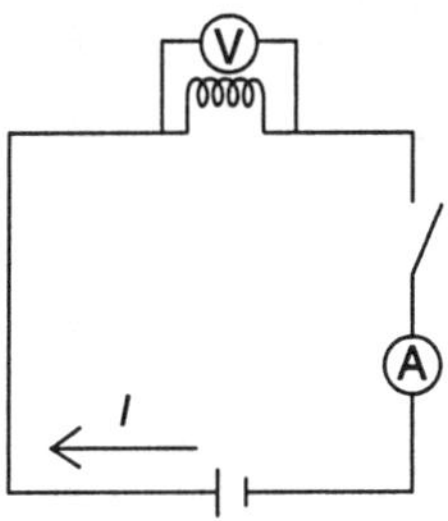

FIGURE 4.

For simple circuits, there will be no observable difference in readings if the ammeter and voltmeter are moved to different locations. However, there is a slight difference in the EMF across the terminals of the battery and the EMF across the bulb. The first EMF reflects the work done by the battery not only to move charge through the circuit but also to move charge across the terminals of the battery. This is sometimes referred to as the *terminal EMF* or *internal EMF*.

ELECTRIC RESISTANCE

In Figure 4, a simple electric circuit is illustrated with measuring devices for voltage and current. If the light bulb is left on for a long time, two observations can be made. First, the bulb will get hot because of the action of the electricity and the filament. The light produced by the bulb is caused by the heat of the filament. Second, the current in the ammeter will begin to decrease.

These two observations are linked to the idea of electric resistance. Friction generates heat. The interaction of flowing electrons and the molecules of a wire (or bulb filament) creates an electric friction called **resistance**. This resistance is temperature-dependent since, from kinetic theory, an increase in temperature increases molecular activity and therefore interferes more with the flow of current.

In the case of a light bulb, it is this resistance that is desired in order for the bulb to do its job. However, resistance along a wire or in a battery is unwanted and must be minimized. In more complicated circuits, a change in current flow is required to protect devices, and so special resistors are manufactured that are small enough to easily fit into a circuit. While resistance is the opposite of conductance, we do not use insulators as resistors since they would stop the flow altogether. Therefore, a range of materials, indexed by *resistivity,* are catalogued in electrical handbooks to assist scientists and electricians in choosing the proper resistor for a given situation.

If the temperature can be maintained at a constant level, then a simple relationship between the voltage and the current in a circuit is revealed. As the voltage is increased, a greater flow of current is observed. This is a direct relationship which was investigated theoretically by a German physicist named Georg Ohm and is called **Ohm's law** (see Figure 5).

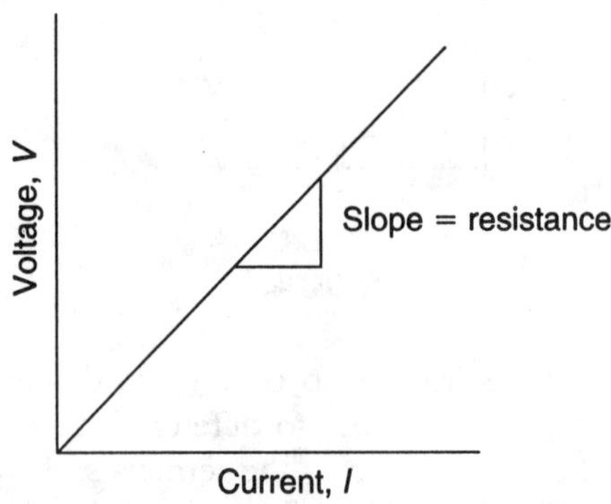

FIGURE 5. *Ohm's law.*

Ohm's law states that in a circuit at constant temperature the ratio of the voltage to the current remains a constant. The slope of the line in Figure 5 is the resistance of the circuit and is measured in ohms (Ω). Algebraically, we can write Ohm's law as $R = V/I$ or $V = IR$. Not all conductors obey Ohm's law, and semiconductors and liquid conductors are some among them.

There are other factors that affect the resistance of a conductor. We have already discussed the effects of temperature and material type (resistivity). Resistivity is designated by the Greek letter ρ (rho). In a wire, electrons try to move through the wire, interacting with the molecules that make up the wire. If the wire has a small cross-sectional area, then the chances of interacting with a bound molecule will increase and thus increase the resistance of the wire. If the length of the wire is increased, then the duration of the interaction time will also increase the resistance of the wire. In summary, these resistance factors all contribute to the overall resistance of the circuit. Algebraically, we may write these relationships (at constant temperature) in the form

$$R = \frac{\rho L}{A}$$

The resistivity is measured in Ω-m and is usually rated at 20°C. In Table 1 the resistivities of various materials are presented. Since the ohm is a standard unit, be sure that all lengths are in meters and all areas are in square meters.

TABLE 1. RESISTIVITIES AT 20°C

Substance	Resistivity, ρ (Ω-m)
Copper	1.69×10^{-8}
Silver	1.59×10^{-8}
Gold	2.44×10^{-8}
Aluminum	2.83×10^{-8}
Tungsten	5.33×10^{-8}
Platinum	10.4×10^{-8}
Nichrome	100×10^{-8}
Carbon	3.5×10^{-5}

ELECTRIC POWER AND ENERGY

Electricity produces energy that can be used to generate light and heat. Electricity can do work to turn a motor. Having measured the voltage and current in a circuit, we can determine the amount of power and energy being produced in the following way. The units of voltage (potential difference) are J/C. This is a measure of the energy supplied to each coulomb of charge flowing in the circuit. The current measures the total number of coulombs per second flowing at any given time. The product of the voltage and the current, VI, is therefore a measure of the total power produced since the units are J/S (watts):

$$P = VI$$

The electric energy expended in a given time t (in seconds) is simply the product of the power and the time:

$$\text{Energy} = Pt = VIt = I^2Rt$$

In the preceding equation we used Ohm's law, $V = IR$, to express the energy in another form. This energy can be related to heat if we consider a direct transfer of energy into some water (for example). The rise in temperature can then be computed. Be careful to note that the electric energy is expressed in joules, while specific heats involve kilojoules and kilograms.

RESISTORS IN SERIES AND PARALLEL CIRCUITS

Series Circuits

A series circuit consists of two or more resistors sequentially placed within one circuit. An example of this type of circuit appears in Figure 6. There are several questions we need to ask. First, what is the effect of adding more resistors in series to the overall resistance in the circuit? Second, what effect does the first condition have on the current flowing in the circuit?

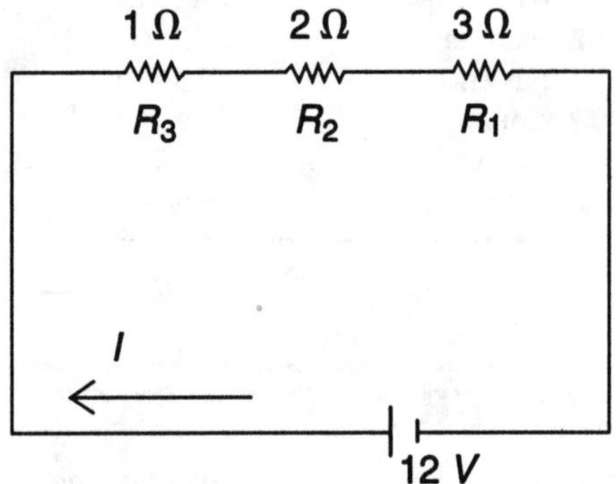

FIGURE 6. *Series circuit*

A look at the series circuit shows that it is still one continuous loop. This means if there is a break in one place (let's say one light bulb goes out), the entire circuit will shut down. Experiments reveal that the current flowing through the entire circuit is the same. This is because we have only one circuit. However, these experiments show that the terminal EMF across the battery is essentially equal to the sum of the EMFs across each resistor. There is a *potential drop* across each resistor.

One way to think about this circuit is to imagine a series of doors, one after the other. As people exit one door, they must wait and open another. This takes energy out of the system and decreases the number of people per second exiting the room. In an electric circuit, adding more resistors in series decreases the current (with the same voltage) by increasing the resistance of the circuit.

Algebraically, all these observations can be summarized as follows. In resistors R_1, R_2, and R_3, we have currents I_1, I_2, and I_3. All three of these currents are equal to each other and to the circuit current I:$I = I_1 = I_2 = I_3$. However, the voltage across each of these resistors is less than the source voltage V. If all three resistors were equal, then each voltage would be equal to one third of the total source voltage. In any case, we have $V = V_1 + V_2 + V_3$. Using Ohm's law, $V = IR$, we can rewrite this expression as $IR = I_1R_1 + I_2R_2 + I_3R_3$. However, all currents are equal, and so they can be canceled out from the expression. This leaves us with

$$R = R_1 + R_2 + R_3$$

In other words, when resistors are added in series, the total resistance of the circuits increases as the sum of each resistance. This explains why the current goes down as more resistors are added. In our example, the total resistance of the circuit is 1 Ω + 2 Ω + 3 Ω = 6 Ω. Since the source voltage is 12 V, Ohm's law provides for a circuit current of $\frac{12}{6}$ = 2 A. The voltage across each resistor can now be determined from Ohm's law since each gets the same current of 2 A. Thus, V_1 = (1)(2) = 2 V; V_2 = (2)(2) = 4 V; V_3 = (3)(2) = 6 V. The voltages in a series circuit add up to the total, and not unexpectedly we have 2 V + 4 V + 6 V = 12 V. It should be noted that when batteries are connected in series (positive to negative), the effective voltage increases additively as well.

Parallel Circuits

A parallel circuit consists of two or more resistors connected across each other with each connected to a common point of potential difference. An example of a parallel circuit is seen in Figure 7. In this circuit, a branch point is reached at which the current is split into I_1 and I_2. Since each resistor is connected to a common potential point, experiments verify that the voltages across each resistor equal the source voltage V. Thus, in this circuit, it is the current that is shared, while the voltage is the same. Another feature of a parallel circuit is the availability of alternative paths. If one part of the circuit is broken, current can

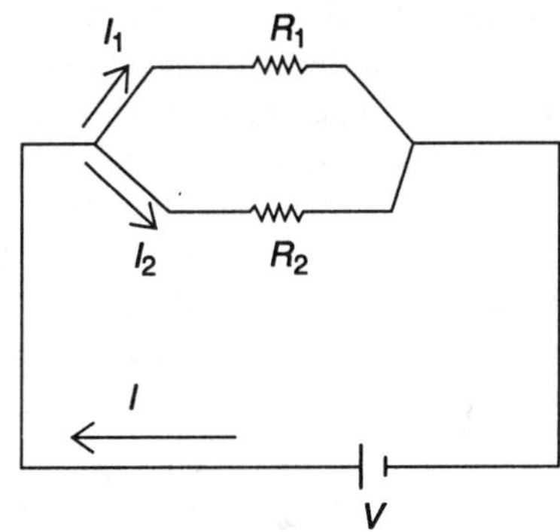

FIGURE 7. *Parallel circuit.*

flow through the other path. While each branch current is less than the total circuit current I, the effect of adding resistors in parallel is to increase the effective circuit current by decreasing the circuit resistance R.

To understand this effect further, imagine a set of doors placed next to each other along a wall in a room. Unlike the series connection (in which the doors are placed one after the other), the parallel circuit analogy involves placing the doors next to each other. Even though each door will have fewer people per second going through it at any given time, the overall effect is to allow more people to exit the room in total. This is analogous to reducing the circuit resistance and increasing the circuit current (at the same voltage).

Algebraically, we can express these observations as follows. We have two resistors R_1 and R_2 with currents I_1 and I_2. Voltmeters placed across each resistor indicate voltages V_1 and V_2 which are essentially equal to the source voltage V. That is, $V = V_1 = V_2$ (in this example). Ammeters placed in the circuit would reveal that the circuit current I is equal to the sum of the branch currents I_1 and I_2. That is, $I = I_1 + I_2$. Using Ohm's law, we see that $V / R = V_1 / R_1 + V_2 / R_2$. Since all the voltages are equal, they can be canceled from the expression, leaving us the following expression for resistors in parallel:

$$\frac{1}{R} = \frac{1}{R_1} + \frac{1}{R_2}$$

This expression indicates that the total resistance is determined *reciprocally*, which in effect reduces the total resistance of the circuit as more resistors are added in parallel. If, for example, $R_1 = 10\ \Omega$ and $R_2 = 10\ \Omega$, the total resistance $R = 5\ \Omega$ (in parallel). It should be noted that connecting batteries in parallel (positive to positive and negative to negative) does not affect the overall voltage of the combination.

COMBINATION CIRCUITS AND NETWORKS

In Figure 8, a circuit consisting of resistors in series and in parallel is presented. The key to reducing each of these circuits is to decide whether it is overall a series or a parallel circuit. In Figure 8, we see that an 8-Ω resistor has been placed in series with a parallel branch containing two 4-Ω resistors. The problem is to reduce the circuit to only one resistor and then determine the circuit current, the voltage across the branch, and the current in each branch.

The procedure for reducing this circuit and finding the circuit resistance R is to first reduce the parallel branch. Thus, if R_e is the equivalent resistance in the branch, then

$$\frac{1}{R_e} = \frac{1}{4} + \frac{1}{4}$$

This implies that $R_e = 2\ \Omega$ of resistance. The circuit can now be thought of as a series circuit between a 2-Ω resistor and an 8-Ω resistor. Thus, $R = 2\ \Omega + 8\ \Omega =$

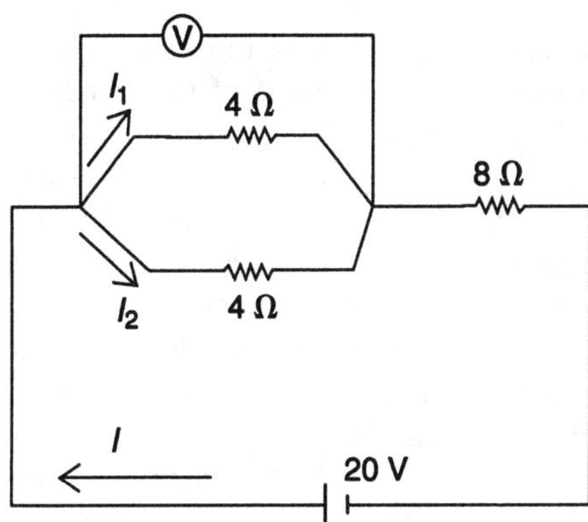

FIGURE 8.

10 Ω. Since the circuit voltage is 20 V, Ohm's law states that the circuit current $I = V/R = \frac{20}{10} = 2$ A.

In a series circuit, the voltage drop across each resistor is shared proportionally since the same current flows through each resistor. Thus, for the branch, $V = IR_e = (2)(2) = 4$ V. To find the current in each part of the parallel branch, recall that in a parallel circuit the voltage is the same across each resistor. Thus, since each resistor is equal, each will get half of the circuit current (since there are two equal resistors; and thus each current is 1 A).

This procedure for reducing a circuit can be used in a large number of combination circuits. However, in Figure 9, we show a circuit in which this procedure will not work because of the presence of a second EMF in a branch. This type of circuit is sometimes called a **network**, and its analysis requires the use of some "laws" stated in the nineteenth century by the physicist Gustav Kirchoff (Kirchoff's laws).

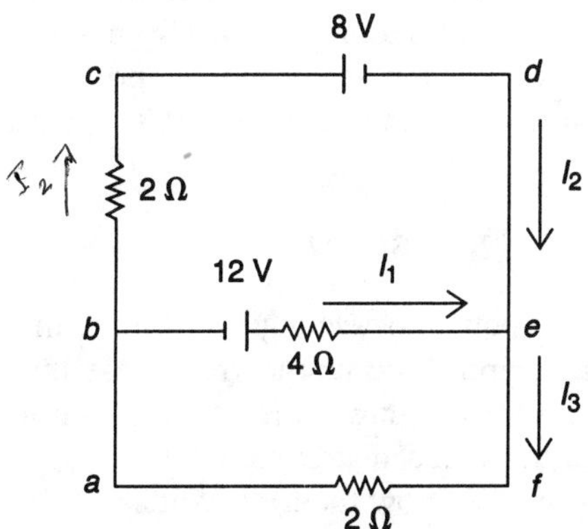

FIGURE 9.

The directions of the currents are given arbitrarily, which has an important effect on the determination of the values of the three currents (which is the problem). Kirchoff's first law states that the sum of all the currents entering a junction point must be equal to the sum of all the currents leaving that junction point. Kirchoff's second law states that the sum of the voltage drops (or rises) across all the elements around any closed loop must equal zero.

Kirchoff's laws are just restatements of the laws of conservation of charge and energy, respectively. An analogous situation is water flowing through a closed system with many branches but no leaks. In our example, we can see that $I_1 + I_2 = I_3$ at junction point *e*.

There are also some useful heuristics to consider as problem-solving strategies. Imagine that we are traversing a loop in the clockwise direction. First, when a resistor is crossed in the direction of the current, the change in electric potential across that resistor is written as $-IR$. Second, when the resistor is crossed in the direction opposite the current, the potential difference is written as $+IR$. Third, when a source of EMF is crossed from $-$ to $+$, we write the EMF as the variable $+V$. Fourth, when the source of EMF is crossed from $+$ to $-$, we state that it is $-V$. Fifth, it is useful to consider the first four rules while imagining that you are traversing each loop in a clockwise direction. Note that the junction rule applies only $(n - 1)$ times if there are n junctions. Since we will be ending up with a system of equations, we must make sure that we have enough equations to equal all the unknowns.

We have applied the junction rule at point *e*, so let's continue the example for loops. Let's go clockwise around loop *abefa*. The EMF is crossed from $-$ to $+$, and the resistor on top is crossed in the direction of I_1 while the resistor on the bottom is crossed in the direction of current I_3. Using Kirchoff's second law, we have

$$12 - 4I_1 - 2I_3 = 0$$

Let's now look clockwise at the loop *bcdeb*. The EMF on top is crossed from $+$ to $-$, the second EMF on the bottom is crossed from $+$ to $-$, the 4-Ω resistor is crossed in the opposite direction as current I_1, and the 2-Ω resistor on the side is crossed in the same direction as current I_2. Using Kirchoff's second law, we have

$$-2I_2 - 8 + 4I_1 - 12 = 0$$

Together with the first equation relating the three currents, we have our necessary three equations and three unknowns. These three equations must be solved simultaneously for the three currents using known algebraic techniques (there are more advanced matrix techniques, such as *Cramer's rule*, which can also be used if you know them). If you solve these equations, you will find that $I_1 = 2.2$ A, $I_2 = -5.6$ A, and $I_3 = -3.4$ A. The negative signs mean that the directions for these currents are opposite the arbitrarily chosen direction.

CAPACITORS IN CIRCUITS

Capacitors in Series

We know from Chapter 16 that the capacitance of a parallel plate capacitor is given by $C = Q / V$. Suppose we have a series connection of capacitors connected to a potential difference V as shown in Figure 10.

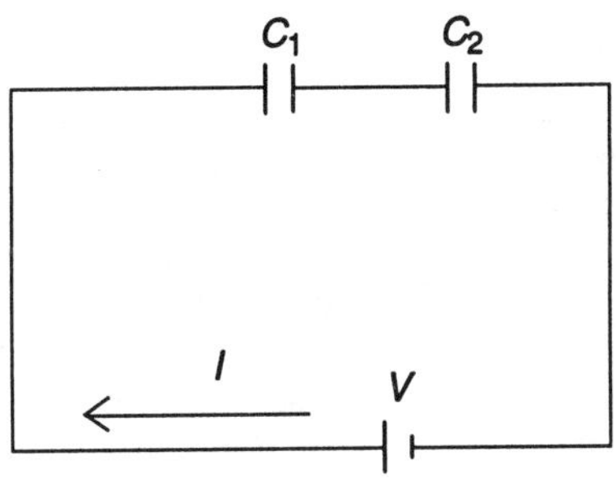

FIGURE 10.

In this circuit, the magnitude of the charge must be the same on all the plates because the left plate of C_1 becomes $+Q$ and the right plate of C_2 becomes $-Q$. By induction, the right plate of C_1 becomes $-Q$ and the left plate of C_2 becomes $+Q$. Additionally, the voltage across each capacitor is shared proportionally such that $V_1 + V_2 = V$ (just like resistors in series). Thus,

$$\frac{Q}{C_1} + \frac{Q}{C_2} = \frac{Q}{C}$$

and

$$\frac{1}{C_1} + \frac{1}{C_2} = \frac{1}{C} \quad \text{or} \quad \text{(for a series combination)}$$

Note that this is the opposite relationship for resistors in series. Be sure not to confuse them!

A parallel combination of capacitors is shown in Figure 11. In this circuit, the current branches off, and so each capacitor is charged to a different total charge Q_1 and Q_2. The equivalent total charge, $Q = Q_1 + Q_2$ and the voltage across each capacitor is the same as the source EMF voltage. Thus,

$$CV = C_1V + C_2V$$

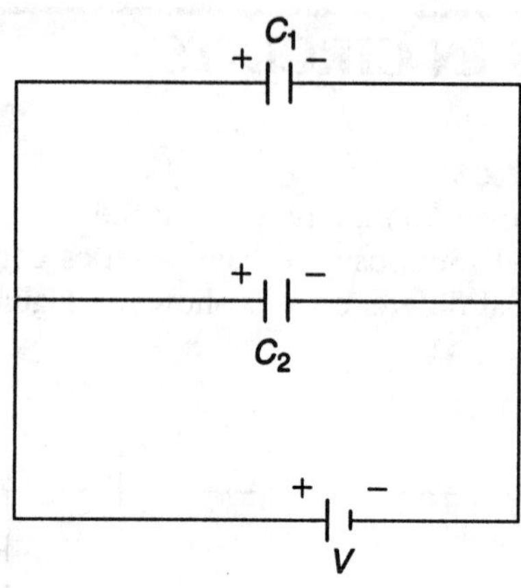

FIGURE 11.

and

$$C = C_1 + C_2 \quad \text{(for a parallel combination)}$$

Note that this is opposite the relationship for resistors in parallel.

ENERGY TRANSFER AND IMPEDENCE MATCHING IN ELECTRIC CIRCUITS

To fully understand the process of electric energy transfer in a simple circuit, let us consider a mechanical example. Suppose we have two spheres about to have an elastic collision with each other. One sphere, M_1, is moving with a velocity $\mathbf{v}_{1i}$, and the second sphere, M_2, is initially at rest. After the elastic collision, the two spheres have velocities $\mathbf{v}_{1f}$ and $\mathbf{v}_{2f}$, respectively.

From our studies of linear momentum we know that

$$M_1\mathbf{v}_{1i} = M_1\mathbf{v}_{1f} + M_2\mathbf{v}_{2f}$$

Now, for an elastic collision,

$$\mathbf{v}_{1i} - \mathbf{v}_{2i} = -(\mathbf{v}_{1f} - \mathbf{v}_{2f})$$

and therefore

$$\mathbf{v}_{1f} = \frac{(M_1 - M_2)\,\mathbf{v}_{1i}}{M_1 + M_2}$$

and

$$\mathbf{v}_{2f} = \frac{2(M_1)(\mathbf{v}_{1i})}{M_1 + M_2}$$

When the collision occurs, kinetic energy is transferred from the first sphere to the second sphere. How does this energy transfer depend on the relative masses? The ratio of kinetic energies is given by

$$\frac{\mathrm{KE}_{2f}}{\mathrm{KE}_{1i}} = \frac{\frac{1}{2}M_2\mathbf{v}_{2f}^2}{\frac{1}{2}M_1\mathbf{v}_{1i}^2} = \frac{4M_1M_2}{(M_1 + M_2)^2} = \frac{4(M_2/M_1)}{(1 + M_2/M_1)^2}$$

If we were to plot this ratio of energy transferred against the mass ratio (Figure 12), we would find that the maximum energy transferred occurs when $M_1 = M_2$. The process of transferring maximum energy through the variation of masses in this elastic collision is called **impedence matching** and is an important concept related to electric circuits and wave energy transfer (resonance).

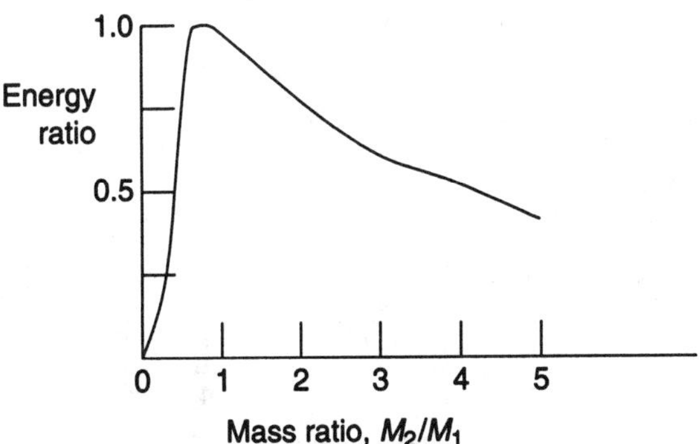

FIGURE 12. *Energy transfer in an elastic collision.*

In an electric circuit, we have the transfer of energy between an EMF V with internal resistance r in series with a load of resistance R. We know that the power associated with the load is given by $P = I^2R$. Thus, using Ohm's law for internal resistance,

$$I = \frac{V}{R + r}$$

and therefore

$$P = \frac{V^2R}{(R + r)^2}$$

The maximum electric power transferred occurs when $R = r$. We therefore have

$$P_{max} = \frac{V^2}{4R}$$

and the ratio of the instantaneous power and the maximum power is given by

$$\frac{P}{P_{max}} = \frac{4R^2}{(R + r)^2} = \frac{4(R^2/r^2)}{(1 + R/r)^2}$$

A plot of the power ratio versus the resistance ratio (Figure 13) is observed to be similar to the one shown in Figure 12. Impedence matching in a circuit is important when considering the effect of transformers, alternating current circuits, and energy transfers in oscillating systems. Additionally, the energy and power curves shown in Figures 12 and 13 look remarkably similar to the binding energy per nucleon curve to be discussed later when we review nuclear physics.

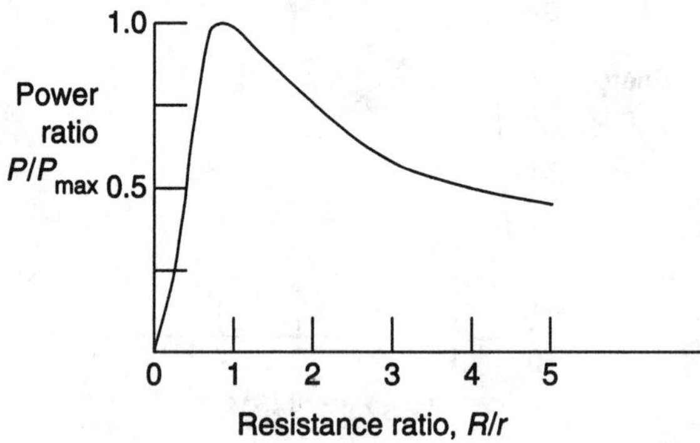

FIGURE 13.

RESISTORS AND CAPACITORS IN CIRCUITS

When we create a circuit consisting of an EMF, an ammeter, a single resistor, and a single capacitor, we notice that a transient current registers in the ammeter. This current starts at an almost instantaneous maximum and then slowly decays to zero. A voltmeter placed across the terminals of the capacitor would record a buildup of potential difference between the two terminals, reaching the same value as the source EMF when the current reaches zero. These observations are illustrated in Figures 14 and 15.

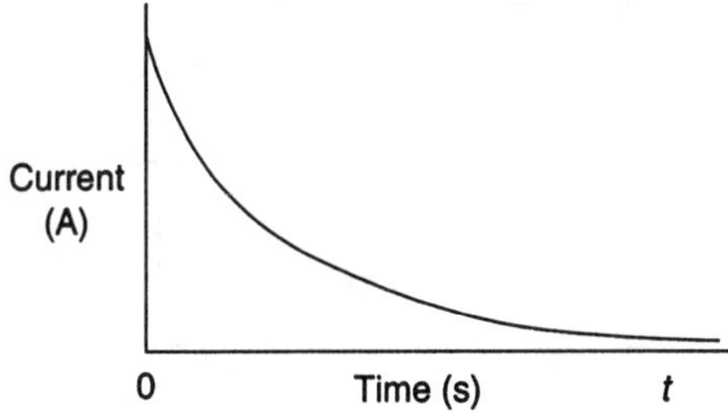

FIGURE 14. *Current decay in a charging capacitor.*

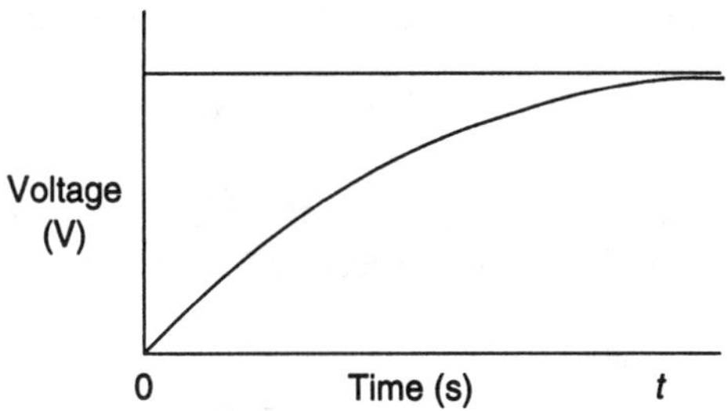

FIGURE 15. *Voltage buildup in a charging capacitor.*

Why should we record a current when the dielectric material in the capacitor is nonconducting? Figure 16 explains this in more detail. When we attach a battery to the capacitor, charge builds up on each plate, and the electric field produced causes electric charges to be repelled to the battery. However, as more charge builds up, so does the potential difference. This makes it more difficult to add more charges on either plate, and so the induced current decays to zero. This curve is logarithmic and provides the basis for understanding this circuit.

To understand the behavior of this circuit, let us consider Kirchoff's law as applied to each element of the circuit. If the EMF of the battery is designated by V, then the sum of all the voltage changes around the circuit must equal zero:

$$\frac{V - IR - q}{C} = 0$$

where q is the charge on the capacitor at any time t and C is its capacitance. Since current is the flow of charge, $I = \Delta q/\Delta t$, and so we can write (with a little rearranging):

$$\frac{\Delta q}{\Delta t} = \frac{-1}{RC}(q - VC)$$

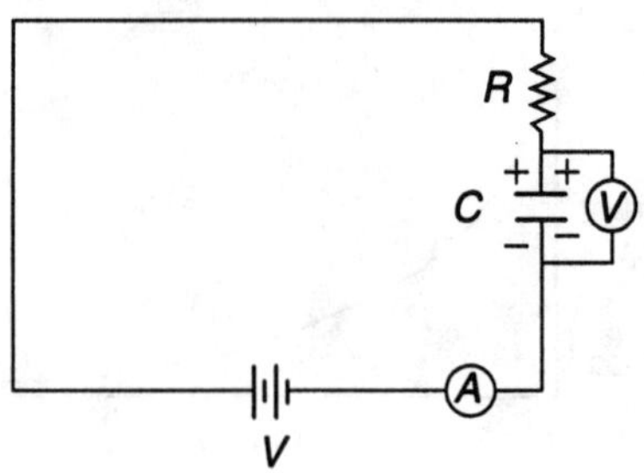

FIGURE 16. *An RC circuit.*

The value of q that satisfies this equation is given by the expression

$$q = VC(1 - e^{-t/RC})$$

If we let V_C represent the voltage across the capacitor at any time t (equal to q/C), then we have the following expression for potential difference across the capacitor:

$$V_C = V(1 - e^{-t/RC})$$

This equation corresponds directly to the graph in Figure 15. As the time value gets very large, the voltage across the capacitor approaches the value of the battery's EMF. Since $I = (V - V_C)/R$, then at any subsequent time t, the current is given by

$$I = \frac{V}{R} e^{-t/RC}$$

This equation corresponds directly to the graph in Figure 14. The time, $\tau = RC$, is known as the **capacitive time constant**.

Sample Problem

How many time constants must elapse before a capacitor reaches 85% of its maximum potential difference?

Solution

Using the formula for the voltage across the plates of a charging capacitor, we can write

$$e^{-t/\tau} = \frac{V - V_C}{V}$$

Since the capacitor is reaching 85% of its maximum EMF, then $V_C = 0.85$ V and therefore

$$e^{-t/\tau} = \frac{1 - 0.85}{1.0} = 0.15$$

$$\frac{-t}{\tau} = \ln(0.15)$$

$$t = 1.897\tau$$

It would therefore take 1.897 time constants to reach 85% of the maximum EMF.

PROBLEM-SOLVING STRATEGIES

Several techniques for solving electric circuit problems have been discussed in this chapter. For resistances with one source of EMF, we use the techniques of series and parallel circuits. Ohm's law plays a big role in measuring the potential drops across resistors and the determination of currents within the circuit.

When working with a combination circuit, try to reduce all subbranches first as well as any series combinations. The goal is to be left with only one equivalent resistor and then to use Ohm's law to identify the various unknown quantities. Drawing a sketch of the circuit (if it is not already provided) is essential. Also, remember that in most college textbooks the direction of current is taken from the positive terminal of a battery. This is opposite the way it is presented in most high school physics textbooks.

In a network with two or more EMFs, Kirchoff's rules must be used. While the directions of the currents are arbitrary, it is important to remember that Kirchoff's rules are based on the laws of conservation of charge and energy. Care must be taken so that enough equations are presented to solve for all the missing information without rendundancy.

In solving a network problem, you can choose to traverse the loop either clockwise or counterclockwise. Working slowly and carefully will minimize careless errors.

PRACTICE PROBLEMS FOR CHAPTER 17

Thought Problems

1. Explain why connecting two batteries in parallel does not increase the effective EMF.
2. You have five 2-Ω resistors. Explain how you can combine them to make equivalent resistances of 1, 3, and 5 Ω.

3. Show that for two resistors in parallel, the equivalent resistance can be obtained from

$$R_{eq} = \frac{R_1 R_2}{R_1 + R_2}$$

Multiple-Choice Problems

1. How many electrons are present in a current of 2 A for 2 s?
   (A) 3.2×10^{-19}
   (B) 6.28×10^{18}
   (C) 4
   (D) 2.5×10^{19}
   (E) 1

2. How much electric energy is consumed by a 100-W light bulb turned on for 5 min?
   (A) 20 J
   (B) 500 J
   (C) 3,000 J
   (D) 15,000 J
   (E) 30,000 J

3. What is the current in a 5-Ω resistor due to a potential difference of 20 V?
   (A) 25 A
   (B) 4 A
   (C) 0.20 A
   (D) 5 A
   (E) 20 A

4. A 20-Ω resistor has 10 A of current in it. The power generated is _____ W.
   (A) 2,000
   (B) 2
   (C) 200
   (D) 4,000
   (E) 20

5. What is the value of resistor R in the circuit shown?

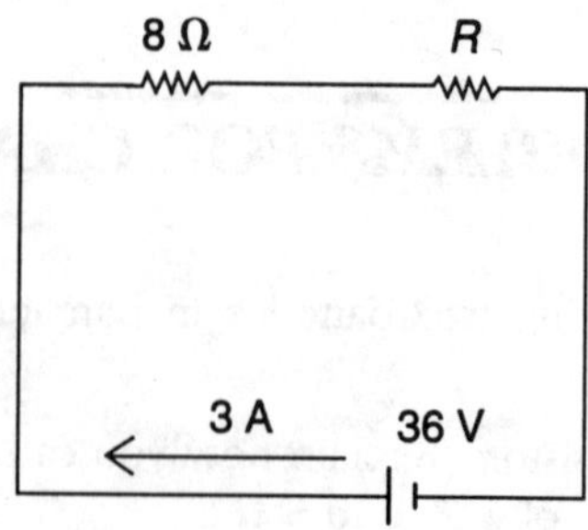

(A) 20 Ω
(B) 2 Ω
(C) 4 Ω
(D) 12 Ω
(E) 10 Ω

6. What is the equivalent capacitance of the circuit shown?

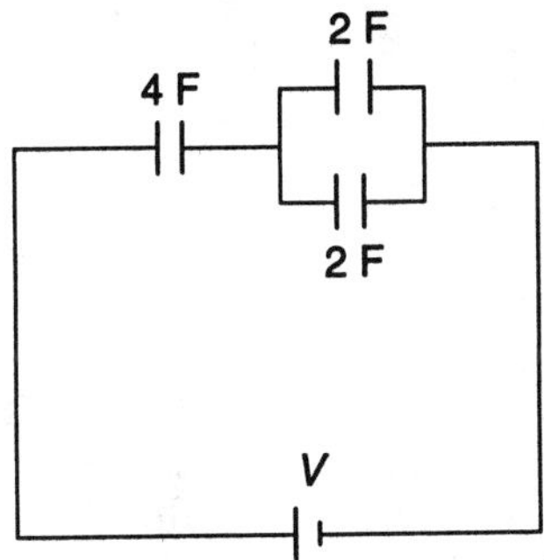

(A) 2 F
(B) 8 F
(C) 5 F
(D) 10 F
(E) 15 F

Answer Problems 7 and 8 based on the circuit shown.

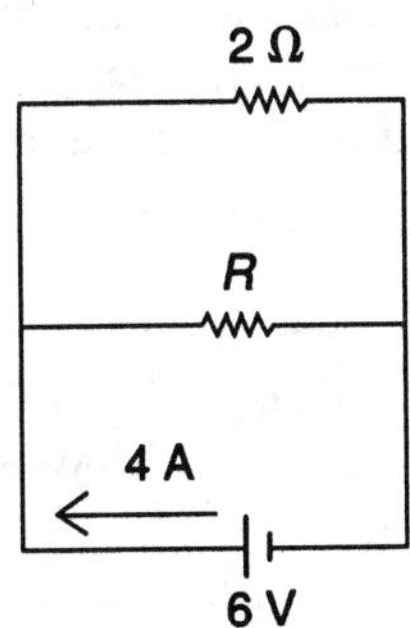

7. If the current in the circuit is 4 A, then the power dissipated by resistor R is _____ W.
(A) 6
(B) 18
(C) 24
(D) 7
(E) 16

8. What is the value of resistor R in the circuit?
   (A) 4 Ω
   (B) 3 Ω
   (C) 5 Ω
   (D) 6 Ω
   (E) 8 Ω

9. A 5-Ω and a 10-Ω resistor are connected in series with one source of EMF of negligible internal resistance. If the energy produced in the 5-Ω resistor is X, then the energy produced in the 10-Ω resistor is _____.
   (A) X
   (B) $2X$
   (C) $X/2$
   (D) $X/4$
   (E) $4X$

10. An RC circuit consists of a battery with an EMF of 24 V and negligible internal resistance, a resistor of 200 Ω, and a capacitor of unknown capacitance, all in series. After 4 s, a current of 50 mA is detected by an ammeter in the circuit. What is the capacitance of the capacitor?
    (A) 0.032 F
    (B) 0.176 F
    (C) 0.17 F
    (D) 0.762 F
    (E) 0.023 F

Free-Response Problems

1. A 30-Ω heating coil is placed into 250 g of water at a temperature of 25°C. If the coil is connected to a 120-V supply and turned on for 2 min, what will be the new temperature of the water? Assume that all the energy is directed into the water and that there are losses to the environment.
2. Answer the following questions based on the circuit shown.

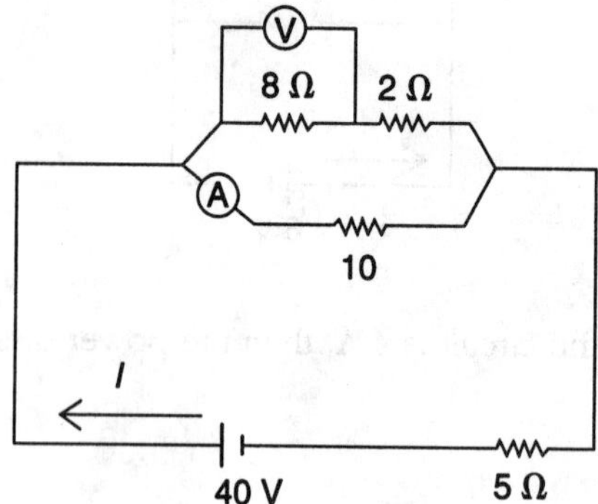

   (a) Determine the equivalent resistance of the circuit.
   (b) Determine the value of the circuit current I.

(c) Determine the value of the reading of ammeter A.
(d) Determine the value of the reading of voltmeter V.

3. Determine the values of the three currents shown in the accompanying network circuit.

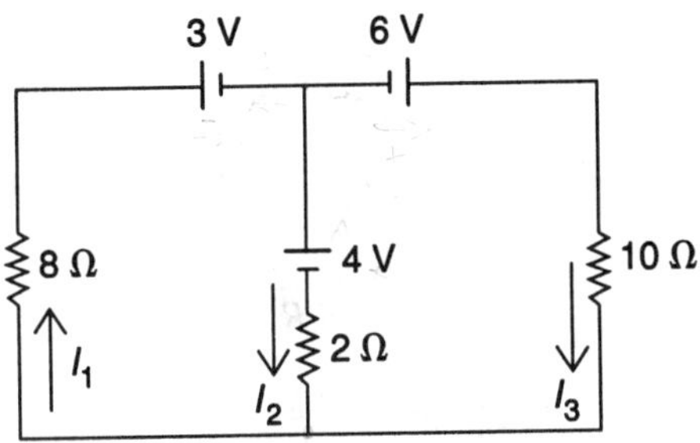

4. A wheatstone bridge is used to measure the resistance of an unknown resistor in the circuit shown below. A galvanometer (a device which detects small currents) is "bridged" across points A and C, and the circuit is "balanced" when the galvanometer reads zero and the voltage across AC is likewise zero. Show that the unknown resistance can be determined from the expression

$$R_x = \frac{R_2 R_3}{R_1}$$

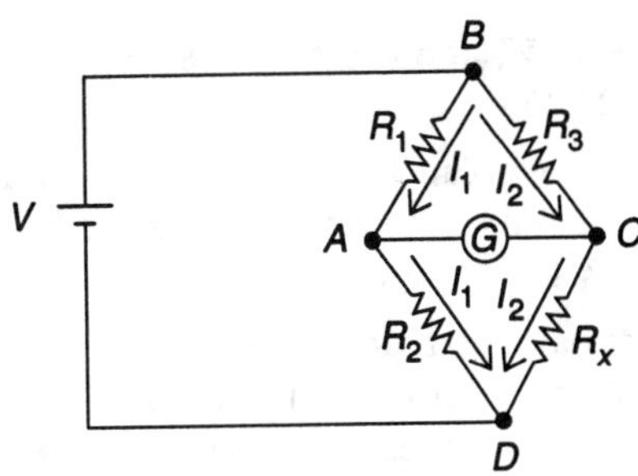

SOLUTIONS TO PRACTICE PROBLEMS

Thought Problems

1. When two batteries are connected in parallel, the EMF of the system is not changed. The storage capacity of the battery system is, however, increased, since the parallel connection of two batteries is similar to the parallel connection of two capacitors.
2. An equivalent resistance of 1 Ω can be achieved by connecting two 2-Ω resistors in parallel. An equivalent resistance of 3 Ω can be achieved by

connecting two 2-Ω resistors in parallel and then connecting the branch in series with another 2-Ω resistor. The addition of another 2-Ω resistor can provide an equivalent resistance of 5 Ω.

3. We start with the regular parallel formula for resistors:

$$\frac{1}{R_{eq}} = \frac{1}{R_1} + \frac{1}{R_2}$$

$$\frac{1}{R_{eq}} = \frac{R_2}{R_2R_1} + \frac{R_1}{R_1R_2}$$

$$\frac{1}{R_{eq}} = \frac{R_1 + R_2}{R_1R_2}$$

$$R_{eq} = \frac{R_1R_2}{R_1 + R_2}$$

Multiple-Choice Problems

1. **D** By definition, 1 A of current measures 1 C of charge passing through a circuit each second. One coulomb is the charge equivalent of 6.28×10^{18} electrons. In this question, we have 2 A for 2 s, which is equivalent to 4 C of charge or 2.5×10^{19} electrons.
2. **E** Energy is equal to the product of power and time, and time must be in seconds. Thus, 5 min = 300 s. Thus, energy = (100)(300) = 30,000 J of energy.
3. **B** Using Ohm's law, $I = V / R = \frac{20}{5} = 4$ A.
4. **A** We know that $P = VI$, and using Ohm's law, $V = IR$, we have $P = I^2R = (100)(20) = 2{,}000$.
5. **C** Using Ohm's law, we can determine the equivalent resistance of the circuit: $\frac{36}{3} = 12$ Ω. Now, the two resistors are in series, and so 12 = 8 + R. Thus, $R = 4$ Ω.
6. **A** We reduce the parallel branch first. In parallel, capacitors add up directly. Thus, $C_1 = 4$ F. Now, this capacitor will be in series with the other 4-F capacitor. In series, capacitors add up reciprocally, and thus the final capacitance is $1 / C = \frac{1}{4} + \frac{1}{4}$. This implies that $C = 2$ F.
7. **A** In a parallel circuit, all the resistors have the same potential difference across them. The branch currents must add up to the source current (4 A in this case). Thus, using Ohm's law, we can see that the current in the 2-Ω resistor is $I = \frac{6}{2} = 3$ A. Thus, the current in resistor R must be equal to 1 A. Since $P = VI$, the power in resistor R is $P = (6)(1) = 6$ W.
8. **D** Using Ohm's law, $R = V / I = \frac{6}{1} = 6$ Ω.
9. **B** In a series circuit, the two resistors have the same current but proportionally shared potential differences. Thus, since the 10-Ω resistor has twice the resistance of the 5-Ω resistor in series, it has twice the

potential difference. With the currents equal, this means the 10-Ω resistor will generate twice as much energy when the circuit is on.

10. **E** The formula for current in an *RC* circuit is given by

$$I = \frac{V}{R} e^{-t/RC}$$

Substituting all given values and using amperes for current, we obtain

$$0.05 = \frac{24}{200} e^{-4/200C}$$
$$0.417 = e^{-0.02/C}$$
$$\ln 0.417 = -0.02/C$$
$$C = 0.023 \text{ F}$$

Free-Response Problems

1. We need to assume that all the electric energy produced is converted into heat. The electric energy is given by $E = VIt$, and using Ohm's law, $I = V/R = \frac{120}{30} = 4$ A. In this case, the time $t = 120$ s. Thus, $E = (120)(4)(120) = 57{,}600$ J. Now, in water, $Q = mc\,\Delta T$. Since the specific heat of water is 4.19 kJ/kg-°C, we need to convert to joules by multiplying by 1,000. Also, $m = 250$ g $= 0.25$ kg. Thus, $57{,}600 = (0.25)(4{,}190)\Delta T$. This implies that $\Delta T = 54.98$ C. Since this is added to the water, the new temperature will be approximately 80°C.

2. (a) The first thing we need to do is to reduce the series portion of the parallel branch. The equivalent resistance there is 8 Ω + 2 Ω = 10 Ω, and this resistor is in parallel with the other 10-Ω resistor. The equivalent resistance is found to be equal to 5 Ω for the entire parallel branch and is in series with the 5-Ω, making a final equivalent resistance of 10 Ω.

   (b) We can determine the source current using Ohm's law: $I = V/R = \frac{40}{10} = 4$ A.

   (c) To determine the reading of ammeter *A*, we first note that the potential drop across the 5-Ω resistor on the bottom is equal to 20 V since $V = IR = (4)(5) = 20$. Now the equivalent resistance of the parallel branch is also 5 Ω, and therefore across the entire branch there is also a potential difference of 20 V (since in a series circuit the voltages must add up—in this case to 40 V). In a parallel circuit, the potential difference is the same across each portion. Thus, the 10-Ω resistor has 20 V across it. Based on Ohm's law, this implies a current reading of 2 A for the ammeter.

   (d) If, from part (c), ammeter *A* reads 2 A, the top branch must be getting 2 A of current as well (since the source current is 4 A). Thus, from Ohm's law, the voltage across the 8-Ω resistor is given by $V = IR = (2)(8) = 16$ V.

3. We begin by applying the junction rule. From the diagram, we see that $I_1 = I_2 + I_3$. Then we apply the loop rule for both loops by going clockwise around each one. For the left loop, we cross the EMFs from the positive side to the negative side, treating them as negatives in our equations. Both resistors are crossed in the direction of each current, and so the potential drop across each of them is taken as a positive value. Thus, we write

$$-3 - 4 - 2I_2 - 8I_1 = 0$$

For the right loop, we cross the EMFs from the negative side, and we write them as positive values. The 10-Ω resistor is crossed in the direction of I_3, and the 2-Ω resistor is crossed in the direction opposite I_2. For this loop we can write

$$4 + 6 - 10I_3 + 2I_2 = 0$$

These three equations must be solved for I_1, I_2, and I_3.

Performing the solutions simultaneously yields $I_1 = -0.55$ A, $I_2 = -1.29$ A, and $I_3 = +0.74$ A.

4. Since the voltage across *AC* is zero, we can use Kirchoff's laws to justify that

$$V_{AB} = V_{BC}$$
$$V_{AD} = V_{DC}$$
$$I_1R_1 = I_2R_3$$
$$I_1R_2 = I_2R_x$$

Taking the ratio of these last two equations leads to the desired result.

18 MAGNETISM

MAGNETIC FIELDS AND FORCES

When two statically charged point objects approach each other, they exert a force on each other that varies inversely with the square of their separation distance. We can also state that one of the charges creates an electrostatic field around itself and that this field transmits the force through space.

When we have a wire carrying current, we no longer consider the effects of static fields. A simple demonstration reveals that when two current-carrying wires are arranged parallel to each other such that the currents are in the same direction, then the wires are attracted to each other with a force that varies inversely with their separation distance and directly as the product of the currents in each wire. If the currents are in opposite directions, then the wires are repelled. Since this is not an electrostatic event, we must conclude that a different force is responsible.

The nature of this force can be further understood when we consider the fact that a statically charged object (like an amber stone) cannot pick up small bits of metal. However, certain naturally occurring rocks called **lodestones** can attract metal objects. Lodestones are called **magnets** and can be used to *magnetize* metal objects (especially those made from iron or steel).

If a steel pin is stroked in one direction with a magnetic lodestone, the pin will become magnetized as well. If an iron rail is placed in the earth for many years, it will also become magnetized. If cooling steel is hammered while lying on a north-south line, it too will become magnetized (heating eliminates the magnetic effect). If the magnetized pin is placed on a floating cork, it will align itself along a general north-south line. If the cork's orientation is shifted, it will oscillate around its original direction and eventually settle down along its original equilibrium direction. Finally, if another magnet is brought near, the "compass" will realign itself toward the new magnet.

Each of these experiences suggests the presence of a magnetic force, and we can therefore consider the actions of this force through the description of a magnetic field. If a magnet is made in the shape of a rectangular bar (see Figure 1), then the orientation of a compass moved around the magnet will demonstrate that the compass aligns itself tangentially to some imaginary field lines that were first discussed by the English physicist Michael Faraday. In general, we label the north-seeking pole of the magnet N and the south-seeking pole S. We also observe that these two opposite poles behave similarly to opposite electric charges; like poles repel, while unlike poles attract.

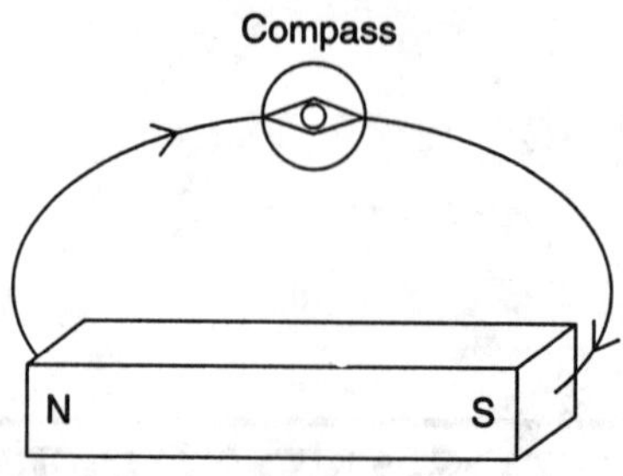

FIGURE 1.

A more fundamental difference between these two effects is that magnet poles never appear in isolation. If a magnet is broken in two pieces, then each new magnet has a pair of poles. This phenomenon continues even to the atomic level. There appear at present to be no magnetic monopoles naturally occurring in nature (new atomic theories have postulated the existence of magnetic monopoles but none as yet have been discovered).

Small pieces of iron (called *filings*) can be used as miniature compasses to illustrate the characteristics of the magnetic field around magnets made of iron or other similar alloys. The ability of a metal to be magnetized is called its **permeability**, and substances like iron, cobalt, and nickel have among the highest permeabilities. These substances are sometimes called **ferromagnets**.

In Figure 2, several different magnetic field configurations are shown. By convention, the direction of the magnetic field is taken to be from the north pole of the magnet to the south pole.

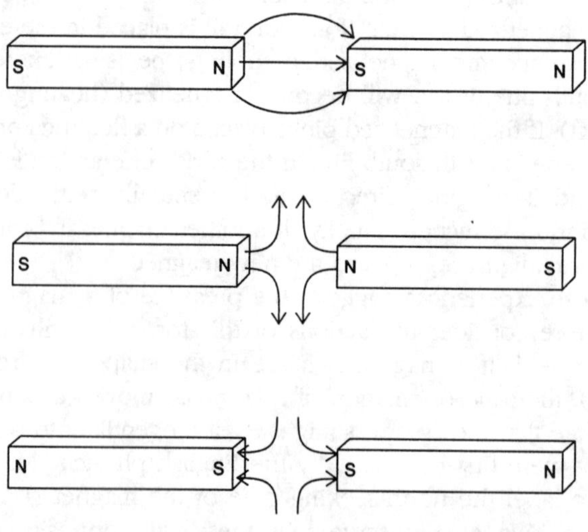

FIGURE 2.

MAGNETIC FORCE ON A MOVING CHARGE

If a moving electric charge enters a magnetic field, it will experience a force that depends on its velocity, charge, and orientation with respect to the field. The force will also depend on the magnitude of the magnetic field strength (designated **B** algebraically). This field strength is a vector quantity just like gravitational and electric field strengths.

Experiments with moving charges in a magnetic field reveal that the resultant force is a maximum when the velocity **v** is perpendicular to the magnetic field, and zero when the velocity is parallel to the field. With this varying angle of orientation θ and a given charge Q, we find that the magnitude of the magnetic force is given by

$$\mathbf{F} = \mathbf{B}Q\mathbf{v} \sin \theta$$

Since the units of force must be newtons, the units for the magnetic field strength **B** must be N-s/C-m. Recall that N-s is the unit for an impulse. One can think of the magnetic field as a measure of the impulse given to 1 C of charge moving a distance of 1 m in a given direction. Additionally, if we have a beam of charges, we effectively have an electric current. The unit s/C can be interpreted as the reciprocal of an ampere, and so the strength of the magnetic field in a current-carrying wire is given in N/A-m.

In the SI system, this combination is called a **tesla** such that 1 T is equal to 1 N/A-m. The earliest experiments on electromagnetism were made by Hans Oersted, who first showed that an electric current can influence a compass. Oersted's discovery, in 1819, established the new science of **electromagnetism**.

The direction of the magnetic force is given by a **right-hand rule** (Figure 3):

> Open your right hand so that your fingers point in the direction of the magnetic field and your thumb points in the direction of the velocity of the charges (or the current if you have a wire). Your open palm will show the direction of the force.

The direction provided by the right-hand rule is the direction that a positive test charge will take. Thus, the direction for a negative charge will be opposite what the right-hand rule predicts. If the velocity and magnetic field are perpendicular to each other, then Figure 4 illustrates that the path of the charge in the field is a circle of radius R.

The circular path arises because the induced force is always at right angles to the deflected path when the velocity is initially perpendicular to the magnetic field. This is similar to the circular path of a stone being swung in an overhead horizontal circle while attached to a string.

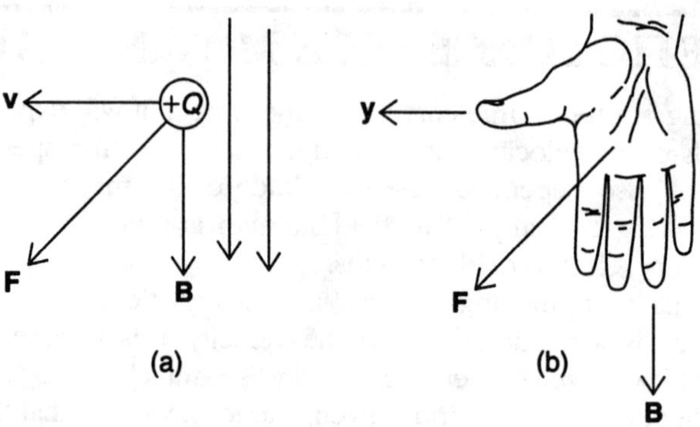

FIGURE 3.

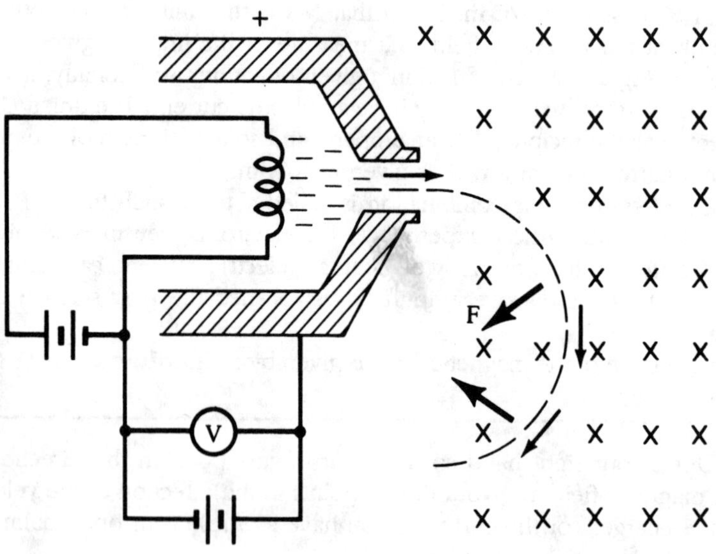

FIGURE 4.

In the frame of reference of the moving charge, the inward magnetic force is apparently balanced by an outward centrifugal force. This relationship is given by

$$\Sigma \mathbf{F} = \mathbf{B}Q\mathbf{v} = \frac{m\mathbf{v}^2}{R}$$

where m is the mass of the charge.

This relationship can be used to determine the radius of the path:

$$R = \frac{m\mathbf{v}}{Q\mathbf{B}}$$

This expression assumes that the velocity and mass of the charge are known independently. Even if you wish to use this expression to find the mass of the charge (as used in a *mass spectrograph*):

$$m = \frac{RQ\mathbf{B}}{\mathbf{v}}$$

you still must know what the velocity is independent of the mass of the charge.

One way to determine the velocity of the charge independently of the mass is to pass the charge through both an electric and a magnetic field (as in a *cathode ray tube*). As shown in Figure 5, if the electric field is horizontal out of the page, then the charges will be attracted or repelled along a horizontal line. If the magnetic field is vertical, then the right-hand rule will force the charges along the same horizontal line. Careful adjustment of both fields (keeping the velocity perpendicular to the magnetic field) can lead to a situation where the effect of one field balances out the effects of the other and the charges remain undeflected. In that case, the magnitudes of the electric and magnetic forces are equal:

$\mathbf{E}Q = \mathbf{B}q\mathbf{v}$, which implies that the velocity $\mathbf{v} = \mathbf{E} / \mathbf{B}$ (independent of mass).

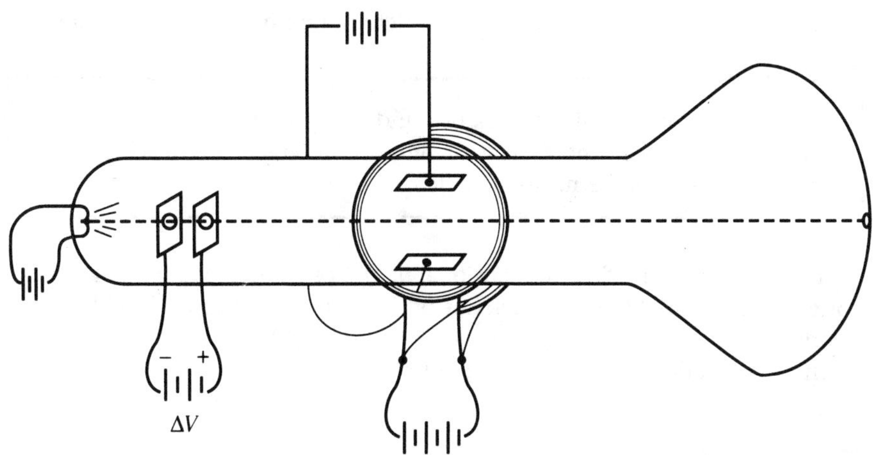

FIGURE 5.

If the velocity is not perpendicular to the magnetic field when it first enters, then the path will be a spiral. This is because the velocity vector will have two components. One component will be perpendicular to the field (and create a magnetic force that will try to deflect the path into a circle), while the other component will be parallel to the field (producing no magnetic force and maintaining the original value of this component). This situation is illustrated in Figure 6.

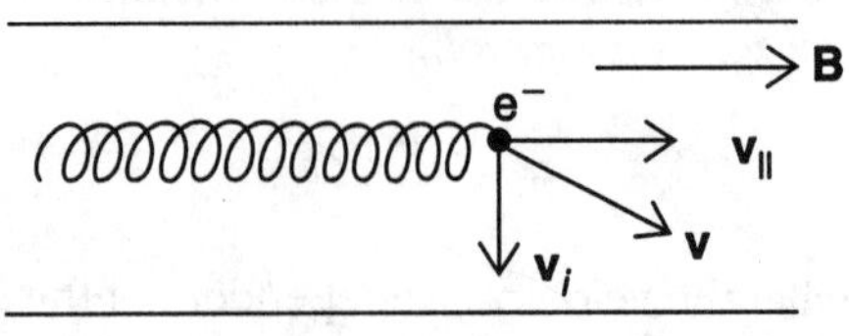

FIGURE 6.

MAGNETIC FIELDS DUE TO CURRENTS IN WIRES

Long, Straight Wires

If a compass is placed near a long, straight wire carrying current, then the magnetic field produced by the current will cause the compass to align tangent to the field. The shape of the magnetic field is a series of concentric circles (Figure 7a) whose rotational direction is determined by a right-hand rule:

> Grasp the wire with your right hand. If your thumb points in the direction of the current, then your fingers will curl around the wire in the same sense as the magnetic field.

A schematic for the three-dimensional nature of the field is sometimes used. For an emerging field, we use a series of dots, and for an inward field we use a series of Xs (see Figure 7b).

The magnitude of the magnetic field strength at some distance *r* from the wire is given by

$$\mathbf{B} = \frac{\mu I}{2\pi r}$$

where μ is the permeability of the material around the wire in N/A^2. In air, the value of μ is approximately equal to the permeability of a vacuum:

$$\mu_0 = 4\pi \times 10^{-7}\ N/A^2$$

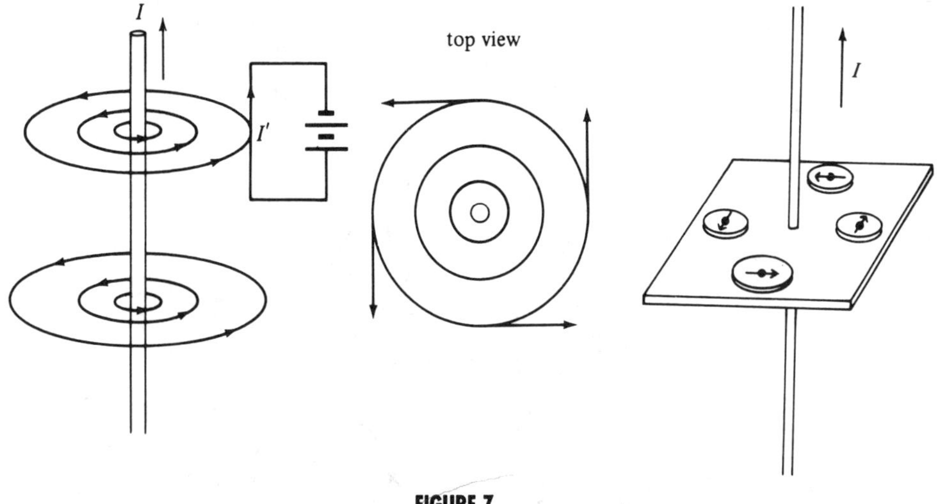

FIGURE 7.

The preceding equation is therefore usually written (for air or a vacuum) in the form

$$\mathbf{B} = \frac{\mu_0 I}{2\pi r}$$

A Current Loop

Imagine a straight current-carrying wire that has been bent into the shape of a loop. Each of the circular magnetic fields now interacts in the center of the loop (of radius *r*). In Figure 8, we see that the effect of all these subfields is to produce one concentrated field at the center pointing inward or outward.

The general direction of the emergent magnetic field is given by a right hand rule (see Figure 9):

> Grasp the loop in your right hand with your fingers curling around it in the same direction as the current. Your thumb will now show the direction of the magnetic field.

If the loop consists of *N* turns of wire, then the magnetic field strength will be increased by *N* times. The magnitude of the field strength at the center of the loop (of radius *r*) is given by

$$\mathbf{B} = \frac{\mu N I}{2r}$$

The strength of the field can also be increased by changing the permeability of the *core*. If the wire is looped around iron, the field will be stronger than if

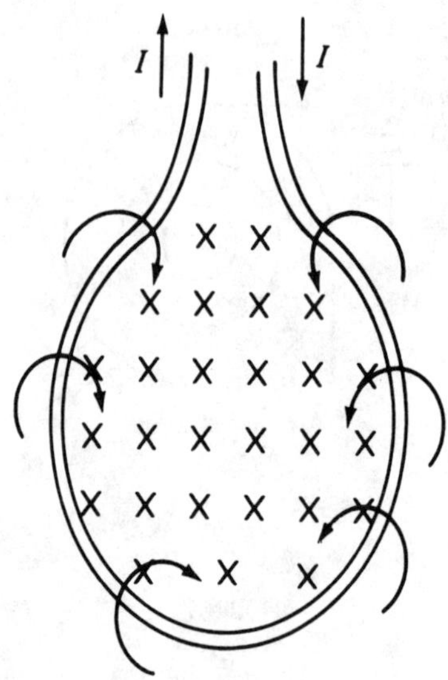

FIGURE 8.

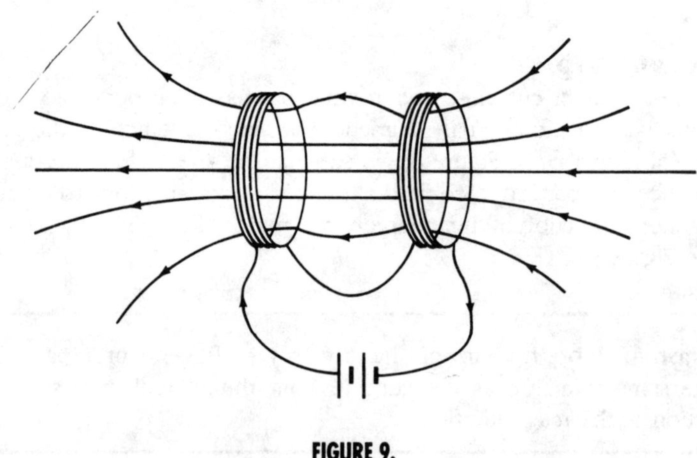

FIGURE 9.

it is looped around cardboard. Also notice that if the loop's radius is smaller, the field will be more concentrated at the center and thus stronger.

The Solenoid

If the loop on page 277 is stretched out to a length L in the form of a spiral, we form what is called a **solenoid** (Figure 10). This is the basic form for an

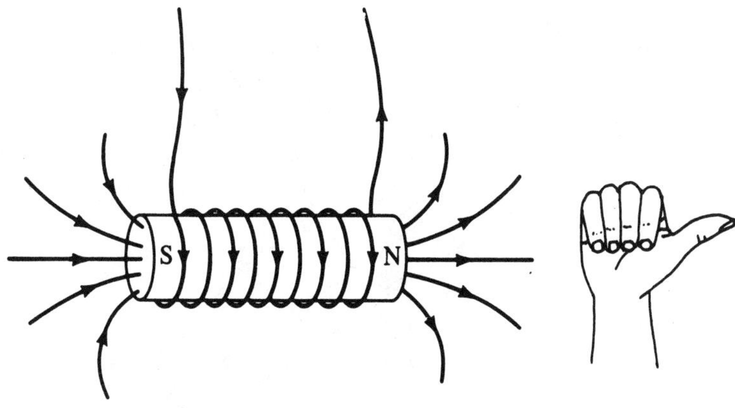

FIGURE 10.

electromagnet in which wire is wrapped around an iron nail and then connected to a battery. The strength of the field increases with the number of turns, the permeability of the core, and the amount of current in the wire:

$$\mathbf{B} = \frac{\mu NI}{L}$$

The direction of the magnetic field is given by another right-hand rule:

> Grasp the solenoid with your right hand so that your fingers curl in the same direction as the current. Your thumb will point in the direction of the emergent magnetic field. The field inside a solenoid is uniform in strength.

MAGNETIC FORCE ON A WIRE

The magnetic force on a charge Q moving in a magnetic field **B** with a velocity **v** is given by $\mathbf{F} = \mathbf{B}q\mathbf{v} \sin \theta$. If we have a collection of charges, then effectively we have an electric current given by $I = Q/t$, where Q is now the total charge present. If the current passes through a length L in time t, then the induced magnetic force is given by

$$\mathbf{F} = \mathbf{B}IL \sin \theta$$

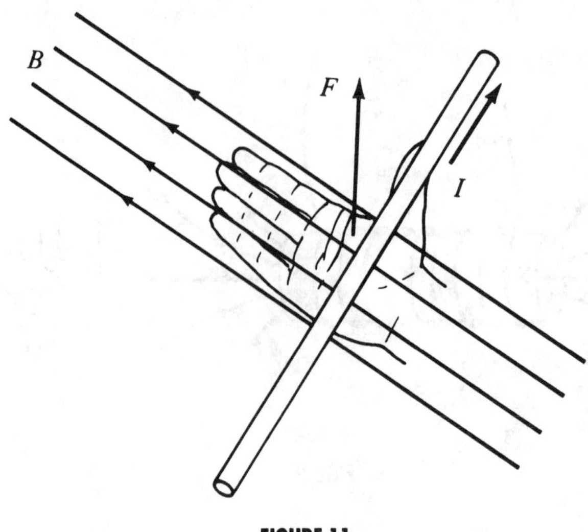

FIGURE 11.

If the charges are contained within a wire and the wire is perpendicular to the magnetic field, then the force is simply given by $\mathbf{F} = \mathbf{B}IL$.

The direction of the force is given by a right-hand rule (Figure 11):

> Point the fingers of your right hand in the direction of the magnetic field such that your thumb points in the direction of the current. The palm of your hand will show the direction of the force.

MAGNETIC FORCE BETWEEN TWO WIRES

Imagine you have two straight wires with currents going in the same direction (Figure 12). Each wire creates a circular magnetic field around itself such that in the region between the wires the net effect of each field is to weaken each other relative to the other side. This difference produces a net inward attractive force. If the current in one wire is reversed, then the force will become repulsive.

The magnitude of the force between them at a distance r can be determined if we consider the fact that one wire sets up the external field that acts on the other. If we want the force on wire 2 due to wire 1, then we can write

$$\mathbf{B}_1 = \frac{\mu_0 I_1}{2\pi r}$$

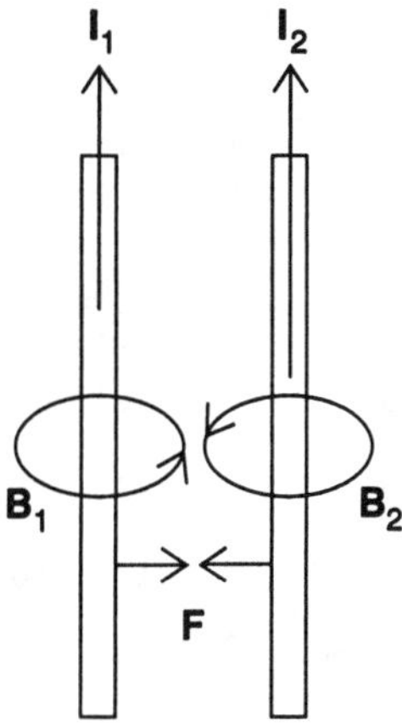

FIGURE 12.

Since $\mathbf{F} = \mathbf{B}IL$, we can write

$$\mathbf{F}_{12} = \frac{\mu_0 I_1 I_2 L}{2\pi r}$$

ELECTRIC MOTORS AND METERS

Torque on a Current Loop of Wire

Imagine a rectangular loop of wire carrying current. If the plane of the loop is parallel to an external magnetic field (see Figure 13), then a net torque will be produced causing the loop to rotate about an axis perpendicular to the field.

The torque can be observed using the right-hand rule. On the right side a downward force is produced, while on the left side there is an upward force. This combination causes the rotation of the loop, and this effect will continue until the plane of the loop is perpendicular to the field. At that point, the net force (and torque) will be zero. With the loop vertical, angular momentum may allow the loop to continue to rotate until it is once again in the original horizontal position.

The Galvanometer

A **galvanometer** is a device that detects small amounts of electric current. A simple galvanometer consists of a coil of wire with a pointer attached. The

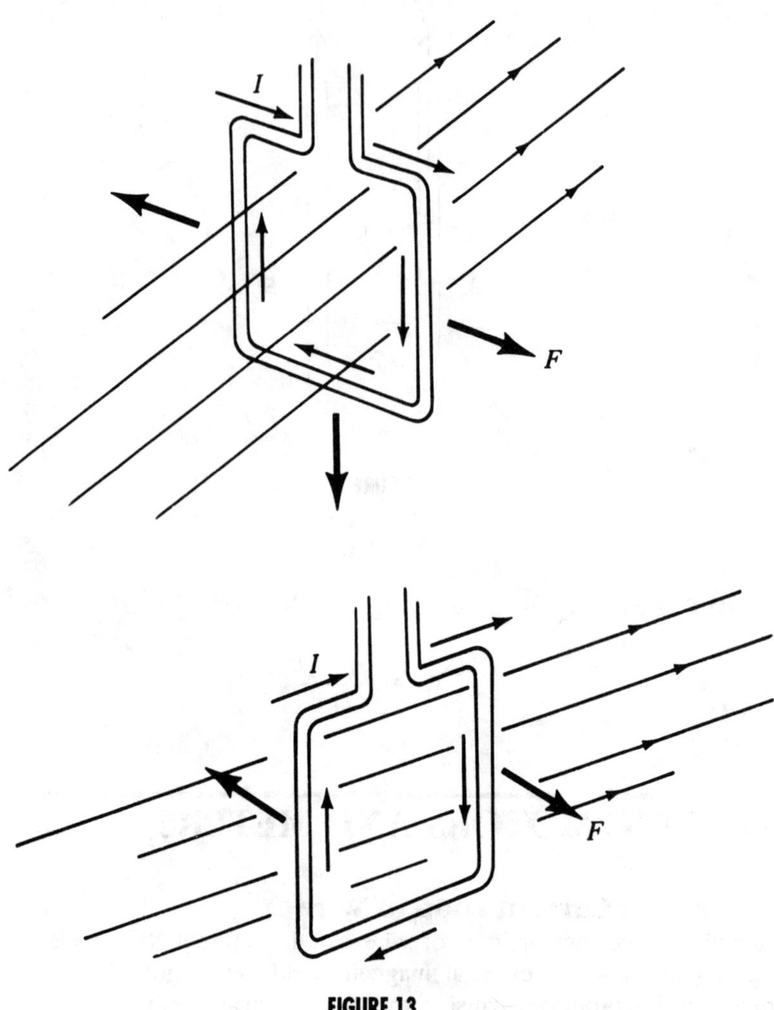

FIGURE 13.

coil is placed in an external magnetic field (see Figure 14) such that when current flows through the coil, the torque produced causes the pointer to deflect along a calibrated scale.

The Voltmeter

When a typically large resistor is placed in series with a coil, the galvanometer becomes a **voltmeter** (see Figure 15). Thus, when placed in parallel within a circuit, the large resistance of the voltmeter implies a low parallel resistance path for current.

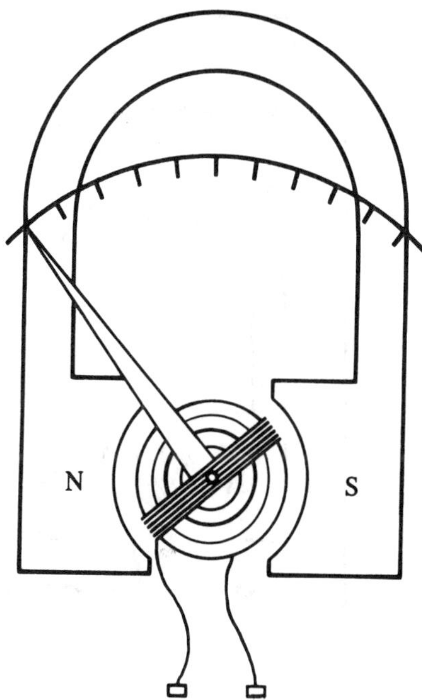

FIGURE 14. *Simple galvanometer.*

The Ammeter

When a small resistor is placed across the coil of a galvanometer (called a *shunt*), the galvanometer becomes an **ammeter** (see Figure 16). This reduces the resistance of the device so that when the ammeter is placed in series within a circuit to measure current, only a small amount of current is used to operate the ammeter.

The Electric DC Motor

On page 278 we saw that a current loop placed in an external magnetic field can experience a torque until its plane is perpendicular to the field. To maintain continuous rotation, the current must be reversed so that the effective torque will maintain a continuous rotational direction. This reversal, with a direct current, is achieved through the use of brushes and a split ring commutator. *Brushes* are metal strips that bring current to the coil, and a *commutator* is a metal band that is alternately connected to the positive and negative terminals of the battery as it rotates through the brushes. The split in the commutator allows for alternation of the current, which creates an alternate torque and maintains the rotation (see Figure 17).

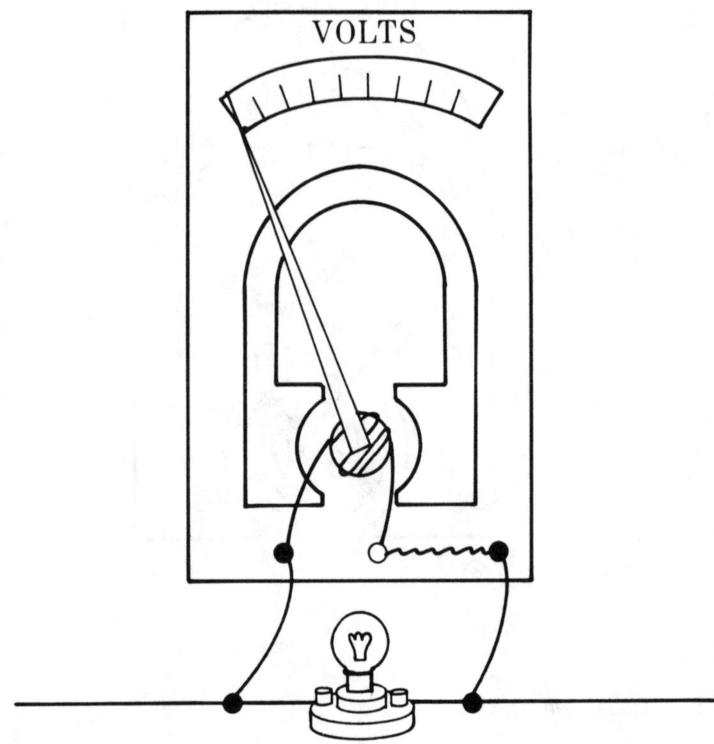

FIGURE 15. *Simple voltmeter.*

MAGNETIC PROPERTIES OF MATTER

When an iron core is placed inside a solenoid, the induced electromagnetic field is greatly increased. Previously, we attributed this observation to the high permeability of iron. In fact, iron, cobalt, and nickel share some atomic properties that enable them to have a high level of magnetization. Collectively, iron, cobalt, and nickel (and their alloys) are called **ferromagnetic** materials. Microscopic photographs reveal that ferromagnetic substances contain groupings of atoms called *domains*.

In an unmagnetized state, the magnetic moments of the atoms are randomly positioned. When placed in an external magnetic field, the moments can be induced to align themselves with the magnetic field, and this effect is called **magnetization**. When no more domains can be aligned, the material is said to be *saturated*.

The magnetization hypothesis can be understood if we accept the idea of an electron orbiting a nucleus. This rotating charge can induce a magnetic field on an atomic scale. In certain elements with an even number of electrons, these magnetic fields can cancel out. This effect is called **diamagnetism** and

FIGURE 16. *Simple ammeter.*

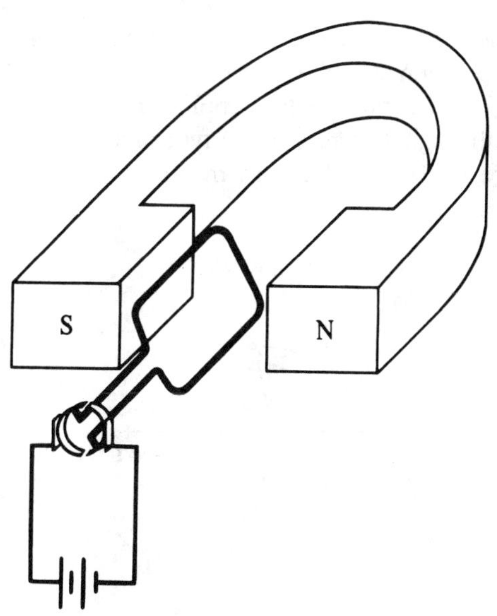

FIGURE 17.

has the tendency to reduce an external field. Elements with an odd number of electrons can have some atomic magnetic fields aligned with an external field, but only weakly. This effect is called **paramagnetism**.

For ferromagnetic materials, the alignment of the domains can be affected by impact or by a change in temperature. Dropping or hitting a ferromagnet can disrupt the domains and decrease the magnetization. Similarly, increasing the temperature of a ferromagnet can disrupt the domains because of the increase in molecular activity.

ENERGY TRANSFER, MAGNETIZATION, AND HYSTERESIS

In Chapter 17, we observed that the energy transfer in a mechanical system or an electric circuit can be maximized if an inertia-dependent variable is matched (*mass* in the mechanical system and *resistance* in the circuit. The opposition to the transfer of energy by a mechanical system is called **impedence**).

In this section, we consider the response of a system to a driving force. The time lag exhibited by a system reacting to changes in the driving forces is called **hysteresis** (from the Greek, meaning "to lag behind"). Additionally, the system demonstrates that the response is dependent on past responses to the same changes (sensitivity to initial conditions).

In magnetism, hysteresis is exhibited in ferromagnetic substances subjected to an external magnetic field. The magnetization process depends on the strength of the field, and once magnetized, ferromagnetic material retains some of the magnetization even after the external field is removed. Hysteresis occurs when the external field is first applied until saturation is achieved and is then reversed, passing through its zero point. A typical ferromagnetic hysteresis loop is shown in Figure 18. Losses of energy may also occur in the process.

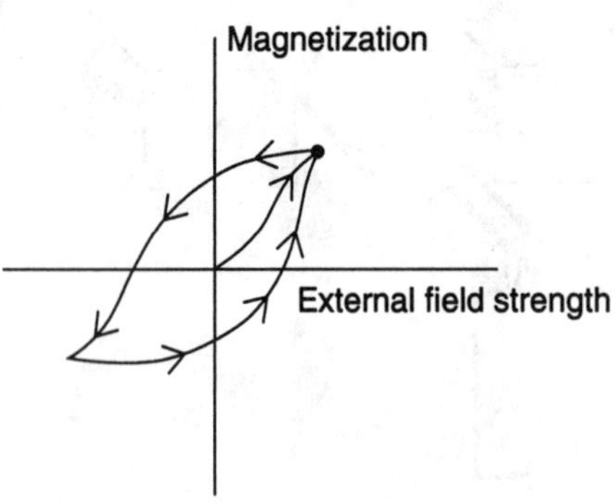

FIGURE 18. *Typical ferromagnetic hysteresis loop.*

PROBLEM-SOLVING STRATEGIES

The key to magnetic and electromagnetic field problems is remembering the right-hand rules. These are heuristics that have been shown to be useful in determining the direction of field interaction forces. Memorize each one and become comfortable with its use. Some rules are familiar in their use of fingers, thumb, and open palm. Make sure, however, that you know what the net result is. Drawing a sketch always helps.

Keeping track of units is also useful. Teslas are SI units and as such require lengths to be in meters and velocities to be in m/s (forces are in newtons). Most force interactions are angle-dependent and have maximum values when two quantities (usually velocity and field or current and field) are perpendicular. Read each problem carefully. Also, remember that the magnetic field **B** is a vector quantity.

Sample Problem

An electron enters a magnetic field perpendicularly as shown here. The strength of the magnetic field is 100 T. If the velocity of the electron is 17,000 m/s, what is the magnitude and direction of the induced magnetic force on the electron?

B

XXXXXXXXXXXXXXXXXX

$e^- \rightarrow$

XXXXXXXXXXXXXXXXXX

XXXXXXXXXXXXXXXXXX

Solution

Using the right-hand rule, we know that if the charge is positive, then the direction of the induced force will be directed upward. Since the electron is negatively charged, the direction is reversed and is downward. The magnitude of the force is given by $\mathbf{F} = \mathbf{B}Q\mathbf{v}$, where $\mathbf{v} = 17{,}000$ m/s, $\mathbf{B} = 100$ T, and $Q = 1.6 \times 10^{-19}$ C. Thus, $\mathbf{F} = 2.72 \times 10^{-13}$ N.

PRACTICE PROBLEMS FOR CHAPTER 18

Thought Problems

1. Describe the actual magnetic polarities in an ordinary compass.
2. Using sketches and the right-hand rule, explain the following observation. A wire coming out of the page is carrying an electric current that is also coming out of the page. It is placed in an external magnetic field directed horizontally from the left to the right-hand side of the page. Explain why a force acting upward is induced on the wire.

3. What happens to the maximum voltage read on a voltmeter as the resistance placed in series with its coil is changed?

Multiple-Choice Problems

1. A charge moves in a circular orbit of radius *R* because of a uniform magnetic field. If the velocity of the charge is doubled, then the orbital radius will become ____.
   (A) 2*R*
   (B) *R*
   (C) *R*/2
   (D) 4*R*
   (E) *R*/4
2. Inside a solenoid, the magnetic field ____.
   (A) is zero
   (B) decreases along the axis
   (C) increases along the axis
   (D) is uniform
   (E) increases and then decreases along the axis
3. A proton enters a magnetic field of 10 T and an electric field of 2,000 N/C such that it passes through both fields undeflected. The velocity of the proton will become____m/s.
   (A) 2,000
   (B) 200
   (C) 200,000
   (D) 20,000
   (E) 2,000,000
4. Which of the following will decrease the strength of the magnetic field of a solenoid?
   (A) an increase in the permeability of the core
   (B) an increase in the temperature
   (C) an increase in the current
   (D) an increase in the number of turns of wire
   (E) none of the above
5. An electron crosses a perpendicular magnetic field as shown. The direction of the induced magnetic force is____.

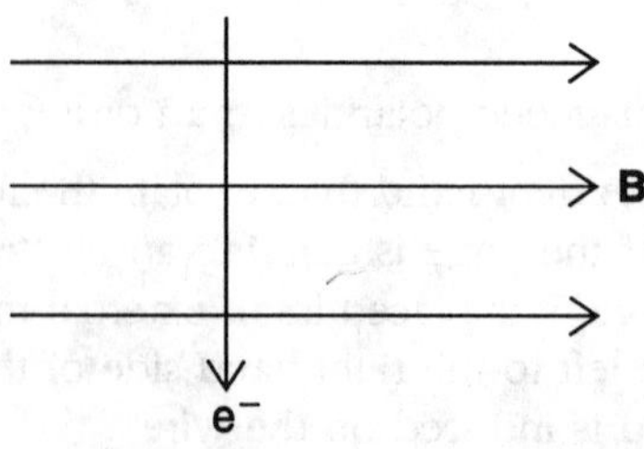

(A) to the right
(B) to the left
(C) up the page
(D) out of the page
(E) into the page

6. Three centimeters from a long, straight wire, the magnetic field produced by the current is determined to be equal to 3×10^{-5} T. The current in the wire must be _____A.
(A) 2.0
(B) 4.5
(C) 1.5
(D) 3
(E) 5.0

7. In which direction does the compass needle point when it is placed near the straight wire carrying current in the direction shown?

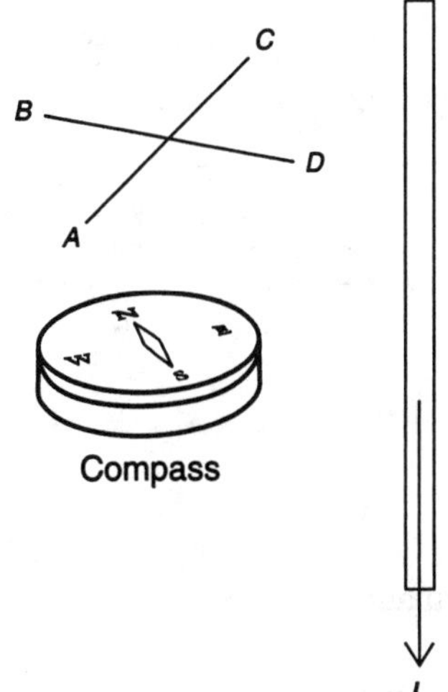

(A) A
(B) B
(C) C
(D) D
(E) the direction varies

8. Magnetic field lines determine _____.
(A) only the direction of the field
(B) the relative strength of the field
(C) both the relative strength and the direction of the field
(D) only the shape of the field
(E) all of these characteristics

9. What is the direction of the magnetic field at point A above the wire carrying current?

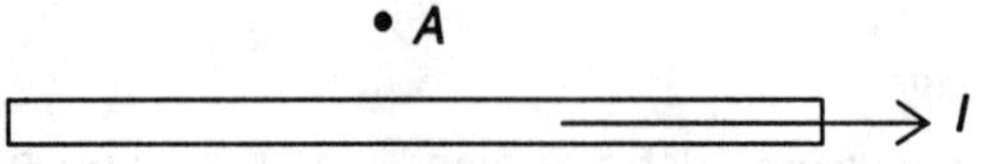

(A) out of the page
(B) into the page
(C) up the page
(D) down the page
(E) to the right

10. In which of the following situations has an electron entered the magnetic field at a nonright angle?

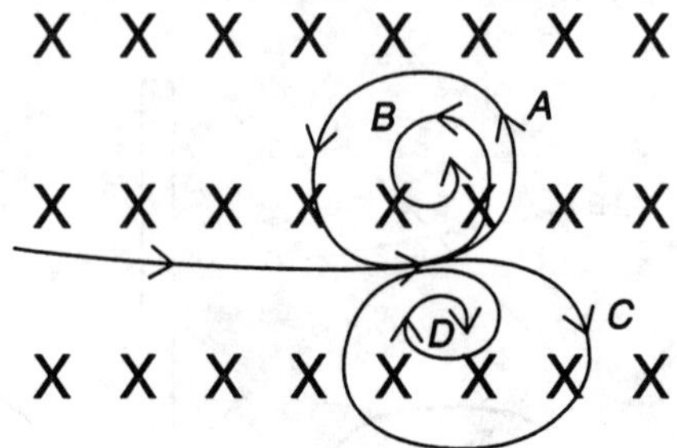

(A) A
(B) B
(C) C
(D) D
(E) none of these situations

Free-Response Problems

1. An electron is accelerated by a potential difference of 12,000 V as shown in the accompanying diagram. The electron enters a cathode ray tube that is 20 cm in length, and the external magnetic field causes the path to deflect along a vertical screen a distance y.

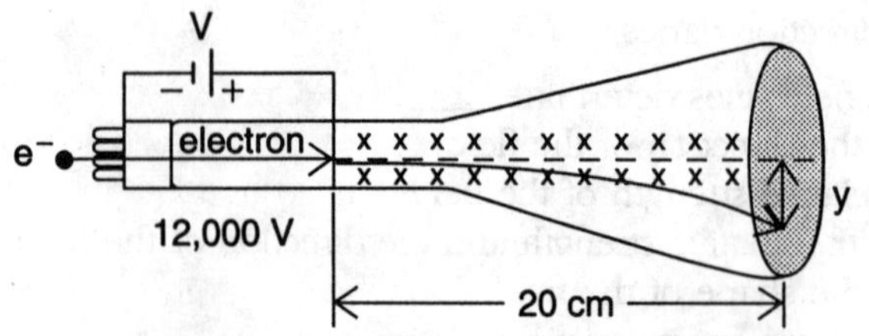

(a) What is the kinetic energy of the electron as it enters the cathode ray tube?
(b) What is the velocity of the electron as it enters the cathode ray tube?
(c) If the external magnetic field has a strength of 4×10^{-5} T, what is the value of y? Assume that the field has a negligible effect on the horizontal velocity.

2. A proton enters a magnetic field perpendicularly with a velocity $\mathbf{v}$.
   (a) Derive an expression for the angular frequency ω in terms of charge, mass, and field strength.
   (b) If the field strength is 2 T, what is the value of the angular frequency?
   (c) If the radius of the path is 2.2 mm, what is the linear velocity of the proton?

3. A straight conductor has a mass of 15 g and is 4 cm long. It is suspended from two parallel identical springs stretching a distance of 0.3 cm. The system is attached to a rigid source of potential difference equal to 15 V, and the overall resistance of the circuit is 5 Ω. When current flows through the conductor, an external magnetic field is turned on, and it is observed that the springs stretch an additional 0.1 cm. What is the strength of the magnetic field?

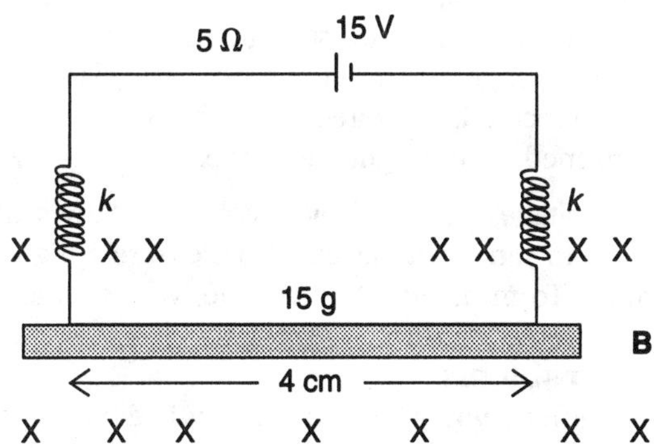

SOLUTIONS TO PRACTICE PROBLEMS

Thought Problems

1. What we generally call the "north" pole of a compass is in actuality a north-seeking pole since only the opposite type of pole can be attracted toward what is referred to as *geomagnetic north*. Likewise, the other end of the compass is a south-seeking pole. Incidentally, geophysicists have long known that the geomagnetic North Pole of the earth does not coincide with the geographic North Pole. Additionally, the geographic North Pole is actually a magnetic South Pole.

2. A diagram of the situation is shown.

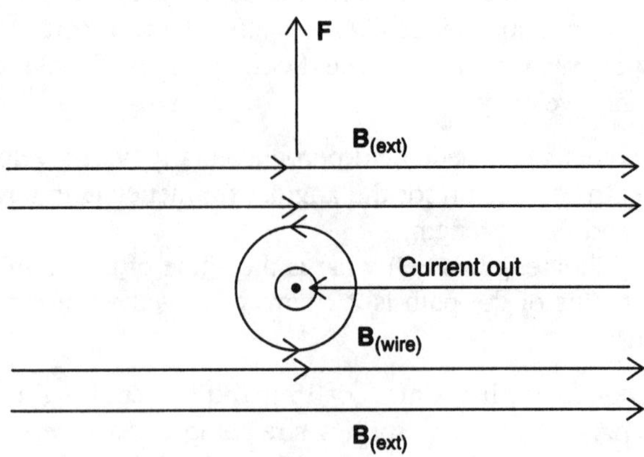

If you examine this diagram closely, you will notice that above the cross section of the wire (a circle), the two fields are interacting in opposite directions. The field of the wire is counterclockwise, using the right-hand rule for a wire, which tends to weaken the resultant field at this location. At the bottom of the wire, the two interacting fields are pointing in the same direction, which tends to strengthen the resultant field at this location. With a stronger field on the bottom relative to the top, the induced force is directed in the direction of the weaker field (i.e., upward), and this is confirmed by the right-hand rule.

3. The maximum voltage measured by a voltmeter varies directly with the resistance placed in series with its coil. To measure a high voltage, we use a high resistance. To measure a low voltage, we use a low resistance.

Multiple-Choice Problems

1. **A** The formula is $R = m\mathbf{v}/q\mathbf{B}$. If the velocity is doubled, so is the radius.
2. **D** Inside a solenoid, the effect of all the coils is to produce a long, uniform magnetic field.
3. **B** The formula is $\mathbf{v} = \mathbf{E} / \mathbf{B} = 2{,}000 / 10 = 200$ m/s.
4. **B** Increasing the temperature of the solenoid increases the resistance of the coils, which decreases the current and hence decreases the strength of the magnetic field.
5. **E** Using the right-hand rule, we place the fingers of the right hand along the line of the magnetic field and point the thumb in the direction of the velocity. The palm points outward. However, since the particle is an electron and the right-hand rule is designed for a positive charge, we must reverse the direction and say that the force is directed *inward*.

6. **B** The formula is

$$\mathbf{B} = \frac{\mu_0 I}{2\pi r} = \frac{2 \times 10^{-7} I}{r}$$

Recalling that $r = 3$ cm $= 0.03$ m, substitution of all the given values yields $I = 4.5$ A.

7. **C** Imagine a circle surrounding the wire. With the right-hand rule, the direction of the circle is clockwise if you are looking down on it. In this orientation, the field goes from right to left. On the left side, it enters the page, and in perspective that is the C direction.

8. **C** Magnetic field lines were introduced by Michael Faraday to determine both the direction of the field and its relative strength (a stronger field is indicated by a greater line density).

9. **A** With the right-hand rule, the circle surrounding the wire will have its field coming out of the page at point *A*.

10. **D** When a charged particle enters a magnetic field at an angle other than 90 degrees, it takes a spiral path. Using the right-hand rule and reversing for the negative charge gives us path D for the desired spiral path (seen in perspective since the field is inward).

Free-Response Problems

1. (a) The kinetic energy of the electron is the product of its charge and potential difference. Thus, KE $= (1.6 \times 10^{-19})(12{,}000) = 1.92 \times 10^{-15}$ J.

   (b) The formula for the velocity, knowing the mass and kinetic energy, is

$$\mathbf{v} = \sqrt{\frac{2\text{KE}}{m}} = \sqrt{\frac{(2)(1.92 \times 10^{-15})}{9.1 \times 10^{-31}}} = 6.5 \times 10^{7} \text{ m/s}$$

   (c) The magnetic force causes a downward uniform acceleration, which can be found from Newton's second law, $\mathbf{F} = m\mathbf{a}$. The force is due to the magnetic force $\mathbf{F} = q\mathbf{v}\cdot\mathbf{B}$. Thus, $\mathbf{a} = q\mathbf{v}\cdot\mathbf{B}/m$. Using the known values, we have $\mathbf{a} = 4.57 \times 10^{14}$ m/s^2. Now, for linear motion downward (recall projectile motion), $y = \frac{1}{2}\mathbf{a}t^2$. The time required to drop y meters is the same time it takes to go 20 cm (0.20 m) (assuming no change in horizontal velocity). Thus, $t = x/\mathbf{v} = 0.20/6.5 \times 10^{7} = 3 \times 10^{-9}$ s. This implies that on substitution $y = 0.00205$ and $m = 2.05$ mm.

2. (a) We know that $\mathbf{v} = q\mathbf{B}r/m$ and $\omega = \mathbf{v}/r$. Thus, $\omega = q\mathbf{B}/m$.

   (b) Substituting the known values gives $\omega = 1.9 \times 10^{8}$ rad/s.

   (c) Now, $\mathbf{v} = \omega$ and $r = (0.0022)(1.9 \times 10^{8}) = 4.2 \times 10^{5}$ m/s.

3. The effect of the conductor is to stretch both springs. Since we have two parallel springs of equal force constant k, we know from our work on

oscillatory motion (Chapter 14) that the effective spring constant is equal to $2k$. Thus,

$$\mathbf{F} = 2k\ \Delta\mathbf{x}$$

where $\Delta\mathbf{x} = 0.003$ m. Thus, with the weight of the conductor, $\mathbf{W} = m\mathbf{g} = (0.015)(9.8) = 0.147$ N and $k = 24.5$ N/m. Now, the magnetic force $\mathbf{F} = \mathbf{B}IL$, is responsible for another elongation $\mathbf{x}$. Since $\mathbf{F} = 2k\mathbf{x}$, we have $\mathbf{B}IL = (2)(24.5)(0.001) = 0.0490$ N. Now, the length of the conductor is 4 cm $= 0.04$ m, and using Ohm's law, we know that the current $I = \frac{15}{5} = 3$ A. Thus, we find that $\mathbf{B}(3)(0.04) = 0.0490$ implies that $\mathbf{B} = 0.4$ T.

19
ELECTROMAGNETIC INDUCTION

INDUCED MOTIONAL EMF IN A WIRE

We know from Oersted's experiments that a current in a wire generates a magnetic field. We also know that the field of a solenoid approximates that of an ordinary bar magnet. Finally, the magnetic properties of materials (called **permeability**) influence the strength of the magnetic field.

A question can now be raised: Can a magnetic field be used to induce a current in a wire? The answer, investigated in the early nineteenth century by the French scientist Andre Ampere, is yes. However, it is not as simple as placing a wire in a magnetic field and having a current "magically" arise.

A simple experimental setup is shown in Figure 1. A horseshoe magnet is arranged with a conducting wire attached to a galvanometer. When the wire rests in the magnetic field, the galvanometer registers zero current. If the wire is moved through the field in such a way that its motion "cuts across" the imaginary field lines, then the galvanomter will register the presence of a small current. If the wire is moved through the field in such a way as to remain parallel to the field, then again no current is present. This experiment also reveals that a maximum current for a given velocity occurs when the wire moves perpendicularly through the field (which suggests a dependency on the sine of the orientation angle).

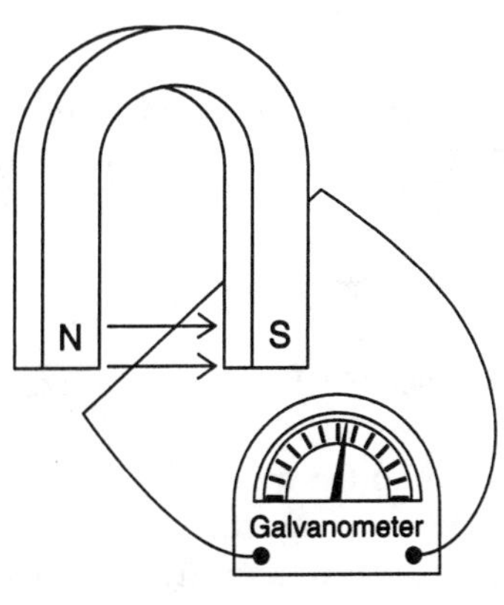

FIGURE 1.

This phenomenon is known as **electromagnetic induction**. The source of the induced current is the establishment of motional EMF as a result of the change in something called the magnetic flux which we will define momentarily.

If the wire is moved up and down through the field in Figure 1, then the galvanometer will show that the induced current alternates back and forth according to a right-hand rule. The fingers of your right hand point in the direction of the magnetic field, and your thumb points in the direction of the velocity of the wire. Finally, your open palm indicates the direction of the induced current.

Now, since a parallel motion through the field implies no induced current, if the wire is moved at any other angle, only the perpendicular component will contribute to the induction process. Additionally, it can be demonstrated that the magnitude of the induced current varies directly with the velocity of the wire, the length of wire in the field, and the strength of the external field. Since current is related to potential difference (EMF) through Ohm's law, $V = IR$, we can write

$$\text{EMF} = \mathbf{B}L\mathbf{v}$$

where $\mathbf{v}$ is the perpendicular velocity component.

The derivation of this expression involves a reconsideration of the induced magnetic force on a charge in an external magnetic field. Recall that if $\mathbf{v}$ is a perpendicular velocity component, then $\mathbf{F} = \mathbf{B}q\mathbf{v}$ is the expression for the magnitude of the induced magnetic force on a charge q. Now, the wire in our situation contains charges that will experience a force $\mathbf{F}$ if we can get them to move through an external field. Physically moving the entire wire accomplishes this task. The direction of the induced force is along the length of the wire (the direction determined by the right-hand rule). If we let Q represent the total charge, then we can write

$$\mathbf{F} = \mathbf{B}Q\mathbf{v}$$

The charges experience a force as long as they are moving in the magnetic field. This occurs for a length L, and the work done by the force is given by $W = \mathbf{F}L$. Thus,

$$\mathbf{W} = \mathbf{B}Q\mathbf{v}L$$

The induced potential difference (EMF) is a measure of the work done per unit charge, W/Q, which means that EMF $= \mathbf{B}L\mathbf{v}$. This EMF is sometimes called **motional EMF** since it is due to the motion of a wire through a field.

An interesting fact about electromagnetic induction is that the velocity in the preceding expression is a *relative velocity*. This means that the induced EMF can be produced whether the wire moves through the field or the field changes over the wire.

In Figure 2, a bar magnet is thrust into and out of a coil of wire. There are N turns of wire in this coil, and the number of coils is related to the overall length L in the original EMF expression. The galvanometer registers the alternating current in the coil as the magnet is inserted and withdrawn. Varying the velocity affects the amount of current as predicted. Moving the magnet around the outside of the coils generates only a weak current. The concept of relative velocity is illustrated if the magnet is held constant and the coil is moved up and down over the magnet.

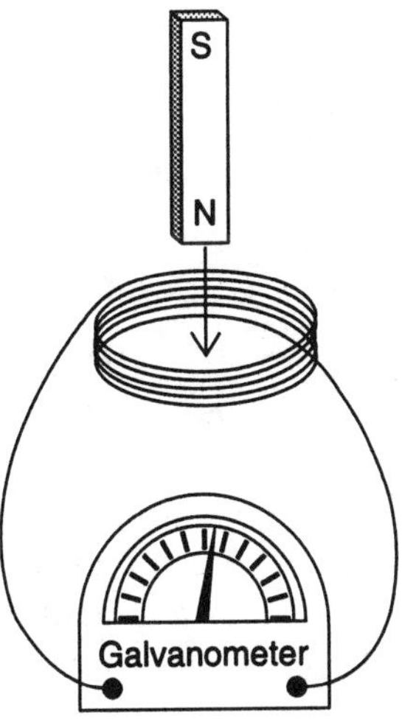

FIGURE 2.

MAGNETIC FLUX AND FARADAY'S LAW OF INDUCTION

In Figure 3, we see a circular region of cross-sectional area A. An external magnetic field **B** passes through the region at an angle θ to the normal of the region, and the perpendicular component of the field is given by $\mathbf{B}_\perp = \mathbf{B} \cos \theta$. The **magnetic flux** Φ is defined to be the product of the perpendicular component of the magnetic field and the cross-sectional area A:

$$\Phi = \mathbf{B}A \cos \theta$$

The unit of magnetic flux is the **weber**, such that 1 Wb = 1 T-m^2. Based on these ideas, the magnetic field strength is sometimes referred to as the **magnetic flux density** and can be expressed in Wb/m^2.

It was the English scientist Michael Faraday who demonstrated that induced motional EMF is due to the rate of change in the magnetic flux. We now

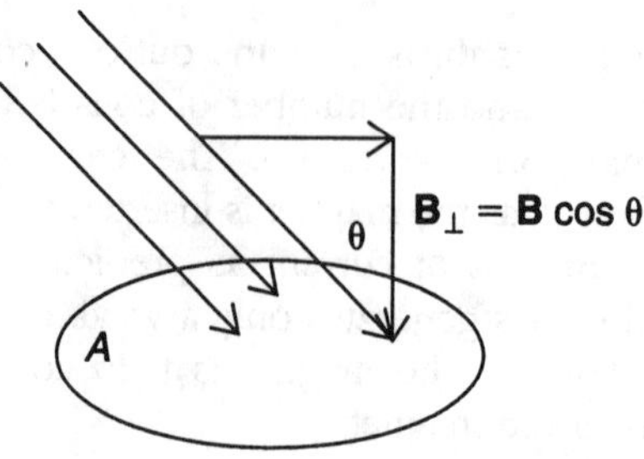

FIGURE 3. *Magnetic flux.*

call this relationship **Faraday's law of electromagnetic induction** and express it as

$$\text{EMF} = -\frac{\Delta\Phi}{\Delta t}$$

The negative sign is used because there is another relationship known as **Lenz's law** which states that an induced current always flows in a direction such that its magnetic field opposes the magnetic field that induced it. Consider the experiment shown in Figure 2. When the magnet is inserted through the coil, an induced current flows. The current in turn produces a magnetic field that is directed either into or out of the coil (along the same axis as the bar magnet).

If the current had a direction such that the "pole" of the coil attracted the pole of the bar magnet, then more energy would be derived from the effect than is possible in nature. Thus, according to Lenz's law, the current in the coil will have a direction such that the new magnetic field will oppose the magnetic field of the bar magnet. The more one tries to overcome this effect, the greater the effect becomes. The opposition of fields in this way prevents violation of the law of conservation of energy and can generate a large amount of heat in the process.

If there are N turns of wire in the coil, then Faraday's law becomes

$$\text{EMF} = -N\,\frac{\Delta\Phi}{\Delta t}$$

Consider the following sample problem.

Sample Problem

A coil is made of 10 turns of wire and has a diameter of 5 cm. The coil is passing out of the field with its axis parallel to the field, and the strength of the field is 0.5 T. What is the change in the magnetic flux? If an average EMF of 2 V is observed, how long was the flux changing?

Solution

The change in the magnetic flux is given by

$$\Delta\Phi = -\mathbf{B}A = -\mathbf{B}\pi r^2 = -(0.5)(3.14)(0.025)^2 = -9.8 \times 10^{-4} \text{ Wb.}$$

The time is therefore given by

$$\Delta t = -N\frac{\Delta\Phi}{\text{EMF}} = \frac{-(10)(-9.8 \times 10^{-4})}{2} = 4.9 \times 10^{-3} \text{ s.}$$

GENERATORS

An electromagnetic generator can be demonstrated using Faraday's law. In Figure 4, wire is shaped into a coil loop and wound around an armature. This armature is connected to a shaft that is easily rotated. Wires from the coil are connected to two slip rings which have no gaps in them, and each ring is then connected to a metallic brush. The remaining wires can be connected to a galvanometer (as shown) or to an electric device.

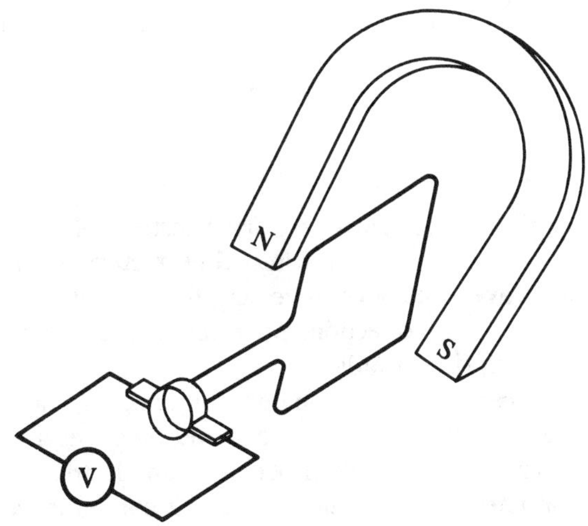

FIGURE 4. *Ac generator.*

The galvanometer registers the presence of a current due to the induced EMF in the coil. The rotation, however, produces an alternating current as the change in the magnetic field reverses direction every half-turn (see Figure 5). This periodic effect results in an **alternating current** which is sinusoidal in shape. The plot of voltage versus time in Figure 5 follows an equation of the form $V = V_{max} \cos 2\pi ft$, where f is the frequency of alternation.

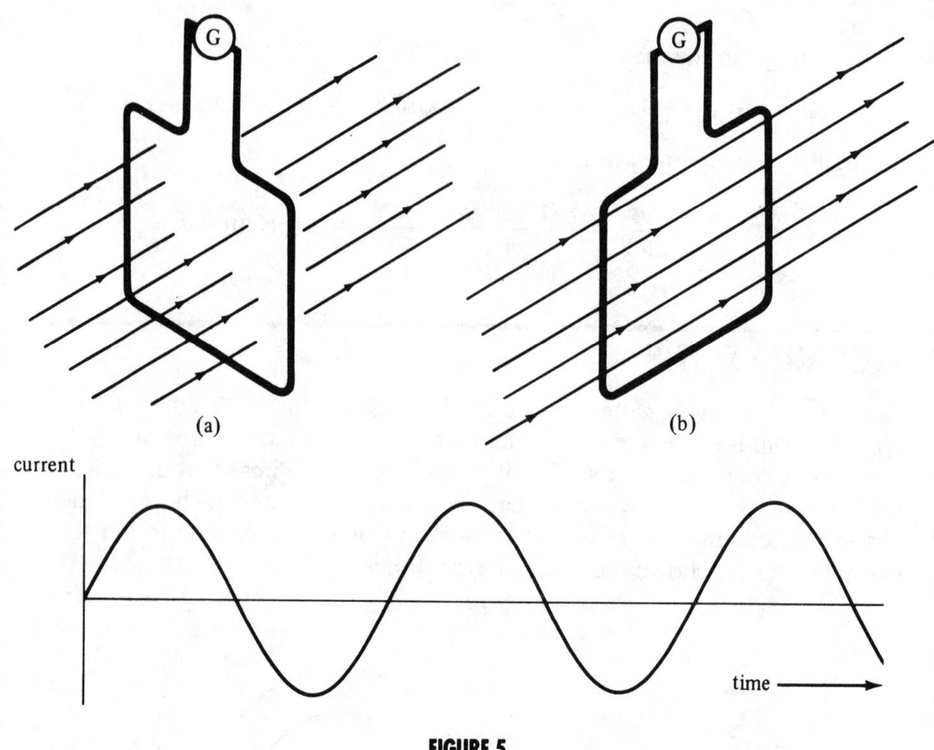

FIGURE 5.

If commutator rings (with gaps) are used instead of slip rings, then the alternating current can be converted into **direct current**. The commutator rings act like **full-wave rectifiers**, reversing the current in one direction so that all of it is in the same direction. An electronic **diode** can also act as a rectifier in an ac circuit (like a radio).

The current induced in the coil in turn produces a magnetic field. This changing magnetic field *self-induces* the coil and produces a back EMF which slows the coil down. This effect often occurs in an electric motor to maintain a constant rotation rate. The back EMF is a direct consequence of Lenz's law.

TRANSFORMERS

Suppose we have an electromagnet consisting of a solenoid with a soft iron core as shown in Figure 6. The solenoid is connected to a switch and a dc battery. There are N turns of wire on the solenoid (now called the *primary*). Near the primary, we have a second solenoid (known as the *secondary*) that is connected to a galvanometer. The secondary consists of n turns of wire

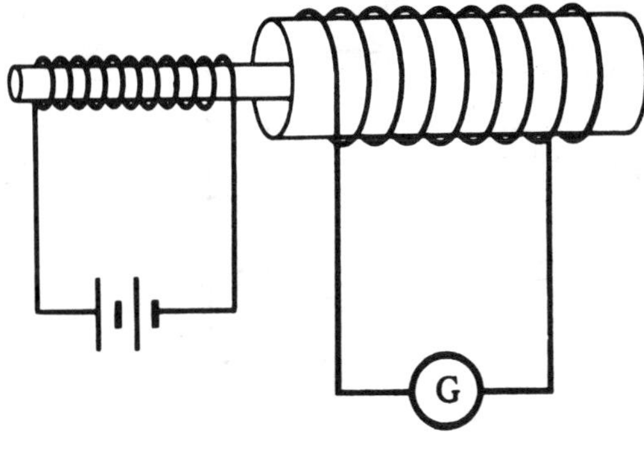

FIGURE 6.

around an iron core. The two solenoids are near each other but are not touching.

When the switch on the primary is initially closed, the galvanometer registers the presence of a current in the secondary that quickly goes to zero. With a steady current flowing through the primary, the galvanometer in the secondary registers zero. However, when the switch on the primary is initially opened, the galvanometer registers a current in the opposite direction and then quickly approaches zero once again.

The ability of one solenoid to induce an EMF in another is called **mutual induction** and is responsible for a phenomenon known as the **transformer effect**. When the switch was initially closed, the current in the primary built up to its maximum value during some interval of time; thus changing current also generates a changing magnetic field. The changing field exists at the location of the secondary and hence induces an EMF there based on the core material and the number of turns of wire. If a specially designed ac voltmeter were attached, the induced EMF could be measured.

Experimentally, it was determined that with a smaller number of turns in the secondary, the voltage reading goes down. If the number of turns in the secondary were greater, then the induced EMF would be greater as well.

The system just described is known as a **transformer**. A simple single-core transformer is shown in Figure 7.

When we induce a larger EMF, we have a **step-up transformer**, and when we induce a lower EMF, we have a **step-down transformer**. In either case, the induced EMF is proportional to the ratio of the number of turns of wire in both solenoids. If we let N_1 be the number of turns in the primary and N_2 be the number of turns in the secondary, then

$$\frac{V_1}{V_2} = \frac{N_1}{N_2}$$

where V_1 and V_2 are the EMFs in the primary and secondary, respectively.

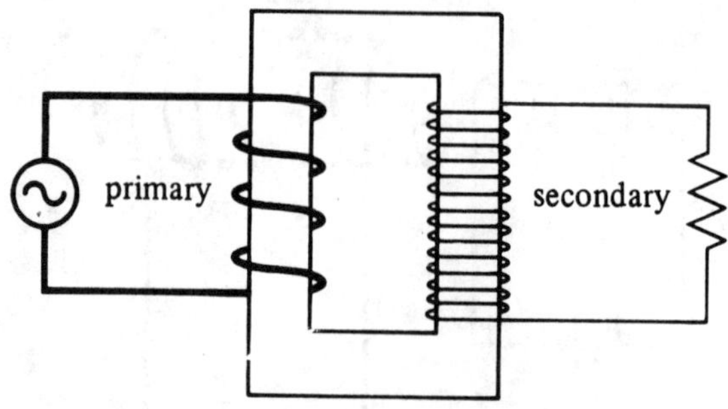

FIGURE 7. *Step-up transformer.*

It should be noted that Lenz's law plays a role here as well. The induced current in the secondary, given by Ohm's law if the resistance of the secondary is known, in turn generates an opposing magnetic field in itself and the primary. This back EMF prevents the buildup of too much energy in violation of the law of conservation of energy.

In an ideal transformer, the power generated in both solenoids is equal:

$$V_1 I_1 = V_2 I_2$$

Using the first equation for induced EMFs, we find that the ratio of currents is inversely proportional to the ratio of turns:

$$\frac{I_1}{I_2} = \frac{N_2}{N_1}$$

Thus, as voltages are stepped up, currents are stepped down. When electricity is brought to your home, the power company steps up the voltage in high-voltage lines to reduce the current. This has the effect of lowering the "joule heating" losses that are proportional to the square of the current. Outside your home, transformers attached to telephone poles step down the voltage, and this is the supply brought to your main electric box. Transformers are used in many electric devices and are most obvious in electric train and racing car sets.

The above-mentioned transformers vary the current output by using variable resistors called *rheostats*. These resistances can be related to the number of turns of wire in the primary and secondary coils using Ohm's law. Since $V = IR$, we can see that

$$I_1^2 R_1 = I_2^2 R_2$$

$$\frac{N_2^2}{N_1^2} = \frac{R_2}{R_1}$$

Sample Problem

A transformer with 1,000 turns on its primary coil is connected to an ac voltage supply of 120 V. The secondary coil has 200 turns. What is the induced secondary EMF?

Solution

Since the ratio of EMFs is directly proportional to the ratio of turns,

$$\frac{V_1}{V_2} = \frac{N_1}{N_2}$$

$$\frac{120}{V_2} = \frac{1{,}000}{200}$$

Thus, $V_2 = 24$ V (ac).

INDUCTORS AND INDUCTANCE

The self-induced EMF is proportional to the rate of change in magnetic flux. Now, since the flux Φ is proportional to the current I, the rate of change in magnetic flux $\Delta\,\Phi/\Delta t$ is likewise proportional to $\Delta I/\Delta t$. Thus, we can state that for a self-induced EMF voltage,

$$V = -L\,\frac{\Delta I}{\Delta t}$$

where the negative sign is needed because of Lenz's law. The proportionality constant L is called the **inductance** and is expressed in **henrys**, such that 1 H is equal to 1 J/A^2.

The schematic symbol for an inductor is shown in Figure 8.

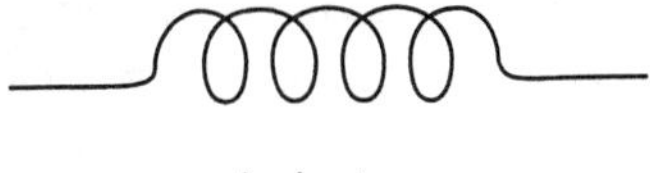

Inductor

FIGURE 8.

The inductance of a coil L depends on the permeability of its core, the number of turns, its length, and its cross-sectional area. To understand these concepts better, consider the following ideas.

From Faraday's law, we know that the induced EMF in an inductor is proportional to the rate of change in magnetic flux and the number of turns of wire in the coil:

$$\text{EMF} = -N\,\frac{\Delta\Phi}{\Delta t}$$

Now, for a fixed geometry, where the field is perpendicular to the plane of the coils so that $\Phi = \mathbf{B}A$, we can therefore write

$$\frac{\Delta\Phi}{\Delta t} = A\,\frac{\Delta \mathbf{B}}{\Delta t}$$

Thus,

$$\text{EMF} = -NA\,\frac{\Delta \mathbf{B}}{\Delta t}$$

For a solenoid, we know that $\mathbf{B} = \mu(N/l)I$, where l is the length of the solenoid. Thus,

$$\text{EMF} = -\mu N^2\,\frac{A}{l}\,\frac{\Delta I}{\Delta t} = -L\,\frac{\Delta I}{\Delta t}$$

$$L = \mu N^2\,\frac{A}{l}$$

When an inductor and resistor are connected in series, some interesting effects occur (Figure 9). From Kirchoff's law, the sum of the EMFs around the loop must equal zero. Thus,

$$IR = \text{EMF} - L\,\frac{\Delta I}{\Delta t}$$

Let the magnitude of the EMF be equal to V. From the preceding equation, we see that if L is large, then the rate of change in the current will be very slow. When the switch is closed, $I = 0$ and $\Delta I/\Delta t = V/L$, which is a maximum. When $IR = V$, then $\Delta I/\Delta t = 0$ and the current is at its maximum value of V/R.

The variation in current with time is exponential and is given by

$$I = \frac{V}{R}(1 - e^{-Rt/L})$$

The value of time $t = L/R$ is called the *inductive time constant* and is designated by τ_L. Figure 10 shows this exponential variation in current with time.

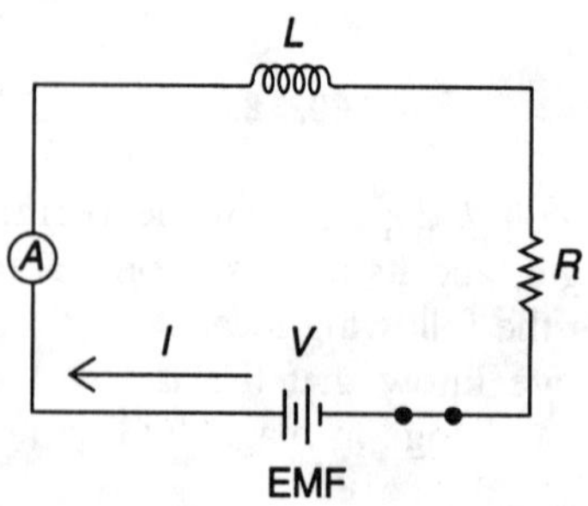

FIGURE 9. *An RL circuit.*

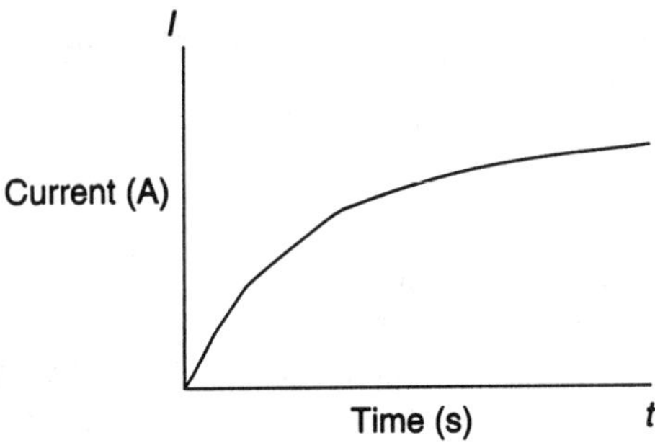

FIGURE 10.

MAGNETIC ENERGY AND ELECTROMAGNETIC OSCILLATIONS

Imagine we have a steady current building up in a different inductance circuit over time. The average current at any time t is given by $I/2$, and the total charge $Q = (I/2)\,\Delta t$. If we let the EMF of the circuit be equal to V, then the energy generated in the circuit is given by $W = QV = (I/2)V\Delta t$. Now, the EMF of the inductance is equal to $-V$, and so $V = L\,\Delta I/\Delta t$. Since the initial current is zero (at time $t = 0$), the total change in current $\Delta I = I$, and so the magnetic energy in the inductor is given by

$$W = \frac{1}{2} IL \frac{I}{\Delta t}\,\Delta t = \frac{1}{2} LI^2$$

When a charged capacitor is connected to an inductor, a situation develops that is remarkably similar to a mass oscillating on a spring. We know from our studies of electric circuits that the energy stored in a charged capacitor is given by

$$W = \frac{1}{2} CV^2 = \frac{1}{2}\frac{Q^2}{C}$$

Now, each of the plates of the capacitor has charge Q, and initially the current is zero and so the inductor has zero energy. The capacitor begins to discharge and current flows through the inductor, giving it energy according to the previous formula. This current generates a magnetic field in the inductor. Eventually, the capacitor is fully discharged, and all the energy resides in the inductor. Now, the magnetic field of the inductor induces a reverse current which begins to charge the capacitor with plates of polarity opposite from before.

This situation is an electromagnetic oscillation and has a definite frequency. The total energy of the circuit is given by

$$W = \frac{1}{2} LI^2 + \frac{1}{2}\frac{Q^2}{C}$$

The frequency of oscillations can be shown to be equal to

$$f = \frac{1}{2\pi\sqrt{LC}}$$

In the middle of the nineteenth century, the theoretical physicist James Clerk Maxwell derived a formula for the oscillations of an electromagnetic circuit. He asserted that the energy would be transmitted into space as a transverse wave with the same velocity as that of light. In the 1890s, the German experimental physicist Heinrich Hertz discovered these "electromagnetic waves" and demonstrated Maxwell's unification of electricity, magnetism, and optics.

Sample Problem

An air core solenoid consists of 100 turns of copper wire with a cross-sectional area of 24 cm^2 and a length of 6 cm. What is the inductance of this coil?

Solution

We first note that the cross-sectional area is equal to 24 cm^2 = 2.4×10^{-3} m^2. The length is equal to 0.06 m, and the permeability μ is very close to the value for the permeability of free space given by $\mu_0 = 1.26 \times 10^{-6}$ T-m/A. Thus, the inductance is given by

$$L = \frac{(1.26 \times 10^{-6})(100)^2(2.4 \times 10^{-3})}{0.06} = 0.000504 \text{ H} = 0.504 \text{ mH}$$

PROBLEM-SOLVING STRATEGIES

When solving electromagnetic problems, be sure to remember that there are vector quantities involved. The directions of the these vectors are usually determined by the right-hand rule. Additionally, remember Lenz's law:

> An induced current always flows in a direction such that its magnetic field opposes the change in the magnetic flux that induced it.

Lenz's law is an immediate consequence of the law of conservation of energy and plays a role in many applications.

The types of effects produced using Faraday's law of electromagnetic induction involve alternating currents. These currents are periodic and have frequencies just as oscillatory motion does.

When solving problems in electromagnetic induction, it is important to remember the right-hand rules. Also, remember that the induced EMF is proportional to the change in the magnetic flux, not in the magnetic field.

PRACTICE PROBLEMS FOR CHAPTER 19

Thought Problems

1. An inductive generator consists of a coil of wire wound around an armature. When the coil is rotated in a magnetic field and connected to an external circuit, an alternating current can be induced in the circuit. A student finds that it is easier to turn the armature when the generator is not connected to the external circuit. Explain why this phenomenon occurs.
2. (a) Show that the units for $\tau = L/R$ are seconds.
   (b) Show that the units for $f = 1/2\pi\sqrt{LC}$ are cycles/s.
3. The cores of commercial transformers are typically made from thin sheets of steel. What is the advantage of this method as opposed to having one solid block of steel with the same thickness?

Multiple-Choice Problems

1. The back EMF of a motor is at a maximum when ____.
   (A) the speed of the motor is constant
   (B) the speed of the motor is at its maximum value
   (C) the motor is first turned on
   (D) the speed of the motor is increasing
   (E) the motor is turned off
2. A bar magnet is pushed through a flat coil of wire. The induced EMF is greatest when ____.
   (A) the north pole is pushed through first
   (B) the magnet is pushed through quickly
   (C) the magnet is pushed through slowly
   (D) the south pole is pushed through first
   (E) the magnet rests in the coil
3. A fully charged 40-pF capacitor is connected to an inductor with an inductance of 100 mH. What is the oscillation frequency of this *LC* circuit?
   (A) 80 kHz
   (B) 120 kHz
   (C) 54 kHz
   (D) 660 MHz
   (E) 35 MHz
4. A flat, 300-turn coil has a resistance of 3 Ω. The coil covers an area of 15 cm^2 such that its axis is parallel to an external magnetic field. How fast must the magnetic field change in order to induce a current of 0.75 A in the coil?
   (A) 0.0075 T/s
   (B) 2.5 T/s
   (C) 0.0005 T/s
   (D) 5 T/s
   (E) 15 T/s

5. When a loop of wire is turned in a magnetic field, the direction of the induced EMF changes every ____
(A) $\frac{1}{4}$ revolution
(B) 2 revolutions
(C) 1 revolution
(D) $\frac{1}{2}$ revolution
(E) never changes

6. A wire of length 0.15 m is passed through a magnetic field with a strength of 0.2 T. What must be the velocity of the wire if an EMF of 0.25 V is to be induced?
(A) 8.3 m/s
(B) 6.7 m/s
(C) 0.0075 m/s
(D) 0.12 m/s
(E) 7.6 m/s

Answer Problems 7 through 9 based on the following information: The primary coil of an ideal transformer has 200 turns, while the secondary coil has 50 turns. The power generated in the primary coil is 600 W when it is connected to a 120-V ac supply.

7. What is the power generated in the secondary coil?
(A) 1,200 W
(B) 60 W
(C) 2,400 W
(D) 150 W
(E) 600 W

8. What is the current in the secondary coil?
(A) 5 A
(B) 10 A
(C) 0.2 A
(D) 20 A
(E) 2 A

9. The induced EMF in the secondary coil is equal to ____.
(A) 60 V
(B) 480 V
(C) 300 V
(D) 120 V
(E) 30 V

10. A loop of wire is rotated through 360 degrees across an external magnetic field. During one period of revolution, the induced EMF in the coil changes in ____.
(A) magnitude only
(B) direction only
(C) both magnitude and direction

(D) neither magnitude nor direction
(E) first magnitude and then direction

Free-Response Problems

1. A horizontal conducting bar is free to slide along a pair of vertical bars as shown. The conductor is of length L and mass M. As it slides vertically downward under the influence of gravity, it passes through an outward-directed uniform magnetic field. The resistance of the entire circuit is R. Find an expression for the terminal velocity of the conductor (neglet any frictional effects due to sliding).

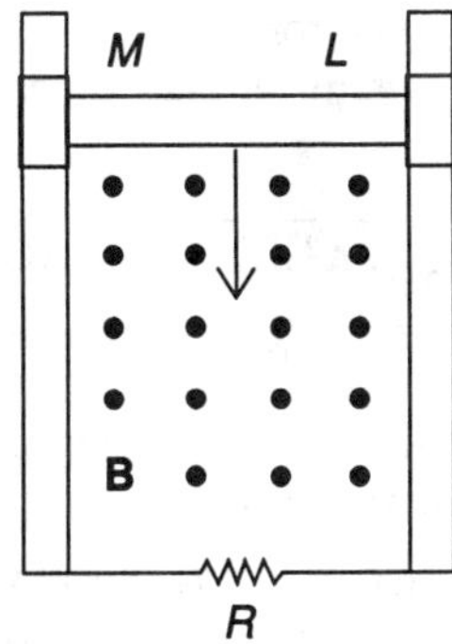

2. A motor has a series of coils that have a total resistance of 20 Ω and is operated with a 120-V power supply. While running at maximum speed, the current in the coils is measured to be 2 A. What is the back EMF at this time?

3. A simple tuned circuit in an old "crystal radio" consists of an antenna and a ground connection attached to a variable capacitor and an inductance coil. The coil was typically wound around a paper core (consider the core to be essentially air) which was a cylinder with a 3-cm radius. Five hundred turns of thin wire, in one layer, made a length of about 15 cm. If the "tuner" receives amplitude-modulated (AM) frequencies in a range of 660 kHz to 1.4 MHz, then what must be the range of the variable capacitor?

4. The total energy in an LC circuit is given by

$$E = \frac{1}{2}LI^2 + \frac{1}{2}\frac{Q^2}{C}$$

Find the frequency of electromagnetic oscillations using the phase space method of analysis outlined earlier in this book. Assume negligible internal resistances. (*Hint*: Define a quantity $H = LI$, in J/A, to derive the phase space ellipse for the variables H and Q.)

SOLUTIONS TO PRACTICE PROBLEMS

Thought Problems

1. When work is done by the current in the circuit, this energy must be supplied by the student. For the open-circuit case, the student needs only to work against friction in the bearings.
2. (a) The units for resistance are ohms, and the units for inductance are henrys. To show that the ratio L/R is in seconds, we need to express these quantities in more fundamental units:

$$\frac{\text{H}}{\Omega} = \frac{\text{J/A}^2}{\text{V/A}} = \frac{\text{J}}{\text{A}^2}\,\frac{\text{A}}{\text{V}} = \frac{\text{V-C}}{\text{A-V}} = \text{seconds}$$

(b) To show that the units for the electromagnetic frequency are in reciprocal seconds, we must first show that the units of $\sqrt{LC}$ are seconds:

$$\sqrt{\text{LC}} = \sqrt{(\text{H})(\text{F})} = \sqrt{(\text{J/A}^2)(\text{C/V})} = \sqrt{(\text{C} - \text{V/A}^2)(\text{C/V})} =$$

$$\sqrt{(\text{C}^2)/(\text{A}^2)} = \sqrt{\text{s}^2} = \text{seconds}$$

Since the frequency involves the reciprocal of this square root, the units are in reciprocal seconds.

3. Thin sheets of steel are easier to magnetically induce. Additionally, there will be mutual induction between each of the sheets, further enhancing the overall strength of the core's magnetic field.

Multiple-Choice Problems

1. **B** The back EMF is proportional to the rate of change in magnetic flux and opposes the ability of the motor to turn. It is at its greatest value when the speed of the motor is a maximum.
2. **B** The motional EMF is equal to $\mathbf{B}lv$, where $\mathbf{v}$ is the velocity of the bar magnet. Thus, the induced EMF is greatest when the magnet is pushed through quickly.
3. **A** The capacitance in farads is equal to 4×10^{-11} F. The inductance in henrys is equal to 0.1 H. Thus, the frequency of electromagnetic oscillations is given by

$$f = \frac{1}{2\pi\sqrt{\text{LC}}} = \frac{1}{6.28\sqrt{(4 \times 10^{-11})(0.1)}} = 79{,}618 \text{ Hz} = 80 \text{ kHz}$$

4. **D** The formula for the rate of change in magnetic flux is EMF $= -N(\Delta\Phi/\Delta t)$. Using the given values and Ohm's law, we find that the magnitude of the flux change is

$$\frac{\Delta\Phi}{\Delta t} = \frac{(0.75)(3)}{300} = 0.0075 \text{ Wb/s}$$

Our question concerns the rate of change in the magnetic field. Recognizing that in a situation in which the axis of the coil is parallel to the magnetic field, then $\Phi = \mathbf{B}A$, where $A = 0.0015$ m^2. Thus, $\Delta\mathbf{B}/\Delta t = 5$ T/s.

5. **D** In an ac generator, the EMF reverses direction every half-turn.
6. **A** The motional EMF is given by EMF = $\mathbf{B}L\mathbf{v}$. Thus, using the given values, $\mathbf{v} = 8.3$ m/s.
7. **E** In an ideal transformer, the power in the primary is equal to the power in the secondary.
8. **D** Since the power in the secondary is equal to 600 W, we only need to find the induced secondary EMF (since $P = VI$, where the voltage V is now the induced EMF). Using the transformer equations,

$$\frac{V_1}{V_2} = \frac{N_1}{N_2}$$

and the given information, the secondary EMF is equal to 30 V. Thus, the current is equal to 20 A.

9. **E** From the preceding analysis, we know that the secondary EMF is equal to 30 V using the transformer equations.
10. **C** While rotating through the magnetic field, the EMF goes through a maximum and a minimum magnitude as well as a reversal in direction.

Free-Response Problems

1. Initially, the bar is accelerated downward by the force of gravity given by $\mathbf{F} = M\mathbf{g}$. The resistance R, in conjunction with the induced current I, produces an EMF = IR. This EMF is also equal to the product $\mathbf{B}L\mathbf{v}$. The magnetic force due to the current I is given by $\mathbf{B}IL$. Putting all these ideas together, we find that at terminal velocity, $M\mathbf{g} = IL\mathbf{B}$ and $I = \mathbf{B}L\mathbf{v}/R$.

   Thus

$$\mathbf{v}_t = \frac{M\mathbf{g}R}{L^2\mathbf{B}^2}$$

2. When the motor is first turned on, the current is given by

$$I = \frac{\text{EMF}}{R} = 6\ \text{A}$$

Now, at maximum speed, the current is 2 A and is given by

$$I = \frac{\text{EMF} - \text{EMF}_{\text{back}}}{R}$$

Thus, $\text{EMF}_{\text{back}} = \text{EMF} - IR = 120 - (2)(20) = 80\text{V}$

3. We first determine the inductance of the coil. The length is 0.15 m, and the permeability (assuming air to be approximately the same as free space)

is $\mu = 1.26 \times 10^{-6}$ T-m/A. The cross-sectional area of a cylinder is a circle. In this case, the area of the circle is equal to $A = \pi R^2 = (3.14159)(0.03)^2 = 2.83 \times 10^{-3}$ m^2. The inductance is therefore equal to

$$L = \frac{\mu N^2 A}{l} = \frac{(1.26 \times 10^{-6})(500)^2 2.83 \times 10^{-3}}{0.15} = 0.005943 \text{ H}$$

Now we can solve for the capacitance in each extreme case:

$$C = \frac{1}{4\pi^2 f^2 L}$$

$$C = \frac{1}{4\pi^2(660{,}000)^2(0.005943)} = 9.785 \times 10^{-12} \text{ F}$$

$$C = \frac{1}{4\pi^2(1.4 \times 10^6)^2(0.005943)} = 2.17 \times 10^{-12} \text{ F}$$

4. By defining a quantity $H = LI$, the equation for total energy can be written in the form

$$E = \frac{H^2}{2L} + \frac{Q^2}{2C}$$

$$1 = \frac{H^2}{2EL} + \frac{Q^2}{2CE}$$

This is our phase space ellipse. The product LI is proportional to the magnetic flux in the coil.

The area of the ellipse is equal to the product of the total energy and the period of oscillation (if no loss due to resistances is accounted for). Thus, we obtain

$$\text{Area} = \pi ab = \pi\sqrt{(2LE)(2CE)} = E2\pi\sqrt{LC}$$

Since the frequency is equal to the reciprocal of the period, we therefore have

$$f = \frac{1}{2\pi\sqrt{LC}}$$

20
ALTERNATING CURRENT CIRCUITS

TIME-VARYING CURRENTS AND VOLTAGES

In a dc electric circuit, the EMF of the battery produces a current that flows in only one direction. An electromagnetic generator can produce a current that alternates its direction periodically in time. In this case the current is said to be *time-varying*. If the time variance is regular, we simply call the current an **alternating current**. The EMF induced by the generator is also time-varying.

The time-varying nature of alternating currents and voltages is sinusoidal, and Figure 1 shows typical ac circuit graphs. The amplitudes correspond to the maximum values I_{max} and V_{max}. The period of alternation T is related to the angular frequency ω:

$$\omega = \frac{2\pi}{T} = 2\pi f$$

where f is the frequency in hertz.

The equations governing these time-varying quantities are

$$I = I_{max} \sin \omega t$$
$$V = V_{max} \sin \omega t$$

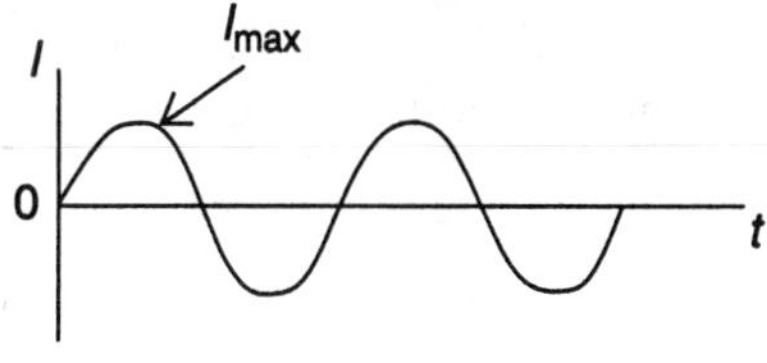

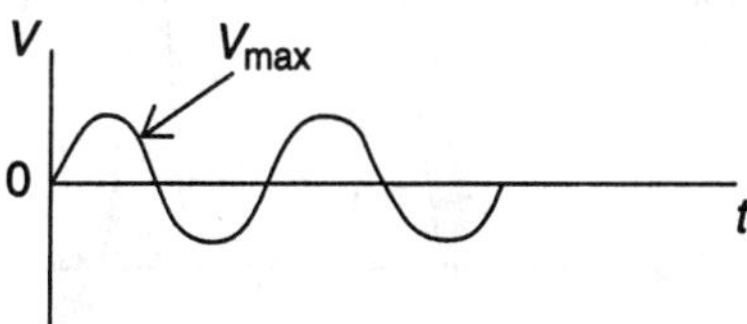

FIGURE 1.

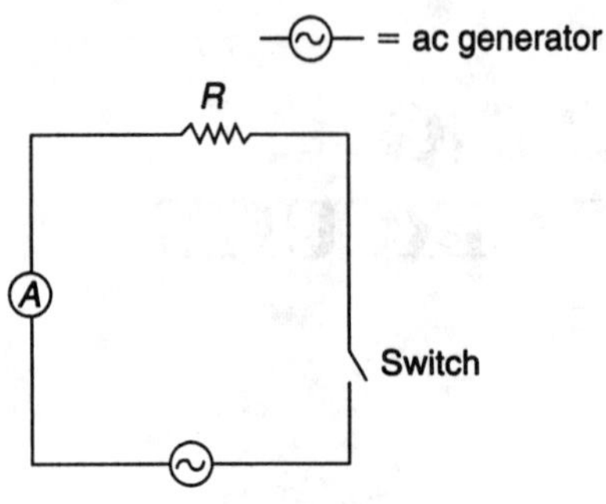

FIGURE 2.

Since these equations are periodic, the concept of *phase* between the current and voltage will soon become a necessity.

When a single resistor is added to the circuit (see Figure 2), then we have the following conditions. The power consumed in the resistor is dissipated as heat (as in the dc circuit) but at a varying rate. Comparisons with dc equations are risky since the maximum current lasts for only a brief interval of time. The instantaneous voltage (which remains in phase with the current through the resistor) is given by Ohm's law:

$$V = I_{max} R \sin \omega t$$

The instantaneous power in the resistor is given by

$$P = I^2R = I^2_{max}R \sin^2 \omega t$$

It is often practical to work with average values. In order to obtain the time-averaged value of the power generated in a resistor, we need to obtain the average value of the square of the sine function. From Figure 1, recall that a sine function when graphed is symmetrical about the "zero" equilibrium position. The average value of $I = I_{max} \sin \omega t$ is equal to zero. However, if we were to square this current equation, the magnitude of the function would always remain positive, as illustrated in Figure 3.

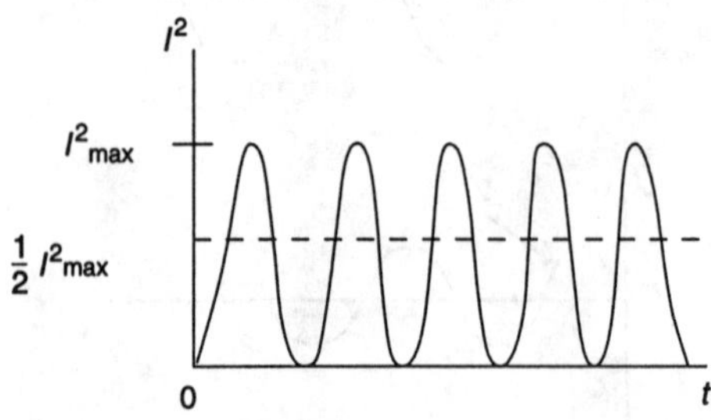

FIGURE 3.

From Figure 3, we can see that the average of the square of the alternating current is given by one half the value of the square of the maximum current. What we need is the *average* or *root mean square* value of this current. That is, we need the square root of the average of the square of the function and not the average of the function itself. We must take the square root to find the root mean square value:

$$\overline{I^2} = \frac{1}{2} I^2_{max}$$

$$I_{rms} = \sqrt{\frac{1}{2} I^2_{max}} = 0.707\ I_{max}$$

The average power is therefore given by

$$P_{avg} = I^2_{rms}R = \frac{1}{2} I^2_{max}R$$

In a similar way, the root mean square value of the alternating EMF can be shown to be equal to

$$V_{rms} = 0.707V_{max}$$

Some textbooks refer to these quantities as "effective" values, and they are equivalent to what we have called root mean square values.

INDUCTORS IN AN ALTERNATING CURRENT CIRCUIT

Consider an ac circuit consisting of a source of EMF and a single inductor. This circuit is illustrated in Figure 4.

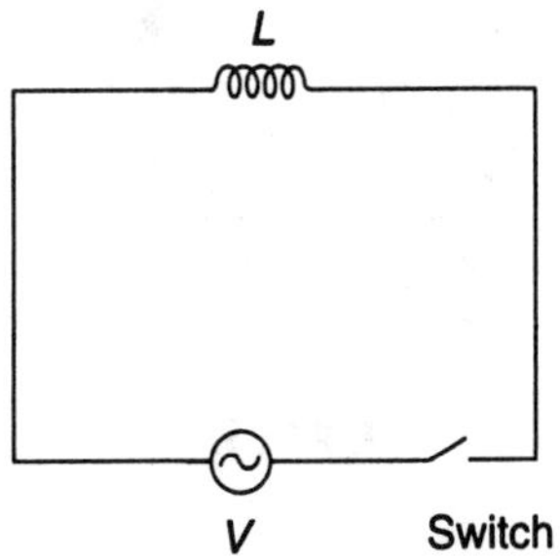

FIGURE 4.

From our discussions of inductors in electric circuits, we know that the EMF is proportional to the rate of change in the current through the inductor (where L is the inductance):

$$V_L = -L\frac{\Delta I}{\Delta t} = -V_{max} \sin \omega t$$
$$V_{source} = -V_L$$

Now, the EMF and the current are both initially zero when the switch is closed. The EMF in the inductor is a maximum when the current is changing at its fastest rate (i.e., when $\Delta I/\Delta t$ is a maximum). This occurs early on since as the current builds in the inductor, its self-inductance causes a back EMF to be produced. This has the effect of reducing the current in the cirucit.

In Figure 1, we can see that the graph of the instantaneous current is changing most rapidly (steepest slope) in the beginning, and as the current approaches its maximum value, the slope tends toward zero. This means that the current lags behind the voltage by one quarter of a cycle (or 90 degrees of phase). This is because the voltage is directly proportional to the rate of change in current. The word *phase* is used to describe how the periodic functions of current and voltage compare in time. These concepts are illustrated in Figure 5 where the current and the change in the current are compared over time. The vertical lines represent fractional portions of complete cycles:

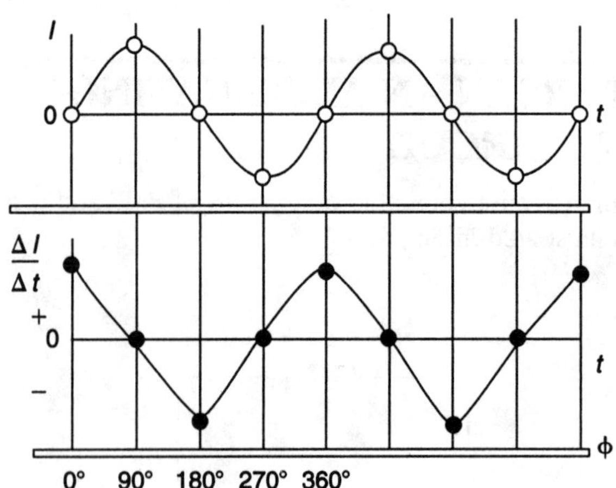

FIGURE 5.

The maximum value for the rate of change in the current can be determined by recalling our work on simple harmonic motion. It was determined at that time that if one plots the time-varying displacement of a mass oscillating on a spring (a sine function), then the maximum rate of change in displacement is equal to the product of the angular frequency and the amplitude:

$$\frac{\Delta \mathbf{x}}{\Delta t_{\max}} = \mathbf{v}_{\max} = \omega A$$

In a similar way, the maximum rate of change in the current oscillating in an ac circuit (also a sine function) is given by the product of the angular frequency and the amplitude:

$$\frac{\Delta I}{\Delta t_{\max}} = \omega A = \omega I_{\max}$$

Additionally, we know that

$$\frac{\Delta I}{\Delta t} = \frac{V_{\max}}{L} \sin \omega t$$

and therefore

$$\frac{\Delta I}{\Delta t_{\max}} = \frac{V_{\max}}{L}$$

Thus, we see that

$$I_{\max} = \frac{V_{\max}}{\omega L}$$

The quantity ωL is called the **inductive reactance** and is represented by X_L (in ohms). We therefore see that $I_{\max} = V_{\max}/X_L$ and that $V_L = I_{\max} X_L \sin \omega t$ is the expression for the EMF across the inductor in an ac circuit.

CAPACITORS IN ALTERNATING CURRENT CIRCUITS

In Figure 6, we show a simple ac circuit consisting of an alternating EMF source and a capacitor. The voltage across the plates of the capacitor is proportional to the the total charge stored and its capacitance:

$$V_C = \frac{Q}{C} = V_{\max} \sin \omega t$$

The instantaneous charge on the capacitor is therefore given by

$$Q = CV_{\max} \sin \omega t$$

When the switch is first closed, the voltage across the plates is zero and the current flows easily. As more charge builds up on the plates, the current slows since the increase in potential difference makes it harder to add more charge. Thus, when the capacitor is fully charged, the voltage across its plates is a

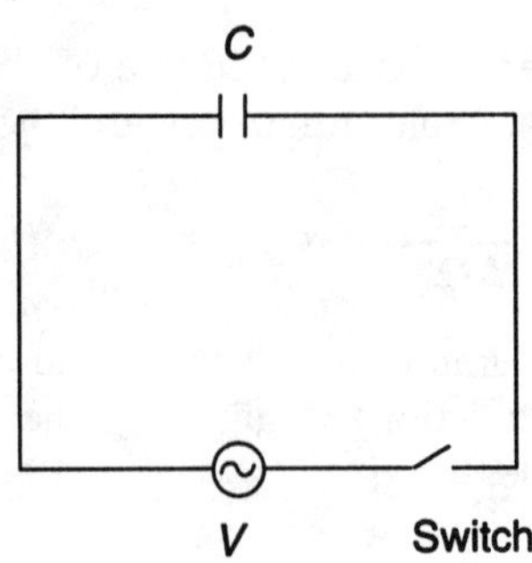

FIGURE 6.

maximum and the current in the circuit is zero. Thus, we say that the current leads the voltage by one quarter of a cycle (or 90 degrees of phase) since the current begins as a maximum while the voltage starts at zero.

Now a graph of charge versus time using the previous equation is sinusoidal with an amplitude CV_{max}, and so the maximum value for the rate of change in the charge (i.e., the current) is again given by the product of the angular frequency and the amplitude:

$$\frac{\Delta Q}{\Delta t_{max}} = \omega A = \omega C V_{max}$$

Since $\Delta Q/\Delta t = I$, we can see that

$$\frac{\Delta Q}{\Delta t} = I_{max} \sin \omega t$$

and that $(\Delta Q/\Delta t)_{max} = I_{max}$. Thus,

$$I_{max} = \omega C V_{max} = \frac{V_{max}}{X_C}$$

where X_C is called the **capacitive reactance** and is given by $X_C = 1/\omega C$. Thus, the voltage across the capacitor plates is given by

$$V_C = X_C I_{max} \sin \omega t$$

THE RLC SERIES CIRCUIT

Consider the circuit shown in Figure 7. It is called an *RLC* circuit because it consists of a resistor, an inductor, and a capacitor in series with an alternating source of EMF.

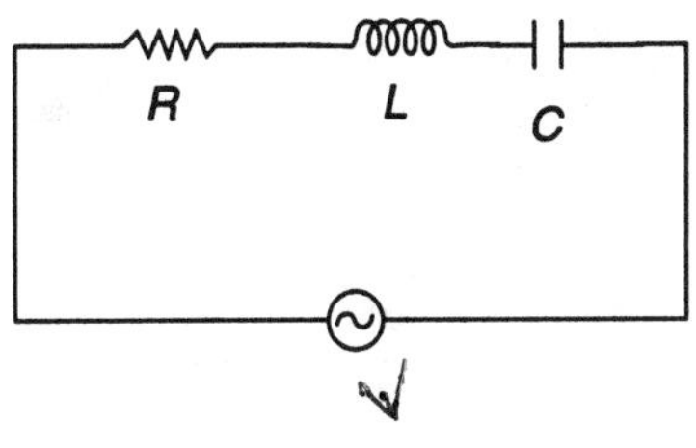

FIGURE 7.

In alternating current circuits, there exist phase relationships between the current through an element of the circuit and the voltage across the element. In general, we can write

$$I = I_{max} \sin(\omega t - \phi)$$

where ϕ is the phase angle.

From our discussions of resistors, capacitors, and inductors in these circuits, we know that in a resistor the current and voltage are in phase. In a capacitor, the current leads the voltage by 90 degrees of phase (or $\pi/2$ rad), and in an inductor, the current lags the voltage by 90 degrees of phase. These relationships can be written algebraically as

$$V_R = V_{R(max)} \sin \omega t$$
$$V_L = V_{L(max)} \sin(\omega t + \frac{\pi}{2}) = V_{L(max)} \cos \omega t$$
$$V_C = V_{C(max)} \sin(\omega t - \frac{\pi}{2}) = - V_{C(max)} \cos \omega t$$

The maximum voltages in each case are given by

$$V_{R(max)} = I_{max}R$$
$$V_{C(max)} = I_{max}X_C$$
$$V_{L(max)} = I_{max}X_L$$

To better visualize these relationships, we use special rotating vectors called *phasors*. Figure 8 illustrates these phase relationships in a rotating frame with frequency ω. Additionally, in an *RLC* series circuit, the maximum current must be the same in all parts of the circuit and have the same phase.

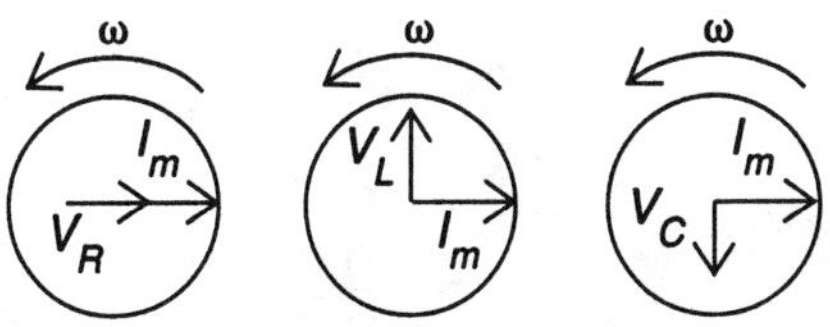

FIGURE 8. *RLC phasor diagrams.*

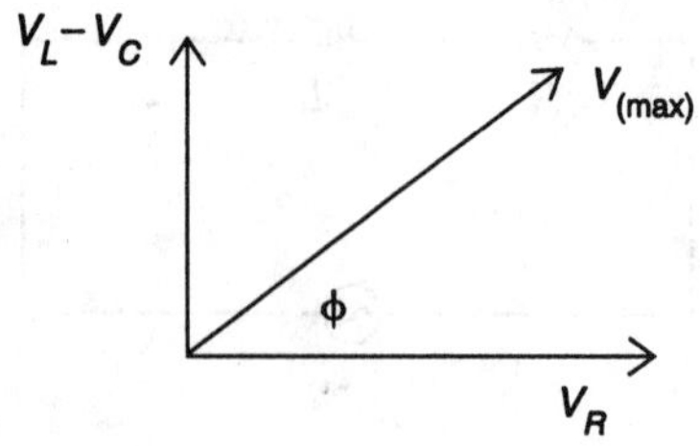

FIGURE 9.

Given these relative orientations, we can summarize the voltage phasors in Figure 9. The angle ϕ is the phase angle from our earlier discussion.

Using the Pythagorean theorem,

$$V_{max} = \sqrt{V_R^2 + (V_L - V_C)^2}$$
$$V_{max} = I_{max}\sqrt{R^2 + (X_L - X_C)^2}$$

where $X_L = \omega L$ and $X_c = 1/\omega C$. The **impedence** Z of the circuit is defined to be equal to

$$Z = \sqrt{R^2 + (X_L - X_C)^2}$$

We can therefore write $V_{max} = ZI_{max}$. This is the ac analog of Ohm's law, where the unit of impedence is the ohm. The phase angle is determined from

$$\tan \phi = \frac{X_L - X_C}{R}$$

RESONANCE IN A SERIES *RLC* CIRCUIT

The instantaneous current in a series *RLC* circuit can be written in the form

$$I = \frac{V_{(max)} \sin \omega t}{Z} = \frac{V_{(max)} \sin \omega t}{\sqrt{R^2 + (X_L - X_C)^2}}$$

The current is a maximum when the inductive reactance is equal to the capacitive reactance.

When this condition is reached, the circuit is said to be in **resonance:**

$$X_L = X_C$$
$$\omega_0 L = \frac{1}{\omega_0 C}$$
$$\omega_0 = \frac{1}{\sqrt{LC}}$$
$$f_0 = \frac{1}{2\pi\sqrt{LC}}$$

The quantity ω_0 is called the **resonant angular frequency**, and f is the **resonant frequency**. This is the same quantity obtained when we considered oscillating dc circuits containing inductors and capacitors in Chapter 19.

The average power consumed by the resistor (capacitors and inductors do not consume power) can now be written as

$$P_{avg} = I_{rms}^2 R = \frac{V_{rms}^2 R}{Z^2} = \frac{V_{rms}^2 R}{R^2 + (X_L - X_C)^2}$$

An alternative version of this formula for the average power is

$$P_{avg} = I_{rms} V_{rms} \cos \phi$$

where the quantity $\cos \phi$ is called the **power factor**.

The average power is dependent on the frequency of oscillations in the ac circuit. To illustrate this fact, we make the following substitutions:

$$(X_L - X_C)^2 = \left(\omega L - \frac{1}{\omega C}\right)^2 = \frac{L^2}{\omega^2}(\omega^2 - \omega_0^2)^2$$

$$P_{avg} = \frac{V_{rms}^2 R \omega^2}{R^2\omega^2 + L^2(\omega^2 - \omega_0^2)^2}$$

This quantity is a maximum when $\omega = \omega_0$ (i.e., at resonance) and is equal to

$$P_{avg(max)} = \frac{V_{rms}^2}{R}$$

In Figure 10, the average power is plotted against frequency in an illustrative example:

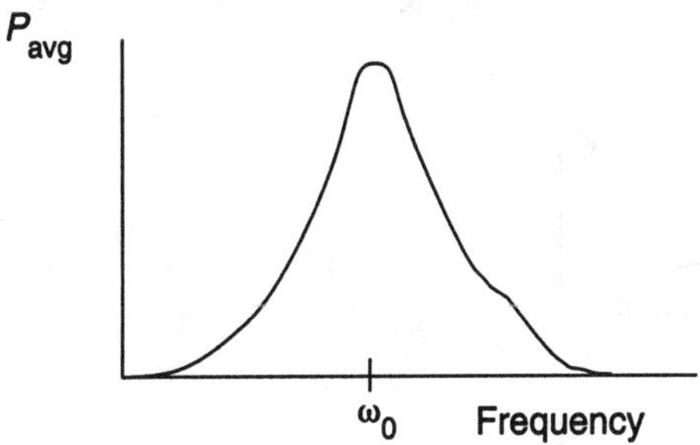

FIGURE 10. *Power resonance curve.*

We can relate the average power at any frequency to the maximum average power at resonance and achieve a result similar to what we discussed in connection with energy transfers in elastic collisions and power transfers in dc circuits. The ratio of average power to maximum average power is

$$\frac{P_{avg}}{P_{avg(max)}} = \frac{R^2\omega^2}{R^2\omega^2 + L^2(\omega^2 - \omega_0^2)^2} = \frac{\omega^2}{(\omega^2 + L^2/R^2)(\omega^2 - \omega_0^2)^2}$$

Now divide both the top and the bottom of the right-hand side by the square of the resonant frequency:

$$\frac{P_{avg}}{P_{avg(max)}} = \frac{\omega^2/\omega_0^2}{\dfrac{\omega^2}{\omega_0^2} + \dfrac{L^2}{\omega_0^2 R^2}(\omega^2 - \omega_0^2)^2}$$

A little bit of factoring brings us to our final form of the relationship:

$$\frac{P_{avg}}{P_{avg(max)}} = \frac{\omega^2/\omega_0^2}{\dfrac{\omega^2}{\omega_0^2} + \dfrac{\omega_0^2 L^2}{R^2}(1 - \dfrac{\omega^2}{\omega_0^2})^2}$$

The quantity $\omega_0 L/R$ is dimensionless and is called the **quality factor**. It is often written in the form

$$Q_0 = \frac{\omega_0 L}{R}$$

The quality factor adjusts the shape of the ratio and resonance curves. For a quality factor of about 0.2, the shape of the ratio curve is similar to that of the two other curves described in Chapter 19 and is illustrated in Figure 11. The maximum transfer of power (resonance) occurs when the ratio of circuit frequency to resonant frequency is 1 as in the other cases.

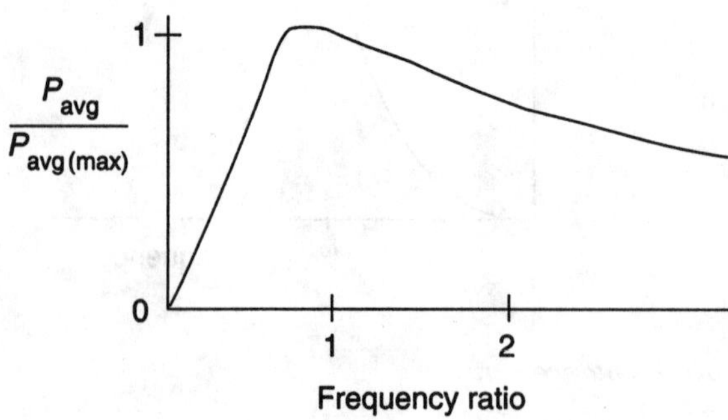

FIGURE 11. *Power ratio curve.*

Sample Problem

A series *RLC* circuit consists of a 100-Ω resistor, a 2-H inductor, and 100-μF capacitor connected to an alternating EMF operating with a maximum voltage of 50 V at a frequency of 60 Hz.

(a) Write an equation for the instantaneous voltage V at any time t.
(b) What is the value of the resonant frequency and angular frequency for this circuit?
(c) What are the values of the inductive and capacitive reactances?
(d) What is the value of the circuit's impedence?
(e) What is the value of the average power generated in this *RLC* circuit?

Solution

(a) In general, $V = V_{max} \sin \omega t$, where $\omega = 2\pi f$ and $\pi = 3.14$. In this problem $f = 60$ Hz, which means that $\omega = 376.80\ s^{-1}$. Thus, $V = \sin(376.80t)$ is an equation for this EMF.

(b) The resonant frequency is given by

$$f_0 = \frac{1}{2\pi\sqrt{LC}} = \frac{1}{6.28\sqrt{(2)(1 \times 10^{-4})}} = 11.26 \text{ Hz}$$

Thus, $\omega_0 = 2\pi f_0 = (6.28)(11.26) = 70.71\ s^{-1}$.

(c) The inductive and capacitive reactances are given by

$$X_L = \omega L = (376.8)(2) = 753.6\ \Omega$$

$$X_C = \frac{1}{\omega C} = \frac{1}{(376.8)(1 \times 10^{-4})} = 26.54\ \Omega$$

(d) The impedence of the circuit is given by

$$Z = \sqrt{R^2 + (X_L - X_C)^2} = \sqrt{(100)^2 + (727.06)^2} = 733.90\ \Omega$$

(e) The average power is given by

$$P_{avg} = I^2_{rms} R = \frac{V^2_{rms}}{Z^2} R = \frac{V^2_{max}}{2Z^2} R = \frac{(2{,}500)(100)}{(2)(53{,}8616.24)} = 0.23 \text{ W}$$

PROBLEM-SOLVING STRATEGIES

Some of the most important ideas to remember when dealing with ac circuits are the various phase relationships between the time-varying voltage and current. The fact that these quantities vary periodically in time means that your familiarity with trigonometric functions is crucial to fully appreciating the subtleties of ac circuits.

The circuit design schematics are similar to those for dc circuits except for the schematic representation of the ac potential source (usually a sine curve inside of a circle). Additionally, the combinations of resistors, inductors, and capacitors leads to behaviors that may not be intuitively obvious to you (such as resonance).

In an ac circuit, we talk about reactances and impedences instead of resistance, and you should recall how we introduced the mechanical analogy of impedence in our earlier discussions of dc power transfers. *Impedence* is a general term that refers to the opposition of the transfer of energy or power. It will appear again in the discussion of the transfer of energy from one pulse in a string to another.

Finally, remember that in a pure resistance circuit, the current and voltage are *in phase*. In a pure inductance circuit, the current *lags* the voltage by one quarter of a cycle. If we have a pure capacitance circuit, then the current *leads* the voltage by one quarter of a cycle.

PRACTICE PROBLEMS FOR CHAPTER 20

Thought Problems

1. Discuss the significance of the ratio R/Z in an RLC series circuit.
2. Show that the units of inductive and capacitive reactance are ohms.
3. Explain why we consider the *root mean square* current instead of the strict *mean* value of the sinusoidally varying alternating current.

Multiple-Choice Problems

1. An ac circuit operates at a frequency of 60 Hz. It consists of a resistor, an inductor, and a capacitor all in series with a source of EMF. If $R = 200\ \Omega$, $L = 1.5$ H, and $C = 120\ \mu$F, what is the value of the phase angle ϕ?
   (A) 71.19 degrees
   (B) 86.37 degrees
   (C) 46.78 degrees
   (D) 57.26 degrees
   (E) 69.78 degrees
2. What is the value of the resonant frequency for the circuit described in Problem 1?
   (A) 11.87 Hz
   (B) 60 Hz
   (C) 123.68 Hz
   (D) 376.2 Hz
   (E) 456.9 Hz
3. An ac circuit has an alternating time-varying voltage given by $V = 110 \sin(175t)$. What is the effective or root mean square voltage of this source?
   (A) 110 V
   (B) 17.5 V
   (C) 67.28 V

(D) 77.77 V
(E) 155.54 V

4. What is the operating frequency of the alternating EMF described in Figure 4?
(A) 175 Hz
(B) 60 Hz
(C) 27.87 Hz
(D) 1,099 Hz
(E) 0.036 Hz

5. In a series *RLC* circuit, resonance occurs when _____.
(A) the voltage and current are in phase
(B) the inductive and capacitive reactances are equal
(C) the operating frequency is 60 Hz
(D) both the inductive and capacitive reactances are zero
(E) the resistance equals zero

6. Which of the following expressions represents the power dissipated as heat in an ac circuit consisting of a pure inductor?
(A) 0
(B) IX_L
(C) I/X_L
(D) I^2X_L
(E) I^2/X_L

7. A series *RLC* circuit contains the following elements: $R = 50\ \Omega$, $X_L = 30\ \Omega$, and $X_C = 8\ \Omega$. What is the value of the impedence for this circuit?
(A) 47.2 Ω
(B) 51.6 Ω
(C) 54.6 Ω
(D) 50 Ω
(E) 62.8 Ω

8. The magnitudes of the voltages measured across the elements of a series *RLC* circuit at a certain instant of time are $V_R = 120$ V, $V_L = 60$ V, $V_C = 20$ V. What is the root mean square voltage for this circuit?
(A) 136 V
(B) 126 V
(C) 120 V
(D) 67 V
(E) 80 V

9. Compared with its resistance at resonance, the impedence of a series *RLC* circuit is _____.
(A) less than *R*
(B) greater than *R*
(C) equal to *R*

(D) any one of these
(E) independent of the resistance

10. A simple inductor is connected to an alternating source of EMF to form a simple ac circuit. At any instant, the product LI is proportional to ____.
(A) the magnetic field strength in the inductor
(B) the magnetic flux through the inductor
(C) the magnitude of the EMF
(D) the energy in the inductor
(E) the power dissipated as heat

Free-Response Problems

1. A simple *RC high-pass* filter is shown here.

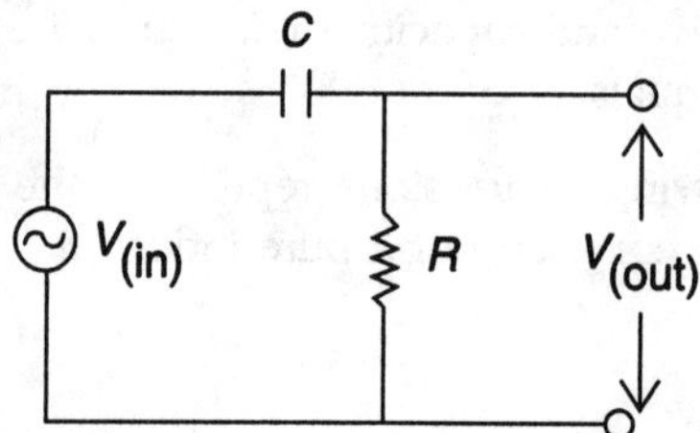

Analyze this circuit and derive an expression for the ratio of the output voltage to the input voltage. Also, explain why this circuit is called a high-pass filter.

2. A simple *RC low-pass* filter is shown here.

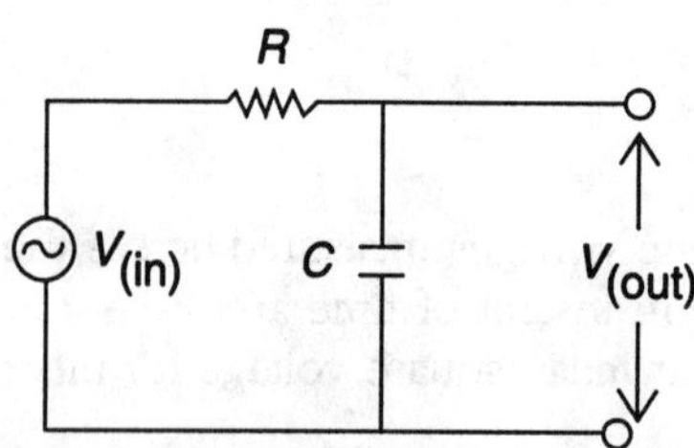

Analyze this circuit and derive an expression for the ratio of the output voltage to the input voltage. Also, explain why this circuit is called a low-pass filter.

3. Based on the derivation of the *RLC* resonance formula for maximum power transfer, show that

$$\left(\omega L - \frac{1}{\omega C}\right)^2 = \frac{L^2}{\omega^2}(\omega^2 - \omega_0^2)^2$$

SOLUTIONS TO PRACTICE PROBLEMS

Thought Problems

1. The ratio R / Z is equal to the power factor cos ϕ, as can be seen in the accompanying phasor diagram.

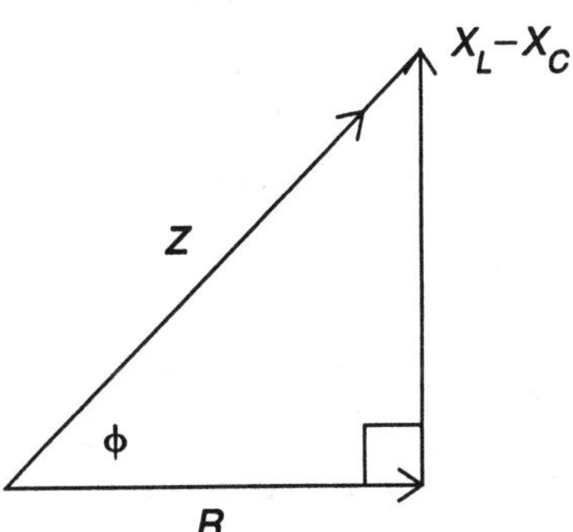

2. We first show that the units for inductive reactance are ohms. The units for inductance are henrys which are equivalent to J/A^2.

   The units for angular frequency are reciprocal second, and so

$$X_L = \omega L = \frac{J}{A^2} \times \frac{1}{s} = \frac{J}{C^2/s^2} \times \frac{1}{s} = \frac{J}{C^2/s} = J/C \times \frac{1}{C/s} = \frac{V}{A} \text{ ohms}$$

   We next show that the units for capacitive reactance are ohms. Recall that capacitance is in farads or C/V. Thus,

$$X_C = \frac{1}{\omega C} = \frac{1}{(1/s)(C/V)} = \frac{1}{A/V} = \frac{V}{A} = \text{ohms}$$

3. The instantaneous current in an ac circuit is given by $I = I_{max} \sin \omega t$. Since the since function is symmetric with respect to the *x* axis, the average value of the current is zero. However, for the electric power through a resistor, we need the square of the current, and thus the square of the sine function is always positive (and hence above the *x* axis). We therefore consider the square root of the average of the square (or root mean square) of the current instead of the strict mean value in order to obtain a meaningful value for the average power dissipated in a resistor.

Multiple-Choice Problems

1. **E** The phase angle is given by

$$\tan \phi = \frac{X_L - X_C}{R}$$

To find the reactances, we need the angular frequency $\omega = 2\pi f = (6.28)(60) = 376.8\ \text{s}^{-1}$.

Thus, we can write

$$X_L = \omega L = (376.8)(1.5) = 565.2\ \Omega$$

$$X_C = \frac{1}{\omega C} = \frac{1}{(376.8)(1.2 \times 10^{-4})} = 22.12\ \Omega$$

$$\tan\phi = \frac{(565.2 - 22.12)}{200} = 2.7154$$

$$\phi = 69.78 \text{ degrees}$$

2. **A** The resonant frequency is given by

$$f_0 = \frac{1}{2\pi\sqrt{LC}} = \frac{1}{2\pi\sqrt{(1.5)(1.2 \times 10^{-4})}} = 11.87\ \text{Hz}$$

3. **D** The effective root mean square voltage is given by

$$V_{rms} = 0.707\ V_{max} = (0.707)(110) = 77.77\ \text{V}$$

4. **C** The frequency is given by $f = \omega/2\pi = 175/6.28 = 27.87$ Hz
5. **B** At resonance, in a series *RLC* circuit the inductive and capacitive reactances are equal.
6. **A** Inductors do not absorb power and dissipate it as heat. The answer is therefore zero.
7. **C** The impedence is given by

$$Z = \sqrt{R^2(X_L - X_C)^2} = \sqrt{(50)^2 + (22)^2} = 54.6\ \Omega$$

8. **B** The effective root mean square voltage is given by

$$V_{rms} = \sqrt{V_R^2 + (V_L - V_C)^2} = \sqrt{(120)^2 + (40)^2} = 126\ \text{V}$$

9. **C** At resonance, the impedence is equal to the resistance in a series *RLC* circuit.
10. **B** The product of the inductance and instantaneous current, LI, is equivalent to N times the magnetic flux in the coil. That is, $LI = N\Phi$.

Free-Response Problems

1. The input voltage is equal to the product of the current and the impedence:

$$V_{in} = IZ = I\sqrt{R^2 + \left(\frac{1}{\omega C}\right)^2}$$

The output voltage is just the product of the current and the resistance:

$$V_{out} = IR.$$

The ratio of the output to the input voltage is therefore equal to

$$\frac{V_{out}}{V_{in}} = \frac{R}{\sqrt{R^2 + (1/\omega C)^2}}$$

At low frequencies, the output voltage is small compared with the input voltage. As the frequency increases toward infinity, the ratio of the output to the input voltage approaches 1. Thus, the circuit preferentially passes high-frequency signals and is thus called a high-pass filter.

2. The output voltage is the voltage across the capacitor:

$$V_{out} = IX_C = \frac{I}{\omega C}$$

The input voltage is equal to the product of the current and the impedence:

$$V_{in} = IZ = I\sqrt{R^2 + \frac{1}{\omega C^2}}$$

The ratio of the output to the input voltage is given by

$$\frac{V_{out}}{V_{in}} = \frac{R}{\sqrt{R^2 + (1/\omega C)^2}}$$

When the frequency is low, the output voltage is practically equal to the input voltage, and when the frequency is low, V_{out} is approximately zero. The circuit therefore preferentially passes low-frequency signals. It is therefore called a low-pass filter.

3. We first expand the given equation and use the definition of the resonant angular frequency:

$$\left(\omega L - \frac{1}{\omega C}\right)^2 = \omega^2 L^2 + \frac{1}{\omega^2 C^2} - \frac{2L}{C}$$

$$\omega_0^2 = \frac{1}{LC}$$

$$\frac{L}{C} = \omega_0^2 L^2$$

$$\left(\omega L - \frac{1}{\omega C}\right)^2 = \omega^2 L^2 + \frac{1}{\omega^2 C^2} - 2\omega_0^2 L^2$$

We will now work exclusively with the right-hand side. The middle term on the right-hand side still includes the capacitance, and it must be eliminated in order to complete our proof. If we return to the definition of the resonant angular frequency, we see that the rest of the solution proceeds with just a little factoring:

$$\omega_0^2 = \frac{1}{LC}$$

$$\frac{1}{C^2} = \omega_0^4 L^2$$

$$\begin{aligned}\omega^2 L^2 + \frac{1}{\omega^2 C^2} - 2\omega_0^2 L^2 &= \omega^2 L^2 + \frac{\omega_0^4 L^2}{\omega^2} - 2\omega_0^2 L^2 \\ &= \frac{L^2}{\omega^2}(\omega^4 + \omega_0^4 - 2\omega_0^2\omega^2) \\ &= \frac{L^2}{\omega^2}(\omega^2 - \omega_0^2)^2\end{aligned}$$

21
WAVES AND SOUND

PULSES

A **pulse** is a single vibratory disturbance in a medium, an example of which is seen in Figure 1. If a string is fixed at both ends, made taut under a tension **T** (in newtons), and given a quick up-and-down snap at the left, then an upwardly pointing pulse will be directed from left to right. The pulse appears to travel with a velocity **v** down the string. In actuality, the energy transferred to the string causes segments to vibrate up and down successively. This effect produces the illusion of a continuous pulse. Only energy is transferred by the pulse, and because the vibration is perpendicular to the apparent direction of motion, the pulse is said to be *transverse*. The **amplitude** of the pulse is the maximum displacement of the disturbance above the level of the string.

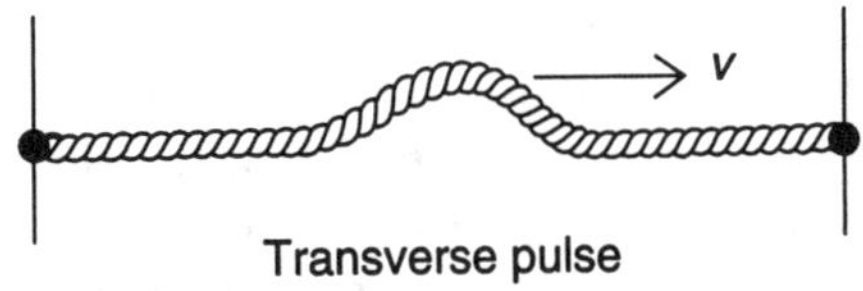

FIGURE 1.

If the tension in the string is increased, then the velocity will increase. If a heavier string is used (greater mass per unit length), then the pulse will move more slowly. Careful measurements of these observations lead to the following result. The velocity of the pulse is given by the equation

$$\mathbf{v} = \sqrt{\frac{\mathbf{T}}{M/L}}$$

where **T** is the tension in newtons, M is the mass in kilograms, and L is the length of the string in meters. The ratio M/L is called the mass density per unit length.

When the pulse reaches one boundary, the energy transferred is directed upward against the wall, creating an upward force. Since the wall is rigid, the reaction to this action is a downward force. The string responds by being displaced downward and is reflected back. The result of this boundary interaction is observed as an inversion of the pulse on reflection and is illustrated in Figure 2. Some energy is lost in the interaction but most is reflected back with the pulse.

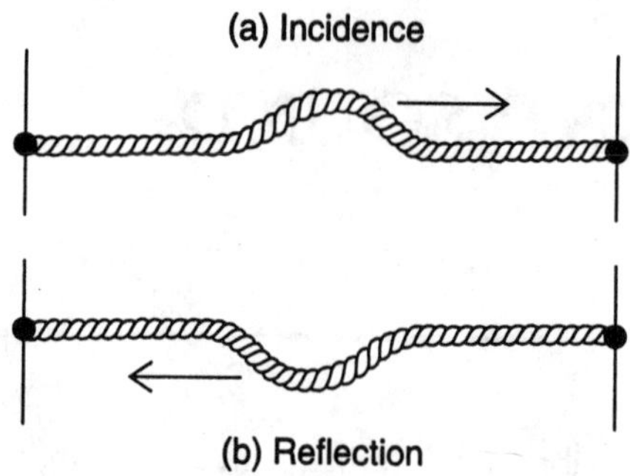

FIGURE 2.

At a nonrigid boundary, say between two strings of different masses (but equal tensions), a different effect occurs. When the pulse reaches the nonrigid boundary, the upward displacement at the boundary causes the new string also to be displaced upward. The magnitude of the new displacement depends on the mass density and per unit length of the new string. Thus, the transmitted pulse has an upward orientation and may travel with a larger or smaller velocity depending on the mass density per unit length (in this case).

The reflected pulse will be upward in orientation since no inversion takes place with a nonrigid boundary unless the difference between the two media (as measured by the new string's inertia) is large enough to behave as a nonrigid boundary. As an example, consider a light string attached to a heavy rope and a pulse traveling from the string to the rope. Figure 3 illustrates both cases.

Since pulses transmit only energy, when two pulses interact, they obey a different set of physical laws than when two pieces of matter interact. Pulses are governed by the **principle of superposition**, which states that when two

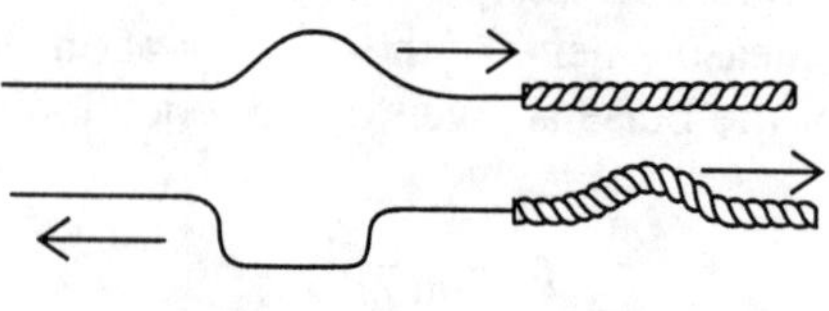

Case I: Nonrigid reflection

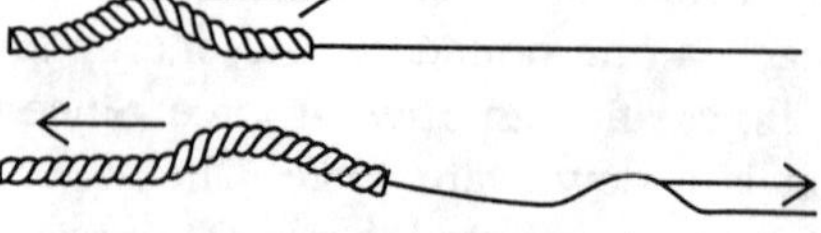

Case II: Nonrigid reflection

FIGURE 3.

pulses interact at the same point and at the same time, the interaction produces a single pulse whose amplitude displacement is equal to the sum of the amplitudes of the original two pulses. After the interaction, however, both pulses continue in their original directions of motion, unaffected by the interaction. Sometimes this interaction is called **interference**.

When a pulse with an upward displacement interacts with another pulse with an upward displacement, the resulting pulse has a displacement larger than that of either one (equal to the numerical sum). This interaction is called **constructive interference**. When a pulse with an upward displacement interacts with a pulse with a downward displacement, the interaction is called **destructive interference**, and the resulting pulse has a smaller amplitude (equal to the numerical difference). Figure 4 illustrates these concepts.

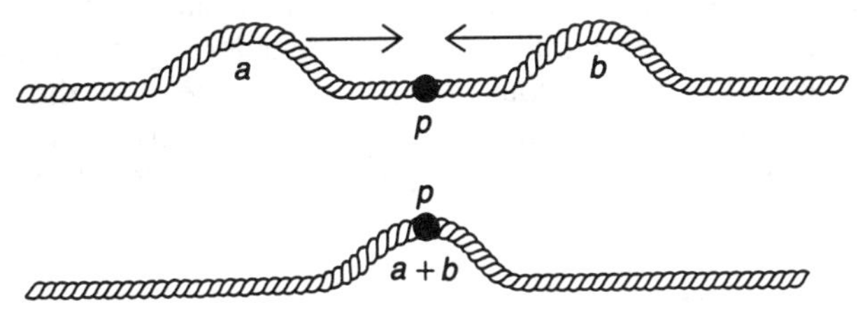

(a) Constructive interference

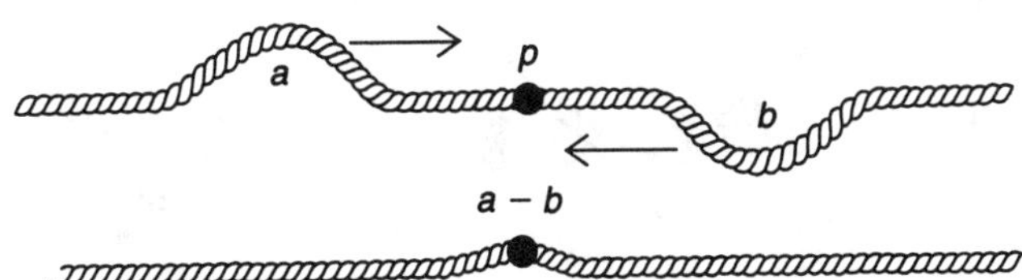

(b) Destructive interference

FIGURE 4.

In the case of destructive interference, it is possible for the interaction to momentarily cancel out both pulses if the amplitudes are equal.

WAVE MOTION

If a continuous up-and-down vibration is given to the string in Figure 1, the resulting set of transverse pulses is called a **wave train** or just a **series of waves**. Whereas a pulse is just an upward- or downward-oriented transverse disturbance, a wave consists of a complete up-and-down segment. Waves are thus periodic disturbances in a medium. The number of waves per second is called the **frequency** designated by the letter f and expressed in cycles/s or **hertz**.

The time it takes to complete one wave cycle is called the **period** and is designated by the letter T (in seconds). The frequency and period of a wave are reciprocals of each other, such that $f = 1/T$.

Just like pulses, waves transmit only energy. The particles in the medium vibrate only up and down as the transverse waves approach and pass a given point. Since there is a continuous series of waves, many points in the medium move up or down in unison. The **phase** of a wave is defined to be the extent to which two successive points on a wave match in position.

The distance between any two successive points in phase is a measure of the **wavelength**, designated by the Greek letter λ (lambda) and measured in meters. The peaks of the wave are called **crests**, while the valleys are called **troughs**. Thus, one wavelength can also be considered the distance between any two successive crests or troughs. The **amplitude** A of the wave is the distance of the maximum displacement (up or down) above the normal rest position in meters. In Figure 5, a typical transverse wave is illustrated.

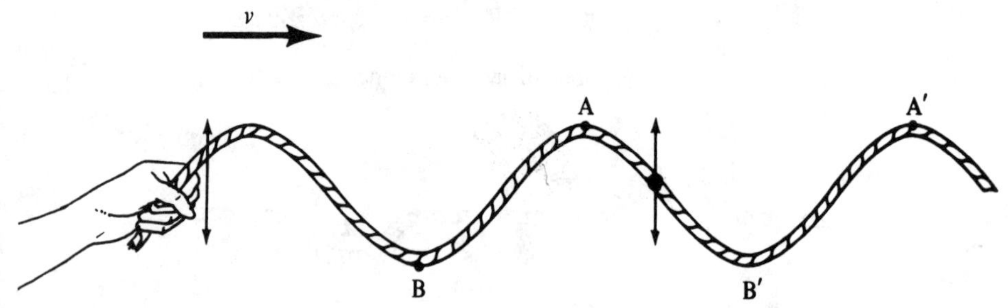

FIGURE 5. *Transverse wave.*

The nature of the wave is sinusoidal, similar to the plot of displacement versus time for a mass on a spring undergoing simple harmonic motion. Recall from Chapter 14 that the maximum energy in a simple harmonic motion system is directly proportional to the square of the amplitude. This is indeed the case with a simple wave motion like the one shown in Figure 5.

In Figure 5, we can see that points a, b, and c are all in phase and that points d and e are also in phase. The distance from d to e can also be used to measure the wavelength. Finally, the velocity of the transverse wave is given by the equation $\mathbf{v} = f\lambda$.

Since each wave carries energy, it should not be surprising that the frequency is also proportional to the energy of a wave. The amplitude is related to what we might consider the wave's **intensity**, which in the case of sound might be called **volume**. The frequency, however, measures the **pitch** of the wave. Waves of higher frequency transmit more energy per second than waves of lower frequency (at the same amplitude).

There are many different kinds of frequencies that interact with human senses. It is therefore worthwhile to review the prefixes used in physics (see Appendix B for more information.) Recall that the prefix *kilo-* represents 10^3, the prefix *mega-* represents 10^6, and the prefix *giga-* represents 10^9.

TYPES OF WAVES

In physics we usually deal with two types of waves: mechanical and electromagnetic. Waves that result from the vibration of a physical medium (a drum, a string, water, etc.) are called **mechanical waves**. Waves that result from electromagnetic interactions (light, x rays, radio waves, etc.) are called **electromagnetic waves**. These electromagnetic waves are special because they do not require a physical medium to carry them.

Electromagnetic waves are also all transverse in nature. Physicists know this based on a behavior analogous to that of mechanical waves. Imagine a long, coiled spring as shown in Figure 6. We can set up transverse vibrations in this spring by shaking it up and down, and the resulting transverse waves can be analyzed as before.

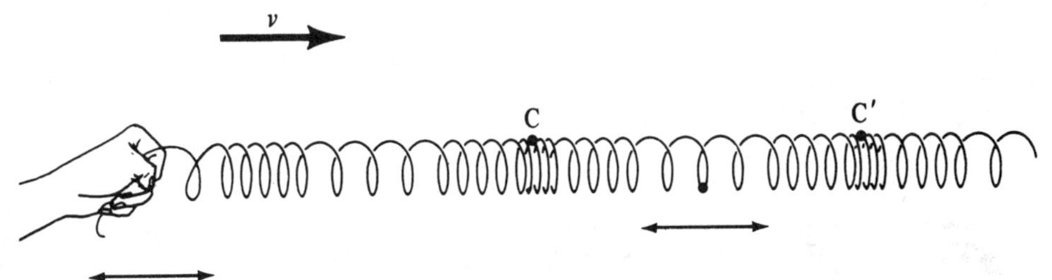

FIGURE 6.

Another way to set up periodic disturbances in the spring is to pull it back and forth along its longitudinal axis. The resulting disturbances consist of regions of compression and expansion that appear to travel parallel to the disturbances themselves. These waves are called **longitudinal** or **compressional** waves, and sound is an example. The vibrations of an object in the air create pressure differences that alternately expand and contract the air which, when impacted on our ears, create the phenomenon known as **sound**.

In a transverse wave, the vibrations must be perpendicular to the direction of apparent motion. However, if we cut a plane containing the directions of vibration, we find that there are many possible orientations in which the vibrations can still be considered perpendicular to the direction of travel. Thus, we can select one or more preferred planes containing a particular vibrational orientation. For example, a sideways vibration is just as "transverse" as an up-and-down or diagonal vibration. This selection process is known as **polarization** (see Figure 7) and applies only to transverse waves.

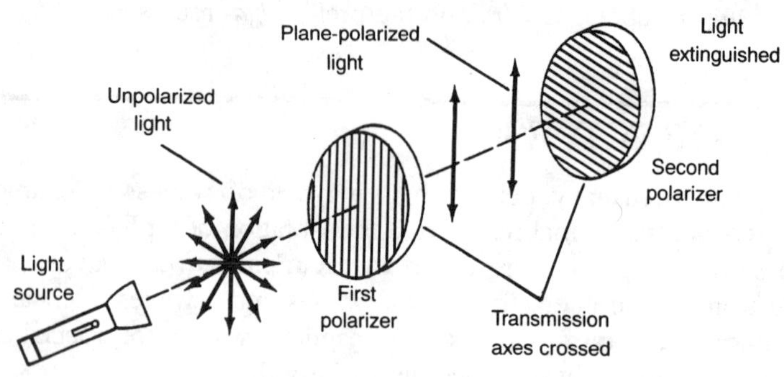

FIGURE 7.

In a longitudinal wave, there is only one way to make the vibrations parallel to the direction of travel. Special devices have been developed to test for polarization in mechanical and electromagnetic waves. Since all electromagnetic waves can be polarized, physicists conclude that they must be transverse waves. We shall not delve into the details of other fundamental considerations concerning the transverse nature of electromagnetic waves. Since sound waves cannot be polarized, physicists conclude that sound is a longitudinal wave.

STANDING WAVES AND RESONANCE

Consider the taut string in Figure 1 again. There are several ways in which we can make it vibrate transversely. One way is to make the entire string move up and down as one unit. This is the easiest frequency mode of vibration possible and is called the **fundamental mode**. When we increase the frequency, something interesting happens. Incoming waves reflect off a boundary inverted and match perfectly the amplitude and frequency of the remaining incoming waves. In this situation, the apparent horizontal motion of a wave stops, and we have a **standing wave**. A standing wave appears to be segmented by **nodal points**, and these nodal points correspond to points where no appreciable displacement takes place.

Figure 8 shows a standing wave in various vibrational modes. In Figure 8a we have the fundamental mode, and in Figure 8b we see the second mode in which one nodal point appears. This point occurs at the one-half-wavelength mark, and as such we have for a string of length L, $\lambda = L$ (at this frequency).

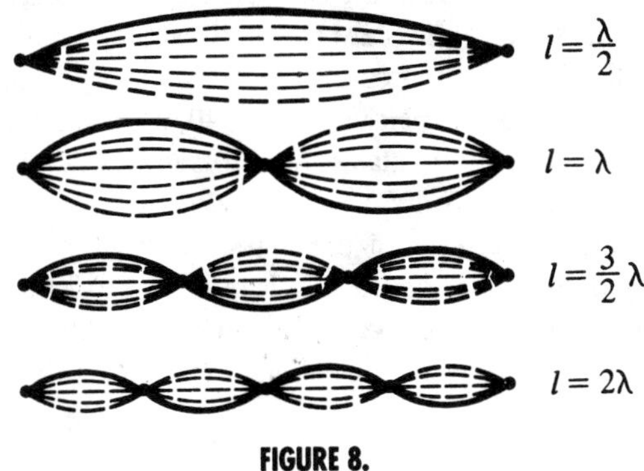

FIGURE 8.

In the fundamental mode, $\lambda = 2L$, while in Figure 7c, $\lambda = 2L/3$ and we have two nodal points.

This standing wave pattern is a form of interference, and with sound the nodal points are determined by zero points in intensity (ideally). With light we observe a darker region relative to the surrounding area. With a string we observe the characteristic segmented pattern shown in Figure 8.

All objects have a natural vibrating frequency. When a glass or a bell is struck, the frequency of vibration heard (actually, the sound is a complex mixture of vibrations) is made up of one fundamental characteristic frequency and superimposed overtones. If two identical tuning forks are held close together, then the phenomenon of resonance can be observed. If one tuning fork is struck, then the waves emanating from the first one strike the second. Since the frequency of these waves matches the natural vibrating frequency of the second tuning fork, it begins to vibrate (see Figure 9). This condition is known as **resonance**.

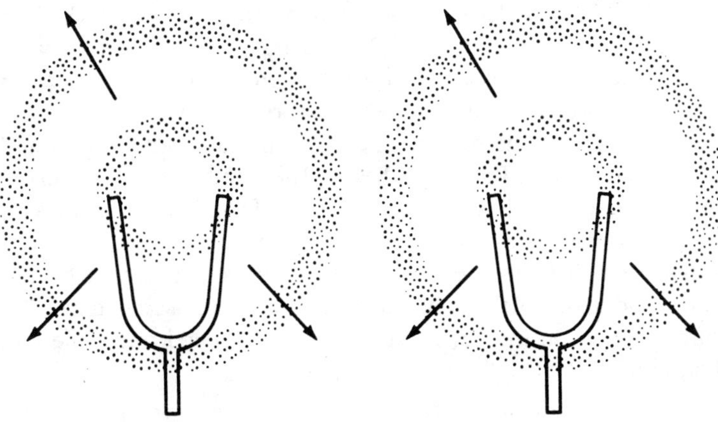

FIGURE 9. *Resonance in two identical tuning forks.*

Resonance also explains how an opera singer can shatter a crystal wine glass. It also explains why soldiers are ordered to "break march" when crossing a bridge. The rhythmic marching can set up resonance vibrations and perhaps cause the bridge to collapse. In November 1940, a resonance vibration caused by a light gale wind generated catastrophic torsional vibrations in the Tacoma Narrows Bridge in Washington State. The violent convulsions were photographed and are shown as one of the classic physics movies illustrating the principle of resonance. Nick-named "Galloping Gertie," the bridge is a testament to why engineers must be very careful when considering the possible effects of resonance.

SOUND

Sound is a longitudinal mechanical wave. Air density is alternately increased and decreased as pressure differences move through the air. When lightning occurs, the expansion of air due to the very high temperature of the lightning bolt creates the violent sound known as thunder.

Since sound is carried by air molecules, the velocity is dependent on the air temperature. At 0°C, the velocity of sound in air is 331 m/s. For each 1°C rise in temperature, the velocity of sound increases by approximately 0.6 m/s. The study of sound is called **acoustics**, and longitudinal waves in matter are sometimes known as *acoustical waves.*

The ability of sound waves to pass through matter depends on the structure of the molecules. Thus, sound travels slower in a gas (in which the molecules have more random motion) than in a liquid or a solid (which has a more rigid structure). In Table 1, the velocity of sound in different substances is presented.

Human hearing can detect sound ranging from 20 to 20,000 Hz. Sound waves exceeding a frequency of 20,000 Hz are called **ultrasonic** waves. The square of the amplitude of a sound wave is proportional to the **intensity** or loudness of the wave, while the frequency relates to the **pitch**. When two sound waves interfere, the regions of constructive and destructive interference produce a phenomenon known as **beats**. The number of beats per second is equal to the frequency difference. The reason for this is as follows. Suppose we have two waves with frequencies f_1 and f_2, where the second frequency is greater than the first. As the two waves interfere, there will be a time of maximum loudness which is heard as a beat. After some time T, another maximum is heard. In this time, one wave has traveled one cycle more than the other, and so

$$f_2T - f_1T = 1 \quad \text{and} \quad f_{\text{beat}} = \frac{1}{T} = f_2 - f_1$$

TABLE 1. VELOCITY OF SOUND

Substance	Velocity (m/s)
Gases (0°C)	
Carbon dioxide	259
Air	331
Helium	965
Liquids (25°C)	
Ethyl alcohol	1,207
Water, pure	1,498
Water, sea	1,531
Solids	
Lead	1,200
Wood	~4,300
Iron and steel	~5,000
Aluminum	5,100
Glass, Pyrex	5,170

Interference between waves is dependent on what is called the **path length difference**. If we have two point sources of sound, *A* and *B*, and a receiver some distance away, such that the distance from source *A* to the receiver is L_A and the distance from source *B* to the receiver is L_B, then the receiver will be at a point of constructive interference if the path length difference, $L_A - L_B$, is equal to a whole multiple of the wavelength, $n\lambda$. Destructive interference will occur if the path length difference is equal to an odd multiple of half-wavelengths, $(n + \frac{1}{2})\lambda$.

Another phenomenon associated with interference is diffraction. When a wave encounters a boundary, it appears to bend around the corners of the boundary. According to one theory, this effect, known as **diffraction**, occurs because the wavefront behaves like a series of point sources and creates circular waves. These waves, because of their shape, reach behind the corners that block the rest of the wavefront. So, at the boundary we have effective wave bending. In Figure 10a, we observe a series of straight waves made in a water tank and the resulting diffraction at a corner. In Figure 10b, we see the diffraction of waves as they pass through a narrow opening. We will discuss the details of interference and diffraction in more detail in connection with light.

The intensity of sound is measured in units called **decibels** (dB). Human hearing does not respond to changes in intensity in a linear fashion. Because of this fact, the decibel system is a logarithmic (base 10) system in which the intensity β is measured in decibels using the relationship

$$\beta = 10 \log \frac{I}{I_0}$$

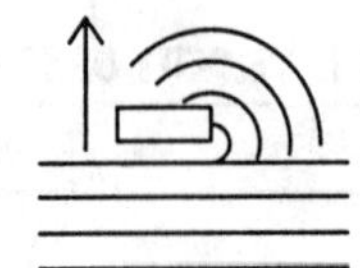

(a) Diffraction around a corner

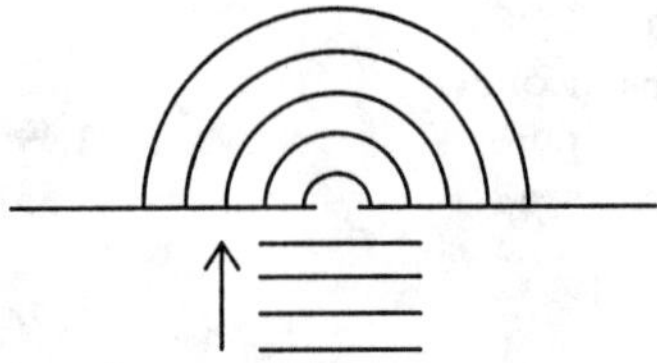

(b) Diffraction through a narrow opening

FIGURE 10.

where I is the intensity in W/m^2. I_0 is taken to be the barely audible intensity of 10^{-12} W/m^2. Some common decibel measurements are presented in Table 2.

TABLE 2. SOUND INTENSITY

Sound	Intensity (dB)
Whisper	30
Conversation	60
Loud	70–90
Deafening	100
Painful	120
Damaging	140

THE DOPPLER EFFECT

Most people have had the experience of hearing a siren pass by. Even though it emits a sound at one frequency, its changing position relative to the observer produces an apparent change in frequency. As it approaches, the pitch increases, and as it passes, the pitch is decreases. This phenomenon is called the **Doppler effect**.

Let us imagine a point source sending out waves in a circular fashion. In the static case shown in Figure 11a, observers at points A and B both experience the same frequency of waves as an observer located at the source S. However, if the source is moving (see Figure 11b), let's say toward point A, then for each period of time between waves, the circular waves will not be

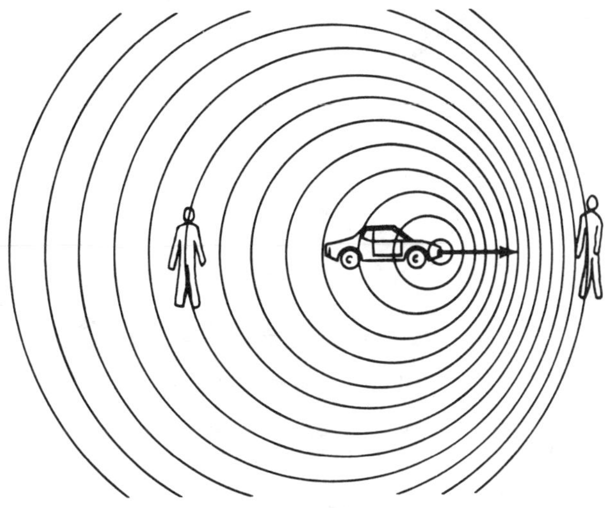

FIGURE 11.

concentric. That is, the spacing between successive waves will be reduced by an amount equal to the distance traveled by the source in the period of time *T*. Thus, an apparent increase in frequency is observed at point *A*, while at point *B* there is an apparent decrease in frequency. The relationship governing this change in frequency is given by the formulas

$$f = f_0 \frac{\mathbf{v}}{\mathbf{v} - \mathbf{v}_s}$$

$$f = f_0 \frac{\mathbf{v}}{\mathbf{v} + \mathbf{v}_s}$$

where $\mathbf{v}$ is the wave velocity in the medium and $\mathbf{v}_s$ is the relative velocity of the source. The first equation is used when the source is approaching a receiver, and the second equation is used when the source is receding from a receiver.

If the source and the observer are both moving relative to each other, then we can write

$$f_0 = f_s \frac{\mathbf{v} + \mathbf{v}_0}{\mathbf{v} - \mathbf{v}_s}$$

The observer's velocity is positive if he or she is moving toward the source, while the source velocity is positive if it is moving toward the observer.

If the source velocity equals the wave velocity, then a very strong interference pattern will build up in front of the source. If the source exceeds the wave velocity, then the source will "outrun" its own waves. This is seen in the

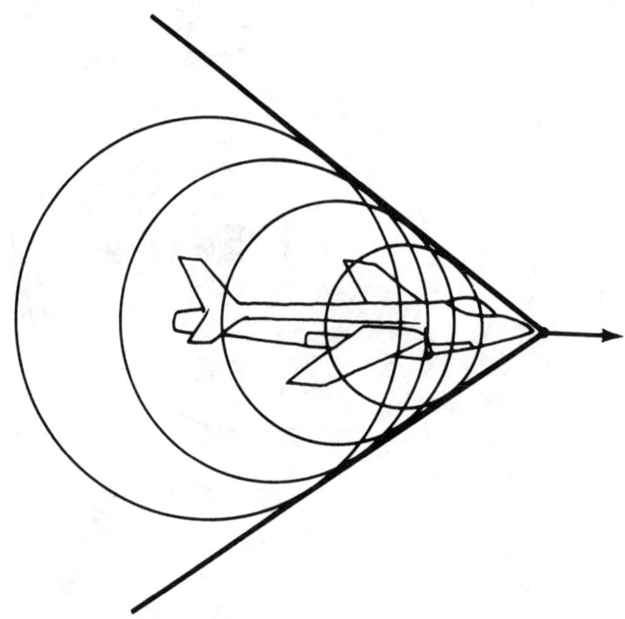

FIGURE 12. *Shock wave pattern*

pattern created by ducks on the water and occurs when airplanes break the "sound barrier." The regions of constructive interference shown in Figure 12 are called **shock waves**. The ratio of the source velocity to the wave velocity is called the **Mach number** after the Austrian physicist Ernst Mach. At Mach 1, the source is traveling at the wave velocity. The **sonic boom** made by supersonic aircraft is an example of these shock waves in air.

PROBLEM-SOLVING STRATEGIES

Solving wave motion problems involves remembering the basic physical concepts. In transverse waves, the only particle motion in the medium is perpendicular to propagation. If a series of waves is presented, you should resist the temptation to view labeled points as beads on a wire. Points that are in phase are either going up at the same time or going down at the same time and are at the same height.

Transverse waves can be polarized, meaning that we can select a preferred axis to contain the vibrations and have them remain perpendicular to the direction of apparent wave motion. Longitudinal waves cannot be polarized, and the vibrations remain parallel to the direction of wave motion.

Sound waves travel through a medium, and their speed is determined by the molecular structure of matter. In air, the speed of sound increases with increasing temperature. Sound wave interference is exhibited by the beat pulsations heard when two frequencies are heard simultaneously.

Sample Problem

A sound wave has a frequency of 350 Hz. If the air temperature is 15°C, what is the wavelength of the sound?

Solution

At 15°C the speed of sound increases from 331 m/s by 0.6 m/s for each degree above 0°C. At 15°C, the speed of sound is 340 m/s. Now, since $\mathbf{v} = \mathbf{f}\lambda$, we find that the wavelength $\lambda = 0.97$ m.

PRACTICE PROBLEMS FOR CHAPTER 21

Thought Problems

1. Explain how the concept of impedence can be applied to a situation where two springs of different sizes are connected together and pulses are sent along one of them.
2. Why are soldiers told to break march when they are going over a bridge?
3. Apartment dwellers are used to hearing the "boom-boom" sound of bass tones when neighbors play their stereos too loudly. What can account for this phenomenon?

Multiple-Choice Problems

1. A stretched string has a length of 1.5 m and a mass of 0.25 kg. What must be the tension in the string in order for pulses in the string to have a velocity of 5 m/s?
   (A) 2.45 N
   (B) 12.5 N
   (C) 4.2 N
   (D) 150 N
   (E) 250 N
2. A stretched string is vibrated in such a way that a standing wave with two nodes appears. The distance between each node is 0.2 m. What is the wavelength of the standing wave?
   (A) 0.4 m
   (B) 0.3 m
   (C) 0.2 m
   (D) 0.6 m
   (E) 0.8 m
3. What is the wavelength of sound produced at a frequency of 300 Hz when the air temperature is 20°C?
   (A) 1.10 m
   (B) 1.30 m

(C) 0.80 m
(D) 1.14 m
(E) 1.2 m

4. Two tuning forks are vibrating simultaneously. One has a frequency of 256 Hz, while the other has a frequency of 280 Hz. The number of pulsational beats per second heard is equal to _____.
(A) 256
(B) 536
(C) 1.1
(D) 0.91
(E) 24

5. In a stretched string with a constant tension **T**, as the frequency of the waves increases, the wavelength _____.
(A) increases
(B) decreases
(C) remains the same
(D) increases and then decreases
(E) decreases and then increases

6. What is the period of a wave that has a frequency of 12,000 Hz?
(A) 36.25 s
(B) 0.000083 s
(C) 0.0275 s
(D) 12,000 s
(E) 0.012 s

7. Sound waves travel fastest in _____.
(A) a vacuum
(B) air
(C) water
(D) wood
(E) not enough information given

8. The amplitude of a sound wave is related to its _____.
(A) pitch
(B) loudness
(C) frequency
(D) resonance
(E) wavelength

9. At 25°C, it is observed that a sound wave takes 3 s to reach a receiver. How far away is the receiver from the source?
(A) 1,038 m
(B) 993 m
(C) 1,215 m
(D) 1,068 m
(E) 887 m

10. How many times more intense is a 100-dB sound compared to a 60-dB sound?
(A) 10^4
(B) 400
(C) 40
(D) 4
(E) 100

Free-Response Problems

1. The noise of a jet engine is coming into an open window 1.0 m wide and 1.2 m high. If the sound level is 140 dB, at what rate does energy enter the window?

2. What must be the tension in a 0.25-m string with a mass of 0.30 kg such that its fundamental frequency mode is 400 Hz?

3. A train whistle has a natural frequency of 1,500 Hz. If the train is approaching a station at a velocity of 30 m/s, what is the frequency perceived by a person standing in the station? (Assume the air temperature is 30°C.)

SOLUTIONS TO PRACTICE PROBLEMS

Thought Problems

1. Impedence involves the transfer of energy from one oscillating system to another. When a pulse encounters a boundary, its ability to be transmitted is a function of the impedence of both media. If the second spring has a higher impedence, then most of the pulse will be reflected (but some will be transmitted). If the impedences are high (as in the case of a spring attached to a wall), then the majority of the pulse's energy will be reflected. However, if the springs are identical, there will be impedence matching and all of the pulse will be transmitted.

2. Soldiers are told to break march when going over a bridge because uniform marching can induce resonance waves in the bridge to the point where its structural integrity might be compromised.

3. The resonant frequency for buildings is quite low, and so the low-frequency bass tones resonate through the walls as the "boom-boom" apartment dwellers are familiar with.

Multiple-Choice Problems

1. **C** The formula for wave velocity in a string is $\mathbf{v} = \sqrt{T/(M/L)}$. Using the given information and solving for tension leads to $\mathbf{T} = 4.2$ N.

2. **A** Standing wave nodes occur at half-wavelength intervals. Thus, $\lambda = 2(0.2) = 0.4$ m.

3. **D** At 20° C, the speed of sound is 331 + (0.6)(20) = 343 m/s. Since $\mathbf{v} = f\lambda$, we obtain a wavelength of λ = 1.14 m.

4. **E** The number of beats per second is equal to the frequency difference: 280 − 256 = 24.

5. **B** Since $\mathbf{v} = f\lambda$ and for a constant tension the velocity is constant, as the frequency increases, the wavelength decreases.

6. **B** Frequency and period are reciprocals of each other. Thus T = 1/12,000 = 0.000083 s.

7. **D** Because of the molecular arrangement, sound travels fastest in solids compared with liquids and gases.

8. **B** Amplitude is related to the intensity of a wave. In sound, wave intensity is perceived as loudness.

9. **A** At 25°C, the speed of sound is 331 + (0.6)(25) = 346 m/s. Since $x = \mathbf{v}t$, then x = (346)(3) = 1,038 m.

10. **A** The decibel scale is defined such that a difference of 40 dB corresponds to a difference (ratio) of 10,000 times the intensity.

Free-Response Problems

1. The formula for sound level in decibels is

$$\beta = 10 \log \frac{I}{I_0}$$

where $I_0 = 10^{-12}$ W/m². In this problem β = 140 dB, and we must solve for the intensity I. If we divide by 10, substitute for I_0, and then "exponentiate" by raising both sides as a power of 10, we find that I = 100 W/m² (after cross-multiplying). The rate of energy per second is the power emitted in watts. Thus, we must multiply 100 W/m² by the area of the window which is 1.2 m². Thus, the energy rate is 120 W.

2. At fundamental frequency, we have $\lambda = 2L$ = (2)(0.25) = 0.5 m. Now, $\mathbf{v} = f\lambda$, which means that $\mathbf{v}$ = (400)(0.5) = 200 m/s. We can now use the formula for the velocity in a fixed string, which is $\mathbf{v} = \sqrt{T/(M/L)}$. Solving for the tension, we obtain $\mathbf{T}$ = 48,000 N.

3. At 30°C, the velocity of sound is 331 + (0.6)(30) = 348 m/s. Now the Doppler formula for a stationary observer and an approaching source is

$$f = f_0\left(\frac{\mathbf{v}}{\mathbf{v} - \mathbf{v}_s}\right)$$

where f_0 = 1,500 Hz, $\mathbf{v}_s$ = 30 m/s, and $\mathbf{v}$ = 348 m/s. Solving for frequency, we have f = 1641.5 Hz.

22
LIGHT

ELECTROMAGNETIC WAVES

In Chapter 19, we reviewed aspects of electromagnetic induction. In one example, we observed how the changing magnetic flux in a solenoid can influence a second solenoid not physically attached to the primary. This **mutual induction** involves the transfer of energy through space by means of an oscillating magnetic field. Experiments in the late nineteenth century by the German physicist Heinrich Hertz confirmed what the Scottish physicist James Clerk Maxwell had asserted theoretically in 1864: Oscillating electromagnetic fields travel through space as transverse waves, and they travel with the same speed as light.

Light is just one form of electromagnetic radiation that travels in the form of transverse waves. We know that these waves are transverse because they can be polarized. The speed of light, designated by the letter **c**, has been determined by many physicists and is approximately equal to 3×10^8 m/s.

Light waves can be coherent if they are produced in a way that maintains a constant phase relationship. For example, lasers produce coherent light of one wavelength. (The word *laser* is an acronym for *l*ight *a*mplification by the *s*timulated *e*mission of *r*adiation.)

Experiments by Sir Isaac Newton in the seventeenth century showed that *white light*, when passed through a prism, contains the colors red, orange, yellow, green, blue, indigo, and violet (abbreviated ROYGBIV). Each color of light is characterized by a different wavelength and frequency, the wavelengths ranging from about 3.5×10^{-7} m for violet light to about 7.0×10^{-7} m. Small wavelengths are sometimes measured in **ångstroms** (Å), where 1 Å = 1×10^{-10} m. In more modern textbooks, **nanometers** are used, where 1 nm= 1×10^{-9} m. Together with other electromagnetic waves (such as radio waves, x rays, and infrared waves), light occupies a special place in the **electromagnetic spectrum** because we can "see" it. A sample electromagnetic spectrum is presented in Figure 1.

Since electromagnetic waves are "waves," they all obey the relationship $\mathbf{c} = f\lambda$, where **c** is the speed of light discussed previously. With this in mind, gamma rays have frequencies in the range of 10^{25} Hz and wavelengths in the range of 10^{-13} m, making them very small but very energetic. They are sometimes referred to as *cosmic rays* since they often originate in the cores of exploding stars and galaxies.

High frequency — Low frequency

Gamma rays	X rays	Ultraviolet	Light Blue Red	Infrared	Micro-waves	Radio waves

Short wavelength — Long wavelength

FIGURE 1. *Electromagnetic spectrum.*

SHADOWS

One property of light that is most useful is its ability to travel in a straight line. We do not always perceive the wave nature of light. If two sticks are placed on the ground, it is possible to mark a straight line by sighting one along the other. This can be done because light travels in a line. If an opaque object is placed in the path of a beam of light, the light will be blocked and a *shadow* will appear in the region behind the object. If a point source of light is used, then the rays of light will emerge radially in all directions. In Figure 2, we see that when an object is placed in the path of the point source, then a clear shadow is formed and a simple geometric relationship exists between the object and the shadow image. If p and q represent the object and image distances from the source, and S_1 and S_2 represent the object and image sizes, then we can write, for a point source of light,

$$\frac{p}{q} = \frac{S_1}{S_2}$$

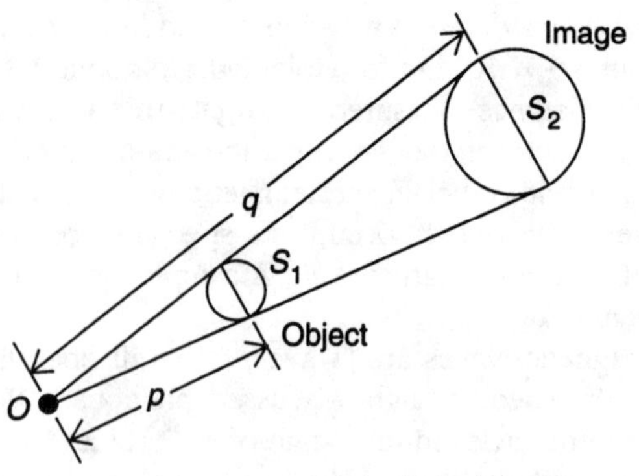

FIGURE 2.

Closer examination reveals that the edges of the shadow are not sharp. There is a distinctive "fuzziness" associated with all shadows. This irregularity was first examined in the seventeenth century by an Italian physicist named F. Grimaldi and attributed to the diffraction of light around the edges of the object. This is evidence for the wave nature of light and was used by Grimaldi, along with a Dutch contemporary named Christian Huygens, to be the basis of a "wave theory of light" advanced in the seventeenth century to counter Isaac Newton's "corpuscular theory of light." Wave interference patterns were first demonstrated for light in 1801 by the English physicist Thomas Young, and the corpuscular theory of light was forgotten until the discovery of the **photoelectric effect** at the end of the nineteenth century.

REFLECTION

Reflection is the ability of light to seemingly bounce off a surface. This phenomenon does not reveal the wave nature of light, and the notion of wavelength or frequency rarely enters into the discussion of reflection. If light is incident on a flat mirror, then the angle of incidence is measured with respect to a line perpendicular to the surface of the mirror called the **normal**. In Figure 3 the **law of reflection** is illustrated, which states that the angle of incidence equals the angle of reflection. Note that the angles of incidence, reflection, and the normal are all coplanar.

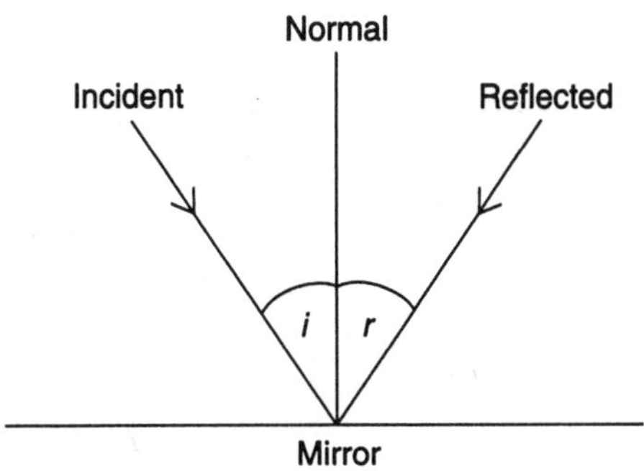

FIGURE 3.

Reflection helps explain the colors of opaque objects. Ordinary light contains many different colors all blended together (called *white light*). A blue object looks blue because of the selective reflection of blue light due the chemical dyes in the painted material. White paper assumes the color of the light incident on it because white reflects all the colors. Black paper absorbs all the colors (and of course nothing can be painted a "pure" color). If light of a single wavelength can be isolated (using a laser for example), then the light is said to be **monochromatic**. If all the waves of light are moving in phase, the light is said to be **coherent**. Laser light is both monochromatic and coherent, which contributes to its strength and energy.

When a concave or convex mirror is used, the law of reflection still holds, but the curved shapes affect the direction of the reflected rays. In Figure 4a, we see that a concave mirror converges parallel rays of light to a **focal point** described as *real* because the light rays really cross. In Figure 4b, we see that a convex mirror causes parallel rays to diverge away from the mirror. If we imagine the rays extended backward, they will appear to originate from a point on the other side of the mirror. This point is called the **virtual focal point** because it is not real. The human eye always traces a ray of light back to its source in a line. This deception of the eye is responsible for images in some mirrors that appear to be on the "other side" of the mirror (called **virtual images**). We will discuss image formation in mirrors in more detail in Chapter 23.

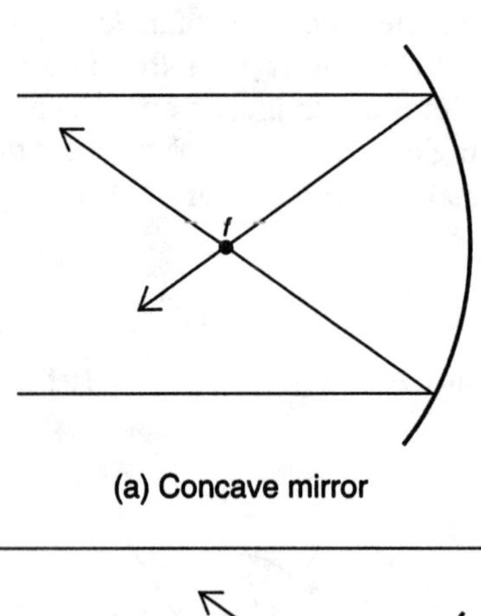

(a) Concave mirror

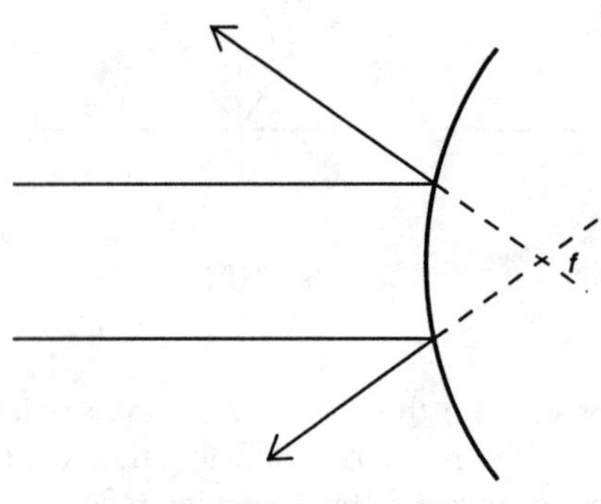

(b) Convex mirror

FIGURE 4.

REFRACTION

Place a pencil in a glass of water and look at it from the side. The apparent bending of the pencil is due to refraction. When light passes obliquely from one transparent medium to another, there is a uniform change in its speed. When this occurs, the light appears to bend toward or away from the normal. If there is no change in speed, then there will be no refraction at any angle (say from benzene to Lucite, for example). Whether or not there will be a change in speed, if the light is incident at an angle of 0 degrees to the normal, then there will again be no refraction (see Figure 5).

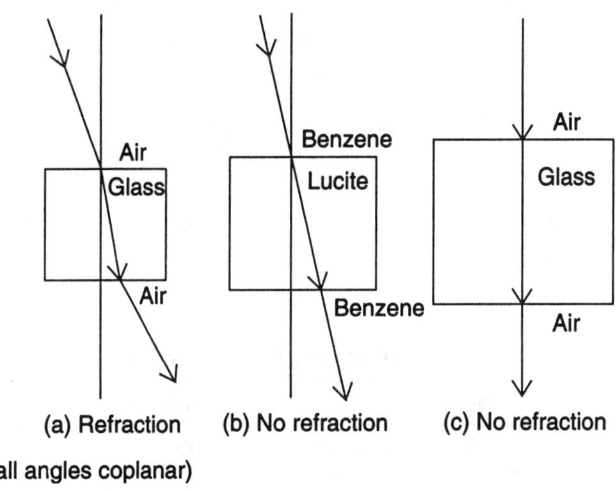

FIGURE 5.

When light travels from one medium to another and slows down at an oblique angle, then the angle of refraction is less than the angle of incidence and we say that the light has been refracted *toward the normal.* In Figure 5c notice that when the light reemerges back into the air, it is parallel to its original direction (but slightly offset). This is explained by the fact that the light speeds up as it goes back into the air. In this case, the angle of refraction is larger than the angle of incidence, and we say that the light has been refracted *away from the normal.* Remember that when the optical properties of both media are the same, there is no refraction since there is no change in the speed of light. In this case, the angle of incidence is equal to the angle of refraction (no deviation). The extent to which a medium is a good refracting medium is measured by how much change there is in the speed of light. This *physical* characteristic is manifested by a *geometric* characteristic, namely, the angle of refraction. The relationship between these quantities is expressed by **Snell's law**. If the angle of incidence is given as i and the angle of refraction is represented by r, then if $\mathbf{v}_1$ is the velocity of light in medium 1 (equal to **c** if medium 1 or 2 is air) and if $\mathbf{v}_2$ is the velocity of light in medium 2, then Snell's law states that

$$\frac{\sin i}{\sin r} = \frac{\mathbf{v}_1}{\mathbf{v}_2} = \frac{n_2}{n_1}$$

The ratio n_2/n_1 is called the **relative index of refraction**. If medium 1 is air, then by definition the **absolute index of refraction** n of air (and the vacuum) is taken to be equal to 1.00 but is actually slightly larger (1.00029; see Table 1). Snell's law can now be written in the form

$$\frac{\sin i}{\sin r} = \frac{c}{\mathbf{v}_2} = n_2$$

Note that when light is refracted, there is no change in the frequency of the light—only a change in the wavelength.

TABLE 1. ABSOLUTE INDICES OF REFRACTION

Substance	Index of Refraction
Air (vacuum)	1.00
Water	1.33
Alcohol	1.36
Quartz	1.46
Lucite	1.50
Benzene	1.50
Glass, crown	1.52
Glass, flint	1.61
Diamond	2.42

The different colors of visible light all have different frequencies. When they are used in a refraction experiment, they produce different angles of refraction since they all travel at different speeds in media other than a vacuum (or air approximately). Substances that allow the different frequencies of light to travel at different speeds are called **dispersive media**. This feature of refraction explains why a prism allows one to see the colored, *continuous*, spectrum and, additionally, why it is red light (at the low-frequency end) that emerges at the smallest angle (see Figure 6).

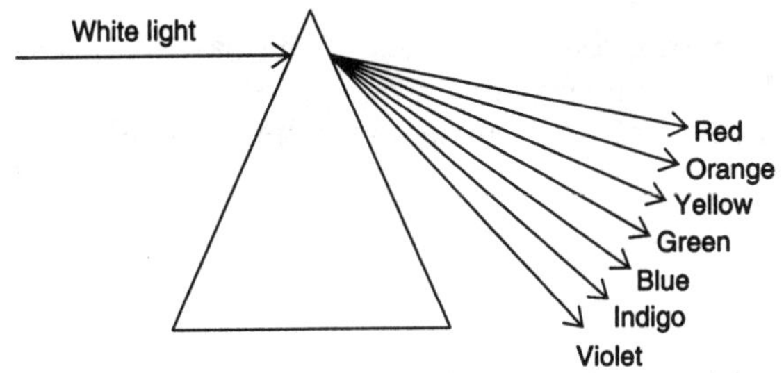

FIGURE 6. *Prismatic dispersion of light.*

APPLICATIONS OF LIGHT REFRACTION

When light is refracted from a medium with a relatively large index of refraction to one with a low index of refraction, the angle of refraction can be quite large. Notice, however, that while the relative index of refraction can be less than 1.00, the absolute index of refraction cannot.

In Figure 7a, we have a situation in which the angle of incidence is at some critical value θ_c such that the angle of refraction is 90 degrees. This can occur only when the relative index of refraction is less than 1.00, as described earlier.

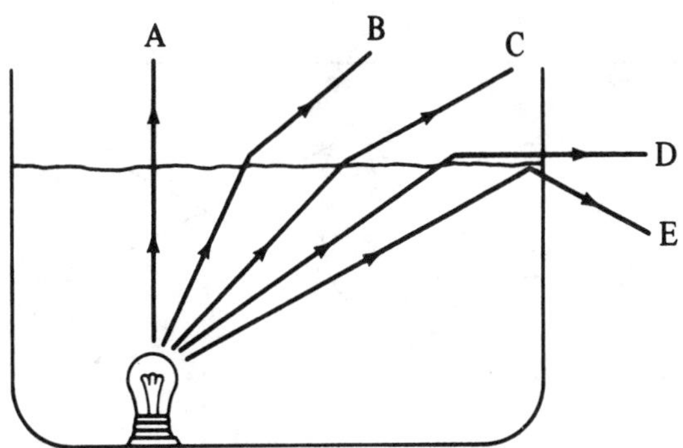

FIGURE 7. *Total internal reflection.*

If the angle of incidence exceeds this critical value (Figure 7b), then the light will be internally reflected. This phenomenon is aptly called **total internal reflection**. The ability of diamonds to sparkle in sunlight is due to total internal reflection and a relatively small critical angle of incidence (resulting from a diamond's high index of refraction). **Fiber optics** communication, in which information is processed along hair-thin glass fibers, works because of total internal reflection.

The critical angle can be determined from the relationship

$$\sin \theta_c = \frac{n_2}{n_1} (n_2 < n_1)$$

As a further example of light refraction, consider the refraction due to a prism. Figure 8a demonstrates what happens when two prisms are arranged base to base and two parallel rays of light incident on them. The ability of the prism to disperse white light does not apply in this example since the light is monochromatic. However, we can observe that the light rays converge to a real focal point some distance away. This situation simulates the effect of refraction by a double convex lens as shown in Figure 8b.

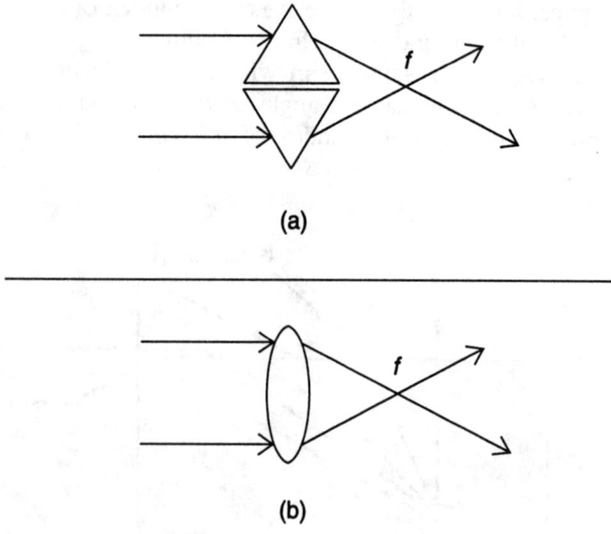

FIGURE 8.

The focal length in each case is dependent on the frequency of the light used, and red light produces a larger focal length than violet light when used with a convex lens.

When the prisms are placed vertex to vertex, as in Figure 9a, the parallel rays of light diverge away from an apparent virtual focal point. As shown in Figure 9b, this simulates the effect of a double concave lens.

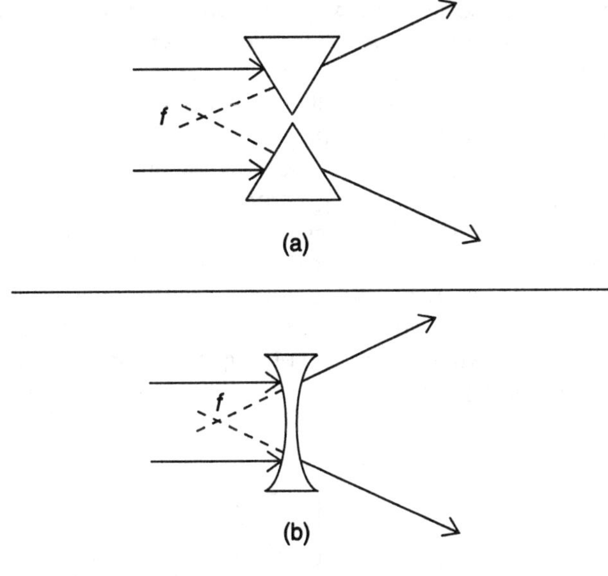

FIGURE 9.

INTERFERENCE AND DIFFRACTION OF LIGHT

The ability of light to be diffracted and to exhibit an interference pattern is evidence for the wave nature of light. Previously, we saw that the diffraction of light occurs in connection with shadows. Light interference can be observed by using two or more narrow slits, and multiple-slit diffraction is achieved using a plastic *grating* onto which over 5,000 lines per centimeter have been scratched.

If white light is used and passed through the grating, then a series of continuous spectra appear, as shown in Figure 10. Interestingly, this continu-

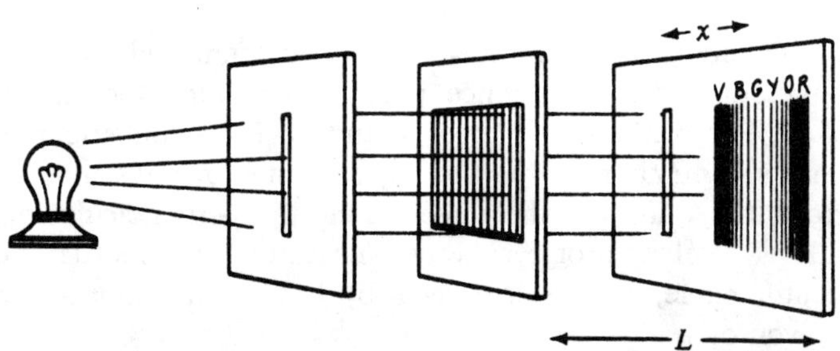

FIGURE 10. *Multiple-slit diffraction.*

ous spectrum is reversed from the way it appears in dispersion. The reason for this difference is that diffraction is wavelength-dependent. When a narrow opening is used, the short-wavelength violet rays are diffracted the least, while the longer-wavelength red rays are diffracted the most.

If monochromatic light is used, then an alternating pattern of bright and dark regions will appear. If a laser is used, the pattern will appears as a series of evenly spaced dots representing regions of constructive and destructive interference (see Figure 11). The origin of this pattern can be understood if we consider a water tank analogy. Suppose we have two point sources vibrating in phase in a water tank. Each source produces circular waves that overlap in the region in front of the sources. Where crests meet crests, we have constructive interference, and where crests meet troughs, we have destructive interference, as indicated in Figure 12.

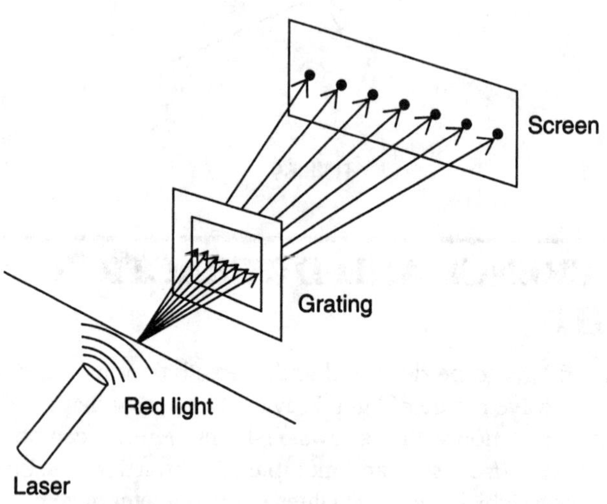

FIGURE 11. *Monochromatic diffraction pattern.*

Notice in Figure 12 that there is a central maximum built up by a line of intersecting constructive interference points lying along the perpendicular bisector of the line connecting sources S_1 and S_2. The symmetric interference pattern consists of numbered *orders* evenly separated by a distance x, and the distance from the sources to the screen along the perpendicular bisector is labeled L. The wavelength of each wave, measured by the spacing between each concentric circle, is of course given by λ. The separation between the sources is given by d.

In Young's "double-slit" experiment, the two explicit sources in the water tank are replaced by two narrow slits in front of a monochromatic ray of light.

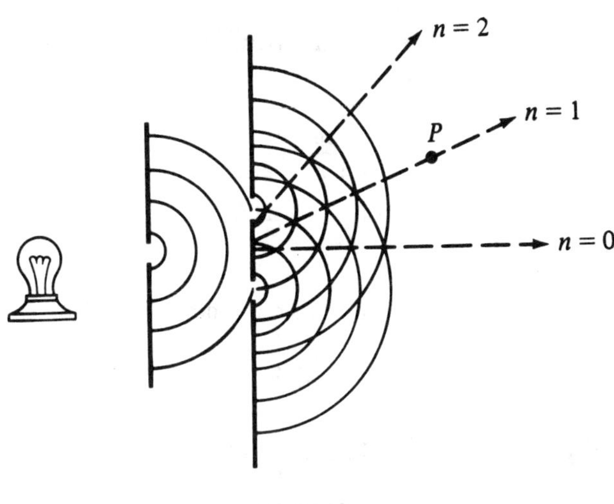

FIGURE 12.

The central maximum is called the "zero-order maximum" because the path length difference from both sources (slits) is equal to zero. The next order is the first order because the path length difference is equal to one wavelength, and so on, as discussed in Chapter 21. Constructive maximum points occur at whole multiples of wavelength, while destructive minimum points occur at odd multiples of half-wavelengths (refer back to Chapter 21 for review).

The relationship governing the variables just discussed is given by the equation

$$n\frac{\lambda}{d} = \frac{x}{L}$$

where n represents the desired order.

If θ is an angle measured from the midpoint between the slits and a particular order, then the ratio x/L is approximately equal to $\sin\theta$ and the diffraction formula becomes

$$n\lambda = d \sin\theta$$

Sample Problem

A ray of monochromatic light (in air) is incident on a diamond (n = 2.42) at an angle of 30 degrees.

(a) What is the angle of refraction in the diamond?

(b) What is the velocity of light in the diamond?

Solution

(a) We use Snell's law to solve this part, recognizing that the index of refraction of air is equal to 1.00. We can then write

$$\frac{\sin 30}{\sin r} = 2.42$$

$$\sin r = 0.2066$$

Therefore, $r = 12$ degrees

(b) The velocity of light in any substance (coming from air) is given by

$$\mathbf{v} = \frac{\mathbf{c}}{n}$$

where n is equal to the absolute index of refraction.

Thus,

$$\mathbf{v} = \frac{3 \times 10^8}{2.42} = 1.24 \times 10^8 \text{ m/s}$$

PROBLEM-SOLVING STRATEGIES

Light is an electromagnetic wave, which means that it can travel through a vacuum. Reflection and refraction do not by themselves verify the wave nature of light. Diffraction and interference are evidence for the wave nature of light. The ability of light to be polarized, using special polaroid filters, is evidence that light is a transverse wave.

When light is refracted, its frequency is not affected. Since the velocity changes, so does the wavelength in the new medium. If the velocity in the medium is frequency-dependent, then the medium is said to be *dispersive.* Light waves can be coherent if they are produced in a fashion that maintains constant phase relationships. Lasers produce coherent light of one wavelength.

When doing refraction problems, remember that light is refracted toward the normal if it enters a medium with a higher index of refraction and is refracted away from the normal if it enters a medium with a lower index of refraction. The absolute index of refraction of a substance can never be less than 1.00.

PRACTICE PROBLEMS FOR CHAPTER 22

Thought Problems

1. Immersion oil is a transparent liquid used in microscopy. It has an absolute index of refraction equal to 1.515. A glass rod attached to the

cap of a bottle of immersion oil is practically invisible when viewed at certain angles (under normal lighting conditions). Explain how this might occur.

2. Explain why a diamond ring sparkles more than a piece of glass of similar shape and size.

3. Explain why total internal reflection occurs at boundaries between transparent media for which the relative index of refraction is less than 1.0.

Multiple-Choice Problems

1. What is the frequency of a radio wave with a wavelength of 2.2 m?
   (A) 3×10^8 Hz
   (B) 1.36×10^8 Hz
   (C) 7.3×10^{-9} Hz
   (D) 2.2×10^8 Hz
   (E) 2.2 Hz

2. A ray of light is incident from a layer of crown glass ($n = 1.52$) on a layer of water ($n = 1.33$). The critical angle of incidence for this situation is equal to ____ degrees.
   (A) 32
   (B) 41
   (C) 49
   (D) 61
   (E) 75

3. The relative index of refraction between two media is 1.20. Compared to the velocity of light in medium 1, the velocity of light in medium 2 ____.
   (A) is greater by 1.2 times
   (B) is reduced by 1.2 times
   (C) is the same
   (D) depends on the two media
   (E) depends on the angle of incidence

4. What is the approximate angle of refraction for a ray of light incident from air on a piece of quartz at a 37-degree angle?
   (A) 24 degrees
   (B) 37 degrees
   (C) 42 degrees
   (D) 66 degrees
   (E) 75 degrees

5. What is the velocity of light in alcohol ($n = 1.36$)?
   (A) 2.2×10^8 m/s
   (B) 3×10^8 m/s
   (C) 4.08×10^8 m/s

(D) 1.36×10^8 m/s
(E) none of these values

6. If the velocity of light in a medium depends on its frequency, the medium is said to be _____.
(A) coherent
(B) refractive
(C) resonant
(D) diffractive
(E) dispersive

7. If the intensity of a monochromatic ray of light is increased while it is incident on a pair of narrow slits, then the spacing between maxima in the diffraction pattern will be _____.
(A) increased
(B) decreased
(C) the same
(D) increased or decreased depending on the frequency
(E) increased or decreased depending on the slit separation

8. In the accompanying diagram, a source of light (at *S*) sends a ray toward the boundary between two media for which the relative index of refraction is less than 1. The angle of incidence is given by *i*. Which ray best represents the path of the refracted light?

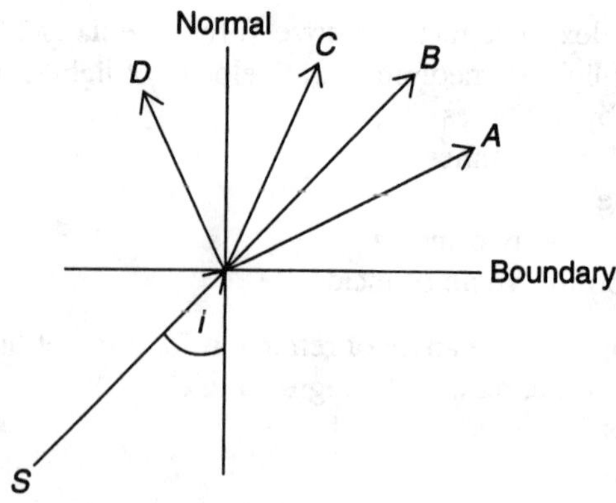

(A) *A*
(B) *B*
(C) *C*
(D) *D*
(E) either *A* or *C*

9. A coin is placed at the bottom of a clear trough which is filled with water (n = 1.33) as shown. Which point best represents the approximate location of the coin as seen by someone looking into the water?

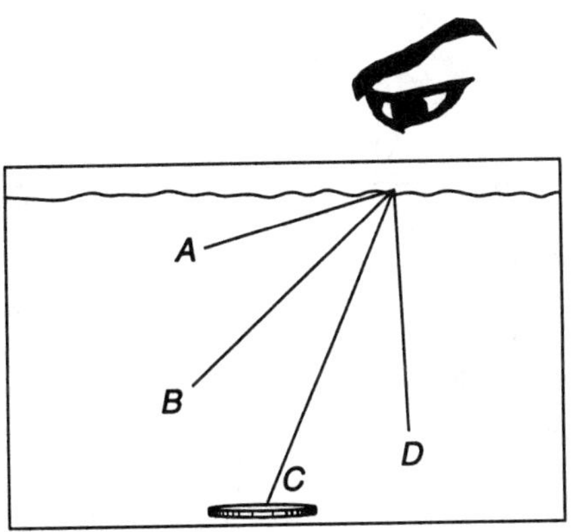

(A) *A*
(B) *B*
(C) *C*
(D) *D*
(E) it depends on the depth of the water

10. If the water in Problem 9 is replaced by alcohol (n = 1.36), then the coin will appear _____.
(A) higher
(B) lower
(C) the same
(D) higher or lower depending on the depth
(E) not enough information given

Free-Response Problems

1. (a) Light of wavelength 700 nm is directed onto a diffraction grating with 5,000 lines per centimeter. What is the angular deviation of the first- and second-order maxima from the central maxima?
   (b) Explain why x rays are used to study crystal structure rather than visible light?

2. Light is incident on a piece of flint glass (n = 1.66) from the air such that the angle of refraction is exactly half the angle of incidence. What are the values of the angles?

3. A ray of light passing through air is incident on a piece of quartz (n = 1.46) at an angle of 25 degrees as shown. The quartz is 1.5 cm thick. Calculate the deviation d of the ray as it emerges back into the air.

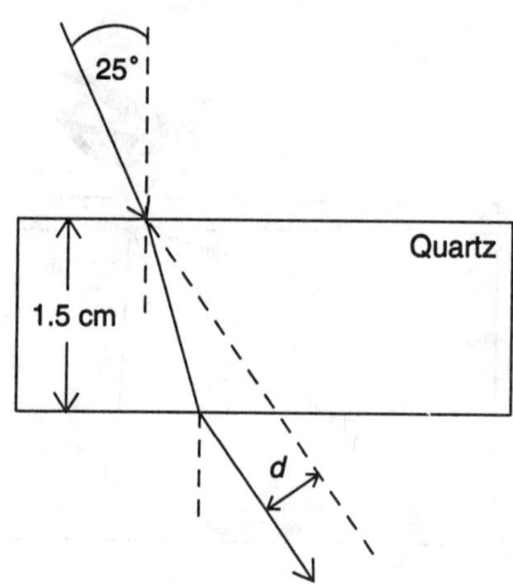

SOLUTIONS TO PRACTICE PROBLEMS

Thought Problems

1. The index of refraction for glass is very close to that for the immersion oil. When the applicator rod is filled with liquid, the light passes through both without refracting and makes the glass rod appear invisible.
2. Diamonds have a lower critical angle of incidence than glass. When a diamond is turned in the sunlight, the rays of light exceed the critical angle and so appear to reflect back up to the eye instead of refracting down toward the ground. The faceted shape causes dispersion of the colors, giving the characteristic sparkle.
3. In order to have total internal reflection, the angle of refraction must exceed 90 degrees. This means that the light must be refracted away from the normal. This occurs only when the light rays speed up as they enter a medium with a lower index of refraction compared to that of the medium they have just left.

Multiple-Choice Problems

1. **B** The velocity of light in air is given by the formula $\mathbf{c} = f\lambda$. Since the wavelength is 2.2 m and the velocity of light in air is 3×10^8 m/s, substituting the known values gives $f = 1.36 \times 10^8$ Hz.
2. **D** The critical angle of incidence is given by the formula $\sin \theta_c = n_2/n_1$, where $n_2 = 1.33$ and $n_1 = 1.52$. Substitution yields a value of 61 degrees for the critical angle.

3. **B** The velocity relationship is given by the formula $\mathbf{v}_1/\mathbf{v}_2 = n_2/n_1$. Since the relative index of refraction is defined to be equal to the ratio n_2/n_1, we see that $\mathbf{v}_1 = 1.2\mathbf{v}_2$. Thus, compared to $\mathbf{v}_1$, $\mathbf{v}_2$ is reduced by 1.2 times.

4. **A** Snell's law in air is given by $\sin i/\sin r = n_2$. The absolute index of refraction for quartz is 1.46. Substitution yields a value of 24 degrees for the angle of refraction.

5. **A** Compared to the velocity of light in air, the velocity of light in any other transparent substance is given by the formula $\mathbf{v} = \mathbf{c}/n$. In this case, $n = 1.36$, and so the velocity of light is equal to 2.2×10^8 m/s.

6. **E** By definition, a medium is said to be dispersive if the velocity of light is dependent on its frequency.

7. **C** The position and separation of interference maxima are independent of the intensity of the light.

8. **A** Since the relative index of refraction is less than 1, the light ray speeds up as it crosses the boundary between the two media. The light therefore bends away from the normal along path *A* (approximately).

9. **B** The human eye traces a ray of light back to its apparent source in a straight line. Thus, if we follow the line from the eye straight back, we reach point *B*.

10. **A** Since alcohol has a higher absolute index of refraction, the light will be bent further away from the normal. Tracing the imaginary line straight back would imply that the image would appear close to the surface (or higher).

Free-Response Problems

1. (a) The general formula for diffraction is $n\lambda = d \sin \theta$, where θ is the angle of deviation from the center. For the first order, $n = 1$, and d is equal to the reciprocal of the number of lines per meter. Thus, we must convert the 5,000 lines per centimeter to 500,000 lines per meter. Now, $\sin \theta = n\lambda/d = (1)(7 \times 10^{-7})(500{,}000) = 0.35$ and $\theta = 20.5$ degrees. For the second-order maximum, we have $n = 2$. Thus, $\sin \theta = (2)(7 \times 10^{-7})(500{,}000) = 0.7$ and $\theta = 44.4$ degrees.

   (b) X rays are used to study crystal structure because of their very small wavelengths. These wavelengths are comparable to the spacings between lattices in a crystal and thus allow the x rays to be diffracted from the different layers. Visible light has wavelengths that are much larger than these spacings and thus is not affected by them. The use of x rays to probe crystals made them one of the first diagnostic tools of atomic physics at the beginning of the twentieth century.

2. We want the angle of refraction to be equal to half the angle of incidence. This means that $i = 2r$. Now, since the light ray is initially in air ($n = 1.00$), Snell's law can be written

$$\frac{\sin i}{\sin r} = n_{\text{glass}}$$

Substituting our requirement that $i = 2r$,

$$\frac{\sin 2r}{\sin r} = 1.66$$

Now, recall the following trigonometric identity: $\sin 2\theta = 2 \sin \theta \cos \theta$. Thus,

$$\frac{2 \sin r \cos r}{\sin r} = 1.66$$

and

$$2 \cos r = 1.66$$

Therefore,

$$\cos r = 0.83$$

and

$$r = 34 \text{ degrees (which means that } i = 68 \text{ degrees)}$$

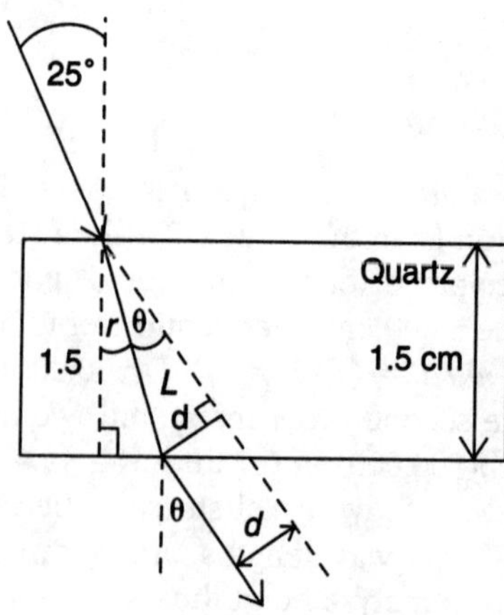

3. The diagram for the problem has been redrawn. From our knowledge of refraction, we know that angle θ must be equal to 25 degrees. Angle r is given by Snell's law:

$$\frac{\sin 25}{\sin r} = 1.46$$

$$\sin r = \frac{\sin 25}{1.46} = 0.2894$$

$$r = 16.8 \text{ degrees}$$

Now angle ϕ is equal to the difference between the angle of incidence and the angle of refraction: $\phi = 8.2$ degrees

Since the quartz is 1.5 cm thick, the length of the light ray in the quartz at the angle of refraction can be determined from $\cos r = 1.5/L$. This implies that $L = 1.57$ cm. Now since we know the length of the diagonal L and the angle ϕ, the deviation d, is just part of the right triangle diagrammed. Thus, $\sin \phi = d/L$ and $d = L \sin \phi = 1.57 \sin 8.2 = 0.224$ cm.

23
GEOMETRIC OPTICS

IMAGE FORMATION IN PLANE MIRRORS

When you look at yourself in a plane mirror, your image appears to be directly in front of you and on the other side of the mirror. Everything about your image is the same as you are except for a left-right reversal. Since no light can be originating from the other side of the mirror, your image is called *virtual*.

The formation of a virtual image in a plane mirror is illustrated in Figure 1. Using an imaginary point object, we construct two rays of light that emerge radially from the object because of ambient light from its environment. Each light ray is incident on the plane mirror at some arbitrary angle and is reflected off at the same angle (relative to the normal). Geometrically, we construct these lines using the law of reflection and a protractor. Since the rays diverge from the object, they continue to diverge after reflection. Your eye, however, perceives the rays as originating from a point on the other side of the mirror and obtained by extending the reflected rays back to an apparent convergence.

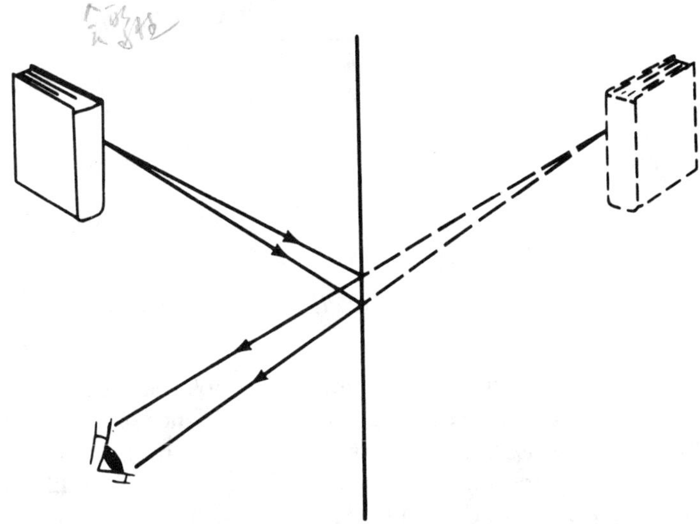

FIGURE 1.

IMAGE FORMATION IN CURVED MIRRORS

Concave Mirrors

In Chapter 22, we saw that if we use a concave or convex mirror, then the shape will cause parallel rays of light to either converge or diverge. In Figure 2, a typical spherical concave mirror is illustrated. The bisecting axis is called the **principal axis**, and its shape should be parabolic in cross section. When the curvature of the mirror is too large, then a defect known as **spherical aberration** occurs and distorts images seen. The real images formed by concave mirrors can be projected onto a screen, and the focal length can be determined using parallel rays of light and observing their converging point. Another, more interesting method is to aim the mirror out the window at distant objects and then focus the images of these objects onto a screen. Since the objects are very far away, they are considered to be "at infinity." At infinity, an object normally sends parallel rays of light and appears as a point. Since we are viewing extended objects, we project smaller images of them, and the distance from the screen to the mirror is the focal length *f*.

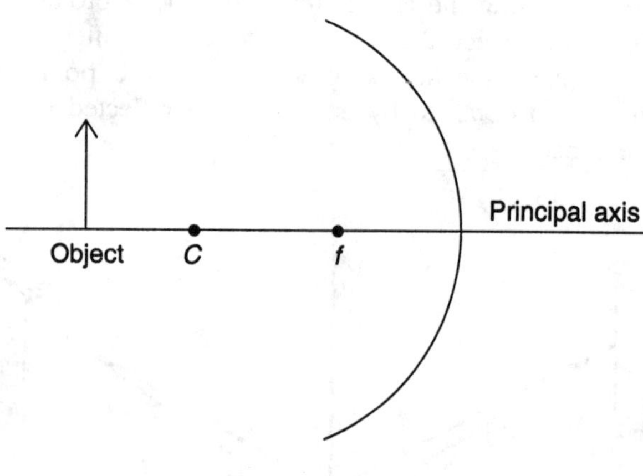

FIGURE 2.

The other point along the principal axis is the point labeled *C*. This point marks the **center of curvature** and is located at a distance of $2f$. By symmetry, Figure 2 can also be used to represent a diverging convex mirror (with the light originating from the right side instead of the left side). In this case, the focal length is *virtual* (as opposed to being *real* in the concave case) and is treated algebraically as $-f$.

To illustrate how to construct an image formed by a concave mirror, we use an arrow as an imaginary object. Its location along the principal axis is measured relative to points f and C. The orientation of the arrow is of course determined by which way it points. We can choose an infinite number of light rays coming off the object as a result of ambient light from its environment. However, for simplicity, we choose two rays emerging from the top of the arrow.

In Figure 3, several different cases of concave mirror constructions are shown. The first light ray is drawn parallel to the principal axis and then reflected through the focal point f. The second light ray is drawn through the focal point and then reflected parallel to the principal axis. The point of intersection (below the principal axis in cases I, II, and III) indicates that the image appears inverted at the location marked in the diagram.

From the constructions in Figure 3, we can see that the real images are always inverted. Additionally, as the object is moved closer to the mirror, the image gets larger and moves further away from the mirror. Notice that when the object is at point C, the image is also at point C and is the same size. When the object is at the focal point, no image can be seen since the light is reflected parallel from all points on the mirror. When the object is moved even closer, we see an enlarged virtual image that is erect. In case IV, we have drawn a light ray toward the center of the mirror and, by the law of reflection, it is reflected below the principal axis at the same angle of incidence.

Convex Mirrors

Convex mirrors are used in a variety of situations. They often appear in stores and elevators because of their ability to reveal (although distorted) images from around corners in aisles. The image in a convex mirror is always virtual and always smaller, which suggests that only one type of construction is necessary to understand image formation in these mirrors. In Figure 4, we show a sample construction, recalling that convex mirrors cause parallel rays to diverge. This divergence, however, is not at any arbitrary angle. The direction of the divergent ray is as though it originated from the virtual focal point.

Algebraic Considerations

We can study the image formed in curved mirrors by means of an algebraic relationship. When we let f represent the focal length (positive in a concave mirror and negative in a convex mirror), p represent the object distance, and q represent the image distance, then for curved mirrors;

$$\frac{1}{f} = \frac{1}{p} + \frac{1}{q}$$

If we let S_o represent the object size and S_i represent the image size, then

$$\frac{S_i}{S_o} = \frac{q}{p}$$

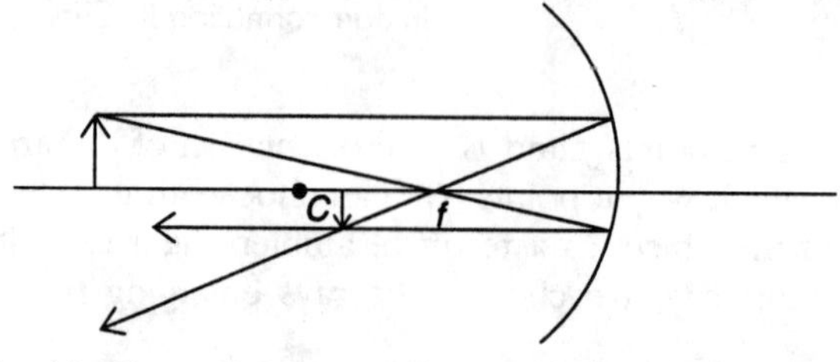

Case I—object beyond *C*; image is real, is between *f* and *C*, is smaller

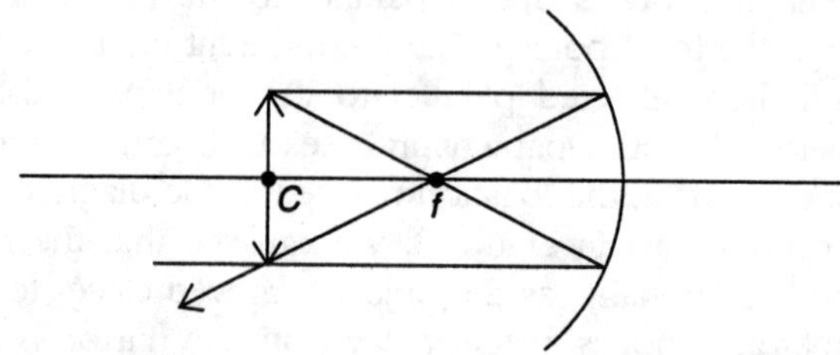

Case II—object at *C*; image is real, is at point *C*, is the same size

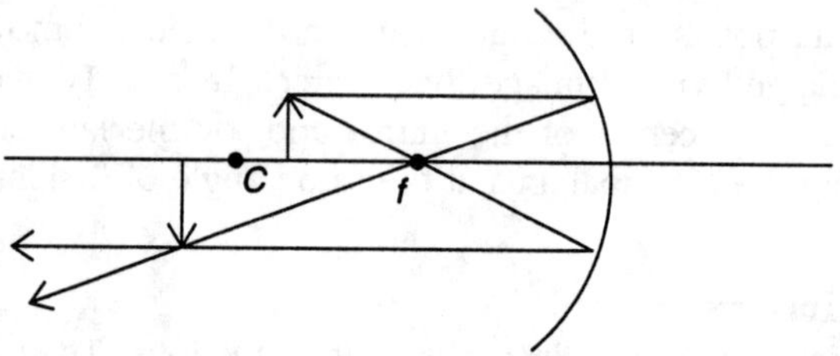

Case III—object between *f* and *C*; image is real, is beyond *C*, is larger

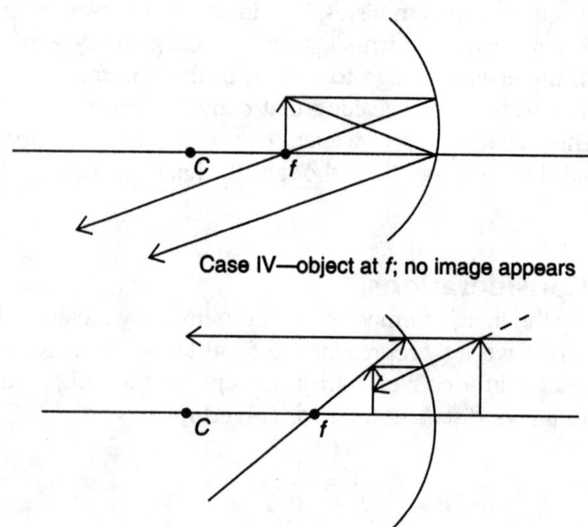

Case IV—object at *f*; no image appears

Case V—object between *f* and the mirror; image is virtual, is larger

FIGURE 3.

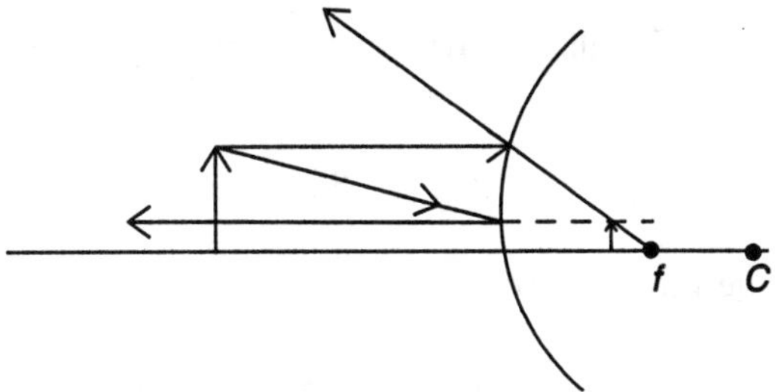

FIGURE 4. *Convex mirror image formation. Image is virtual, erect, and smaller.*

The ratio of image size to object size is called the **magnification**. We can derive this expression quite easily. In Figure 5, we show a case I situation. Now, if we assume that the curvature of the mirror is small, then the line segment AB will be approximately equal to DO (assumed to be vertical and perpendicular to the principal axis). Likewise, we assume that line segment GH is approximately equal to OE (also assumed to be vertical and perpendicular to the principal axis).

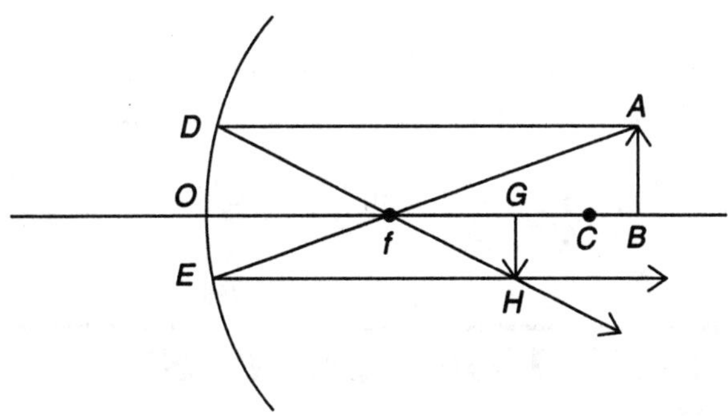

FIGURE 5. *Mirror equation derivation.*

From the geometry, we can see that triangle fBA is similar to triangle fOE, and the corresponding sides are therefore in proportion:

$$\frac{fB}{fO} = \frac{BA}{OE}$$

We can see that $BA = S_o$ (the size of the object) and $OE = GH = S_i$ (the size of the image). The line segment $fB = p - f$, while $fO = f$. Thus, we can write

$$\frac{p - f}{f} = \frac{S_o}{S_i}$$

Now, triangle *fOD* is similar to triangle *fGH*, and so we have the following proportions:

$$\frac{fO}{FG} = \frac{OD}{GH}$$

This can be rewritten in the form

$$\frac{f}{q - f} = \frac{S_o}{S_i}$$

If we equate these two expression for S_o / S_i, cross-multiply, and divide by the common factor *fpq*, we will arrive at the desired formula.

Sample Problem

A 10-cm-tall object is placed 20 cm in front of a concave mirror with a focal length of 8 cm. What is the location of the image and what is its size?

Solution

Using the formula just presented we can write

$$\frac{1}{8} = \frac{1}{20} + \frac{1}{q}$$

Solving for the image distance, we obtain $q = 13.3$ cm. This is consistent with case I. The image size can be obtained from the expression

$$S_i = S_o\left(\frac{q}{p}\right) = 10\left(\frac{13.3}{20}\right) = 6.7 \text{ cm}$$

IMAGE FORMATION IN LENSES

Converging Lenses

When discussing the formation of images using lenses, we usually invoke what is called **thin lens approximation**. The top and bottom of the lens are tapered like a prism, and a defect known as **chromatic aberration**, which is due to dispersion, can sometimes occur. This defect causes the different colors of light to disperse in the lens and focus at different places because of their different frequencies and velocities in the lens material.

Figure 6 presents a series of cases in which a double convex lens is constructed as a straight line using thin lens approximation. The symmetry of the lens creates two real focal points on either side. Our object is again an arrow drawn along the principal axis (which bisects the lens).

From Chapter 22, we know that a light ray parallel to the principal axis will be refracted through the focal point. Therefore, a light ray originating from the

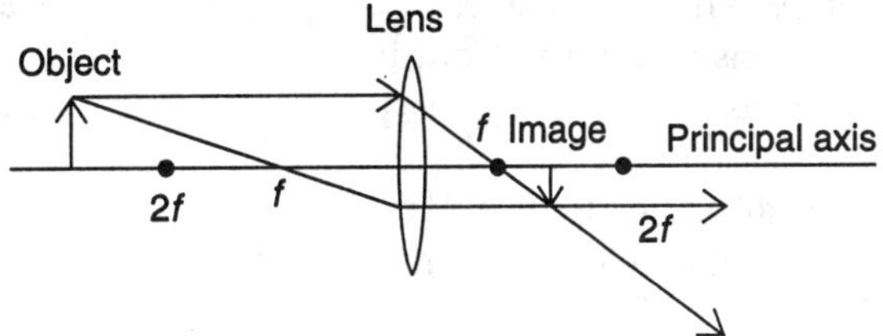

Case I—object beyond 2*f*; image is between *f* and 2*f*, is real, is smaller

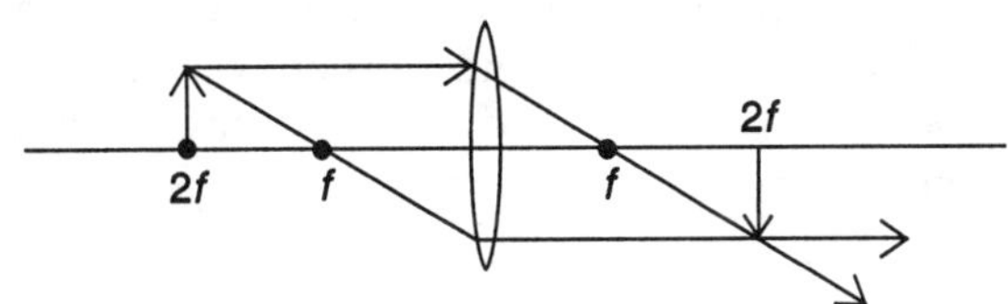

Case II—object at 2*f*; image is at 2*f*; is real, is same size

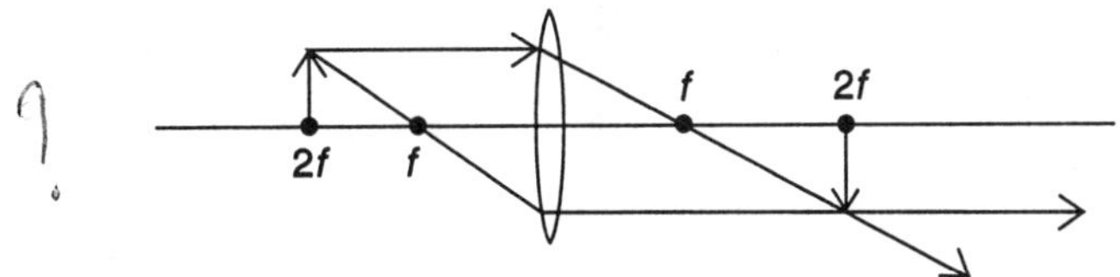

Case III—object between *f* and 2*f*; image is beyond 2*f*, is real, is larger

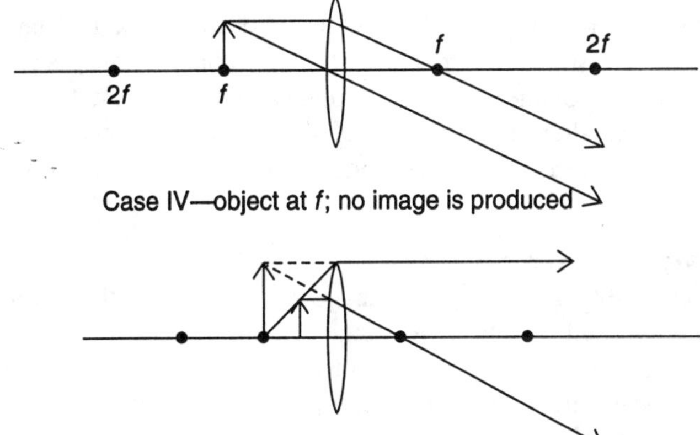

Case IV—object at *f*; no image is produced

Case V—object between *f* and the lens; image is behind object, is virtual, is larger

FIGURE 6.

focal point will be refracted parallel to the principal axis. We shall use these two light rays to construct most of our images.

Notice that in case IV no image is produced and a different light ray is shown passing through the optical center of the lens (since the object is situated at the focal point). In case V, an enlarged virtual image is produced on the same side as the object, as in the case of a simple magnifying glass.

Diverging Lenses

A concave lens causes parallel rays of light to diverge away from a virtual focal point. Figure 7 illustrates a typical ray construction for a diverging lens. Again, if we assume the thin lens approximation, we will draw the lens itself as a line for ease of construction. The context of the problem tells us that it represents a concave lens instead of a convex lens.

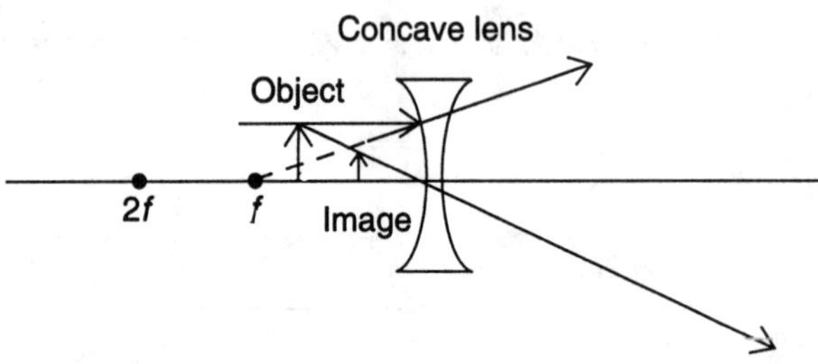

FIGURE 7. *Concave lens.*

In order to construct the image we have drawn a parallel light ray which then diverges away from the focal point. The second light ray is drawn through the optical center of the lens and is undeviated. The resulting virtual image will be erect, smaller, and in front of the object. This will be true for any location of the object, and so only one sample construction is necessary.

Algebraic Considerations

The relationship governing the image formation in a thin lens is the same as it is for curved mirrors. Positive image distances occur when the image forms on the other side of the lens. A negative image distance implies a virtual image. A positive focal length implies a convex lens, while a negative focal length implies a concave lens. Therefore, as before, we can write

$$\frac{1}{f} = \frac{1}{p} + \frac{1}{q}$$

$$\frac{S_i}{S_o} = \frac{q}{p}$$

If two converging lenses are used in combination, then the combined magnification of the system is given by

$$m = m_1 m_2$$

The lens equation can be derived from the case I situation shown in Figure 8.

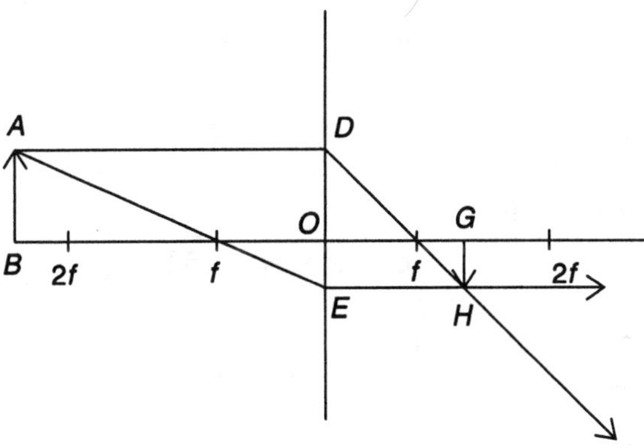

FIGURE 8. *Lens equation derivation.*

From the geometry, we can see that triangle *fOD* is similar to triangle *fGH*. Thus, we can write

$$\frac{f}{q - f} = \frac{fO}{fG} = \frac{OD}{GH} = \frac{S_o}{S_i}$$

Additionally, we know that triangle *fBA* is similar to triangle *fOE*, and so we can also write

$$\frac{p - f}{f} = \frac{fB}{fO} = \frac{BA}{OE} = \frac{S_o}{S_i}$$

This is the same situation as with the spherical mirrors, and so a similar algebraic conversion will result in the desired result.

PROBLEM-SOLVING STRATEGIES

Solving ray diagram problems requires you to remember light rays and how they are reflected and refracted using mirrors and lenses. Numerically, remember that positive focal lengths imply concave mirrors and convex lenses.

Memorize all the cases and remember that real images are always inverted. If a given problem does not require a construction, draw a sketch anyway.

Additionally, remember that the focal length of a lens is dependent on the frequency of light used and the material the lens is made of.

PRACTICE PROBLEMS FOR CHAPTER 23

Thought Problems

1. Why do passenger-side mirrors on cars have a warning that states: Objects are closer than they appear?

2. Two lenses have identical sizes and shapes. One is made of quartz (n = 1.46), and the other is made of glass (n = 1.5). Which lens will make a better magnifying glass?

3. Why do lenses produce chromatic aberration but spherical mirrors do not?

Multiple-Choice Problems

1. Which material will produce the converging lens with the longest focal length?
   (A) Lucite
   (B) crown glass
   (C) flint glass
   (D) quartz
   (E) diamond

2. An object is placed in front of a converging lens such that the image produced is inverted and larger. If the lens is replaced by one with a larger index of refraction, then the size of the image will _____.
   (A) increase in size
   (B) decrease in size
   (C) increase or decrease depending on the degree of change
   (D) remain the same size
   (E) increase or decrease depending on the wavelength of light

3. You wish to make an enlarged reproduction of a document using a copy machine. When you push the enlargement button, the lens inside the machine moves to a point _____.
   (A) equal to f
   (B) equal to $2f$
   (C) between f and $2f$
   (D) beyond $2f$
   (E) less than f

4. A negative image distance means that the image formed by a concave mirror will be _____.
   (A) real
   (B) erect

(C) inverted
(D) smaller
(E) both B and D

5. Real images are always produced by ____.
   (A) plane mirrors
   (B) convex mirrors
   (C) concave lenses
   (D) convex lenses
   (E) both B and C

6. The focal length of a convex mirror with a radius of curvature of 8 cm is ____ cm.
   (A) 4
   (B) −4
   (C) 8
   (D) −8
   (E) 16

7. An object appears in front of a plane mirror as shown. Which of the following diagrams represents the reflected image of this object?

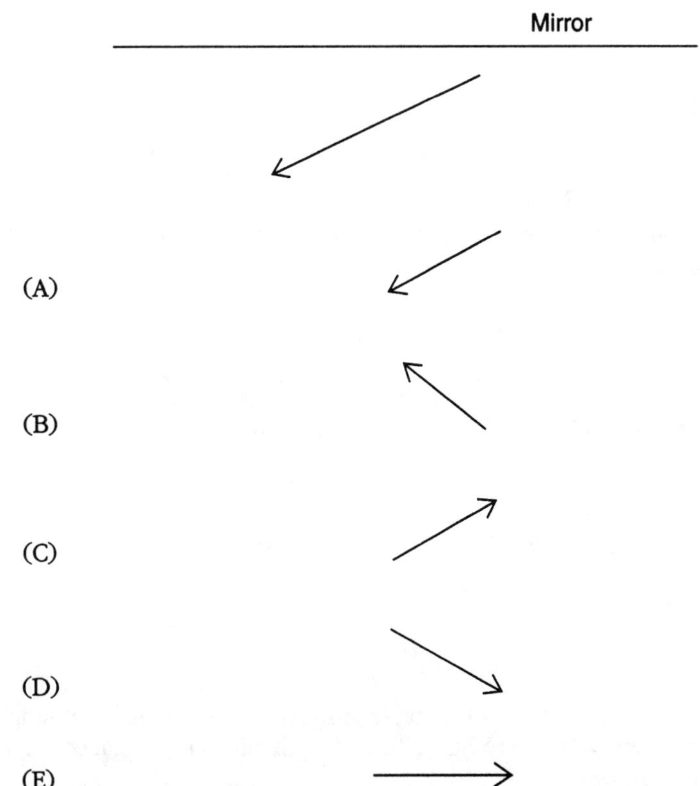

8. An object is placed in front of two mirrors hinged at right angles to each other. How many images will be seen of an object placed exactly between them and viewed from a point directly behind the object?
   (A) one
   (B) two
   (C) three
   (D) four
   (E) five

9. An object is located 15 cm in front of a converging lens. An image twice the size of the object appears on the other side of the lens. The image distance must be _____ cm.
   (A) 15
   (B) 30
   (C) 45
   (D) 60
   (E) 75

10. A 1.6-m-tall person stands 1.5 m in front of a vertical plane mirror. The height of his image is _____ cm.
   (A) 0.8 m
   (B) 2.6 m
   (C) 3.2 m
   (D) 4.5 m
   (E) 1.6 m

Free-Response Problems

1. Using a geometric ray diagram construction, prove Newton's version of the lens equation:

$$f^2 = d'd$$

   where f is the focal length of the lens, d is the distance from the object to the focal point, and d' is the distance of the image from the other focal point.

2. Prove that if two thin convex lenses are touching, then the effective focal length of the combination is given by the formula

$$\frac{1}{f} = \frac{1}{f_1} + \frac{1}{f_2}$$

3. An object is 25 cm in front of a converging lens with a 5-cm focal length. A second converging lens with a focal length of 3 cm is placed 10 cm behind the first one.

(a) Locate the image formed by the first lens.
(b) Locate the image formed by the second lens if the first image is used as an "object" for the second lens.
(c) What is the combined magnification of this combination of lenses?

SOLUTIONS TO PRACTICE PROBLEMS

Thought Problems

1. Right-hand-side passenger mirrors are convex in shape, which means that they cause reflected light to diverge. This distortion makes the images appear to be smaller and further away than they actually are (recall the ray diagram construction for diverging mirrors).

2. Based on the ray diagram construction for lenses (case V), we can see that if we have a shorter focal length, we will produce greater magnification of the enlarged virtual image. The shorter focal length is achieved by using a material with a relatively large index of refraction. In our comparison, the choice would be glass over quartz.

3. Chromatic aberration is caused by the dispersion of white light. Since dispersion is a phenomenon associated with refraction, reflecting spherical mirrors do not produce dispersion and hence do not exhibit chromatic aberration. Spherical mirrors can exhibit spherical aberration based on their shape.

Multiple-Choice Problems

1. **D** The longest focal length will be produced by the material that is refracted the least. This material will have the smallest index of refraction. Given the four choices, quartz has the lowest index of refraction.

2. **B** The lens formula can be rewritten as $q = pf/(p - f)$. If the object remains in the same position relative to the lens, then using a larger index of refraction will imply a smaller focal length. The image will move closer to the lens and consequently will be smaller in size.

3. **C** To produce an enlarged real image, the object must be located between f and $2f$.

4. **B** In a concave mirror, a negative image distance implies a virtual image which is enlarged and erect. Of the choices given, only (B) is correct.

5. **D** Only a convex lens or a concave mirror can produce a real image.

6. **B** The radius of curvature of a spherical mirror is twice the focal length. However, for a convex mirror, the focal length is taken to be negative.

7. **B** The reflected image must point away from the mirror but on the other side, flipped over.

8. **C** A sketch of the situation appears here. Each mirror will have one image, but an image will also be seen at the vertex point.

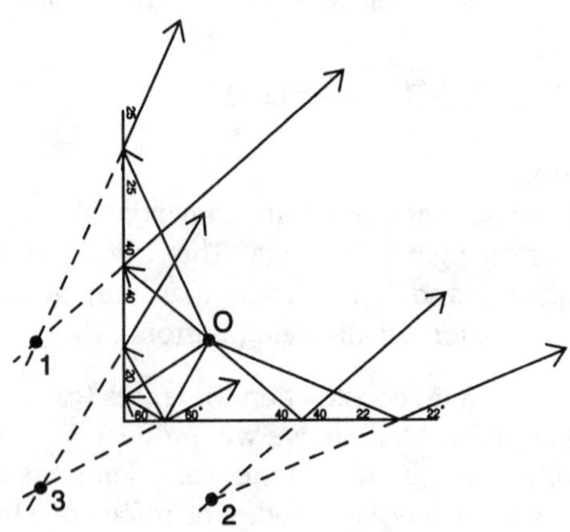

9. **B** The equation governing magnification in a converging lens is

$$m = \frac{q}{p}$$

Since $m = 2$ and $p = 15$ cm, the image distance must be $q = 30$ cm.

10. **E** A plane mirror produces a virtual image that is the same size as the object in all cases.

Free-Response Problems

1. A drawing of the situation is shown here. We have used the thin lens approximation. $AF = d$ and $FO = f$ (focal length) on the left side of the lens. On the right side, we again have $FO = f$ and $FC = d'$. Since we have a thin lens, $AB = OH$ and $OI = CD$. Now we have the following similar

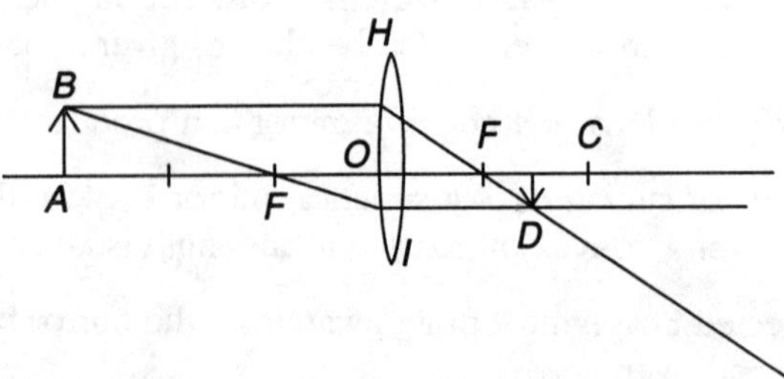

triangles: ΔFAB is similar to ΔFOI and ΔFOH is similar to ΔFCD. Thus, the following sides are in proportion:

$$\frac{AB}{OI} = \frac{AF}{FO}$$

$$\frac{OH}{CD} = \frac{FO}{FC}$$

Thus, $f/d = d'/f$ using the relationships defined here. Therefore, $f^2 = dd'$, as desired.

2. If we have two lenses in contact, the image of the first lens becomes a "negative" object for the second lens. For the first lens, we have

$$\frac{1}{f_1} = \frac{1}{p} + \frac{1}{q_1}$$

where p is the object distance for the combination. For the second lens, we must now have

$$\frac{1}{f_2} = \frac{1}{q_1} + \frac{1}{q_2}$$

If we add both these equations, we obtain for the combination,

$$\frac{1}{f_1} + \frac{1}{f_2} = \frac{1}{p} + \frac{1}{q_2}$$

Since the two thin lenses are in contact, q_2 can be considered the image distance from the first lens also. Therefore, we have

$$\frac{1}{p} + \frac{1}{q_2} = \frac{1}{f}$$

Then the combination of two thin lenses in contact behaves equivalently as a single lens with a focal length f such that

$$\frac{1}{f} = \frac{1}{f_1} + \frac{1}{f_2}$$

3. (a) For the first lens, we have $f = 5$ cm and $p = 25$. Thus,

$$q = \frac{pf}{p - f} = \frac{(25)(5)}{20}$$

and $q = 6.25$ cm.

(b) Since $d = 10$ cm for the second lens (its distance from the first lens) and $q = 6.25$ for the first lens, the image q now becomes a new object at a distance $p' = 10 - 6.25 = 3.75$. The focal length of the second lens is 3 cm, and therefore the final image distance q' is q' = (3.75)(3)/(0.75) = 15 cm.

(c) The combined magnification of the system is given by the product of each separate magnification. Since $m = q/p$, in general we have $m_1 = 6.25 / 25 = 0.25$ and $m_2 = 15 / 3.75 = 4$. Therefore, the combined magnification $m = m_1 m_2 = (0.25)(4) = 1$.

24
QUANTUM THEORY

BLACKBODY RADIATION

Quantum theory was born on December 14, 1900, when the theoretical physicist Max Planck delivered a paper to the German Academy of Science on a revolutionary way to explain the distribution of energy emitted by hot bodies.

Radiation is one process by which heat can be transported. After a metal is heated, it begins to glow when its temperature reaches a certain value. First, it appears red and then, as its temperature increases, its color changes through the electromagnetic spectrum (red to yellow to blue). This energy is transported by waves and is called **thermal radiation**.

The radiating objects do not necessarily have to be metals. Glowing gases (like stars) emit thermal radiation related to their temperatures. The yellow color of the sun is due to its surface temperature of 5,800 K. Objects that would theoretically be perfect emitters of radiation are called **blackbody radiators**. Stars are only approximately blackbodies, and astrophysicists have been working on many different theories to explain the types of radiation emitted by stars. Figure 1 illustrates some typical blackbody curves which are called *distribution spectra* because the energy emitted over a given surface is compared over all wavelengths (or frequencies). The intensity of distribution peaks at certain wavelengths is based on the given absolute temperature of the blackbody.

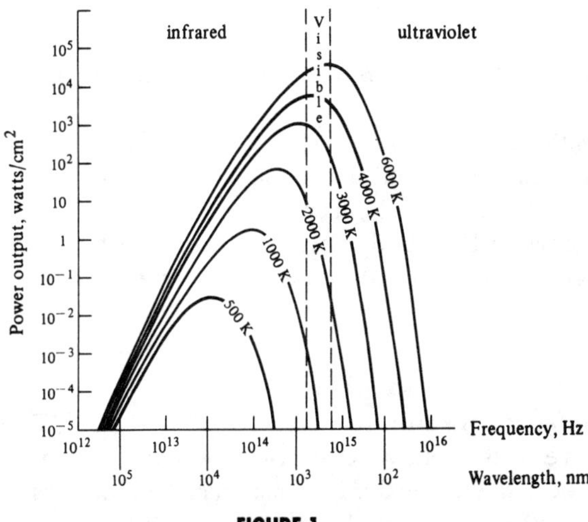

FIGURE 1.

The area under this type of curve measures the power of the radiation emitted per square meter (or centimeter in some textbooks). The analysis of these curves about 100 years ago helped establish the science of quantum theory as well as a method for analyzing the light emitted by stars (the science known as astrophysics).

One of the earliest attempts to empirically study blackbody radiation was made by the experimental physicist Max Wien. He established a relationship known as **Wien's displacement law** which states that the peak or maximum wavelength emitted by a blackbody is related to its absolute temperature by the formula

$$\lambda_{max}T = \text{constant} = 0.002898 \text{ m-K}$$

Sample Problem

What is the absolute temperature of a blackbody emitting radiation at a peak wavelength of 550 nm?

Solution

Using Wien's displacement law, we see that

$$\lambda_{max}T = 0.002898$$

$$T = \frac{0.002898}{5.5 \times 10^{-7}} = 5{,}084.2 \text{ K}$$

Theoretical attempts to understand the nature of the blackbody spectrum and Wien's law began with a theory developed by Lord Rayleigh and Sir James Jeans in England. They predicted that the intensity distribution (as a function of wavelength and absolute temperature) should be given by

$$I(\lambda, T) = \frac{2\pi \mathbf{c} kT}{\lambda^4}$$

where **c** is the velocity of light, k is Boltzmann's constant, and T is the absolute temperature of the blackbody.

The *Rayleigh-Jeans law,* as it was called, suffered from an "ultraviolet catastrophe" in which a theoretical intensity was predicted at short wavelengths but was contradicted by actual observations.

Further analysis of blackbody curves showed that the total power emitted per square meter is proportional to the fourth power of the absolute temperature:

$$P = \sigma T^4$$

where σ is **Stefan's constant** and is equal to 5.7×10^{-8} W/m^2-K^4.

An alternative to the ultraviolet catastrophe hypothesis was developed by Max Planck who treated blackbody radiators as harmonic oscillators with one major assumption. He postulated that the possible energy emitted by blackbody radiators is directly proportional to some whole multiple of the frequency of the emitted light:

$$E_n = nhf$$

where h is a constant equal to 6.63×10^{-34} J/Hz which we refer to as **Planck's constant**. According to Planck, blackbody radiation is therefore **quantized** into discrete packets (now called **photons**). Planck's constant is expressed equivalently as the product of energy and time (called **action**) and is therefore sometimes referred to as the **elementary quantum of action**.

Using his quantum hypothesis, Planck reformulated the blackbody radiation law and correctly predicted the distribution spectrum for various absolute temperatures using the equation

$$I(\lambda, T) = \frac{2\pi hc^2}{\lambda^5(e^{hc/\lambda kT} - 1)}$$

PHOTOELECTRIC EFFECT

Light has a very interesting effect on certain metals. Imagine a piece of zinc placed on top of an electroscope and charged negatively by contact (in a vacuum) as shown in Figure 2. It has been determined that if the zinc is exposed to various frequencies of light, light in the ultraviolet region will cause the electroscope to discharge. This phenomenon does not occur when the zinc and the electroscope are positively charged. The conclusion is that ultraviolet light causes zinc to emit electrons, which causes the electroscope to discharge, a phenomenon known as the **photoelectric effect**.

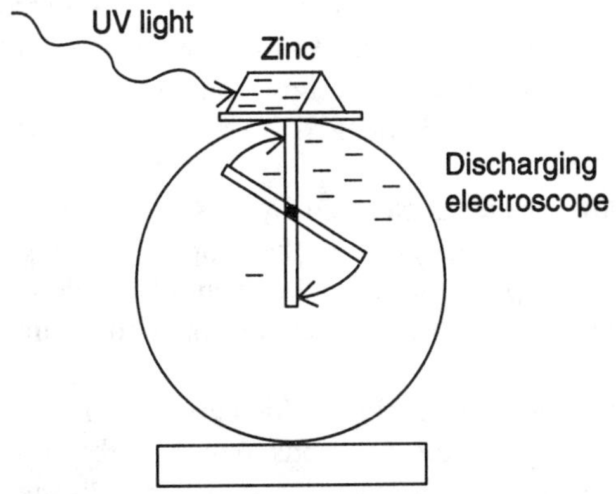

FIGURE 2.

Many metals exhibit the photoelectric effect, which can be studied in a more quantitative way. In Figure 3 we see an evacuated tube containing a photoemissive material. On the other side of the tube is a collecting plate. The tube is connected to a microammeter, which can be used to measure the current generated by the emitted electrons. When exposed to light of suitable frequency, the ammeter implicitly measures the number of electrons emitted per second. This number is sensitive only to the intensity of the electromagnetic wave applied, provided it is above a certain minimum frequency. The minimum frequency at which electrons are emitted is called the **threshold frequency** and is designated by f_0.

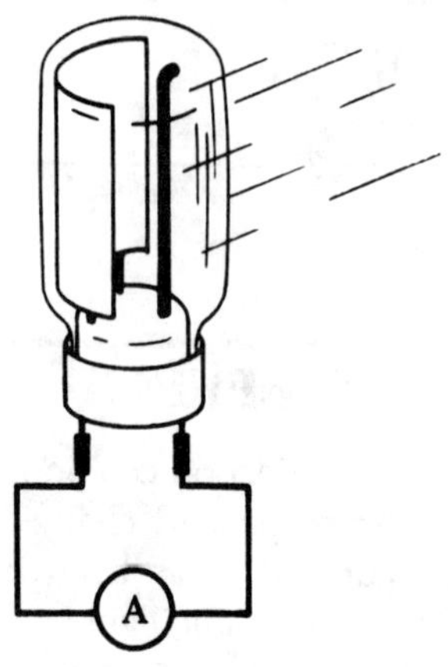

FIGURE 3.

To measure the energy of the emitted electrons, we add a source of potential difference to the circuit such that the negative electrons approach a negatively charged side (see Figure 4). This creates a braking force, and if we select the voltage properly, we can stop the current completely. The voltage that performs this feat is called the **stopping voltage** or **stopping potential** and is designated by V_0.

The energy of the electrons stopped by this voltage is determined by the product of the electron's charge and the stopping voltage eV_0. Since this stopping voltage is independent of the intensity of the light used (a violation of the wave theory of light), it measures the maximum kinetic energy of the emitted electrons.

As the frequency of light is slowly increased, there is a change in the maximum kinetic energy of the electrons. The fact that this energy is independent of the intensity of the light was used by Albert Einstein to demonstrate that light consists of discrete or quantized packets of energy called photons. The energy of a given photon is directly proportional to its frequency. To

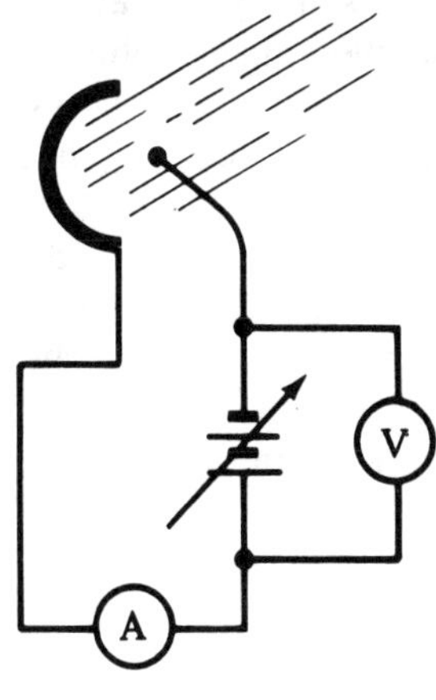

FIGURE 4.

further clarify this relationship, Figure 5 illustrates a typical plot of maximum kinetic energy versus frequency.

Starting with the threshold frequency, the curve is a straight line. For any frequency point f the slope of the line is given by

$$\text{Slope} = \frac{KE_{max}}{f - f_0} = h$$

which is exactly equal to Planck's constant.

The preceding equation can be solved for the maximum kinetic energy:

$$KE_{max} = hf - hf_0$$

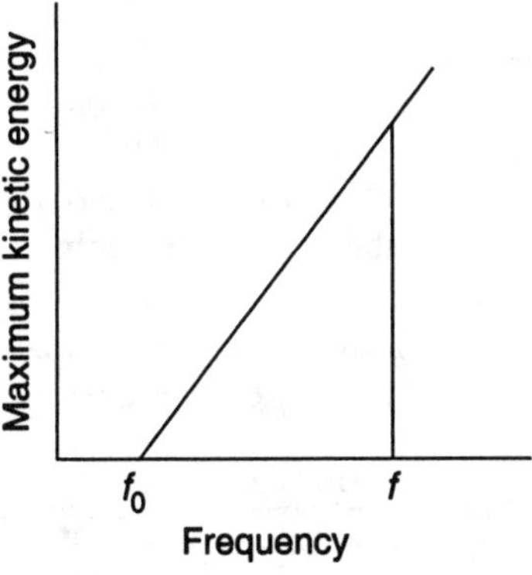

FIGURE 5.

This expression is known as the **photoelectric equation**. The quantity hf represents the energy of the original photon, while the quantity hf_0 is a property of the metal and is called the **work function**. This quantity is a measure of the minimum amount of energy needed to free an electron and is designated in some textbooks by the Greek letter phi (ϕ) or by W_0. In this book, we shall use the latter designation for the work function.

Experiments with other metals yield some interesting results. Even though different metals have different threshold frequencies, they all obey the same photoelectric equation and have the same slope h. A typical comparison graph of several metals is shown in Figure 6.

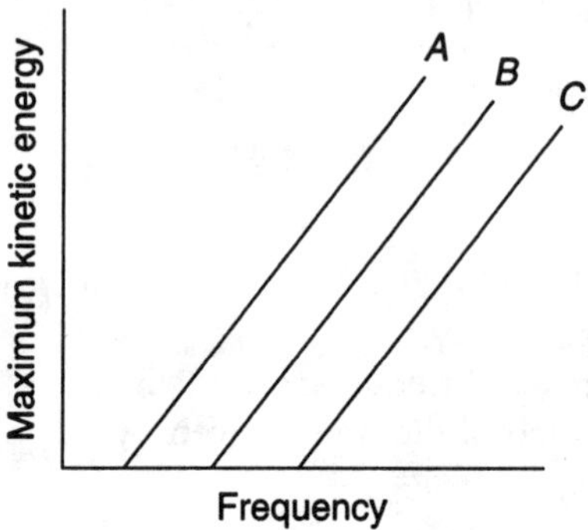

FIGURE 6.

Einstein was awarded the Nobel Prize for Physics in 1921 for his theoretical work on the quantum theory of light in 1905. (This was the same year his celebrated paper on special relativity appeared, but controversy over relativity forced the Nobel committee to award the prize on the basis of his uncontroversial work on the photoelectric effect.) The idea that light behaves as packets of quantized energy (photons), as suggested by the photoelectric effect, was in direct conflict with the prevailing wave theory of light (as evidenced by interference and diffraction). Thus, physicists in the early twentieth century were presented with a *dual nature of light.*

In the wave theory of light, the intensity of light is related to its amplitude and measured with a **photometer**. The intensity of light energy varies inversely as the square of the distance from the source. In the quantum theory, light intensity is related to the number of photons:

$$N = \frac{\text{detected energy}}{\text{energy of a photon}}$$

In studies of the photoelectric effect, when the intensity was varied, only the number of photons varied, not their kinetic energy. Modern instruments used to measure light intensity now employ **photoelectric photometers** to measure how many photons are entering the device. In telescopes and video cameras, image detectors use **charge coupled devices** (CCDs) to work with small numbers of photons.

Sample Problem

Light with a frequency of 2×10^{15} Hz is incident on a piece of copper.

(a) What is the energy of the light in joules and electronvolts?
(b) If the work function for copper is 4.5 eV, what is the maximum kinetic energy (in electronvolts) of the emitted electrons?

Solution

(a) The energy is given in joules by $E = hf$. Thus, $E = hf = (6.63 \times 10^{-34})(2 \times 10^{15}) = 1.326 \times 10^{-18}$ J. Since 1 eV $= 1.6 \times 10^{-19}$ J, we have $E =$ 8.28 eV.

(b) To find the maximum kinetic energy, we simply subtract the work function from the photon energy: $KE_{max} = 8.28 - 4.5 = 3.79$ eV.

COMPTON EFFECT AND PHOTON MOMENTUM

In the early 1920s, the American physicist Arthur Compton conducted experiments with x rays and graphite. He demonstrated that when an x-ray photon collides with an electron, the collision obeys the law of conservation of momentum. The scattering of photons due to this momentum interaction is known as the **Compton effect** (Figure 7), and the scattered photon has a lower frequency than the incident photon.

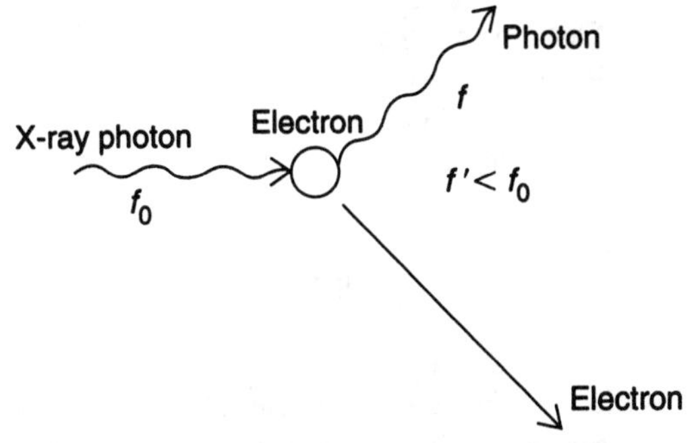

X-ray photon f_0 — Electron — Photon f — $f' < f_0$ — Electron

FIGURE 7.

The momentum of the photon is given by the relationship $\mathbf{p} = h/\lambda$. Thus, x rays possess a large photon momentum because of their very high frequency. Even though we state that photons can have momentum, they do not possess what is called **rest mass**. If light can have a momentum as well as a wavelength, the question now raised is, Can particles have a wavelength as well as a momentum?

MATTER WAVES

If a beam of electrons is incident on a narrow slit (Figure 8), then the pattern that emerges is a typical wave diffraction pattern. Electron diffraction is one example of the dual nature of matter. Under certain conditions, matter exhibits a wave nature, and the circumstances can be determined by experiment. When you perform an interference experiment with light, you will demonstrate its wave properties. But if you do a photoelectric effect experiment, light reveals its particle-like properties.

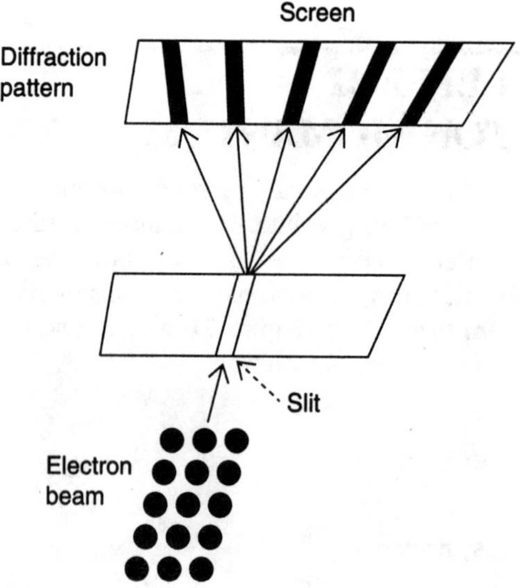

FIGURE 8.

In the 1920s, the French physicist Louis DeBroglie used the dual nature of light to suggest that matter can have a wavelike nature as evidenced by its **DeBroglie wavelength**. Since a photon's momentum is given by $\mathbf{p} = h/\lambda$, the wavelength can be transposed to give $\lambda = h/\mathbf{p}$.

DeBroglie suggested that if the momentum of a particle, $\mathbf{p} = m\mathbf{v}$, has a magnitude suitable for overcoming the small magnitude of Planck's constant, then a significant wavelength can be observed. The criterion, since mass is in the denominator of the expression for wavelength, is for the mass to be extremely small (on an atomic scale). Thus, while electrons, protons, and neutrons might have wave characteristics, a falling stone, because of its large mass, would not exhibit any wavelike effects. The equation for the DeBroglie wavelength of a particle is

$$\lambda = \frac{h}{m\mathbf{v}}$$

Sample Problem

Find the DeBroglie wavelength for

(a) a 10-g stone moving with a velocity of 20 m/s
(b) an electron moving with a velocity of 1×10^7 m/s.

Solution

(a) $m = 10$ g $= 0.01$ kg, and $\lambda = h/m\mathbf{v}$. Therefore,

$$\lambda = \frac{6.63 \times 10^{-34}}{(0.01)(20)} = 3.315 \times 10^{-33} \text{ m}$$

(b) We have

$$\lambda = \frac{h}{m\mathbf{v}} = \frac{6.63 \times 10^{-34}}{(9.1 \times 10^{-31})(1 \times 10^{7})} = 7.3 \times 10^{-11} \text{ m}$$

This aspect of particle motion on a small scale leads to the **Heisenberg uncertainty principle**. Developed by Werner Heisenberg, again in the 1920s, it states that because of the wave nature of matter, it is impossible to determine precisely both the simultaneous position and momentum of a particle. Any experimental attempt to observe the particle will result in an inherent uncertainty because of this wave nature of matter. Stated mathematically, the Heisenberg uncertainty principle can be written as

$$\Delta x \, \Delta \mathrm{p} \geq \frac{h}{2\pi}$$

where h is Planck's constant.

SPECTRAL LINES

When a hot source emits white light, it is said to be **incandescent**. When analyzed with a prism, its light is broken up into a continuous spectrum of colors. A **spectroscope** is a device that uses either a prism or a grating to create a spectrum. If a colored filter is used over the incandescent source, then light of only one color will be emitted and the spectrum will appear as a continuous band of that color.

If hydrogen gas is contained at low pressure in a tube and electrically sparked, it will emit a bluish light. If this light is analyzed with a spectroscope, something different appears. Instead of a continuous band of color, the spectroscope reveals a discrete series of colored lines ranging from red to violet. When sparked, other chemical elements show characteristic *emission line spectra* that have been useful for chemical analysis. Figure 8 illustrates a typical hydrogen spectral series in the visible range of the electromagnetic spectrum. This series is often called the **Balmer series** after Jakob Balmer, a Swiss mathematician who empirically studied the hydrogen spectrum in the late nineteenth century.

A hydrogen tube is excited and its light passed through a narrow slit to provide a line source, as shown in Figure 9. This light is then passed through a prism (or grating). While many emission lines are produced, we have illustrated only the first four. They are designated Hα (red), Hβ, Hγ, and Hδ (violet). The wavelengths range from about 656 nm to about 400 nm.

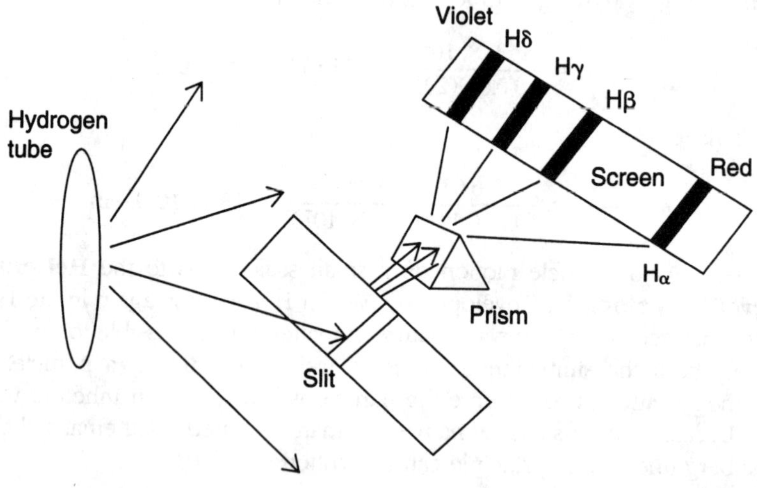

FIGURE 9.

It was Balmer, in 1884, who first demonstrated that the visible hydrogen spectrum obeys the following empirical (nontheoretical) relationship in ångstroms:

$$\lambda\ (\text{Å}) = 3{,}645.6\,\frac{n^2}{n^2 - 4}$$

where $n = 3, 4, 5, \ldots$

Balmer did not understand the significance of the integral index n, but it was clear that the light emitted by atomic compounds was not continuous but discrete (quantized). An alternative form of the Balmer formula allows for some additional insight. If we solve for the **wavenumber** $1/\lambda$, then we will have a relationship in which it is directly proportional to the emitted energy of the photons. For the Balmer series, this expression takes the form

$$\frac{1}{\lambda} = R\left(\frac{1}{2^2} - \frac{1}{n^2}\right)$$

The constant R is known as the **Rydberg constant** and has a value of $1.097 \times 10^7\ \text{m}^{-1}$.

Several other emission line series for hydrogen were discovered in the ultraviolet region (Lyman series) and in the infrared region (Paschen series). In each case, a similar version of the Balmer formula is found.

$$\text{Lyman series: } \frac{1}{\lambda} = R\left(\frac{1}{1^2} - \frac{1}{n^2}\right) \qquad n = 2,3,4,\ldots$$

$$\text{Paschen series: } \frac{1}{\lambda} = R\left(\frac{1}{3^2} - \frac{1}{n^2}\right) \qquad n = 4,5,6,\ldots$$

PROBLEM-SOLVING STRATEGIES

The photoelectric effect demonstrates the particle nature of light and is independent of the intensity of the electromagnetic waves used. The number of photons, however, is proportional to the intensity of energy.

In some problems, the energy is expressed in electronvolts. This is convenient because the maximum kinetic energy is proportional to the stopping potential. Additionally, wavelengths are sometimes expressed in ångstroms and need to be converted to meters. Be careful with units and make sure equations involving energies and other quantities employ SI units.

PRACTICE PROBLEMS FOR CHAPTER 24

Thought Problems

1. Explain why electrons are diffracted through a crystal.
2. Explain why increasing the intensity of electromagnetic radiation on a photoemissive surface does not affect the kinetic energy of the ejected electrons. Why is the kinetic energy referred to as the *maximum kinetic energy*?
3. Some stars are red, and some stars are blue. Explain these color differences assuming that stars radiate like blackbodies. Some red stars are very bright

and very far away, and some red stars are very faint and relatively close to the earth. Explain these differences in terms of blackbody radiation.

Multiple-Choice Problems

1. How many photons are associated with a beam of light with a frequency of 2×10^{16} Hz and a detectable energy of 6.63×10^{-15} J?
   (A) 25
   (B) 500
   (C) 135
   (D) 8
   (E) 80

2. A photoelectric experiment reveals a maximum kinetic energy of 2.2 eV for a certain metal. The stopping potential for these emitted electrons is ___ V.
   (A) 1.2
   (B) 1.75
   (C) 3.5
   (D) 4.2
   (E) 2.2

3. The work function for a certain metal is 3.7 eV. What is the threshold frequency for this metal?
   (A) 9×10^{14} Hz
   (B) 2×10^{15} Hz
   (C) 7×10^{14} Hz
   (D) 5×10^{15} Hz
   (E) 3.5×10^{15} Hz

4. Which of the following photons has the greatest momentum?
   (A) radio
   (B) microwave
   (C) red light
   (D) ultraviolet
   (E) x ray

5. What is the momentum of a photon associated with yellow light that has a wavelength of 5,500 Å?
   (A) 1.2×10^{-27} kg-m/s
   (B) 1.2×10^{-37} kg-m/s
   (C) 1.2×10^{-17} kg-m/s
   (D) 1.2×10^{-30} kg-m/s
   (E) 5,000 kg-m/s

6. What is the DeBroglie wavelength for a proton ($m = 1.67 \times 10^{-27}$ kg) with a velocity of 6×10^{7} m/s?
   (A) 1.5×10^{14} m
   (B) 1.5×10^{-14} m
   (C) 4.8×10^{-11} m

(D) 6.6×10^{-15} m
(E) 3×10^{8} m

7. What is the velocity of an electron with a DeBroglie wavelength of 3×10^{-10} m?
(A) 3×10^{8} m/s
(B) 2.4×10^{6} m/s
(C) 7.28×10^{4} m/s
(D) 1.5×10^{7} m/s
(E) 5.7×10^{5} m/s

8. Which of the following is a formula for photon momentum?
(A) $\mathbf{p} = E\mathbf{c}$
(B) $\mathbf{p} = E\mathbf{c}^2$
(C) $\mathbf{p} = E/\mathbf{c}$
(D) $\mathbf{p} = \mathrm{m}\mathbf{c}^2$
(E) $\mathbf{p} = E + \mathbf{c}$

9. Which of the following statements is correct in reference to emission line spectra?
(A) all the lines are evenly spaced
(B) all elements in the same chemical family have the same spectra
(C) only gases emit emission lines
(D) all lines result from discrete energy differences
(E) all of these statements are correct

10. Which of the following is equivalent to the units of Planck's constant, J-s?
(A) kg-m/s
(B) kg-m/s^2
(C) kg-m^2/s
(D) kg-m^2/s^2
(E) kg/s

Free-Response Problems

1. (a) Starting with the Balmer formula for wavelength in ångstroms,

$$\lambda(\text{Å}) = 3{,}645.6\frac{n^2}{n^2 - 4}$$

derive the alternative version of the Balmer formula in reciprocal meters

$$\frac{1}{\lambda} = R\left(\frac{1}{4} - \frac{1}{n^2}\right)$$

where $R = 1.097 \times 10^{7}\ \text{m}^{-1}$.

(b) What is the wavelength of the photon associated with $n = 4$?

2. (a) Explain the implication of Planck's constant being larger than its present value.
(b) Explain the basic theory behind electron diffraction.

3. The threshold frequency for calcium is 7.7×10^{14} Hz.
   (a) What is the work function of calcium in joules?
   (b) If light of wavelength 2.5×10^{-7} m is incident on calcium, what will be the maximum kinetic energy of the emitted electrons?
   (c) What is the stopping potential for the electrons emitted under the conditions in part (b)?
4. (a) Show, using the Planck blackbody radiation formula, that for long wavelengths the formula reduces to the Rayleigh-Jeans formula:

$$I(\lambda, T) = \frac{2\pi ckT}{\lambda^4}$$

   (*Hint*: Use a series expansion $e^x \approx 1 + x$.)
   (b) Show, using the Planck blackbody radiation formula, that for short wavelengths the Planck formula reduces to Wien's radiation law:

$$I(\lambda, T) = \frac{2\pi hc^2}{\lambda^5} e^{hc/\lambda kT}$$

SOLUTIONS TO PRACTICE PROBLEMS

Thought Problems

1. Waves are diffracted through a grating when the spacings are comparable to the wavelength. The DeBroglie wavelength of a low-energy electron is on an atomic scale and is comparable to the spacing between the lattice layers of a crystal.
2. According to quantum theory, the energy of a photon is proportional to its frequency, and the intensity of light is proportional to the number of photons. In the photoelectric effect, one photon interacts with one electron. Increasing (or decreasing) the intensity changes the photocurrent but not the kinetic energy of the electrons emitted. The kinetic energy is the maximum kinetic energy because energy may be lost to other processes.
3. The colors of stars indicate their temperatures. Red stars are cooler than blue stars. However, the energy emitted per second per unit of area depends on the size of the star. Stars known as *red giants* are much cooler than the sun but are hundreds of times larger. Thus, even over the vast distances of space (covering trillions of miles), these red giant stars are relatively bright because of their large surface emitting area.

Multiple-Choice Problems

1. **B** If E is the total detectable energy and E_p is the energy of a photon, then

$$N = \frac{E}{E_p} = \frac{E}{hf} = \frac{6.63 \times 10^{-15}\ \text{J}}{1.326 \times 10^{-17}\ \text{J}} = 500$$

2. **E** By definition, 1 eV is the amount of energy given to one electron placed in a potential difference of 1 V. Thus, if $KE_{max} = 2.2$ eV, the stopping potential must be equal to 2.2 V.

3. **A** The formula for the work function is $W_0 = hf_0$ and it must be expressed in joules before dividing by Planck's constant:

$$f_0 = \frac{(3.7)(1.6 \times 10^{-19})}{6.63 \times 10^{-34}} = 8.9 \times 10^{14} \text{ Hz}$$

4. **E** Since $\mathbf{p} = h/\lambda$, the waves with the smallest wavelength will have the greatest photon momentum. Of the five choices, an x-ray photon has the smallest wavelength.

5. **A** The wavelength must first be converted to meters, and then we can use $\mathbf{p} = h/\lambda$:

$$p = \frac{6.63 \times 10^{-34}}{(5500)(1 \times 10^{-10})} = 1.2 \times 10^{-27} \text{ kg-m/s}$$

6. **D** The formula for the DeBroglie wavelength is $\lambda = h/m\mathbf{v}$. Substituting the values given yields

$$\lambda = \frac{6.63 \times 10^{-34}}{(1.67 \times 10^{-27})(6 \times 10^{7})} = 6.6 \times 10^{-15} \text{ m}$$

7. **B** Using the DeBroglie formula for wavelength gives $\mathbf{v} = h/m\lambda$. Substituting values,

$$\mathbf{v} = \frac{6.63 \times 10^{-34}}{(9.1 \times 10^{-31})(3 \times 10^{-10})} = 2.4 \times 10^{6} \text{ m/s}$$

8. **C** First we have $\mathbf{p} = h/\lambda$ and $E = hf = h\mathbf{c}/\lambda$. Thus, $E = p\mathbf{c}$, and so $\mathbf{p} = E/\mathbf{c}$.

9. **D** All spectral emission lines are different for every atom or molecule, and the lines are not evenly spaced. They do, however, arise from energy difference transitions (see Chapter 26) as discussed in this chapter.

10. **C** The units of Planck's constant are J-s, and since 1 J = 1 kg-m^2/s^2, the units for h are kg-m^2/s. These units are also the units for angular momentum.

Free-Response Problems

1. (a) To derive the *Rydberg formula*, we need to rewrite the Balmer formula for meters:

$$\lambda \text{ (m)} = 3.6456 \times 10^{-7} \frac{n^2}{n^2 - 4}$$

Now take the reciprocal of both sides:

$$\frac{1}{\lambda} = \frac{n^2 - 4}{3.6456 \times 10^{-7} n^2}$$

Distribute the denominator into the two factors in the numerator:

$$\frac{1}{\lambda} = \frac{1}{3.6456 \times 10^{-7}}\left(\frac{1}{1} - \frac{4}{n^2}\right)$$

Finally, factor out a 4:

$$\frac{1}{\lambda} = \frac{4}{3.6456 \times 10^{-7}}\left(\frac{1}{4} - \frac{1}{n^2}\right) = 1.097 \times 10^7\left(\frac{1}{4} - \frac{1}{n^2}\right) = R\left(\frac{1}{4} - \frac{1}{n^2}\right)$$

(b) To find the wavelength associated with $n = 4$, just substitute for $n = 4$ to obtain $1/\lambda = 2{,}056{,}875\ \text{m}^{-1}$, which means that $\lambda = 4.861 \times 10^{-7}$ m.

2. (a) If Planck's constant were larger than its established value, then an implication would be that the uncertainty principle, $\mathbf{\Delta p}\ \Delta x \geq h/2\pi$, applies to a larger magnitude domain and we would see quantum effects occurring at larger sizes as well.

(b) Electron diffraction is based on the wave theory of matter. Since an electron has a small enough mass, its DeBroglie wavelength is large enough for it to be diffracted when small openings are used and thus exhibit seemingly common wave properties. Scanning electron microscopes, although much more complex in theory, use this basic principle as well.

3. (a) The work function is given by $W_0 = hf_0 = (6.63 \times 10^{-34})(7.7 \times 10^{14}) = 5.1 \times 10^{-19}$ J.

(b) The frequency associated with this wavelength is given by

$$f = \frac{\mathbf{c}}{\lambda} = \frac{3 \times 10^8}{2.5 \times 10^{-7}} = 1.2 \times 10^{15}\ \text{Hz}$$

The energy of the associated photon is equal to $E = hf = 7.956 \times 10^{-19}$ J. The maximum kinetic energy of the ejected electrons is equal to the difference between this energy and the work function: $KE_{max} = 7.956 \times 10^{-19} - 5.1 \times 10^{-19} = 2.856 \times 10^{-19}$ J.

(c) This maximum kinetic energy is equal to the product of the electron charge and the stopping potential. Thus,

$$V_0 = \frac{KE_{max}}{e} = \frac{2.856 \times 10^{-19}}{1.6 \times 10^{-19}} = 1.785\ \text{V}$$

4. (a) If we use the given series expansion, then we can write for long wavelengths:

$$e^x \approx 1 + x$$

$$e^{hc/\lambda kT} \approx 1 + \frac{hc}{\lambda kT}$$

Making this substitution in the Planck formula, we obtain (after canceling out the factor 1)

$$I(\lambda, T) = \frac{2\pi ckT}{\lambda^4}$$

(b) For short wavelengths, the exponential term is much larger than 1. We can therefore effectively neglect the subtraction in the denominator and simply write (for short wavelengths)

$$I(\lambda, T) = \frac{2\pi hc^2}{\lambda^5}\, e^{hc/\lambda kT}$$

25
THE ATOM

ATOMIC STRUCTURE AND RUTHERFORD'S MODEL

In 1896, the French physicist Henri Becquerel was studying the phosphorescent properties of uranium ores. Certain rocks glow after being exposed to sunlight, and many physicists at that time were studying the light emanations from these ores.

One day while engaged in such a study, Becquerel placed one of these uranium ores in a drawer on top of some sealed photographic paper. To his surprise, several days later he found that the paper had been exposed even though no light was incident on it. His conclusion was that radiation emitted by the uranium had penetrated the sealed envelope and exposed the paper.

Subsequent work by Marie and Pierre Curie discovered that this natural radioactivity (as they called it) came from the uranium itself and was a consequence of its instability as an atom. Ernest Rutherford discovered (see Figure 1) that the radiation emitted yielded particles that were useful for further atomic studies.

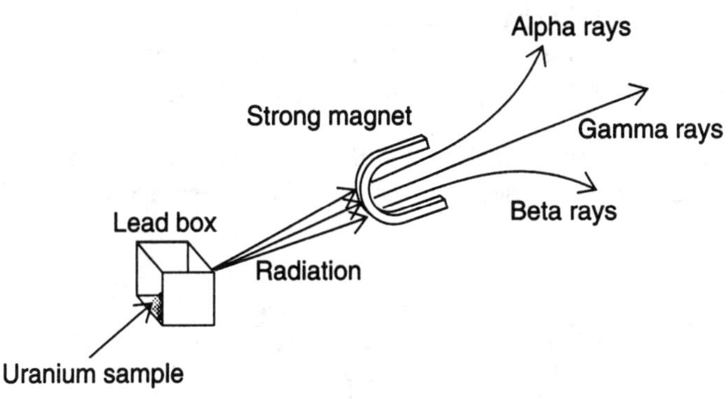

FIGURE 1.

When passed through a magnetic field, the emitted radiation split into three separate parts. These were called **alpha rays**, **beta rays**, and **gamma rays**. Analysis of their trajectories lead to the conclusion that alpha particles are positively charged (later identified to be helium nuclei), beta particles are negatively charged (later identified to be electrons), and gamma rays are electrically neutral (later identified to be photons).

J. J. Thomson, using the data available at the end of the nineteenth century, devised an early model of the atom based on the fact that all atoms are electrically neutral. He concluded that the atom was positively charged and that electrons were likewise associated with it. Thomson's model proposed that the atom consisted of a relatively large uniformly distributed positive mass with negative charged electrons embedded in it like raisins in a pudding (see Figure 2).

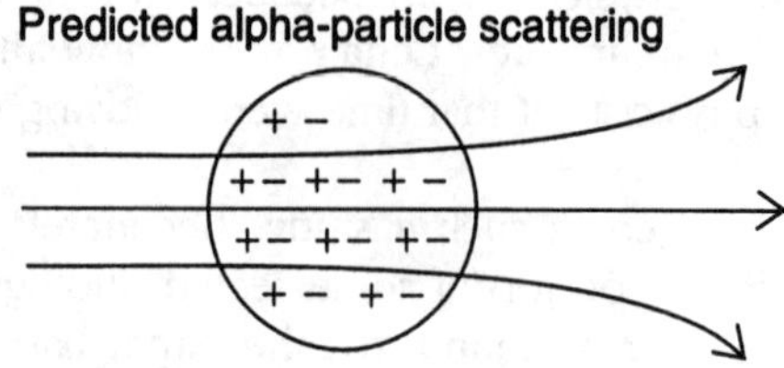

FIGURE 2. *Thomson's model of the atom.*

In 1911 Rutherford and his coworkers decided to test Thomson's model using alpha particle scattering. In the Thomson model, the positive charge was uniformly distributed. Since the negative electrons were embedded in the nucleus, the Coulomb force of attraction would weaken the deflecting force on a passing positively charged alpha particle.

Rutherford's idea was to use alpha particles, aiming them at a thin metal foil (gold was used) observing the scattering pattern on a zinc sulfide screen (Figure 3).

When the results were analyzed, Rutherford realized that most of the alpha particles had not been deflected. A few had been strongly deflected in

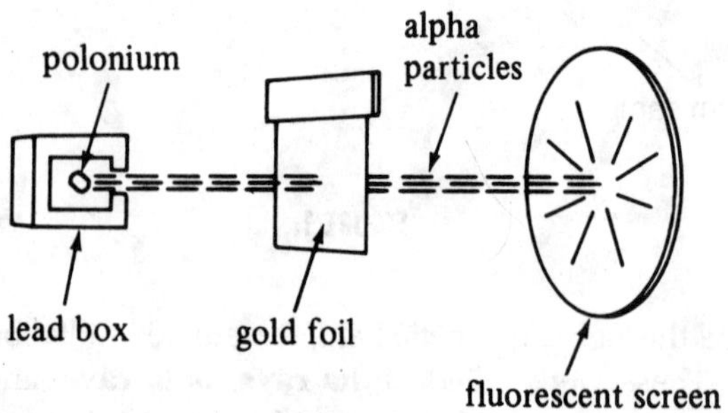

FIGURE 3.

hyperbolic paths as a result of the Coulomb force of repulsion. Since these forces were strong, Thomson's model was incorrect. Some alpha particles were even scattered back close to 180 degrees (Figure 4).

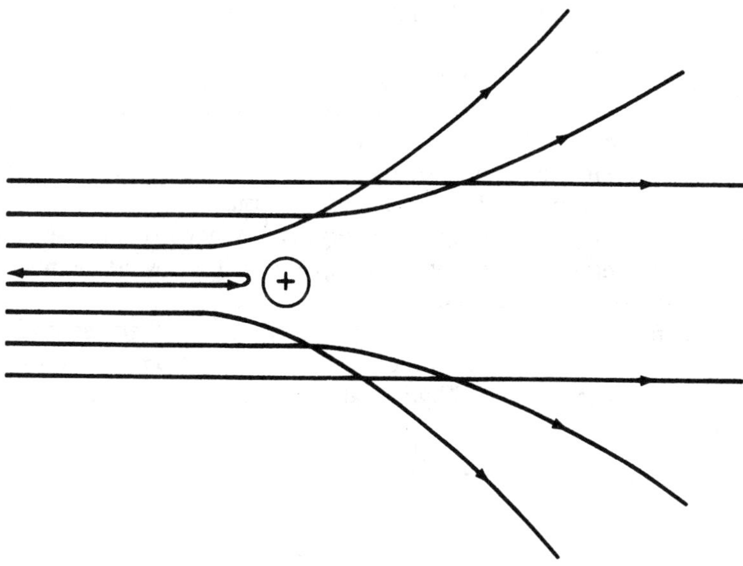

FIGURE 4. *Alpha particle scattering trajectories.*

Rutherford's conclusion was that the atom consisted mostly of empty space and that the nucleus was a very small, densely packed positive charge. Electrons, he proposed, orbited around the nucleus (like planets orbiting the sun).

Rutherford's model had several major problems. Among them was the fact that it could not account for the appearance of discrete emission line spectra. The model had electrons continuously orbiting around the nucleus, and this circular motion was accelerated motion which should produce a continuous band of electromagnetic radiation but did not. Additionally, the predicted orbital loss of energy would result in an atom disintegrating in a very short period of time and thus breaking apart all matter. This phenomenon also did not occur.

THE BOHR MODEL

Niels Bohr was a Danish physicist who in 1913 decided to study the hydrogen spectral lines again. Einstein had proposed a quantum theory of light years earlier, developing the idea of a photon and completely explaining the photoelectric effect.

Bohr observed that the Balmer formula for the hydrogen spectrum could be viewed in terms of energy differences. He then realized that these energy differences involved a new set of postulates about atomic structure. The main ideas of his theory were that

1. Electrons do not emit electromagnetic radiation while in a given orbit (energy state).
2. Electrons cannot remain in any arbitrary energy state. They can remain only in discrete or quantized energy states with an angular momentum proportional to Planck's constant.
3. The angular momentum of the electron is quantized (by an integer index n) and proportional to Planck's constant.
4. Electrons emit electromagnetic energy when they make transitions from higher energy states to lower energy states. The lowest energy state is known as the **ground state**.
5. Absorption spectra (dark lines appearing on an otherwise continuous color spectrum) occur because of transitions from lower to higher energy states as emitted light is absorbed by a cooler gas in front and reradiated.

Recall from Chapter 24 that for the visible hydrogen spectrum we have the formula

$$\frac{1}{\lambda} = R\left(\frac{1}{4} - \frac{1}{n^2}\right)$$

Now the energy of the emitted photon can be determined if we multiply both sides by the quantity hc:

$$E_p = \frac{hc}{\lambda} = \frac{hcR}{4} - \frac{hcR}{n^2}$$

Recall that R is the Rydberg constant and that (in this case) n is an integer equal to 3, 4, 5, ...

For the ultraviolet hydrogen spectrum the term $hcR/4$ is replaced by $hcR/1$, and for the infrared spectrum we have the term $hcR/9$. In these cases, acceptable values for the *principal quantum number n*, likewise change.

Since electrons are "bound" to a nucleus (as a result of attractive Coulomb forces), forces designate the energy levels as increasingly negative as the electron gets closer to the nucleus (with $n = 1$ being the ground state). Thus, we can qualitatively write

$$E_n = -\frac{hcR}{n^2}$$

The energy states are then given by $E_1 = -13.6$ eV, $E_2 = -3.4$ eV, $E_3 = -1.5$ eV, $E_4 = -0.84$ eV, and so on. Negative signs are used to allow for the binding of the electron to the nucleus (much like a stone dropped into a well, which has negative potential energy with respect to ground level).

These energy states have been derived qualitatively on the basis of the appearance of emission line spectra. It was the genius of Niels Bohr to begin with the electric Coulomb forces and the postulated quantized angular momentum to theoretically derive the same relationship for the hydrogen spectrum as has been determined empirically.

Bohr knew from Rutherford's experiments that the average radius of a hydrogen atom is 5.3×10^{-11} m. The circumference of its orbit would therefore be equal to

$$C = 2\pi r = (6.28)(5.3 \times 10^{-11}) = 3.3 \times 10^{-10} \text{ m}$$

To better understand how the angular momentum is quantized, let us use an argument that came after Bohr's theory but is more illustrative. Recall that an electron has a wavelength inversely proportional to its momentum $\lambda = h/m\mathbf{v}$.

Physicists suspected, using Rutherford's model, that the orbital velocity of an electron was about 2.2×10^6 m/s. The wavelength associated with this velocity is given by

$$\lambda = \frac{h}{m\mathbf{v}} = \frac{6.63 \times 10^{-34}}{(9.1 \times 10^{-31})(2.2 \times 10^6)} = 3.3 \times 10^{-10} \text{ m}$$

which is exactly the same as the circumference derived previously.

If we state that the wavelength corresponds to a closed standing wave (see Figure 5) that is a multiple of the circumference (a two-dimensional standing wave), then we can write

$$n\lambda = 2\pi r = \frac{nh}{m\mathbf{v}}$$

If we transpose some of the variables, we can rewrite this expression in the form

$$\mathbf{L} = m\mathbf{v}r = \frac{nh}{2\pi}$$

The quantity $m\mathbf{v}r$, designated by the letter **L**, is the angular momentum of the electron (in a classical sense). The preceding expression is Bohr's statement

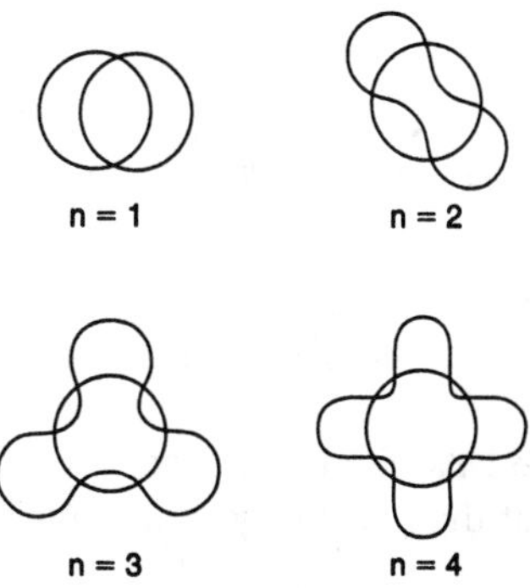

FIGURE 5. *Electron standing waves.*

about the quantization of the angular momentum. The angular momentum is proportional to Planck's constant divided by 2π. Recall that the units of Planck's constant are consistent with those for angular momentum. Some examples of these orbital standing waves are shown in Figure 5.

Bohr was able to derive that the energy levels are determined by

$$E_n = -\frac{2\pi^2 k^2 m e^4}{n^2 h^2} = -\frac{13.6 \text{ eV}}{n^2}$$

where k is Coulomb's constant, m is the mass of an electron, e is the charge of an electron, h is Planck's constant, and n is the principal quantum number equal to the energy level (ground state and excited states).

Bohr's expression completely matched the empirical and qualitative energy level values suggested by the emission line spectra in the absence of any quantum theory (except that $E = hf$ for photons). According to the Bohr theory, the energy of a photon is equal to the amount of energy difference between two energy states (Figure 6).

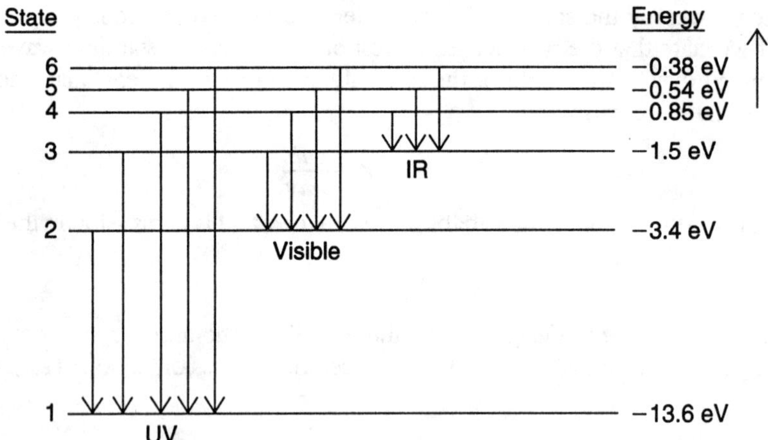

FIGURE 6. *Energy level diagram for hydrogen.*

The different spectral series arise from transitions from higher energy states to specified lower energy states. Thus, the ultraviolet Lyman series occurs when electrons fall to level 1. The visible Balmer series occurs when electrons fall to level 2. The infrared Paschen series occurs when electrons fall to level 3. Thus, we can write that the frequency of an emitted photon is given by

$$f = \frac{E_i - E_f}{h}$$

Figure 6 is the energy level diagram for hydrogen. The energy in a given state represents the amount of energy (called **ionization energy**) needed to free an electron from an atom. A charged atom is called an **ion**. Electron ionization energies were confirmed by experiment in the 1920s.

Even though hydrogen has only one electron, the many hydrogen atoms in a sample of hydrogen gas, coupled with the different probabilities of electrons being in any one particular state, produce the multiple lines that appear in the visible spectrum when the gas is "excited."

Sample Problem

What is the wavelength of a photon emitted when an electron makes a transition in a hydrogen atom from level 5 to level 3?

Solution

The transition from level 5 to level 3 involves an energy difference of

$$E = -0.54 \text{ eV} - (-1.50 \text{ eV}) = 0.96 \text{ eV} = 1.536 \times 10^{-19} \text{ J}$$

Now the frequency of the photon is equal to $f = (1.536 \times 10^{-19})/(6.63 \times 10^{-34}) = 2.3 \times 10^{14}$ Hz. The wavelength of this photon is given by $\lambda = \mathbf{c}/f = \lambda = \mathbf{c}/f = 3 \times 10^{8}/2.3 \times 10^{14} = 1.29 \times 10^{-6}$ m (which is in the infrared part of the electromagnetic spectrum).

It is important to remember that an electron will make a transition from a lower to a higher energy state only when it absorbs a photon equal to a given possible (allowed) energy difference. This is the nature of the quantization of the atom. The electron can exist only in a specified energy state, and to change states, it must either absorb or emit a photon whose energy is proportional to Planck's constant and equal to the energy difference between two energy states.

The uncertainty principle, discussed in Chapter 24, introduces the function of probability in describing the nature of electron orbits and energy states. This subject is studied in the science of quantum mechanics developed by Erwin Schrödinger and Werner Heisenberg.

QUANTUM MECHANICS AND THE ELECTRON CLOUD MODEL

The success of the Bohr model was limited to the hydrogen atom and certain hydrogen-like atoms (atoms with one electron in the outermost shell; ionized helium is such an example). The wave nature of matter lead to a problem of predicting the behavior of an electron in a strict Bohr energy state.

In 1926 the physicist Erwin Schrödinger developed a new mechanics of the atom. It later became known as **quantum mechanics**, but at the time it was called **wave mechanics**. In the wave mechanics of Schrödinger, the electron's behavior was determined by a mathematical function called the **wave func-**

tion, which was related to the probability of finding an electron in any one energy state. A **wave equation** called the **Schrödinger equation** was developed using ideas about time-dependent motion in classical mechanics.

The energy states of Bohr were then replaced by a probability "cloud." In this **cloud model** the electrons are not necessarily limited to specified orbits. A cloud of uncertainty is produced with the densest regions corresponding to the highest probability of an electron being in a given state.

The use of probability to explain the structure of the atom lead to intense debates between the proponents of the new quantum mechanics and one of its fiercest opponents, Albert Einstein. Einstein, who had helped to develop the quantum theory of light 20 years earlier, was not convinced that the universe could be determined by probability. To paraphrase, Einstein said he could not believe that God would "play dice" with the universe. "God is subtle," he said, "but not malicious."

At about the same time that Schrödinger introduced his wave mechanics, Werner Heisenberg developed his **matrix mechanics** for the atom. Instead of using a wave analogy, Heisenberg developed a purely abstract mathematical theory using matrices. Working with Heisenberg, Wolfgang Pauli developed the **Pauli exclusion principle**, which states that two electrons having the same spin orientation cannot occupy the same quantum state at the same time.

The new quantum mechanics was born when the two rival theories were shown to be equivalent. Electron spin was experimentally verified, and the theory proved to bc morc successful than the older Bohr theory in explaining atomic and molecular structures, leading to new ideas like solid-state physics and the subsequent development of lasers.

PROBLEM-SOLVING STRATEGIES

Understanding Bohr's postulates is a key part of successively solving atomic structure and spectra problems. Semiclassical ideas about Coulomb's law, energy, and uniform circle motion were used by Bohr.

Remember that the levels are inversely proportional to the square of the principal quantum number n and directly proportional to the factor -13.6 eV. However, in a frequency or wavelength calculation, these energies must be converted back to joules.

PRACTICE PROBLEMS FOR CHAPTER 25

Thought Problems

1. In Rutherford's model of the atom, what would happen to the frequency of the emitted electron radiation as the electron spiraled in its orbit?

2. If a source of light from excited hydrogen gas were moving toward or away from an observer, what changes, if any, would be perceived in the emission spectral lines?

3. A hydrogen atom consists of only one electron. Explain why the emission spectrum of excited hydrogen gas shows the presence of many different spectral lines.

Multiple-Choice Problems

1. How much energy is needed to ionize a hydrogen atom in the $n = 4$ state?
   (A) 0.85 eV
   (B) 13.6 eV
   (C) 12.75 eV
   (D) 10.2 eV
   (E) 3.4 eV

2. Which of the following electron transitions will emit a photon with the greatest frequency?
   (A) $n = 1$ to $n = 4$
   (B) $n = 5$ to $n = 2$
   (C) $n = 3$ to $n = 1$
   (D) $n = 7$ to $n = 3$
   (E) $n = 1$ to $n = 3$

3. An electron in the ground state of a hydrogen atom can absorb a photon with all the following energies except _____ eV.
   (A) 10.2
   (B) 12.1
   (C) 12.5
   (D) 12.75
   (E) 13.06

4. What is the frequency of a photon emitted in an electron transition from $n = 3$ to $n = 1$ in a hydrogen atom?
   (A) 1.83×10^{34} Hz
   (B) 2.92×10^{15} Hz
   (C) 1.7×10^{15} Hz
   (D) 4.7×10^{14} Hz
   (E) 3.3×10^{15} Hz

5. Which of the following statements about the atom is correct?
   (A) Orbiting electrons can sharply deflect passing alpha particles.
   (B) The nucleus of the atom is electrically neutral.
   (C) The nucleus of the atom deflects alpha particles into parabolic trajectories.
   (D) The nucleus of the atom contains most of the atomic mass.
   (E) None of these statements is correct.

6. Which of the following statements about the Bohr theory reflects how it differs from classical predictions about the atom.
   I: An electron can orbit without a net force acting.
   II: An electron can orbit about a nucleus.
   III: An electron can be accelerated without radiating energy.
   IV: An orbiting electron has a quantized angular momentum.
   (A) I and II
   (B) I and III
   (C) III and IV
   (D) II and IV
   (E) I and IV

7. What is the maximum angle at which an alpha particle can be deflected by a nucleus?
   (A) 30 degrees
   (B) 60 degrees
   (C) 90 degrees
   (D) 135 degrees
   (E) 180 degrees

8. An electron makes a transition from a higher energy state to the ground state in a Bohr atom. As a result of this transition, _____.
   (A) the total energy of the atom is increased
   (B) the force on the electron is increased
   (C) the energy of the ground state is increased
   (D) the charge on the electron is increased
   (E) the electron's energy remains the same

9. How many different photon frequencies can be emitted if an electron is in excited state $n = 4$ in a hydrogen atom?
   (A) one
   (B) three
   (C) five
   (D) six
   (E) eight

10. A radioactive atom emits a gamma ray photon, and as a result, _____.
   (A) the energy of the nucleus is decreased
   (B) the charge in the nucleus is decreased
   (C) the ground state energy is decreased
   (D) the force on an orbiting electron is decreased
   (E) none of these statements is correct

Free-Response Problems

1. (a) Using Coulomb's law and an assumed circular orbit, derive an expression for the orbital velocity of an electron in the ground state of a hydrogen atom.

(b) Use the Bohr quantized standing wave hypothesis along with the results from part (a) to show that in the ground state,

$$r = \frac{h^2}{4\pi^2 km\, e^2}$$

(c) Using all known values of the quantities in part (b), calculate the classical radius of a ground state hydrogen atom.

2. (a) The electric potential energy for an orbiting ground state electron (in a hydrogen atom) is given by $U = ke^2/r$. Use the result from part (a) of Problem 1 to show that the equation for the total energy of the atom is

$$E = -\frac{ke^2}{2r}$$

(b) The total energy for the ground state hydrogen atom is -13.6 eV. Using this information, derive the ground state radius for the atom based on this expression and compare it with the results from Problem 1.

3. (a) Using the information in Problems 1 and 2, show that the energy levels in a hydrogen atom are given by the formula

$$E_n = -\frac{2\pi^2 k^2 me^4}{h^2 n^2}$$

(b) Using the known values in the preceding expression, show that this equation reduces to

$$E_n = -\frac{13.6 \text{ eV}}{n^2}$$

SOLUTIONS TO PRACTICE PROBLEMS

Thought Problems

1. According to the Rutherford model, electrons emit continuous radiation while orbiting a nucleus. The appearance of emission spectral lines refutes this. Additionally, the loss of energy would make the electrons spiral into the nucleus. This would be accompanied by a shift in the frequency of the light since as the electron falls inward, its orbital frequency would increase, creating a "blue shift." These phenomena do not occur in nature.

2. If a source of light were moved toward or away from an observer, the position of the spectral lines would shift as a result of the Doppler effect. If the source were moving toward the observer, a blue shift (in which the lines are shifted toward the blue end) would be detected. If the source were going away from the observer (in both cases, along the line of sight),

a red shift would be seen. Astronomers use Doppler shifts to determine the *radial velocities* of stars and galaxies.

3. Hydrogen contains only one electron. In confined excited hydrogen gas, collisions between atoms cause different electrons to exist momentarily in a variety of excited states. If the temperature and pressure are suitable, the probability of these electrons falling to the $n = 2$ state (producing visible light) from higher states will be greater than the probability of their falling to the ground state. At very high temperatures, hydrogen emits ultraviolet light, and at even higher temperatures, it can be ionized.

Multiple-Choice Problems

1. **A** In the $n = 4$ state, the energy level is -0.85 eV. Thus, to ionize the electron in this state, $+\ 0.85$ eV must be supplied.
2. **C** Transitions to level 1 produce ultraviolet photons and have frequencies greater than those produced by transitions to other levels.
3. **C** An electron in the ground state can absorb a photon with energy equal to the energy difference between the ground state and any other level. Only 12.5 eV is not such an energy difference.
4. **B** The energy difference from level 3 to level 1 is 12.1 eV. The emitted frequency is $f = E/h = (12.1)(1.6 \times 10^{-19})/(6.63 \times 10^{-34}) = 2.92 \times 10^{15}$ Hz.
5. **D** The nucleus of an atom contains most of its mass.
6. **C** The Bohr theory challenged classical theories by stating that an electron can accelerate without radiating energy and that its angular momentum is quantized and proportional to Planck's constant.
7. **E** Rutherford's experiments revealed that some alpha particles were backscattered through angles as large as 180 degrees.
8. **B** As an electron gets closer to the nucleus, the force of attraction on the electron increases.
9. **D** There are six possible photon frequencies emitted from state $n = 4$: (1) $n = 4$ to $n = 1$ (2) $n = 4$ to $n = 3$ (3) $n = 4$ to $n = 2$ (4) $n = 3$ to $n = 2$ (5) $n = 3$ to $n = 2$ (6) $n = 2$ to $n = 1$
10. **A** Since photons do not have any rest mass, their emission releases only energy.

Free-Response Problems

1. (a) Assuming a circular orbit for the electron, we set the Coulomb force equal to the centripetal force:

$$\frac{ke^2}{r^2} = \frac{m\mathbf{v}^2}{r}$$

Solving for the velocity **v**, we obtain

$$\mathbf{v} = e\sqrt{\frac{k}{mr}}$$

(b) Using the standing wave hypothesis for $n = 1$, we have

$$\lambda = 2\pi r = \frac{h}{m\mathbf{v}} = \frac{h}{me\sqrt{\frac{k}{mr}}}$$

Squaring both sides we obtain

$$4\pi^2 r^2 = \frac{h^2 mr}{m^2 e^2 k}$$

Finally, solving for r we have

$$r = \frac{h^2}{mk\ e^2 4\pi^2}$$

(c) Using the known values for the physical quantities we have

$$r = \frac{(6.63 \times 10^{-34})^2}{(9.1 \times 10^{-31})(9 \times 10^9)(1.6 \times 10^{-19})^2 4(3.14)^2} = 5.3 \times 10^{-11}\ \text{m}$$

2. (a) The total energy is equal to the sum of the kinetic and potential energies and is given by:

$$E = \frac{1}{2}m\mathbf{v}^2 - \frac{ke^2}{r}$$

Substituting for the velocity expression in part (a) of Problem 1, we get

$$E = \frac{1}{2}m\left(\frac{e^2k}{mr}\right) - \frac{ke^2}{r} = \frac{ke^2}{2r} - \frac{2ke^2}{2r} = -\frac{ke^2}{2r}$$

(b) We know that for $n = 1$, $E = -13.6$ eV, and we need to convert this energy to joules:

$$E\ (\text{J}) = -(13.6)(1.6 \times 10^{-19}) = -2.176 \times 10^{-18}\ \text{J}$$

Solving for the radius r gives

$$r = \frac{(9 \times 10^9)(1.6 \times 10^{-19})^2}{2(2.176 \times 10^{-18})} = 5.3 \times 10^{-11}\ \text{m}$$

3. (a) If we go back to the standing wave hypothesis, then for any quantum number n,

$$\lambda = \frac{nh}{m\mathbf{v}}$$

$$r = \frac{n^2h^2}{mk\ e^2 4\pi^2}$$

Now if we substitute this radius formula into the total energy formula, we have

$$E = -\frac{ke^2}{2r} = -\frac{ke^2}{2}\left(\frac{me^2k4\pi^2}{n^2h^2}\right) = -\frac{2\pi^2mk^2e^4}{n^2h^2}$$

(b) If we substitute all known values for the quantities (except n), we obtain

$$E = -\frac{2.167 \times 10^{-18}}{n^2} = -\frac{13.54\ \text{eV}}{n^2}$$

26
THE NUCLEUS

NUCLEAR STRUCTURE AND STABILITY

The experiments of Hans Geiger and Ernest Marsden (under Ernest Rutherford's supervision) in 1911 confirmed that the nucleus of the atom was compact and positively charged. Experiments on nuclear masses using a **mass spectrograph**, which deflects the positively charged nuclei into curved paths with different radii, confirmed that the estimate of nuclear mass was too low. The nucleus could not, therefore, be made entirely of positively charged protons. Physicists theorized that another electrically neutral particle, called the **neutron**, must exist.

Thus, the question of nuclear stability depended on the particular structure of the nucleus. The work on radioactive decay by Rutherford and by Pierre and Marie Curie, confirmed that the radioactive disintegration of an atom is a nuclear phenomenon and is tied to stability as well.

The basic structure of the atom, using the Bohr-Rutherford planetary model, consists of a nucleus with a certain number of protons and neutrons (together called **nucleons**). The number of protons (Z) characterizes the element since no two elements have the same number of protons. The **atomic number** Z also indicates the relative charge on the nucleus, and the **mass number** A is the total number of nucleons (protons plus neutrons) in the nucleus. Both of these numbers are whole numbers, while the actual nuclear mass is not a whole number.

Since an atom is normally electrically neutral, the initial number of protons and electrons is the same. Schematically, an atom is designated with a capital letter (sometimes followed by one additional lowercase letter), and the two nuclear numbers A and Z are written as superscripts and subscripts on the left-hand side. That is, for any element X we write ^A_ZX. For example, ^1_1H means that the element is hydrogen with one proton and no neutrons. All elements have **isotopes**. These are nuclei with the same number of protons but different numbers of neutrons. For example, helium 4 is written as ^4_2He, which indicates two protons and two neutrons. An isotope of this atom is helium 3, written as ^3_2He. Some examples of nuclear and atomic structure are shown in Figure 1.

A neutron is slightly more massive than a proton. The proton mass is approximately $m_p = 1.673 \times 10^{-27}$ kg, while the mass of a neutron is approximately $m_n = 1.675 \times 10^{-27}$ kg. Additionally, nuclear masses are usually

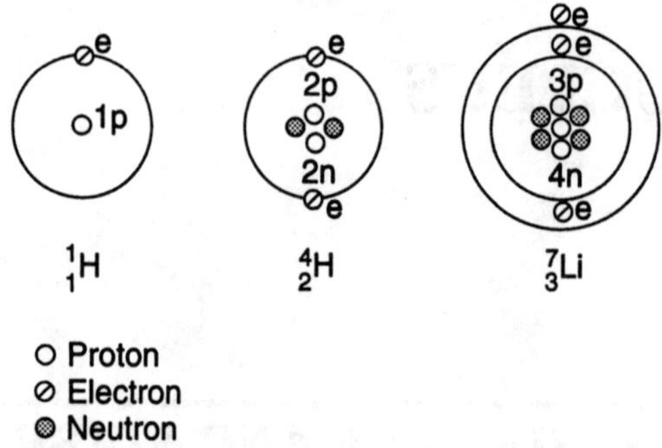

FIGURE 1. *Nuclear structure.*

expressed in **atomic mass units** (u). One atomic mass unit is defined to be equal to $\frac{1}{12}$ the mass of a carbon-12 atom:

$$1 \text{ u} = 1.66 \times 10^{-27} \text{ kg}$$

On this scale, the proton's mass is given as 1.0078 u and the neutron's mass is given as 1.0087 u.

If one looks at all stable isotopes of known nuclei, something qualitatively interesting emerges. Experiments show that lighter nuclei have about the same number of protons as neutrons; however, heavier stable nuclei have more neutrons than protons.

This suggests that the neutrons must somehow help overcome the repulsive forces in a tightly packed positive nucleus. Unstable radioactive nuclei lie off the stability curve (see Figure 2), and as they decay, alpha and beta emissions **transmute** the original nucleus into a more stable one.

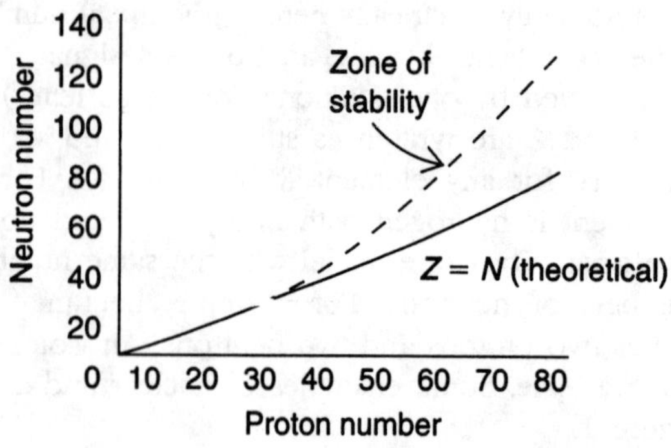

FIGURE 2. *Proton-neutron plot.*

BINDING ENERGY

If you were to look at a particular isotope, say ^4_2He, and determine its total mass based on the number of nucleons, then you would have a nucleus that was more massive than is measured. For example, the helium-4 nucleus contains two protons (each with a mass of 1.0078 u) and two neutrons (each with a mass of 1.0087 u). The total predicted mass would be 4.0330 u, but experimentally the mass of the helium-4 nucleus is found to be 4.0026 u. The difference of 0.0304 u is called the **mass defect** and is proportional to the **binding energy** (BE) of the nucleus.

Using Einstein's special theory of relativity (Chapter 27), we know that mass can be converted to energy. The energy equivalent of 1 u is equal to 931.5 MeV. The binding energy of any nucleus is therefore determined from the relationship

$$\text{BE} = \text{mass defect} \times 931.5 \text{ MeV/u}$$

In our example, the binding energy of the helium-4 nucleus is

$$\text{BE} = 0.0304 \times 931.5 = 28.3 \text{ MeV}$$

It is useful to think about the average binding energy per nucleon, which is equal to BE/A. For helium 4, A = 4..., and

$$\text{BE}_{\text{avg}} = \frac{28.3}{4} = 7.08 \text{ MeV per nucleon}$$

In Figure 3, average binding energy per nucleon is plotted against the atomic number Z.

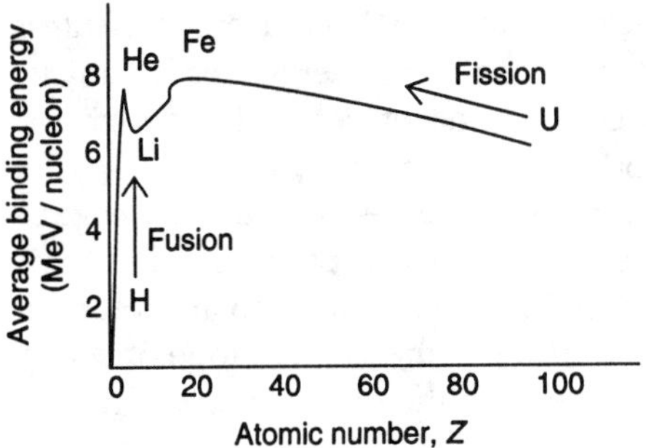

FIGURE 3.

As can be seen from the graph, a maximum binding energy per nucleon is reached at Z = 26 (iron). The greater the average binding energy per nucleon, the greater the stability of the nucleus. Thus, since hydrogen isotopes have less stability than helium isotopes, the **fusion** of hydrogen into helium increases the stability of the nucleus through the release of energy. Likewise, the unstable uranium isotopes yield more stable nuclei if they are split (**fission**) into lighter nuclei (also releasing energy).

RADIOACTIVE DECAY

Radioactivity was discovered in 1895 by Henri Becquerel, and the relationship between radioactivity and nuclear stability was studied by Pierre and Marie Curie. Ernest Rutherford pioneered the experiments on nuclear bombardment and introduced the word **transmutation** into the nuclear physics vocabulary. The discovery of alpha and beta particle emission from nuclei led to their use in probing the structure of the nucleus in the first decades of the twentieth century. Subsequent research showed the existence of positively charged electrons (called **positrons**) and even negatively charged protons. We have the following designations for some of these particles:

Proton (hydrogen nucleus): ${}^{1}_{1}H$
Beta particle (electron): ${}_{-1}^{0}e$
Neutron: ${}^{1}_{0}n$
Alpha particle (helium nucleus): ${}^{4}_{2}He$
Positron (positive electron): ${}^{0}_{1}e$
Photon: ${}^{0}_{0}\gamma$

When an isotope of uranium, ${}^{238}_{92}U$, decays, it emits an alpha particle according to the following equation: ${}^{238}_{92}U \rightarrow {}^{234}_{90}Th + {}^{4}_{2}He$. When the radioactive thorium isotope decays, it emits a beta particle according to the following equation: ${}^{234}_{90}Th \rightarrow {}^{234}_{91}Pa + {}_{-1}^{0}e + \overline{\nu}$. In the first example, both the atomic number and the mass number change as the uranium transmuted into thorium. However, in the decay of the thorium nucleus, only the atomic number changes. This occurs because in beta decay a neutron in the nucleus is converted to a proton, an electron, energy, and a particle called the **antineutrino** ($\overline{\nu}$). There is currently no evidence that neutrinos have any rest mass, although some theories require that they do.

In addition to alpha and beta decay, there are several other scenarios for nuclear disintegration. In **electron capture**, an isotope of copper captures an electron into its nucleus and transmutes into an isotope of nickel: ${}^{64}_{29}Cu + {}_{-1}^{0}e \rightarrow {}^{64}_{28}Ni$. In **positron emission**, the same isotope of copper decays by emitting a positron and transmutes into the same isotope of nickel (plus a neurtino, ν): ${}^{64}_{29}Cu \rightarrow {}_{-1}^{0}e + {}^{64}_{28}Ni + \nu$.

The rate of nuclear decay is characterized by the element's **half-life**. This is the time it takes for half of a given amount of a radioactive substance to decay. If we have a sample with N nuclei, the number ΔN that decay in a time Δt will be proportional to the number N:

$$\frac{-\Delta N}{\Delta t} \approx \lambda N$$

where λ is the **decay constant**. The number of remaining undecayed nuclei is given by

$$N = N_0 e^{-\lambda t}$$

where N_0 is the number of nuclei present when $t = 0$ and e is the base of the natural logarithm system and is approximately equally to 2.718.

The meaning of the constant λ is as follows. In a time $\tau = 1/\lambda$ (called the **characteristic time**), the number of nuclei is equal to $1/e$ or 37% of the original amount. The half-life is related to the characteristic time. Let T represent the half-life, and then

$$T = (\ln\ 2)\tau = 0.693\tau$$

The *activity* of a radioactive substance is the number of decays per second:

$$\text{Activity} = R \approx -\frac{\Delta N}{\Delta t} \approx \lambda N \approx \left(\frac{0.693}{T}\right)N_0$$

The SI unit of decay rate (decays per second) is the **becquerel** (Bq). A graph of radioactive decay is shown in Figure 4.

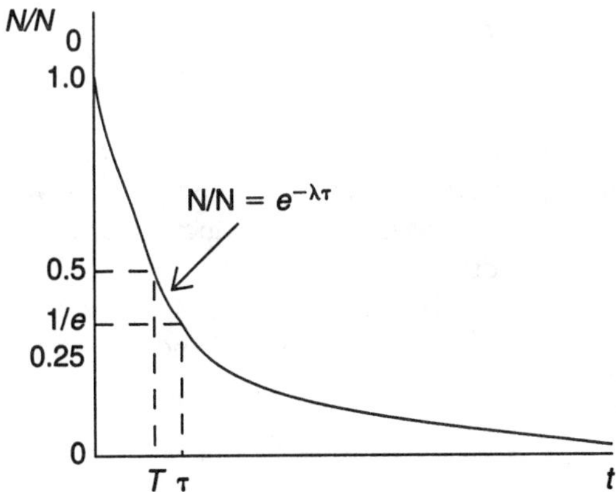

FIGURE 4.

An alternative way of expressing the decay relationship is as follows. Let T represent the half-life, let t be any time (in the same units), and let m_0 be the original mass at time $t = 0$. The mass remaining after time t can therefore be given by

$$m = \frac{m_0}{2^{t/T}} \rightarrow N = \frac{N_0}{2^{t/T}}$$

In this expression, we have eliminated the need to use natural logarithms, but common logarithms will still be necessary. We have also eliminated the necessity of knowing the decay constant and the characteristic time. Consider the following sample problem.

Sample Problem

Radon, $^{222}_{86}Rn$, has a half-life of 3.8 days. Suppose we start with 0.03 kg of radon. How long will it take to have 0.001 kg remaining?

Solution

Our starting formula is

$$\frac{m}{m_0} = \frac{1}{2^{t/T}}$$

Now take the logarithm of both sides:

$$\log \frac{m}{m_0} = \log \frac{1}{2^{t/T}} = \log \left(\frac{1}{2}\right)^{t/T} = \left(\frac{t}{T}\right) \log \frac{1}{2}$$

We need to solve for the time t:

$$t = T\,\frac{\log (m/m_0)}{\log 0.5} = (3.8)\,\frac{\log 0.333}{\log 0.5} = 18.7 \text{ days}$$

The quantity of energy required to balance a nuclear equation is called the *Q value*. It can be expressed in terms of the energy equivalent of the mass of the reactants minus the total mass of the products or as a measure of the total kinetic energy of the products.

Given a nuclear reaction between nucleus X and a subatomic particle a resulting in the production of a new nucleus Y and a subatomic particle b such that

$$a + \mathrm{X} = \mathrm{Y} + b$$

then the Q value is given by (with the masses in kilograms):

$$Q = (m_a + M_\mathrm{X} - M_\mathrm{Y} - m_b)c^2$$

In an alpha decay (from the uranium disintegration series, for example) the Q value is

$$Q = (M_\mathrm{X} - M_\mathrm{Y} - M_\alpha)(931.5) \text{ MeV/u}$$

where M_X is the mass of the original nucleus expressed in atomic mass units, M_Y is the mass of the product nucleus, and α represents the alpha particle emitted. Q values are positive for exothermic reactions and negative for endothermic reactions.

FISSION

Nuclear fission was discovered in 1939 by German physicists Otto Hahn and Fritz Strassman. When an isotope of uranium, $^{235}_{92}U$, absorbs a slow-moving neutron, the resulting instability splits the nucleus into two smaller (but still radioactive) nuclei and releases a large amount of energy. Recall from the binding energy curve that the binding energy per nucleon for uranium is lower than for other lighter nuclei. The increase in stability (an increase in binding energy per nucleon) is accompanied by a loss of kinetic energy. One possible uranium fission reaction could occur as follows:

$$^{235}_{92}U + ^{1}_{0}n \rightarrow ^{140}_{54}Xe + ^{94}_{38}Sr + 2\,^{1}_{0}n + 200 \text{ MeV}$$

In controlled-fission reactions, water or graphite is used as a **moderator** to slow down the neutrons because they cannot be slowed by electromagnetic processes. The production of two additional neutrons means that a chain reaction is possible. In a fission reactor, cadmium control rods adjust the reaction rates to desirable levels.

The production of plutonium as a fissionable material is accomplished using another isotope of uranium, $^{238}_{92}U$. First, the uranium absorbs a neutron to form a rare element called neptunium, which then decays to form plutonium:

$$^{238}_{92}U + ^{1}_{0}n \rightarrow ^{239}_{92}U \rightarrow ^{239}_{93}Np + ^{0}_{-1}e$$

$$^{239}_{93}Np \rightarrow ^{239}_{94}Pu + ^{0}_{-1}e$$

FUSION

The binding energy curve indicated that if hydrogen can be fused into helium, the resulting nucleus will be more stable and energy can be released. In a fusion reaction, isotopes known as deuterium and tritium are used. The following equations represent one kind of hydrogen fusion reaction:

$$^{1}_{1}H + ^{1}_{1}H \rightarrow ^{2}_{1}H + ^{0}_{1}e + 0.4 \text{ MeV} + \nu$$

$$^{1}_{1}H + ^{2}_{1}H \rightarrow ^{3}_{2}He + 5.6 \text{ MeV}$$

$$^{3}_{2}He + ^{3}_{2}He \rightarrow ^{4}_{2}He + 2\,^{1}_{1}H + 12.9 \text{ MeV}$$

In the preceding reaction, hydrogen is first fused into deuterium, which is then fused with hydrogen to form helium 3. The helium-3 isotope is then fused into stable helium 4.

In a different reaction, the isotope tritium is fused with deuterium to form stable helium 4 in one step:

$$^{2}_{1}H + ^{3}_{1}H \rightarrow ^{4}_{2}He + ^{1}_{0}n + 17.6 \text{ MeV}$$

PROBLEM-SOLVING STRATEGIES

Solving nuclear equations requires you to remember certain rules. First, make sure that you have conservation of charge. That means that all the atomic numbers, which are the numbers of protons (and hence the charges), must add up to the same total on each side of the equation. However, to find the actual charge, you must multiply the atomic number by the actual proton charge in coulombs.

The second rule to consider is the conservation of mass. The atomic mass numbers in the superscripts should add up, and any missing mass should be accounted for by the appearance of energy (sometimes designated by the letter Q). The mass energy equivalence is that 1 u is equal to 931.5 MeV. This second rule applies for problems involving binding energy and mass defect. Remember that an actual nucleus has less mass than the total sum of its nucleons.

PRACTICE PROBLEMS FOR CHAPTER 26

Thought Problems

1. Why do heavier stable nuclei have more neutrons than protons?
2. Why are the conditions for the fusion of hydrogen into helium favorable in the core of a star?
3. Explain the radioactive disintegration series of uranium 238 into stable lead 206 in terms of the neutron-proton stability plot.

Multiple-Choice Problems

1. How many neutrons are in a nucleus of $^{213}_{84}Po$?
   (A) 84
   (B) 129
   (C) 213
   (D) 297
   (E) 546
2. The nitrogen isotope $^{14}_{7}N$ emits a beta particle as it decays. The new isotope formed is _____.
   (A) $^{14}_{7}N$
   (B) $^{13}_{6}C$
   (C) $^{13}_{8}O$
   (D) $^{14}_{8}O$
   (E) not enough information given
3. The half-life of a radioactive material is 5 years. After 20 years, how much of a 75-g sample of the material remains?
   (A) 4.68 g
   (B) 3.75 g
   (C) 18.75 g

(D) 15 g
(E) 22.4 g

4. Which type of radiation has the greatest ability to penetrate matter?
(A) x ray
(B) alpha particle
(C) gamma ray
(D) beta particle
(E) positron

5. As a sample of radioactive material decays, its half-life _____.
(A) increases and then decreases
(B) decreases and then increases
(C) increases
(D) decreases
(E) remains the same

6. As a radioactive material undergoes nuclear disintegration, _____
(A) its atomic number must always remain the same
(B) its atomic mass number never increases
(C) its atomic mass number never decreases
(D) it always emits an alpha particle
(E) none of these statements is correct

7. If all the following particles had the same velocity, which one would have the smallest wavelength?
(A) alpha particle
(B) beta particle
(C) neutron
(D) proton
(E) positron

8. Which of the following statements about nuclear isotopes is correct?
(A) They all have the same binding energy.
(B) They all have the same atomic mass number.
(C) They all have the same number of nucleons.
(D) They all have the same number of protons.
(E) All these statements are correct.

9. In the following nuclear reaction, an isotope of aluminum is bombarded by alpha particles:

$$^{27}_{13}\text{A1} + ^{4}_{2}\text{He} \rightarrow ^{30}_{15}\text{P} + \text{Y}$$

Quantity Y must be a(n)_____.
(A) neutron
(B) electron
(C) positron
(D) photon
(E) proton

10. When a gamma ray photon is emitted from a radioactive nucleus, which of the following occurs?
    (A) The nucleus goes to an excited state.
    (B) The nucleus goes to a more stable state.
    (C) An electron goes to an excited state.
    (D) The atomic number of the nucleus decreases.
    (E) The number of nucleons decreases.

Free-Response Problems

1. (a) A 75-g radioactive sample decays to 30 g in 3.5 yr. What is the half-life of this substance?
   (b) Make a chart of mass versus time in units of half-lives for a total of six half lives.
   (c) Make a graph (appropriately scaled) of time (in half-lives) versus mass on the set of axes shown.

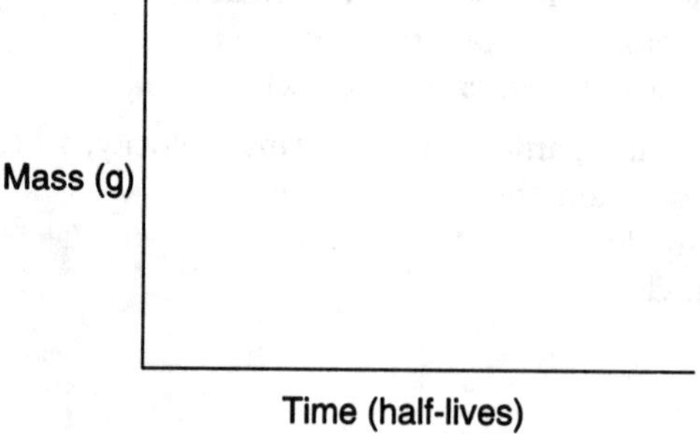

2. Given the isotope $^{56}_{26}Fe$, which has an actual mass of 55.934939 u,
   (a) determine the mass defect of the nucleus in atomic mass units
   (b) determine the average binding energy per nucleon in MeV per nucleon.

3. Show that if the Q value for an alpha decay is equal to the sum of the kinetic energies of the emitted alpha particle and the daughter nucleus, then the kinetic energy of the alpha particle can be written as

$$KE_{\alpha} = \frac{Q}{1 + M_{\alpha}/M_{d}}$$

where M_{α} is the mass of the alpha particle (in kilograms) and M_d is the mass of the daughter nucleus. (*Hint*: Use the fact that the magnitudes of the momentum of the alpha particle and daughter nucleus are equal.)

SOLUTIONS TO PRACTICE PROBLEMS

Thought Problems

1. Heavier nuclei have more protons, and this increases the coulomb force repulsion between them. To maintain stability, the nucleus contains more neutrons to help the nuclear forces counter this repulsive tendency.
2. The density, pressure, and temperature at the center of a star are so great that the fusion of hydrogen (a new star's most abundant element) into helium occurs.
3. Radioactive uranium 238 is unstable and off the neutron-proton stability curve. Through a process of alpha and beta decays, the nucleus transmutes into nuclei that bring it closer to that stability zone. Finally, at lead 206, the nucleus falls within the zone of stability and radioactivity stops.

Multiple-Choice Problems

1. **B** The number of neutrons $N = A - Z = 213 - 84 = 129$.
2. **C** The emission of a beta particle involves the transformation of a neutron into a proton and an electron. Thus, the total number of nucleons Z remains the same but the number of protons increases by 1. The new isotope is therefore oxygen 13.
3. **A** Twenty years corresponds to four half-lives. If we take just half of the given amount, compounded over 20 yr (counting every fifth year), we end up with 4.68 g.
4. **C** Gamma rays have the greatest energy with the greatest penetration power.
5. **E** The half-life is "preset" by nature. It is not affected by the decay process itself.
6. **B** In nuclear disintegration, the nucleus always either loses mass or releases energy in the form of photons.
7. **A** We know that at the same velocity the particle with the largest mass will have the smallest wavelength. Of the four choices, the alpha particle, being a helium nucleus, has the largest mass.
8. **D** Isotopes, by definition, are of the same chemical element. Hence they must have the same number of protons (but different numbers of neutrons and different binding energies).
9. **A** We must have conservation of charge and mass. The subscripts add up to 15 on both sides already, and so we are left with a subscript of 0 (the particle is therefore neutral). The superscripts must add up to 31. To balance, the missing particle needs a superscript equal to 1. This makes particle Y a neutron.
10. **B** After nuclei emit photons, they settle down to a more stable state since no mass has been lost.

Free-Response Problems

1. (a) We use the formula

$$m = \frac{m_0}{2^{t/t_h}}$$

If we transpose some variables, we can rewrite this expression in the form

$$2^{t/t_h} = \frac{m_0}{m}$$

Now taking the logarithm of both sides and using the rules of exponents, we have

$$\frac{t}{t_h} \log (2) = \log \frac{m_0}{m}$$

We now substitute all the given values: $t = 3.5$ yr, $m_0 = 75$ g, and $m = 30$ g. Solving for the half-life, we obtain 2.648 yr.

(b) Knowing the half-life, we can get a sense of the decay rate. The half-life is the time it takes for half of a given amount of radioactive material to decay. We start with 75 g when $t = 0$ and obtain the data in the accompanying table for six half-lives.

t (yr)	m (g)
0	75
2.648	37.5
5.296	18.75
7.944	9.375
10.592	4.6875
13.240	2.343
15.888	1.170

(c) We plot mass versus time (in half-lives) as shown.

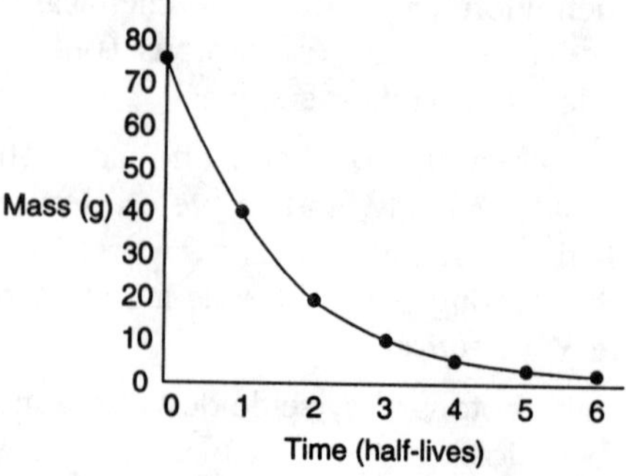

2. (a) The actual mass of iron 56 is 55.934939 u. The total additive mass is determined by noting that this isotope of iron has 26 protons and 30 neutrons. Each photon has a mass of 1.0078 u, and each neutron has a mass of 1.0087 u. Thus, the total constituent mass is

$$26(1.0078) = 26.2028 \text{ u}$$
$$30(1.0087) = 30.2610 \text{ u}$$
$$\text{Total } M = 56.4638 \text{ u}$$

The mass defect is therefore the difference between this mass and the actual mass:

$$\text{Mass defect} = 0.528861 \text{ u}$$

(b) The binding energy is given by BE = (0.528861)(931.5) = 492.63402 MeV
The average binding energy is BE_{avg} = 492.63402/56 = 8.797 MeV per nucleon.

3. We are given

$$Q = \frac{1}{2} M_\alpha \mathbf{v}_\alpha^2 + \frac{1}{2} M_d \mathbf{v}_d^2$$
$$Q = \text{KE}_\alpha + \frac{1}{2} M_d \mathbf{v}_d^2$$

Now, we use the fact that the magnitudes of the momentum of the alpha particle and the daughter nucleus are equal, such that

$$M_\alpha \mathbf{v}_\alpha = M_d \mathbf{v}_d$$

which means that

$$\mathbf{v}_d = \frac{M_\alpha \mathbf{v}_\alpha}{M_d}$$

and

$$M_d = \frac{M_\alpha \mathbf{v}_\alpha}{\mathbf{v}_d}$$

Making some substitutions, we can see that

$$Q = \text{KE}_\alpha + \frac{1}{2}\left(\frac{M_\alpha \mathbf{v}_\alpha}{\mathbf{v}_d}\right)\mathbf{v}_d^2$$
$$= \text{KE}_\alpha + \frac{1}{2} M_\alpha \mathbf{v}_\alpha \mathbf{v}_d$$

If we now use the substitution for $\mathbf{y}_d$, we can arrive at the desired result:

$$
\begin{aligned}
Q &= \mathrm{KE}_\alpha + \frac{1}{2} M_\alpha \mathbf{v}_\alpha \frac{M_\alpha \mathbf{v}_\alpha}{M_d} \\
&= \mathrm{KE}_\alpha + \frac{1}{2} M_\alpha \mathbf{v}_\alpha^2 \frac{M_\alpha}{M_d} \\
&= \mathrm{KE}_\alpha + \mathrm{KE}_\alpha \frac{M_\alpha}{M_d} \\
&= \mathrm{KE}_\alpha \left(1 + \frac{M_\alpha}{M_d}\right) \\
\mathrm{KE}_\alpha &= \frac{Q}{1 + M_\alpha / M_d}
\end{aligned}
$$

27
SPECIAL RELATIVITY

FRAMES OF REFERENCE AND THE POSTULATES OF RELATIVITY

We end our review of physics exactly where we began. All motion is relative. However, when we discussed relative motion back in Chapter 4, we restricted ourselves to inertial frames of reference moving at velocities much less than the velocity of light (300,000 km/s). This type of restricted motion is known as **Galilean relativity**.

In classical physics, if an inertial frame of reference has a velocity **v** relative to another inertial frame of reference (one considered fixed), then the velocity of an object (velocity **u**) moving within the first frame of reference will be predicted to have a velocity equal to **u** + **v**.

Do the same transformations apply to light? In the nineteenth century, that was a valid research question. Physicists who believed in the electromagnetic wave theory of light had speculated that a material in space known as "ether" was responsible for transmitting electromagnetic radiation (including light) through space. Similar to the way in which water undulates to transmit water waves, the ether supposedly moved in a transverse manner to carry light. As the earth revolved about the sun in orbit, it was expected that the velocity of light might be different in directions parallel and perpendicular to the orbital velocity.

As an analogy, consider the situation shown in Figure 1. Imagine two identical swimmers of equal ability. Both can swim with maximum velocities **c**. A course

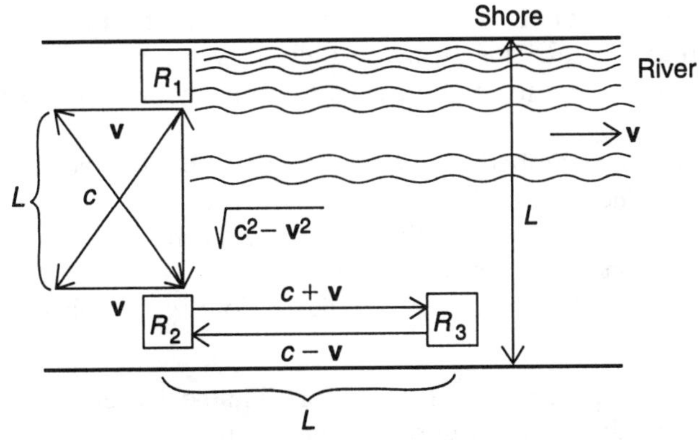

FIGURE 1.

is set up in a river with a current velocity **v**. Three rafts are set up at equal distances from each other. One swimmer will go across the river and swim back, and the other swimmer will swim downstream to a raft and then upstream back to the start. Both swimmers will cover the same effective distance $2L$.

Let's consider the first person who swims parallel to the current. When she travels with the current, her relative velocity is $\mathbf{c} + \mathbf{v}$. When she travels against the current, her relative velocity is $\mathbf{c} - \mathbf{v}$. Since her round trip distance is $2L$, her total travel time is given by

$$t_1 = \frac{L}{\mathbf{c} + \mathbf{v}} + \frac{L}{\mathbf{c} - \mathbf{v}} = \frac{2L\mathbf{c}}{\mathbf{c}^2 - \mathbf{v}^2}$$

Now, let's consider the person who swims across the current. In order to head straight across, she must swim slightly upstream at her maximum velocity **c**. Her relative velocity across the river is therefore given by $\sqrt{\mathbf{c}^2 - \mathbf{v}^2}$. The total distance she travels is again equal to $2L$. Thus, her round trip time is given by

$$t^2 = \frac{2L}{\sqrt{\mathbf{c}^2 - \mathbf{v}^2}}$$

Now, the question is, who wins the race? The answer can determined by looking at the ratio of times:

$$\frac{t_1}{t_2} = \frac{2L\mathbf{c}/(\mathbf{c}^2 - \mathbf{v}^2)}{2L/\sqrt{\mathbf{c}^2 - \mathbf{v}^2}} = \frac{1}{\sqrt{1 - \mathbf{v}^2/\mathbf{c}^2}}$$

Since this ratio is greater than 1, it means that $t_1 > t_2$, and so the second swimmer wins the race.

In 1887 an experiment was performed by two American physicists, Albert Michelson and Edward Morley. They were attempting to establish a difference between the velocity of light in the direction of the earth's revolution in its orbit about the sun and the velocity of light in a direction perpendicular to the earth's revolution. To their surprise, the "race" turned out to be a tie! The velocity of light was found to be the same in all directions. In fact, the velocity of light is an **invariant**. This means that all observers, no matter where they are or what they are doing, will measure the velocity of light in a vacuum to be the same value.

In 1905 Einstein published his theory under the title "On the Electrodynamics of Moving Bodies." Two physicists, Hendrik Lorentz and George Fitzgerald, had independently theorized that there might be a contraction of length in the direction of motion by a factor of $\sqrt{1 - \mathbf{v}^2/\mathbf{c}^2}$.

In his 1905 paper, Einstein stated the following postulates of special relativity:

1. The laws of physics are the same for all frames of reference moving at constant velocity with respect to one another.
2. The velocity of light in a vacuum is the same for all observers regardless of their state of motion or the motion of the source of light.

COORDINATE TRANSFORMATIONS

Galilean Transformations

Imagine an inertial frame of reference in which the x,x' axes are colinear, the y,y' and z,z' axes are coplanar, and the primed system is moving to the right with a constant velocity **v** relative to the unprimed system.

If the inertial frame of reference is moving in one dimension with a velocity **v**, then the coordinates of a point x in the new frame of reference are given by (see Figure 2)

$$x' = x - \mathbf{v}t$$
$$y' = y$$
$$z' = z$$
$$t' = t$$

In classical physics, the time measured in one frame of reference is the same as the time measured in the other.

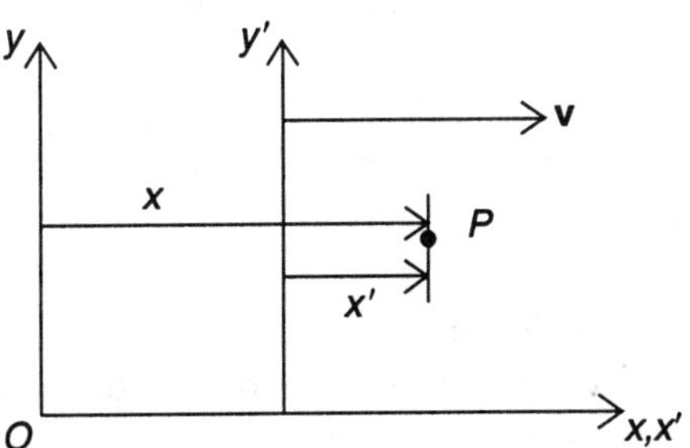

FIGURE 2.

Now consider two intervals of time t_1 and t_2 and their corresponding positions of the time intervals in the original frame (x_1 and x_2). In the new frame of reference, we have

$$x_1' = x_1 - \mathbf{v}t_1$$
$$x_2' = x_2 - \mathbf{v}t$$

Thus, the difference in these positions can be written

$$\Delta x' = x_2' - x_1' = (x_2 - \mathbf{v}t_2) - (x_1 - \mathbf{v}t_1) = \Delta x - \mathbf{v}\ \Delta t$$

The average velocity can be written as a transformation in the form

$$\frac{\Delta x'}{\Delta t} = \frac{\Delta x}{\Delta t} - \mathbf{v}$$

or

$$\mathbf{u}' = \mathbf{u} - \mathbf{v}$$

which is the usual Galilean velocity transformation.

If we consider the object accelerating uniformly in a time Δt, then it is easy to show that

$$\frac{\Delta \mathbf{u}'}{\Delta t} = \frac{\Delta \mathbf{u}}{\Delta t}$$

where **u** is the velocity of the object. To derive this concept, think of two velocities $\mathbf{u}_1'$ and $\mathbf{u}_2'$ and consider their difference:

$$\Delta \mathbf{u}' = \mathbf{u}'_2 - \mathbf{u}'_1 = (\mathbf{u}_2 - \mathbf{v}) - (\mathbf{u}_1 - \mathbf{v}) = \mathbf{u}_2 - \mathbf{u}_1 = \Delta \mathbf{u}$$

We simply divide both sides by Δt to obtain the desired result.

This equation simply states that the acceleration of the object is the same in both inertial frames. Since $\mathbf{F} = m\mathbf{a}$, we see that Newton's laws remain invariant under a Galilean transformation.

The special theory of relativity does not allow these tranformations to apply at all. They are only an approximation valid at low velocities. Instead, we use **Lorentz transformations**.

Lorentz Transformations

Hendrik Lorentz was the first to develop the coordinate transformations that Einstein would incorporate into his special theory of relativity. If we consider an object moving with a velocity **v** relative to a fixed observer (in the x direction only), then we can write

$$x' = \frac{x - \mathbf{v}t}{\sqrt{1 - (\mathbf{v}^2/\mathbf{c}^2)}}$$

$$y' = y$$

$$z' = z$$

$$t' = \frac{t - (\mathbf{v}/\mathbf{c}^2)x}{\sqrt{1 - (\mathbf{v}^2/\mathbf{c}^2)}}$$

Notice that with Lorentz transformations, following the postulates of Einstein, the dimension of time is no longer an absolute quantity. We shall discuss this aspect of relativistic time in connection with the concept of *simultaneity* later on.

Lorentz transformations can be inverted (by interchanging primes and substituting $-\mathbf{v}$ for $\mathbf{v}$) based on the chosen frame of reference:

$$x = \frac{x' + \mathbf{v}t'}{\sqrt{1 - (\mathbf{v}^2/\mathbf{c}^2)}}$$

$$y = y'$$

$$z = z'$$

$$t = \frac{t' + (\mathbf{v}/\mathbf{c}^2)x'}{\sqrt{1 - (\mathbf{v}^2/\mathbf{c}^2)}}$$

LENGTH CONTRACTION

Suppose we have a ruler that is moving with a frame of reference O' such that it is fixed in that frame. The ruler is used to measure the length of an object in the frame. The coincidence of two points on the ruler with the linear boundary of the object defines its "proper" length L_o. If we assume that the measurement in O' occurs simultaneously as observed by an observer O, then we can write

$$X_2' - X_1' = \frac{(X_2 - X_1) + \mathbf{v}(t_2 - t_1)}{\sqrt{1 - (\mathbf{v}^2/\mathbf{c}^2)}}$$

Now, if O observes the placement of the ruler at both ends as a simultaneous event, then $t_2 - t_1 = 0$, and since $x_2 - x_1 = L$, we can write an equation that relates how observer O measures the length of the object in O':

$$L = L_0\sqrt{1 - \frac{\mathbf{v}^2}{\mathbf{c}^2}}$$

Since $L < L_o$, we say that the length of the object in O' has been contracted relative to the observed length in O. This phenomenon is called **length contraction** in the special theory of relativity.

SIMULTANEITY AND TIME DILATION

Two events are simultaneous to an observer if he experiences them as occurring at the same time in his frame of reference. In Galilean relativity, if one observer records simultaneous events, then all observers will likewise record simultaneous events. This is not true in special relativity.

Suppose we have two events measured by O' in two places. The time interval $\Delta t'$ in the O' frame will not be equal to the time interval Δt observed in O:

$$\Delta t = t_2 - t_1 = \frac{(t_2' - t_1') + (\mathbf{v}/\mathbf{c}^2)(X_2' - X_1')}{\sqrt{1 - (\mathbf{v}^2/\mathbf{c}^2)}}$$

If O' observes two events as being simultaneous, then $\Delta t'$ is equal to zero and we can write

$$\Delta t = \frac{(\mathbf{v}/\mathbf{c}^2)(X_2' - X_1')}{\sqrt{1 - (\mathbf{v}^2/\mathbf{c}^2)}}$$

Now if O' observes the simultaneous events occurring at the same spatial location in O', then $\Delta X'$ is equal to zero and both O and O' will observe simultaneous events (since $\Delta t = 0$). However, if $\Delta X' \neq 0$, then observer O will *not* record the events in O' as being simultaneous.

Special relativity allows for the fact that time is not an absolute quantity. Using Lorentz transformations, we can see that if Δt_0 is the "proper" time measured by a clock moving with observer O (in their frame of reference), then the observed time interval in O' is given by

$$\Delta t' = \frac{\Delta t_0}{\sqrt{1 - (\mathbf{v}^2/\mathbf{c}^2)}}$$

Since the denominator is less than 1, the time interval, as measured in O', is dilated. This phenomenon is known as **time dilation** in the special theory of relativity.

Confirmation of these predictions can be found in cosmic ray particles called **muons**, which are created in the nuclear collisions of particle accelerators on the earth or in the ionosphere and magnetosphere surrounding the earth. In a laboratory, muons have a lifetime of approximately 2.2 μs (2.2×10^{-6} s). At velocities of about 0.998**c**, a particle travels only 0.66 km before decaying. How is it possible for them to be detected on earth if they are created high in the atmosphere? The answer is time dilation. Using Einstein's formula and the velocity of muons as measured on earth, we find that while the frame of reference moving with a muon observes a lifetime of 2.2 μs, we on earth observe a lifetime of

$$t = \frac{2.2 \times 10^{-6}}{\sqrt{1 - (0.998\ \mathbf{c})^2/\mathbf{c}^2}} = 34.8 \times 10^{-6}\ \text{s} = 34.8\ \mu\text{s}$$

Sample Problem

Imagine a meter stick moving relative to the earth with a velocity of 0.85**c**. By what factor does its length appear to be contracted?

Solution

Using the length contraction formula, we have

$$\frac{L}{L_0} = \sqrt{1 - (0.85\ \mathbf{c})^2/\mathbf{c}^2} = 0.53$$

which is approximately one half its rest length.

RELATIVISTIC VELOCITY TRANSFORMATIONS

In Galilean relativity, if a person is walking toward the front of a moving train at 2 m/s (relative to the train) and the train has a velocity of 20 m/s relative to the earth, then the velocity of the person relative to the earth is 22 m/s. One might think that if the train turns on its headlights, then the velocity of the photons emitted will also be additive in the same way that the person's velocity is.

This observation, in fact, does not take place. The second postulate of relativity states that the velocity of light is invariant. It remains the same for all inertial observers. At velocities close to the velocity of light, the transformation equation that governs the relative velocity observed is given by

$$\mathbf{v}' = \frac{\mathbf{v} + \mathbf{u}}{1 + \mathbf{uv}/\mathbf{c}^2}$$

where $\mathbf{u}$ is the velocity of the frame of reference, $\mathbf{v}$ is the velocity relative to the frame of reference, and $\mathbf{v}'$ is the velocity observed by a observer. If the object were moving in the opposite direction, we would use $-\mathbf{v}$ instead of $\mathbf{v}$. This equation governs all bodies and is approximated by the familiar sum of velocities $\mathbf{u} + \mathbf{v}$ if $\mathbf{v} << \mathbf{c}$.

Sample Problem

What is the observed velocity of a photon emitted by the light of a spacecraft (in the direction of the craft's motion) that has a velocity of 0.95**c** relative to the earth?

Solution

We know that the answer should be equal to **c**, but let's prove it. In this problem, $\mathbf{v} = \mathbf{c}$ and $\mathbf{u} = 0.95\mathbf{c}$. This means that the observed velocity $\mathbf{v}'$ is given by

$$\mathbf{v}' = \frac{\mathbf{c} + 0.95\,\mathbf{c}}{1 + (0.95\ \mathbf{c})(\mathbf{c})/\mathbf{c}^2} = \frac{1.95\,\mathbf{c}}{1 + 0.95} = \mathbf{c}$$

RELATIVISTIC MASS AND ENERGY

When we introduced Newton's laws of motion, we discussed the meaning of inertia. **Inertia** is the property of matter that resists a net force changing its state of motion. In Newtonian mechanics, the inertia of an object is its *mass*. In all Newtonian interactions, the mass of an object remains the same unless something physical happens to it (as in an explosion or a rocket using up fuel).

In his paper on special relativity, Einstein demonstrated that inertial mass varies with velocity. The experimental evidence for this phenomenon is observed in the mass of high-speed electrons accelerated by magnetic fields. According to special relativity, the mass m varies according to the formula

$$m = \frac{m_0}{\sqrt{1 - \mathbf{v}^2/\mathbf{c}^2}}$$

where m_0 is the rest mass (a constant).

We can see now why it is impossible for a mass to be accelerated at the velocity of light in a vacuum. As the velocity $\mathbf{v}$ approaches the velocity of light $\mathbf{c}$, the mass of the object increases toward infinity and is undefined if $\mathbf{v} = \mathbf{c}$. The relativistic momentum, $\mathbf{p} = m\mathbf{v}$, can now be expressed as

$$\mathbf{p} = m\mathbf{v} = \frac{m_0\mathbf{v}}{\sqrt{1 - \mathbf{v}^2/\mathbf{c}^2}}$$

In his analysis of mechanics, Einstein also introduced the most famous equation of the twentieth century. Any object with a rest mass of m_0 has an equivalent rest energy given by

$$E_0 = m_0\mathbf{c}^2$$

This equation accounts for the binding energy factor of 931.5 MeV per atomic mass unit.

If the object is in motion, then it has an equivalent energy given by

$$E = m\mathbf{c}^2 = \frac{m_0\mathbf{c}^2}{\sqrt{1 - \mathbf{v}^2/\mathbf{c}^2}}$$

This energy is the sum of the object's kinetic energy and its rest energy. Thus, the kinetic energy is given by

$$\text{KE} = m\mathbf{c}^2 - m_0\mathbf{c}^2 = \frac{m_0\mathbf{c}^2}{\sqrt{1 - \mathbf{v}^2/\mathbf{c}^2}} - m_0\mathbf{c}^2$$

For low velocities, this equation reduces to the familiar classical expression (in approximation)

$$\text{KE} = \frac{1}{2} m_0\mathbf{v}^2$$

We can link the relativistic energy to the momentum in the following way. Recall that in classical mechanics, $\text{KE} = \frac{1}{2} m_0\mathbf{v}^2 = \mathbf{p}^2/2m_0$. Using the relativistic forms for the total energy E and the momentum $\mathbf{p}$, we can show that (see Problem 2 at the end of this chapter)

$$E^2 = \mathbf{p}^2\mathbf{c}^2 + (m_0\mathbf{c}^2)^2$$

Sample Problem

What is the rest energy for a 60-kg student?

Solution

We use the formula

$$E_0 = m_0\mathbf{c}^2 = (60)(9 \times 10^{16}) = 5.4 \times 10^{18} \text{ J}$$

PROBLEM-SOLVING STRATEGIES

Solving relativity problems can be somewhat confusing. The key is to remember the various frames of reference involved.

Be sure to follow the equations carefully and have them memorized for use on the examination. Try not to let your intuition about classical mechanics influence the concepts of relativity. Time and space are all relative concepts, and even the notion of simultaneity is suspect since time measurements depend on the frame of reference. Thus, two events that may appear to be simultaneous in one frame may not be simultaneous in another.

PRACTICE PROBLEMS FOR CHAPTER 27

Thought Problems

1. The special theory of relativity restricts frames of reference to motion at a constant velocity. What would be the implication of an accelerated frame of reference?
2. An object undergoing uniform circular motion has an average velocity equal to zero (assuming that it is averaged over an integral multiple of periods). Would you expect there to be any relativistic mass increase in this case?
3. How is the density of an object affected by special relativity?

Multiple-Choice Problems

1. Which of the following statements is a postulate of special relativity?
   (A) All motion is relative.
   (B) Objects can never go faster than the velocity of light in any medium.
   (C) Accelerated frames of reference produce fictitious forces.
   (D) The laws of physics are the same in all inertial frames of reference.
   (E) All laws of physics reduce to Newton's laws at low velocity.
2. What is the perceived length of a meter stick if it is moving with a velocity of 0.5**c** relative to the earth?
   (A) 0.87 m
   (B) 0.75 m
   (C) 0.50 m
   (D) 0.25 m
   (E) 0.35 m

3. What is the energy equivalent of an electron at rest?
   (A) 931.5 MeV
   (B) 0.017 MeV
   (C) 0.512 MeV
   (D) 535 MeV
   (E) 675 MeV
4. The observed mass of a moving object is three times its rest mass. What is its velocity relative to the observer?
   (A) $(2\sqrt{2})\mathbf{c}/3$
   (B) $\mathbf{c}$
   (C) $3\mathbf{c}$
   (D) $(2\sqrt{2})\mathbf{c}$
   (E) $\mathbf{c}/3$
5. An observer is moving with a velocity of 0.95**c** in a direction perpendicular to a rod of length *L*. The observer will measure the length of the rod to be _____.
   (A) equal to *L*
   (B) less than *L*
   (C) greater than *L*
   (D) zero
   (E) maybe more or less than *L* depending on the velocity
6. A subatomic particle called a *pion* has an observed laboratory lifetime of 2.6×10^{-8} s. If a pion is observed to have a velocity of 0.93**c** relative to the laboratory, then by what factor is the lifetime of the pion observed to increase?
   (A) 1.35
   (B) 3.6
   (C) 1.75
   (D) 2.72
   (E) 0.56
7. An observer on the earth sees a cosmic particle coming toward him at a velocity of 0.96**c** when it is at an altitude of 1,850 m. What is the altitude of the particle as measured from a frame of reference moving with the particle?
   (A) 130 m
   (B) 518 m
   (C) 768 m
   (D) 6,607 m
   (E) 9,833 m
8. A pendulum is set swinging inside a spaceship moving at a velocity of 0.91**c** relative to an observer. The period is measured to be 2.5 s inside the spaceship. What is the period observed by a person on the earth?
   (A) 3 s
   (B) 6.03 s

(C) 1.03 s
(D) 4.27 s
(E) 2.3 s

9. A person is standing on a platform that is moving with a velocity of 0.75**c** relative to an observer. The person throws a ball in the forward direction with a velocity of 0.60**c** relative to herself. What is the velocity of the ball relative to the observer?
(A) 0.15**c**
(B) 1.35**c**
(C) 0.93**c**
(D) 0.82**c**
(E) 0.75**c**

10. For the situation described in Problem 9, if the mass of the ball relative to the thrower is 150 g, then what is the mass relative to the observer?
(A) 408 g
(B) 187.5 g
(C) 226.7 g
(D) 334.2 g
(E) 150 g

Free-Response Problems

1. (a) What is the relativistic kinetic energy of an electron moving with a velocity of 0.9**c**?
(b) Use the binomial expansion series to show that for low velocities the relativistic formula for kinetic energy reduces (as an approximation) to the classical expression $\frac{1}{2} m_0\mathbf{v}^2$.

2. (a) Starting with the expressions for the relativistic total energy E and momentum $\mathbf{p}$, show that $E^2 = \mathbf{p}^2\mathbf{c}^2 + (m_0\mathbf{c}^2)^2$.
(b) If a photon has no rest mass, derive an expression for the momentum of the photon in terms of its energy E.
(c) Explain why the expression $E^2 - \mathbf{p}^2\mathbf{c}^2$ is a relativistic invariant (i.e., it has the same value in any frame of reference), while the quantities E and $\mathbf{p}$ are not.

3. A proton has a kinetic energy of 3.5×10^{-11} J.
(a) What is the velocity of the proton?
(b) By what factor has the mass of the proton been increased?

4. An observer in a frame O records that two events in their frame are linearly separated in space by 750 m and separated in time by 5×10^{-7} s. How fast must an observer O' be traveling (as a fraction of the speed of light **c**) so that the two events appear to be simultaneous?

SOLUTIONS TO PRACTICE PROBLEMS

Thought Problems

1. In an accelerated frame of reference, there appears to be another force in the direction opposite the acceleration. This inertial force is equivalent to gravity. In the general theory of relativity, there is an equivalence principle which states in part that accelerated frames of reference are indistinguishable from gravitational fields.

2. The relativistic mass formula depends on the instantaneous velocity of the object and not the average velocity. Thus, even though an object undergoing uniform circular motion has zero average velocity (assuming it has been averaged over an integral number of periods), it does not have zero instantaneous velocity. There will therefore be an observed mass increase in this case.

3. Since density is directly dependent on the mass of an object, the observed density of an object increases with motion as well as the observed mass. There is also a volume decrease with length contraction.

Multiple-Choice Problems

1. **D** "The laws of physics are the same in all inertial frames of reference" is one of Einstein's postulates of relativity.

2. **A** We use the length contraction formula with $L_0 = 1.0$ m:

$$L = L_0\sqrt{1 - \frac{\mathbf{v}^2}{\mathbf{c}^2}} = 1.0\sqrt{1 - \frac{(0.5\mathbf{c})^2}{\mathbf{c}^2}} = 0.87 \text{ m}$$

3. **C** We use the mass-energy relationship for a stationary electron:

$$E_0 = m_0\mathbf{c}^2 = (9.1 \times 10^{-31})(9 \times 10^{16}) = 8.14 \times 10^{-14} \text{ J} = 0.512 \text{ MeV}$$

4. **A** We use the relativistic mass relationship

$$m = \frac{m_0}{\sqrt{1 - \mathbf{v}^2/\mathbf{c}^2}}$$

$$3m_0 = \frac{m_0}{\sqrt{1 - \mathbf{v}^2/\mathbf{c}^2}}$$

$$\sqrt{1 - \frac{\mathbf{v}^2}{\mathbf{c}^2}} = \frac{1}{3}$$

Now square both sides and solve for the velocity **v**:

$$1 - \frac{\mathbf{v}^2}{\mathbf{c}^2} = \frac{1}{9}$$

$$\frac{\mathbf{v}^2}{\mathbf{c}^2} = \frac{8}{9}$$

$$\mathbf{v} = \frac{(2\sqrt{2})\mathbf{c}}{3}$$

5. **A** If the rod is laid along the x axis and the motion is in the y direction, there will be no observed changes in length.
6. **D** We use the time dilation formula and solve for the ratio t/t_0:

$$\frac{t}{t_0} = \frac{1}{\sqrt{1 - (0.93\mathbf{c})^2/\mathbf{c}^2}} = 2.72$$

7. **B** We use the length contraction formula since an observer moving with the particle will perceive a shorter distance:

$$L = 1{,}850\sqrt{1 - \frac{(0.96\mathbf{c})^2}{\mathbf{c}^2}} = 518 \text{ m}$$

8. **B** We use the time dilation formula to measure the observed period:

$$t = \frac{2.5}{\sqrt{1 - (0.91\mathbf{c})^2/\mathbf{c}^2}} = 6.03 \text{ s}$$

9. **C** We use the relativistic velocity formula for both objects moving in the same direction relative to the observer:

$$\mathbf{v} = \frac{0.75\mathbf{c} + 0.60\mathbf{c}}{1 + (0.75\mathbf{c})(0.60\mathbf{c})/\mathbf{c}^2} = 0.93\mathbf{c}$$

10. **A** We need to use the relativistic mass formula but with the relative velocity measured in Problem 9 ($\mathbf{v} = 0.93\mathbf{c}$):

$$m = \frac{150}{\sqrt{1 - (0.93\mathbf{c})^2/\mathbf{c}^2}} = 408 \text{ g}$$

Free-Response Problems

1. (a) To find the kinetic energy of a moving electron, we use the formula

$$\text{KE} = \frac{m_0\mathbf{c}^2}{\sqrt{1 - \mathbf{v}^2/\mathbf{c}^2}} - m_0\mathbf{c}^2$$

$$= \frac{(9.1 \times 10^{-31})(9 \times 10^{16})}{\sqrt{1 - (0.9/\mathbf{c})^2/\mathbf{c}^2}} - (9/.1 \times 10^{-31})(9 \times 10^{16}) = 1.06 \times 10^{-13} \text{ J}$$

(b) To show what happens for low velocities, $\mathbf{v} << \mathbf{c}$, we use the binomial expansion

$$(1 - x^2)^{-1/2} \approx 1 + \frac{x^2}{2} + \cdots \quad (x \ll 1)$$

In our expression, $x^2 = \mathbf{v}^2/\mathbf{c}^2$, and if $\mathbf{v} << \mathbf{c}$, we can write

$$\left(1 - \frac{\mathbf{v}^2}{\mathbf{c}^2}\right)^{-1/2} \approx 1 + \frac{\mathbf{v}^2}{2\mathbf{c}^2}$$

$$\mathrm{KE} \approx m_0\mathbf{c}^2\left(1 + \frac{\mathbf{v}^2}{2\mathbf{c}^2}\right) - m_0\mathbf{c}^2 \approx 1/2 m_0\mathbf{v}^2$$

2. (a) The relativistic expressions for energy and momentum are

$$E = \frac{m_0\mathbf{c}^2}{\sqrt{1 - \mathbf{v}^2/\mathbf{c}^2}}$$

$$\mathbf{p} = \frac{m_0\mathbf{v}}{\sqrt{1 - \mathbf{v}^2/\mathbf{c}^2}}$$

Now we need to square both the E and $\mathbf{pc}$ terms:

$$E^2 = \frac{(m_0\mathbf{c}^2)^2}{1 - \mathbf{v}^2/\mathbf{c}^2}$$

$$\mathbf{p}^2\mathbf{c}^2 = \frac{m_0^2\mathbf{c}^2\mathbf{v}^2}{1 - \mathbf{v}^2/\mathbf{c}^2}$$

To the second equation we need to add the quantity $(m_0\mathbf{c}^2)^2$:

$$\mathbf{p}^2\mathbf{c}^2 + (m_0\mathbf{c}^2)^2 = \frac{m_0^2\mathbf{c}^2\mathbf{v}^2}{1 - \mathbf{v}^2/\mathbf{c}^2} + (m_0\mathbf{c}^2)^2$$

$$= \frac{m_0^2\mathbf{c}^2\mathbf{v}^2}{1 - \mathbf{v}^2/\mathbf{c}^2} + \frac{(m_0\mathbf{c}^2)^2(1 - \mathbf{v}^2/\mathbf{c}^2)}{1 - \mathbf{v}^2/\mathbf{c}^2}$$

$$= \frac{m_0\mathbf{c}^2\mathbf{v}^2 + (m_0\mathbf{c}^2)^2(1 - \mathbf{v}^2/\mathbf{c}^2)}{1 - \mathbf{v}^2/\mathbf{c}^2}$$

$$= \frac{(m_0\mathbf{c}^2)^2}{1 - \mathbf{v}^2/\mathbf{c}^2} = E^2$$

(b) If the particle has no rest mass, then the preceding expression reduces to $E = \mathbf{p} \cdot \mathbf{c}$. For a photon, this is consistent with the expression $\mathbf{p} = h/\lambda$ since $E = hf$.

(c) The energy and momentum are relativistic quantities that depend on the frame of reference and its relative velocity. However, by verifying that the expression $E^2 = \mathbf{p}^2 \cdot \mathbf{c}^2 + (m_0\mathbf{c}^2)^2$ is valid, the expression $E^2 - \mathbf{p}^2 \cdot \mathbf{c}^2$ depends only on the rest mass (which is the same in any frame of reference) and the square of the velocity of light (which is of course an invariant).

3. (a) To find the velocity of the proton, we need to use the relativistic kinetic energy formula

$$\text{KE} = \frac{m_0\mathbf{c}^2}{\sqrt{1 - \mathbf{v}^2/\mathbf{c}^2}} - m_0\mathbf{c}^2$$

$$3.5 \times 10^{-11} = \frac{(1.67 \times 10^{-27})(9 \times 10^{-16})}{\sqrt{1 - \mathbf{v}^2/\mathbf{c}^2}} - (1.67 \times 10^{-27})(9 \times 10^{16})$$

$$1.853 \times 10^{-10} = \frac{1.503 \times 10^{-10}}{\sqrt{1 - \mathbf{v}^2/\mathbf{c}^2}}$$

$$\mathbf{v} = 0.585\mathbf{c}$$

(b) Now that we know the velocity, we can use the relativistic mass formula to find the ratio of mass increase:

$$m = \frac{m_0}{\sqrt{1 - \mathbf{v}^2/\mathbf{c}^2}}$$

$$\frac{m}{m_0} = \frac{1}{\sqrt{1 - (0.585\mathbf{c})^2/\mathbf{c}^2}} = 1.23$$

The proton's mass increases by 1.23 times.

4. We use the Lorentz transformation for time to express $\Delta t'$ in terms of unprimed quantities:

$$\Delta t' = \frac{(t_2 - t_1) - (\mathbf{v}/\mathbf{c}^2)(X_2 - X_1)}{\sqrt{1 - (\mathbf{v}^2/\mathbf{c}^2)}}$$

Now, if O' observes the events to be simultaneous, then $\Delta t'$ equals zero:

$$0 = \frac{5 \times 10^{-7} - (\mathbf{v}/\mathbf{c})(750)}{3 \times 10^8}$$

$$\frac{\mathbf{v}}{\mathbf{c}} = 0.2$$

$$\mathbf{v} = 0.2\mathbf{c}$$

SUPPLEMENTARY PROBLEM SETS

In each supplementary problem set there are 25 short-answer exercises as well as 6 additional free-response problems that are more difficult. These problem sets can serve as additional practice for general review or final examinations. They cover a wide array of areas in physics and are not meant to be all-inclusive. Topics therefore vary from set to set. Full solutions and explanations are provided after each problem set.

PROBLEM SET 1

Exercises

1. A rock is thrown horizontally off the top of a building that is 7.5 m above the ground. It is observed that the rock lands 22 m away from the base of the building. With what velocity was the rock thrown?

2. A student swings a 3.5-g rubber stopper on a string in a horizontal circle over her head. The length of the string is 0.6 m. The stopper is observed to complete 10 revolutions in 11.7 s. What is the tangential velocity of the stopper at any time *t*?

3. A projectile is launched at a 25-degree angle to the ground. The projectile lands 1,650 m away along a flat plane. What was the initial velocity of the projectile?

4. In the situation shown, a 2-kg mass is attached by a light string over a frictionless pulley to a 5-kg mass hanging below. The 2-kg mass rests on a frictionless surface. If the system is released, what will be the acceleration of both masses?

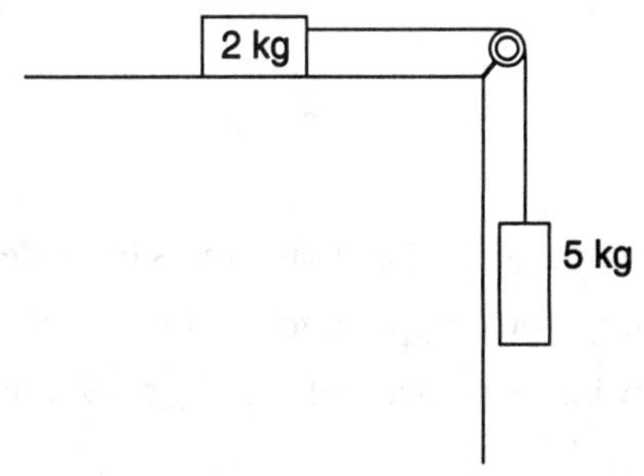

5. A 45-kg boy is sliding down a frictionless hemispherical surface with radius R as shown in the diagram. The boy loses contact with the surface when he makes an angle θ with the horizontal. What is the magnitude of the normal force acting on the boy at that instant?

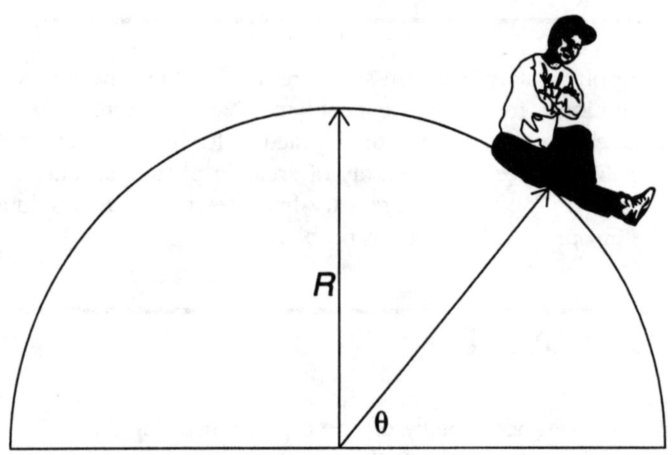

6. A crate weighing 15 N is moving along a horizontal rough surface with a constant velocity **v**. The coefficient of kinetic friction between the crate and the surface is 0.17. What is the magnitude of the horizontal force maintaining the constant velocity?

Questions 7 through 9 are based on the plot of force versus time shown here for a 15-kg mass.

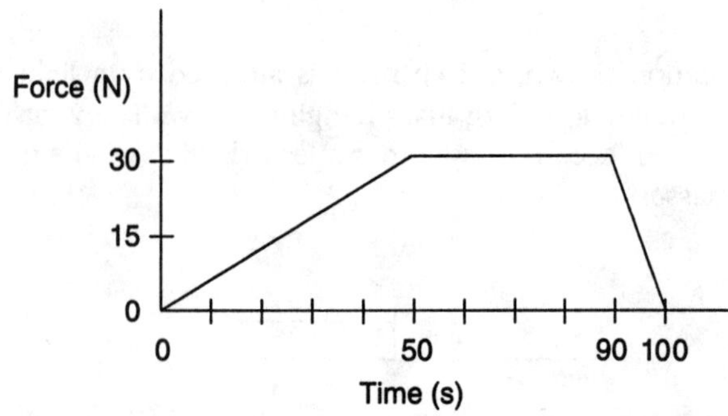

7. What was the total impulse applied to the mass for the entire 100-s interval?
8. What was the average force applied to the mass during the first 50 s?
9. If the mass had an initial velocity of 4 m/s, what was its velocity at the end of 50 s?

10. A 10-kg object has a velocity of 2 m/s to the right. It is struck by a 0.05-kg wad of putty moving to the left at 10 m/s, and the putty sticks to the object. What percentage of initial kinetic energy is lost in this inelastic collision?

11. A spring with a force constant of $k = 500$ N/m is compressed horizontally by a 10-kg mass along a rough surface as shown. The spring is compressed 0.3 m, and the coefficient of kinetic friction between the mass and the surface is 0.15. If the system is released, what will be the velocity of the mass when the spring is no longer compressed?

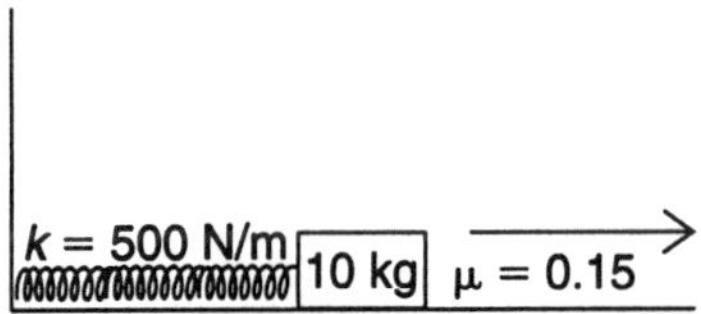

12. A meter stick is balanced at the 50-cm mark as shown, and four masses are suspended at the positions indicated. What is the magnitude of the missing mass?

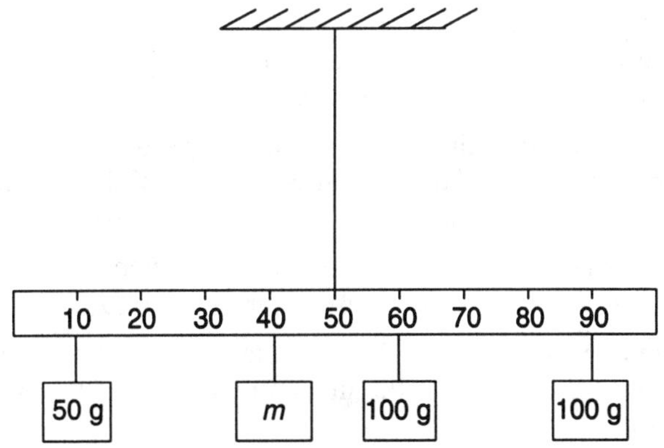

Questions 13 and 14 are based on the following information: A wheel is rolling to the right with a constant velocity **v**.

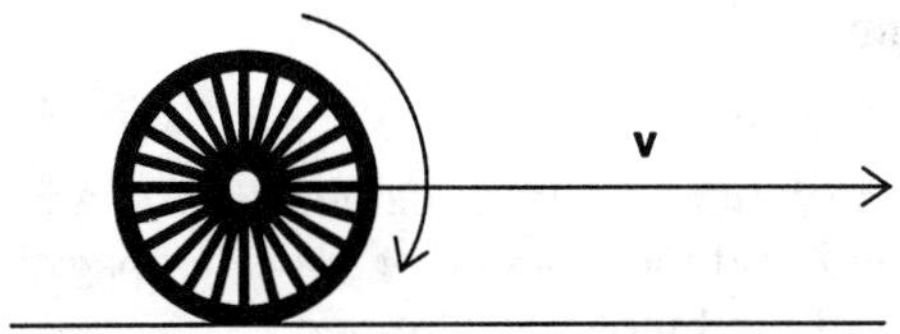

13. What is the direction of the angular velocity vector **ω**?
14. If the radius of the wheel is 0.45 m and the velocity is 5 m/s, what is the magnitude of the angular velocity **ω**?
15. In the accompanying diagram, a meter stick is attached to the wall with a pivot that it is free to rotate. If the meter stick is initially horizontal and at rest and is then released, what will be the magnitude of the initial angular acceleration of the meter stick? (The moment of inertia for the stick about this axis is equal to $\frac{1}{3}\ mL^2$.)

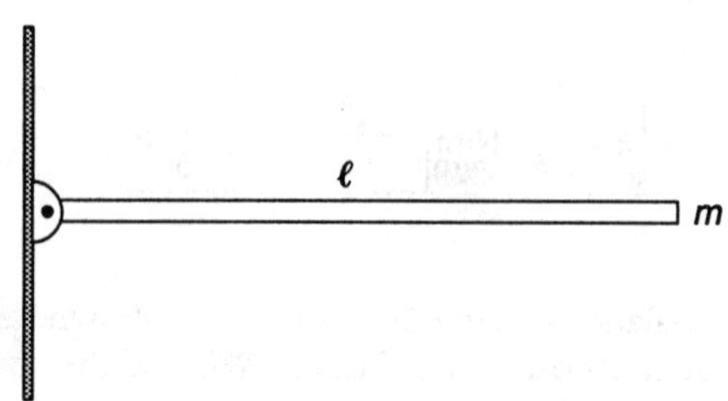

16. An insulated tank contains 2.5 m^3 of an ideal gas under a pressure of 0.5 atm. If the pressure is raised to 3 atm while the temperature remains constant, what will be the new volume of the gas?
17. An insulated tank contains carbon dioxide gas (CO_2 = 28 g/mol) and nitrogen gas (N_2 = 14 g/mol). What is the ratio of the root mean square velocity of the nitrogen molecules to the carbon dioxide molecules?
18. An ideal gas is kept under constant pressure at a temperature of 50°C. What change in Celsius temperature must occur in order to double its volume?
19. An insulated container holds 50 g of ice at 0°C. Water is added at 20°C. If all the ice is to melt, what is the minimum amount of water necessary to achieve this?

Questions 20 and 21 are based on the following information: A mass is attached to a horizontal spring with a force constant k and oscillates with simple harmonic motion. The amplitude of the motion is 0.2 m, and the period is 0.5 s.

20. Derive an expression that represents the displacement of the mass as a function of time.
21. What is the magnitude of the force constant k if a 0.4-kg mass is used?
22. A taut string is vibrating in its fundamental resonance mode. If the length of the string is L and the velocity of wave propagation is **v**, then what is the frequency of the fundamental mode?

23. What is the equivalent resistance of the circuit shown here?

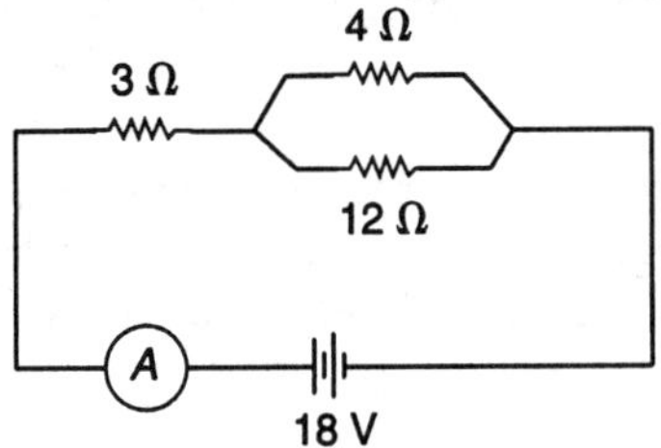

24. Which of the following electric circuits has the same equivalent capacitance?

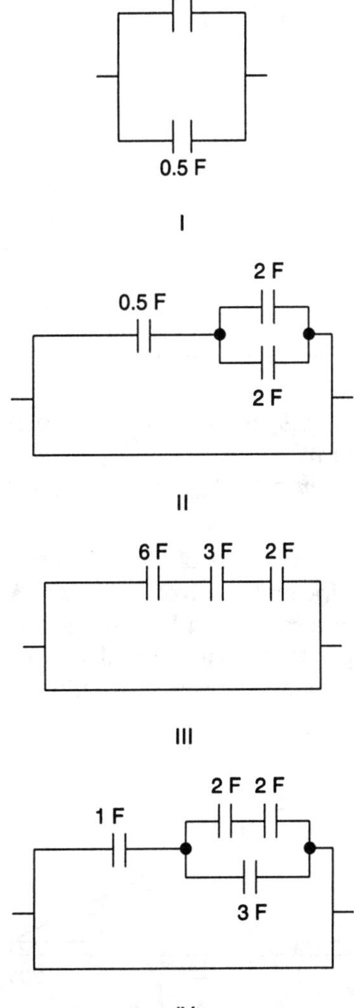

25. What is the momentum of a photon that has a wavelength of 1,000 Å?

Free-Response Problems

1. In the arrangement shown, a spring with a force constant k is attached to a vertical wall. The other end of the spring is attached to a mass m resting on a rough horizontal surface. The coefficient of kinetic friction for the mass and the surface is μ. A light string is attached to m and passes over a frictionless pulley to an overhanging mass M. The system is initially at rest. If the mass M is released, it will fall a distance h before coming to rest.

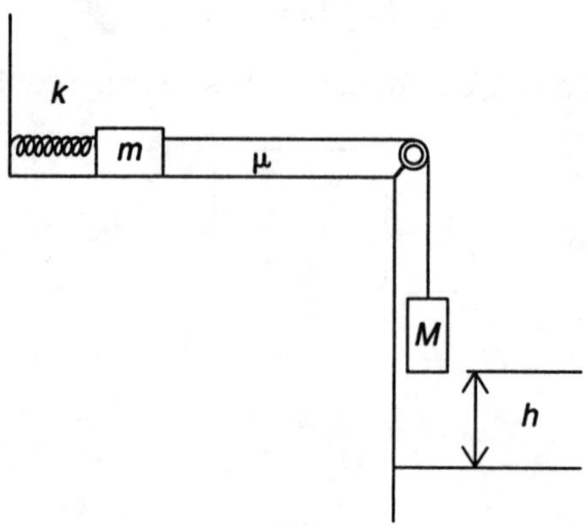

   (a) Write an expression for the conservation of energy as it applies to this situation.
   (b) In terms of m, M, g, h, and k, derive an expression for the coefficient of kinetic friction μ.
   (c) If $k = 1{,}000$ N/m, $m = 0.5$ kg, $M = 1.5$ kg, and $h = 0.04$ m, calculate the value of μ.

2. A uniform rigid rod with a mass of 20 kg and a length L is attached to a wall by means of a frictionless pivot as shown. A mass of 5 kg is hung from the middle of the rod, and a light string is attached to the free end of the rod. The string is attached to the wall, making an angle of 25 degrees with the horizontal. The system is in equilibrium.

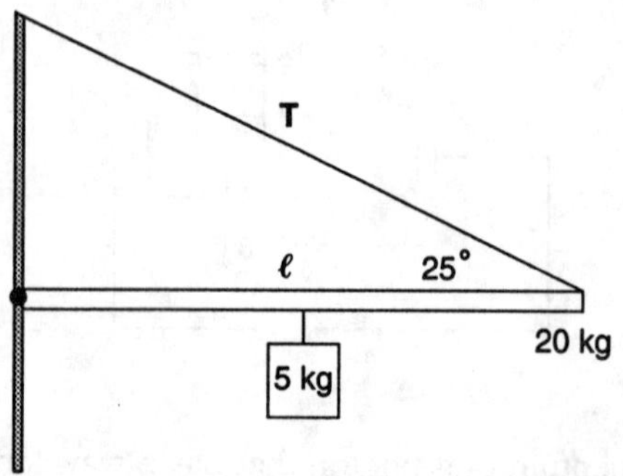

(a) Draw a free body diagram for this situation and label all the forces involved.
(b) Determine the magnitude of the tension **T**.
(c) Determine the magnitude and direction of the reaction force **R** acting at the pivot.

3. A 3-m^3 sample of argon gas is at a temperature of 40°C in an insulated container. The pressure of the gas is 2 atm.
   (a) How many moles of argon gas are present?
   (b) How much energy is required to raise the temperature to 60°C if no work is to be done by the gas?
   (c) What is the change in internal energy for the gas?

4. For the electric circuit shown:

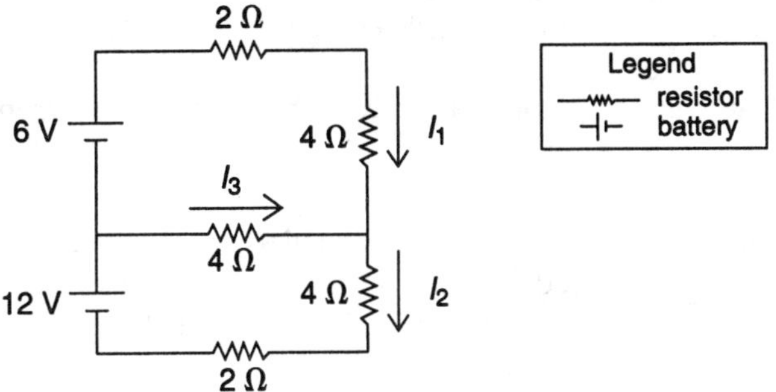

   (a) State Kirchoff's rules for electric circuits.
   (b) Determine the magnitude of each current identified in the circuit.

5. An electron is accelerated across a 1,200-V potential difference as shown in the accompanying diagram. On reaching the other side, the electron enters a downward electric field of uniform strength **E** = 1,000 N/C. The length of the second region containing the electric field is 1.2 m.

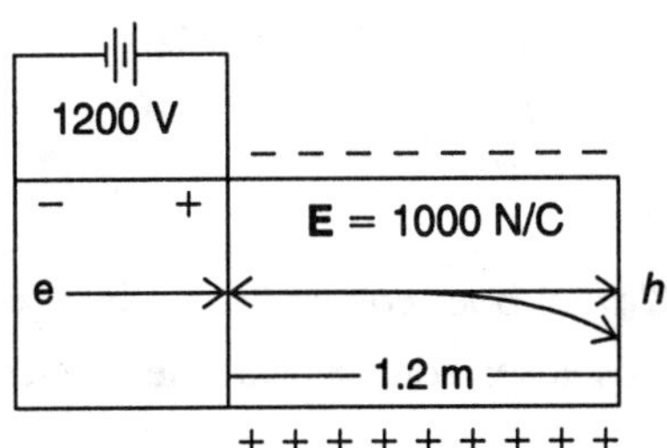

   (a) How much work is done in moving the electron across the potential difference region?
   (b) What is the velocity of the electron as it enters the electric field region?

(c) Neglecting any gravitational effects, the electron is deflected away from its straight path. Derive an expression for the deflected distance h in terms of the mass of the electron, its charge, the length of the region L, and its entrance velocity **v**. Assume constant horizontal velocity.

(d) Using the known values of each quantity, calculate the magnitude of the deflected height h.

6. The threshold frequency for a photoelectric metal is 2.5×10^{15} Hz. Electromagnetic radiation with a wavelength of 500 Å is incident on the surface of the metal.
   (a) What is the frequency of the radiation photons?
   (b) If the total energy of the radiation is equal to 9.945×10^{-16} J, how many photons are associated with this radiation?
   (c) What is the work function for this metal?
   (d) What is the maximum kinetic energy for the emitted electrons in joules and electronvolts?
   (e) What is the stopping potential for these emitted electrons?

Solutions to Exercises

1. First, we must determine the time required to fall 7.5 m. This will be the same as the time needed to go out the 22-m distance. The time is given by $t = \sqrt{2y/\mathbf{g}} = \sqrt{15/9.8} = 1.24$ s. The horizontal velocity **v** is therefore equal to 22 m/1.24 s = 17.74 m/s.

2. The tangential velocity **v** is given by $\mathbf{v} = \boldsymbol{\omega}\, r = 2\pi f r = (6.28)(10/11.7)(0.6) = 3.22$ m/s.

3. The formula for the range of the projectile is $R = \mathbf{v}_0^2 \sin 2\,\theta/\mathbf{g}$. Substituting the given values yields a value of 145 m/s for the initial velocity.

4. To find the acceleration of both masses (since they are coupled) we set up Newton's second law of motion in both directions. The tension in the string caused by the weight of the hanging mass is the same tension that accelerates the 2-kg mass to the right. In the y direction we can write

$$\mathbf{T} - (5)(9.8) = -5\mathbf{a}$$

   In the x direction we can write

$$\mathbf{T} = 2\mathbf{a}$$

   Thus, eliminating **T** from both equations, we have $\mathbf{a} = 7$ m/s^2.

5. At the instant the boy loses contact with the surface, the normal force is zero.

6. The force necessary to maintain a constant velocity is equal to the force of kinetic friction and is given by $\mathbf{f} = \mu\mathbf{N} = (0.17)(15) = 2.55$ N.

7. The total impulse for the entire 100-s interval is given by the total area which equals 2,100 N-s.

8. For the first 50 s, the impulse applied was to be equal to 750 N-s by calculating the area of that triangular segment. The average force is equal to the impulse divided by the time of 50 s and is 15 N.

9. During the 50-s interval, the impulse was equal to 750 N-s. This is equal to the change in momentum given by $m\,\Delta\mathbf{v}$. Since the mass is equal to 15 kg, the velocity changed by 50 m/s. Since the force was directed to accelerate the mass, the new velocity is 54 m/s.

10. The initial kinetic energy of the block is 20 J, and the initial kinetic energy of the putty is 2.5 J. The total initial kinetic energy is therefore 22.5 J. To find the kinetic energy after the collision, we need to know the final velocity. This is an inelastic collision, but momentum is still conserved. Therefore,

$$(10)(2) - (0.05)(10) = (10 + 0.05)\mathbf{v}$$
$$\mathbf{v} = 1.94 \text{ m/s}$$

The final kinetic energy of the combined 10.05-kg mass is thus 18.9 J. The percentage of energy lost is found to be 16% since 18.9/22.5 = 0.84.

11. Conservation of energy requires that the potential energy stored in the spring be converted into kinetic energy and the work done against friction:

$$\frac{1}{2}kx^2 = \frac{1}{2}m\mathbf{v}^2 + W_f = \frac{1}{2}m\mathbf{v}^2 + \mu m\mathbf{g}x$$

Substituting the given values yields $\mathbf{v}$ = 1.9 m/s.

12. We need to show that the torques on both sides of the balance point are equal. Since all masses would be multiplied by $\mathbf{g}$, we can leave it out without any loss of generality: (50)(40) + m(10) = (100)(10) + (100)(40), which implies that m = 300 g.

13. According to the right-hand rule for rotation, the direction of the angular velocity vector $\boldsymbol{\omega}$ is into the page.

14. We use the formula $\boldsymbol{\omega} = \mathbf{v}/r$ = 5/0.45 = 11 rad/s.

15. The torque $\boldsymbol{\tau} = I\boldsymbol{\alpha}$, where I is the moment of inertia for the given axis and $\boldsymbol{\alpha}$ is the angular acceleration. Thus,

$$\frac{M\mathbf{g}L}{2} = \frac{1}{3}ML^2\alpha$$
$$\alpha = \frac{3}{2}\mathbf{g}L$$

where L = 1 m.

16. At constant temperature, we use Boyle's law for an ideal gas:

$$P_1V_1 = P_2V_2$$
$$(0.5)(2.5) = (3)V_2$$
$$V_2 = 0.42 \text{ m}^3$$

17. The ratio of the root mean square velocities is equal to the square root of the inverse ratio of molecular masses: $\mathbf{v}_{\text{ratio}} = \sqrt{28/14} = \sqrt{2}$.

18. First, we must change the 50° Celsius to 323 K. To double the volume at constant pressure, we use Charles's law and simply double the Kelvin temperature to 646 K. This corresponds to a new temperature of 373°C. Thus, we must change the temperature by 323°C to double the volume.

19. In order to melt ice at 0°C, we must add 80 cal/g. Thus, 4,000 cal of heat must be supplied by the warm water. Since the water is at 20°C, we must have 200 g of water (since the specific heat of water is 1 cal/g°C).

20. We have as a general expression $x = A \sin \omega t$, where A is the amplitude in meters and $\omega = 2\pi/T$. Thus, $\omega = 4\pi$ in this case and $A = 0.2$ m.

21. In simple harmonic motion, $\omega^2 = k/m$. Substituting yields $k = 63$ N/m.

22. Since $\mathbf{v} = f\lambda$, $f = \mathbf{v}/\lambda$. For a string vibrating in its fundamental mode, $\lambda = 2L$.

23. We first reduce the resistance in the parallel branch. $1/R = \frac{1}{4} + \frac{1}{12}$ and $R = 3\ \Omega$. This resistance is added, in series, to the other 3-Ω resistor, making a total of 6 Ω.

24. The equivalent capacitance of both I and III is 1 F.

25. We use the formula $\mathbf{p} = h/\lambda$. Thus, $\mathbf{p} = (6.63 \times 10^{-34})/(1 \times 10^{-7}) = 6.63 \times 10^{-27}$ kg-m/s.

Solutions to Free-Response Problems

1. (a) In this situation the potential energy lost by the falling mass M is converted into stretching the spring and moving the mass m horizontally by doing work against friction. Thus, ΔGPE = ΔEPE + W (friction), where GPE = gravitational potential energy and EPE = elastic potential energy.

   (b) We write out each of the terms in part (a):

$$\frac{1}{2}kh^2 = Mgh + \mu mgh$$

$$\mu = \frac{\frac{1}{2}kh - Mg}{mg}$$

   Notice that the elongation of the spring $\mathbf{x}$ is equal to the downward displacement h.

(c) We make all the substitutions with the given values to obtain

$$\mu = \frac{(0.5)(1{,}000\ \text{N/m})(0.04\ \text{m}) - (1.5\ \text{kg})(9.8\ \text{m/s}^2)}{(0.5\ \text{kg})(9.8\ \text{m/s}^2)} = 1.08$$

2. (a) The free body diagram is shown here.

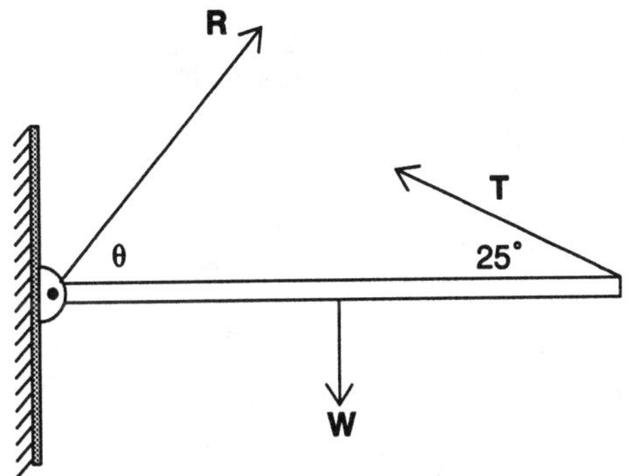

We take the action of the weight of the rod to be at its center of mass ($L/2$). There is also the weight of the 5-kg mass acting at the center of the rod, making for a combined weight **W**.

(b) To find the tension **T** we choose the equilibrium point to be about the pivot. The reaction force **R** does not contribute to any torques about this point. The upward component of the tension, **T** sin θ, is a counterclockwise torque acting at a length L. The total weight, **W** = (25 kg)(9.8 m/s²) = 245 N, acts at the distance $L/2$ and is a clockwise torque. Thus,

$$\Sigma\tau = 0$$

$$(\mathbf{T}\sin\theta)L = \frac{\mathbf{W}L}{2}$$

$$\mathbf{T}\sin 25 = 245\ \text{N}/2$$

$$\mathbf{T} = 290\ \text{N}$$

(c) To find the reaction force, note that **R** has perpendicular components $\mathbf{R}_x$ and $\mathbf{R}_y$. Now, consider the torques about the far right edge of the rod. The tension **T** and $\mathbf{R}_x$ do not contribute to any torques about this axis. However, the total weight **W** acts at a distance of $L/2$ from the right edge, and the component force $\mathbf{R}_y$ acts at a distance L. In equilibrium, the sum of these torques must be equal to zero:

$$\Sigma\tau = 0$$

$$\mathbf{R}_y L = \frac{\mathbf{W}L}{2}$$

$$\mathbf{R}_y = 245\ \text{N}/2 = 122.5\ \text{N}$$

To find $\mathbf{R}_x$, we note that in equilibrium the sum of all forces must equal zero as well: $\Sigma \mathbf{F} = 0$. This means that $\mathbf{R}_x$ must balance the horizontal component of the tension, which equals $\mathbf{T} \cos \theta$:

$$\mathbf{R}_x = (290 \text{ N})(\cos 25) = 262.83 \text{ N}$$

The magnitude of the reaction force is given by the Pythagorean theorem:

$$\mathbf{R} = \sqrt{(122.5 \text{ N})^2 + (262.83 \text{ N})^2} = 290 \text{ N}$$

The direction of $\mathbf{R}$ is given by

$$\tan \theta = \frac{\mathbf{R}_y}{\mathbf{R}_x} = \frac{122.5 \text{ N}}{262.83 \text{ N}} = 0.466$$

$$\theta = 25 \text{ degrees}$$

3. (a) The number of moles n is given by

$$n = \frac{PV}{RT} = \frac{(2)(101{,}000)(3)}{(8.31)(313)} = 232.98 \text{ mol}$$

(b) From the first law of thermodynamics, $\Delta Q = \Delta U + \Delta W$. If no work is done by the gas, then $\Delta W = 0$ and $\Delta Q = \frac{3}{2} n R \Delta T$ (where $\Delta T = 20$ K in this case).

$$\Delta Q = (1.5)(233.87 \text{ mol})[(8.31 \text{ J/(mol–K)}](20 \text{ K}) = 58{,}303.79 \text{ J}$$

(c) Since no work is done by the gas, $\Delta U = \Delta Q = 58{,}303.79$ J

4. (a) (i) The sum of any currents entering a junction in an electric circuit must equal the sum of the currents leaving that junction.
(ii) The algebraic sum of the changes in potential across any element in a closed loop must equal zero.

(b) Applying Kirchoff's rules to the circuit given, we take the two given loops and move clockwise around them using the conventions discussed in the text:

$$I_2 = I_1 + I_3$$
$$6 - 6I_1 + 4I_3 = 0$$
$$12 - 4I_3 - 6I_2 = 0$$

Then we substitute for I_2 in the third equation to obtain

$$6 - 6I_1 + 4I_3 = 0$$
$$12 - 6(I_1 + I_3) - 4I_3 = 0$$

Eliminating I_1 from the two equations yields $I_3 = 0.429$ A. Substitution back into either equation yields $I_1 = 1.286$ A. Finally, substitution back into the first equation yields $I_2 = 1.714$ A.

5. (a) The work done is equal to the product of the electron's charge and the potential difference:

$$W = \text{eV} = (1.6 \times 10^{-19}\ \text{C})(1{,}200\ \text{J/C}) = 1.92 \times 10^{-16}\ \text{J}$$

(b) The work done is equal to the change in the kinetic energy. Thus,

$$\mathbf{v} = \sqrt{\frac{2\text{KE}}{m}} = \sqrt{\frac{(2)(1.92 \times 10^{-16}\ \text{J})}{9.1 \times 10^{-31}\ \text{kg}}} = 2.05 \times 10^{7}\,\text{m/s}$$

(c) Since the electron enters the electric field with an assumed constant horizontal velocity, the downward acceleration will cause a parabolic projectile path. Thus, since the electric field is uniform, we know that $\mathbf{F} = e\mathbf{E}$ (where e is the magnitude of the electron charge). Also, we know that $\mathbf{F} = m\mathbf{a}$, and thus, $\mathbf{a} = e\mathbf{E}/m$. Now, with a constant horizontal velocity $\mathbf{v}$ and a distance L, the time within the electric field is given by $L/\mathbf{v}$. This is the same time needed to "fall" the deflected distance h:

$$h = \frac{1}{2}\mathbf{at}^2 = \frac{e\mathbf{E}L^2}{2m\mathbf{v}^2}$$

(d) Using the known values from the problem, we find that h is equal to

$$h = \frac{e\mathbf{E}L^2}{2m\mathbf{v}^2} = \frac{(1.6 \times 10^{-19}\ \text{C})(1{,}000\ \text{N/C})(1.2\ \text{m})^2}{(2)(9.1 \times 10^{-31}\ \text{kg})(2.05 \times 10^{7}\ \text{m/s})^2} = 0.30\ \text{m}$$

6. (a) The frequency of the radiation is given by $f = \mathbf{c}/\lambda = (3 \times 10^8\ \text{m/s})/(5 \times 10^{-8}\ \text{m}) = 6.0 \times 10^{15}$ Hz.

(b) The number of photons $N = E_{\text{total}}/E_{\text{photon}}$. Thus,

$$N = \frac{(9.945 \times 10^{-16}\ \text{J})}{(6.0 \times 10^{15}\ \text{s}^{-1})(6.63 \times 10^{-34}\ \text{J-s})} = 250$$

(c) The work function is equal to the product of the threshold frequency and Planck's constant:

$$W_0 = hf_0 = (6.63 \times 10^{-34}\ \text{J-s})(2.5 \times 10^{15}\ \text{s}^{-1}) = 1.6575 \times 10^{-18}\ \text{J}$$

(d) The maximum kinetic energy is equal to the difference between the energy of one photon and the work function. Therefore, $\text{KE}_{\text{max}} = \text{E}_\text{P} - \text{W}_0 - (3.978 \times 10^{-18}\ \text{J}) - (1.6575 \times 10^{-18}\ \text{J}) = 2.32 \times 10^{-18}\ \text{J} =$ 14.5 eV.

(e) Since 1 eV is the energy given to one electron accelerated through 1 V, the stopping potential will be equal to 14.5 V in this case.

PROBLEM SET 2

Exercises

1. A particle has a position time function given by $x = 2t^2 + t + 2$. What was the average speed of the object from $t = 0$ to $t = 2$ s?
2. A stone is dropped from a height of 49 m. Two seconds later, a similar stone is thrown from the same height such that the two stones hit the ground at the same time. What was the initial velocity of the second stone?
3. A pendulum of length L is swinging in an arc AB as shown. It has a velocity **v** at point O. What is the tension in the string at point O?

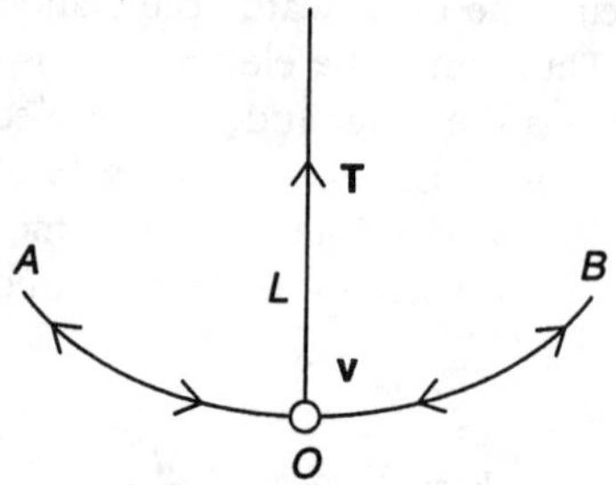

4. A conical pendulum consists of a mass M attached to a light string of length L. The mass swings around in a horizontal circle, making an angle θ with the vertical. What is the tension **T** in the string?

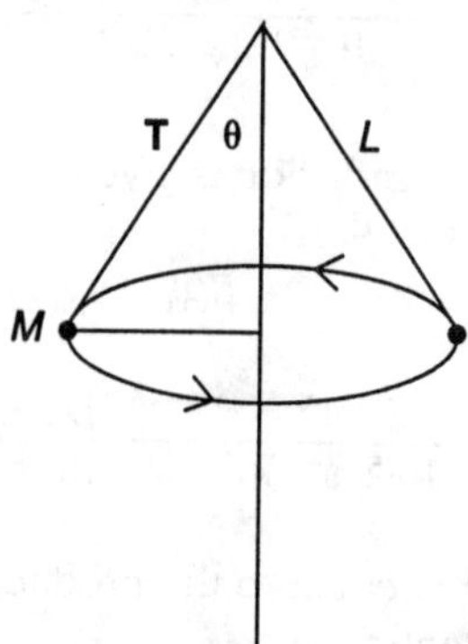

5. A 10-N force is applied to two masses in contact as shown. The horizontal motion is along a frictionless plane, and the masses are 4 kg and 2 kg, respectively. What is the magnitude of the contact force between the two masses?

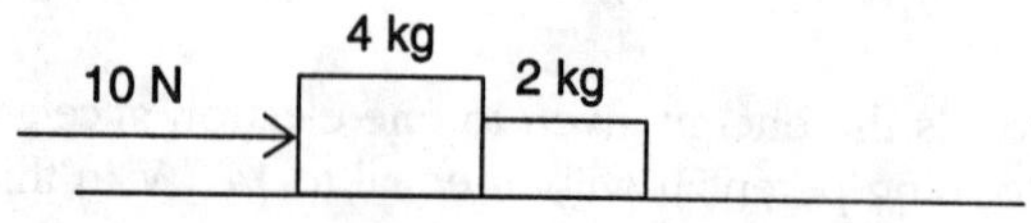

6. A mass M rests on a turntable that has a coefficient of static friction μ. The turntable is rotating with a constant angular frequency ω and has a radius r. The mass is placed at the edge of the turntable. If the angular frequency is at the magnitude where the mass is just about to fly off, which expression allows one to determine the magnitude of μ?

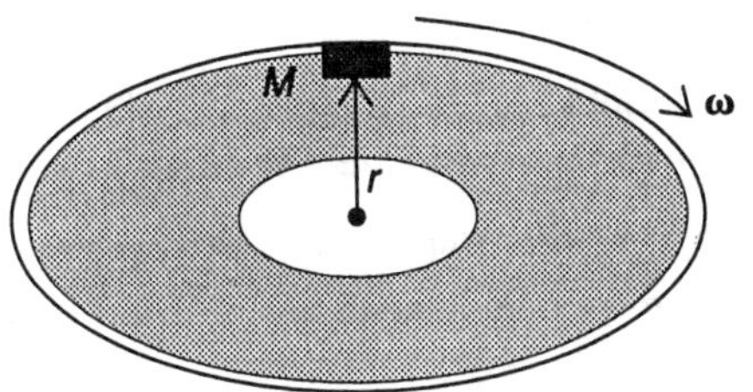

7. A 0.2-kg wad of putty is moving horizontally with a velocity of 15 m/s. It strikes a stationary sphere of radius R at a distance above the ground equal to $\frac{4}{5}R$. What will be the angular momentum of the putty-sphere system after the collision has taken place?

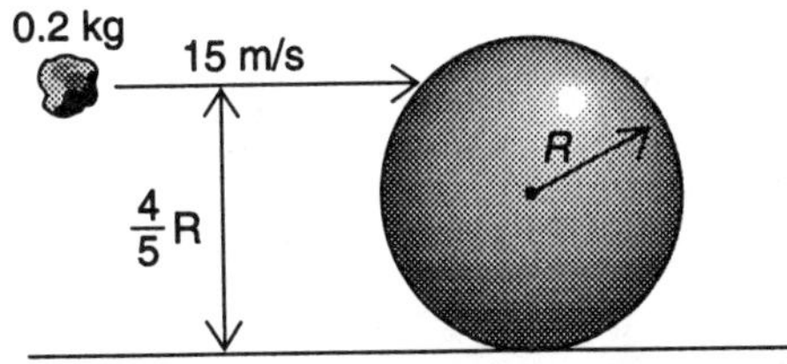

8. A 200-kg cart rests on top of a frictionless hill as shown.

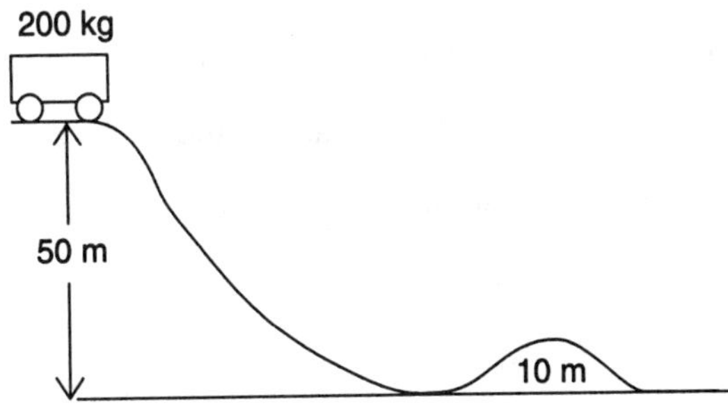

What will be the kinetic energy of the cart when it is at the top of the 10-m hill?

Questions 9 and 10 are based the following information: The magnitude of the one-dimensional momentum of a 2-kg particle obeys the relationship $\mathbf{p} = 2t + 3$.

9. What was the velocity of the particle at $t = 1$ s?

10. What was the average force applied to the particle from $t = 1$ s to $t = 3$ s?

11. A 50-kg box is pushed up a 1.5-m incline with an effort of 200 N. The top of the incline is 0.5 m above the ground. How much work was used to overcome friction?

12. The incline in Question 11 is an example of a simple machine. Based on the information provided, what is the efficiency of this machine?

13. A 200-g metal block absorbs 1,500 J of heat, and its temperature changes by 150°C. What is the specific heat capacity of this metal?

14. How many moles of an ideal monatomic gas are there if it occupies a volume of 2.5 L at sea level and its temperature is 400 K?

Questions 15 and 16 are based on the following circuit:

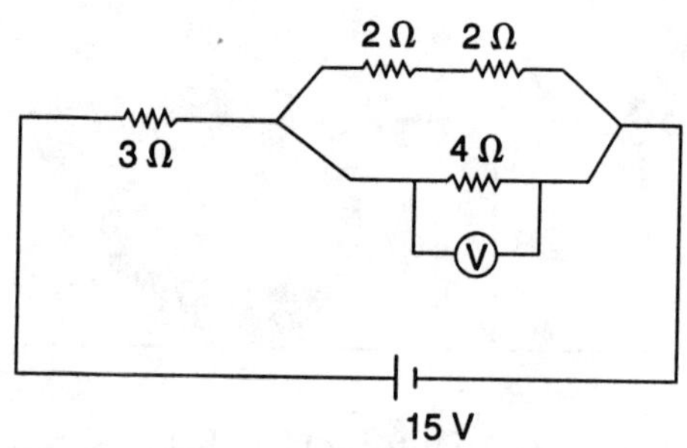

15. What is the equivalent resistance of the circuit?

16. What is the reading of the voltmeter across the 4-Ω resistor?

17. What is the equivalent capacitance of the circuit shown?

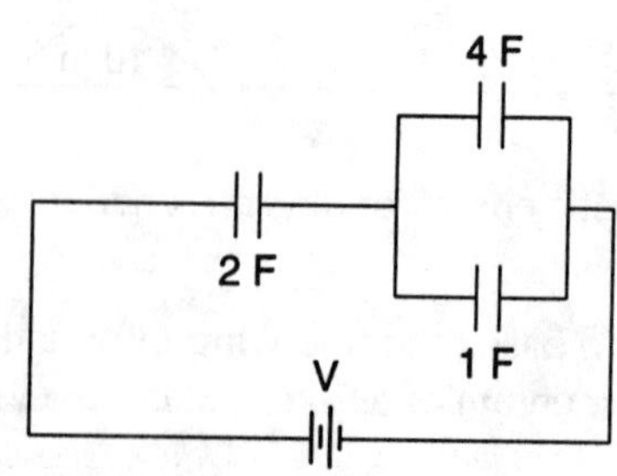

Questions 18 and 19 are based on the following circuit:

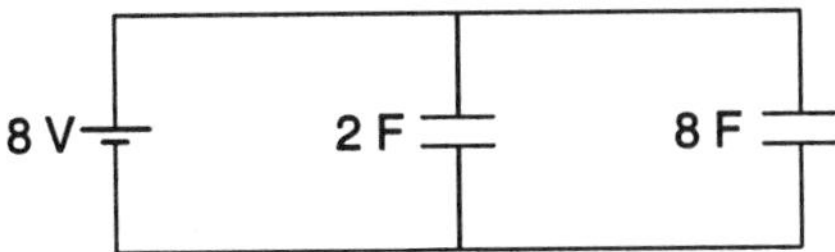

18. What is the charge stored in the 2-F capacitor?
19. What is the energy stored in the 8-F capacitor?
20. What is the distance to a sound source if it takes 3 s to detect the sound after it is produced when the air temperature at sea level is 25°C?
21. A 340-Hz tuning fork sets an air column vibrating in fundamental resonance as shown. A hollow tube is inserted in a column of water, and the height of the tube is adjusted until a strong resonance is heard. In this situation, the air column is 25 cm in length. What is the velocity of the sound?

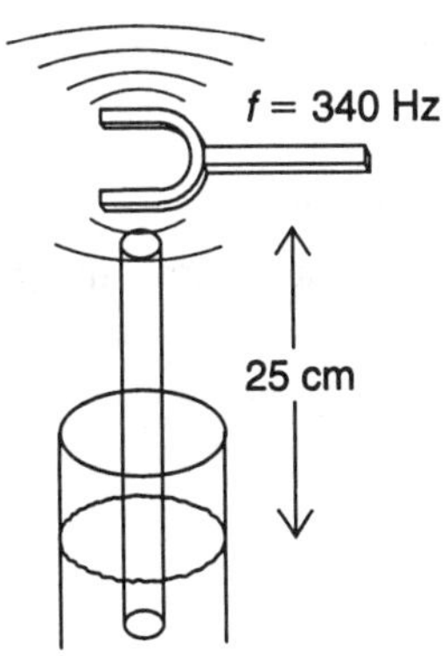

22. What is the approximate critical angle of incidence for a light ray incident on a crown glass (n = 1.51) to water (n = 1.33) interface?
23. A 5-cm-tall object is located 10 cm in front of a converging lens that has a 6-cm focal length. What is the size of the image produced?
24. In a photoelectric effect experiment, the emitted electrons can be stopped with a retarding potential of 12 V. What was the maximum kinetic energy of these electrons?
25. What is the energy associated with a 150-nm photon?

Free-Response Problems

1. A mass m is suspended by a light string and attached to a ring stand as shown. The ring stand sits on top of a rough platform that can rotate with

an angular frequency $\boldsymbol{\omega}$ such that the mass is pulled outward, making an angle θ with the vertical.

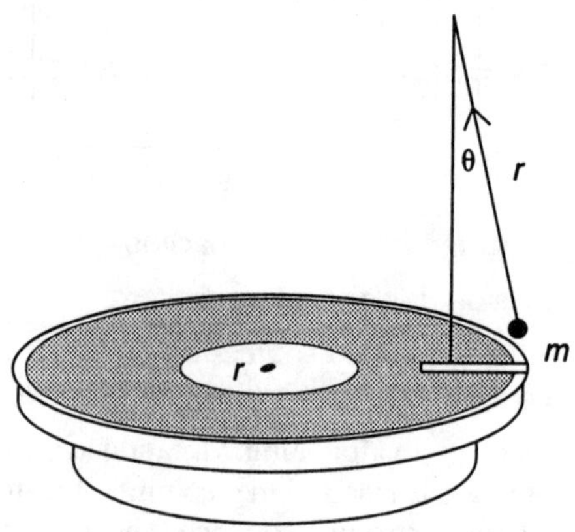

(a) Draw a free body diagram for the forces acting on the mass and label each force appropriately.
(b) In terms of $\mathbf{g}$, r, and $\boldsymbol{\omega}$, derive an expression for the magnitude of the angle θ.
(c) If $r = 0.25$ m, $m = 0.15$ kg, and the period of rotation is 0.45 s, what is the value of the angle θ?

2. A mass m is resting at a height h above the ground. When released, it can slide down a frictionless track to a loop-the-loop of radius R as shown.

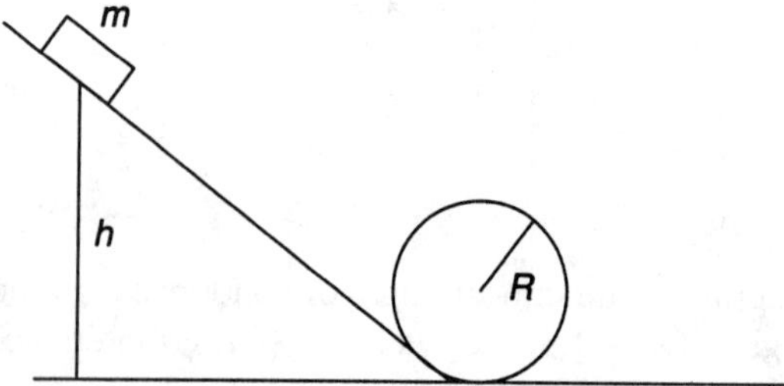

(a) Draw a free body diagram for the mass when it is at the top and at the bottom of the loop.
(b) Derive expressions for the reaction force of the track on the mass when it is at the top and at the bottom.
(c) At what height h must the mass be released such that the reaction force of the track on the mass when it is at the top of the loop is exactly equal to its weight?

3. The 501.5-nm line of helium is observed at an angle of 45 degrees in the second-order spectrum of a diffraction grating.

(a) How many lines per millimeter does this diffraction grating have?
(b) Calculate the angular deviation for the 667.8-nm helium line in the first-order spectrum for the same grating.

4. A cooler is made from a hollow rectangular box that is 50 cm long, 20 cm wide, and 20 cm high. The walls are 3 cm thick and made of a thermal material that has a conductivity of k = 0.045 W/(m-K). The internal temperature is maintained at 0°C using ice, and the outside temperature is 30°C.
   (a) Write an expression representing the amount of heat transferred per unit time across the walls of the cooler.
   (b) Show that the units on both sides of your equation reduce to J/s.
   (c) How much ice melts each hour based on the equation and conditions given?

5. An electron is projected at an angle of 35 degrees above the horizontal with a velocity of 300,000 m/s. It encounters an upward electric field of strength 600 N/C. Neglecting all gravitational effects,
   (a) Determine the time it takes the electron to return to its initial height.
   (b) Determine the maximum height reached by the electron.
   (c) Determine the horizontal displacement when it has reached its maximum height.

6. The following information is given: proton mass = 1.0078 u, neutron mass = 1.0087 u, mass of $^{226}_{88}$Ra = 226.0244 u.
   (a) Determine the mass defect for this isotope.
   (b) Determine the binding energy per nucleon for this isotope.
   (c) Radium 88 undergoes alpha decay via the following equation:

$$^{226}_{88}\text{Ra} \rightarrow \, ^{222}_{86}\text{Rn} + \, ^{4}_{2}\text{He} + Q$$

   If the mass of the radon-86 isotope is 222.0165 u and the mass of the helium-4 isotope is 4.0026 u, then how much energy (in MeV), is associated with the quantity Q?

Solutions to Exercises

1. The average velocity is given by the change in distance over time. At t = O s, the position is x = 2 m. At t = 2 s, x = 12 m. Thus, Δx = 10 m and Δt = 2 s. The average velocity is therefore 5 m/s.

2. It takes a stone 3.16 s to fall 49 m. The second stone is thrown downward 2 after the first one is dropped, and therefore it takes only 1.16 s to hit the ground. The vertical displacement is still −49 m. Thus,

$$y = \mathbf{v}_{0y}t - \frac{1}{2}\mathbf{g}t^2$$

$$-49m = \mathbf{v}_{0y}(1.16 \text{ s}) - (4.9 \text{ m/s}^2)(1.16 \text{ s})^2$$

and

$$v = -47.92 \text{ m/s}$$

3. At the bottom of the swing, we set

$$\mathbf{F} = m\mathbf{a} = \frac{m\mathbf{v}^2}{r} = \mathbf{T} - m\mathbf{g}$$

Thus,

$$\mathbf{T} = \frac{m\mathbf{v}^2}{r} + m\mathbf{g}$$

4. From the diagram we can see that the upward component of the tension, in magnitude, is given by $\mathbf{T} \cos \theta = m\mathbf{g}$. Thus, $\mathbf{T} = m\mathbf{g}/\cos \theta$.
5. Since both masses are in contact, the force $\mathbf{F} = 10$ N gives them the same acceleration: $\mathbf{a} = 10$ N/6 kg $= 1.67$ m/s^2. If we isolate the 2-kg mass in a free body diagram, then the only force acting on it is $\mathbf{P}$, the force exerted by the 4-kg mass. However, in magnitude, $\mathbf{P} = m\mathbf{a}$, where $m = 2$ kg and $\mathbf{a}$ is the acceleration of the system. Thus, we find that $\mathbf{P} = 3.34$ N.
6. In the frame of reference of the mass, the centripetal force is balanced by the force of friction. The normal force in this case is equal to the weight $m\mathbf{g}$. Thus,

$$m\omega^2 r = \mu m\mathbf{g}$$

$$\mu = \frac{\omega^2 r}{\mathbf{g}}$$

7. The instantaneous angular momentum is given by $\mathbf{L} = m\mathbf{v}r = (0.2 \text{ kg})(15 \text{ m/s})(\frac{4}{5}Rm)$. Thus, the magnitude of the angular momentum is equal to $\frac{12}{5}R$ kg-m^2/s.
8. Using conservation of energy we see that the total energy at the start is $E = m\mathbf{g}h$ which is 98,000 J. At the top of the second hill, the kinetic energy is equal to the difference in potential energies: 98,000 J − 19,600 J = 78,400 J.
9. At $t = 1$ s, the momentum $\mathbf{p} = 5$ kg-m/s. With a mass of 2 kg, that implies an initial velocity of 2.5 m/s.
10. The average force $\mathbf{F} = \Delta\mathbf{p}/\Delta t$. In this situation, $\Delta t = 2$ s and $\Delta\mathbf{p} = 4$ kg-m/s. Thus, $\mathbf{F} = 2$ N.
11. The work done by friction is the difference between the work input and the work output. In this problem $W_{in} = (200 \text{ N})(1.5 \text{ m}) = 300$ J, and we can take $\mathbf{g} = 9.8$ N/kg. Then $W_{out} = (50 \text{ kg})(9.8 \text{ N/kg})(0.5 \text{ m}) = 245$ J. Thus, $W_{friction} = 55$ J.
12. The efficiency is given by the ratio of the work output to the work input in a simple machine. Thus, $e = 245 \text{ J}/300 \text{ J} \times 100\% = 82\%$.

13. The specific heat capacity $c = \Delta Q/m\,\Delta T = 1{,}500\ \text{J}/(200\ \text{g})(150°\text{C}) = 0.05\ \text{J}/(\text{g}°\text{-C})$.

14. The number of moles $n = 12.2PV/T$ if P is in atmospheres, V is in liters, and T is in kelvins. We are at sea level, and so $P = 1$ atm and hence $n = 0.076$ mole after substitution.

15. First reduce the series portion in the parallel branch: $R = 2\ \Omega + 2\ \Omega = 4\ \Omega$. Now, this 4-Ω resistance is in parallel with the other 4-Ω resistor. Recalling that

$$\frac{1}{R} = \frac{1}{R_1} + \frac{1}{R_2}$$

we find that the next equivalent resistance is 2 Ω. Finally, this resistance is in series with the 3 Ω, giving a final equivalent resistance of $3\ \Omega + 2\ \Omega = 5\ \Omega$.

16. With a 5-Ω equivalent resistance, Ohm's law states that the circuit current will be equal to $I = 15\ \text{V}/5\ \Omega = 3$ A. In a series circuit, the potential drop across each resistor is given by $V = IR$. Now, the 4-Ω resistor is part of a branch that has an equivalent resistance of 2 Ω. Thus, the potential drop across that entire branch is 6 V. The voltmeter recording the potential difference across the 4-Ω resistor will register 6 V also.

17. Capacitors add up directly in parallel and add up reciprocally in series. This is the opposite of what resistors do. Thus, in the parallel branch the equivalent capacitance is 5 F. For the two series capacitors, we have $1/C = \frac{1}{2} + \frac{1}{5}$, which implies $C = \frac{10}{7}$ F.

18. Both capacitors have the same potential difference of 8 V. Thus, $Q = CV = 16$ C.

19. The energy stored in a charged capacitor is given by $E = \frac{1}{2}QV = 256$ J.

20. The velocity of sound increases by approximately 0.6 m/s for each 1°C rise in air temperature. At 25°C, the velcoity of sound in air is 346 m/s. At 3 s of travel, the distance to the source is 1,038 m.

21. In a closed-end tube, the standing wave resonance point occurs at a length $l = \lambda/4$. Thus, $\lambda = 1.0$ m, and since $\mathbf{v} = f\lambda$, $\mathbf{v} = 340$ m/s in this case.

22. The critical angle of incidence is given by $\sin\theta = n_2/n_1 = 1.33/1.51$, and $\theta = 62$ degrees.

23. To find the image size we must first locate its position:

$$\frac{1}{f} = \frac{1}{p} + \frac{1}{q}$$

$$\frac{1}{6} = \frac{1}{10} + \frac{1}{q}$$

$$q = 15\ \text{cm}$$

The magnification of the image is given by the ratio $q/p = 1.5$, and so $S_i = 5 \times 1.5 = 7.5$ cm.

24. The stopping potential of 12 V implies a maximum kinetic energy of 12 eV for each electron since $KE_{max} = eV$ and for one electron, $KE_{max} = (12\text{ V})(1$ electron) which equals 12 eV.

25. $E = hf$, and for 150 nm, the frequency is $f = c/\lambda = (3 \times 10^8 \text{ m/s})/(1.5 \times 10^{-7} \text{ m}) = 2 \times 10^{15} \text{ s}^{-1}$. Multiplying this by Planck's constant yields $E = hf = (6.63 \times 10^{-34} \text{ J-s})(2 \times 10^{15} \text{ s}^{-1}) = 1.326 \times 10^{-18}$ J.

Solutions to Free-Response Problems

1. (a) A free body diagram for the mass m is shown here.

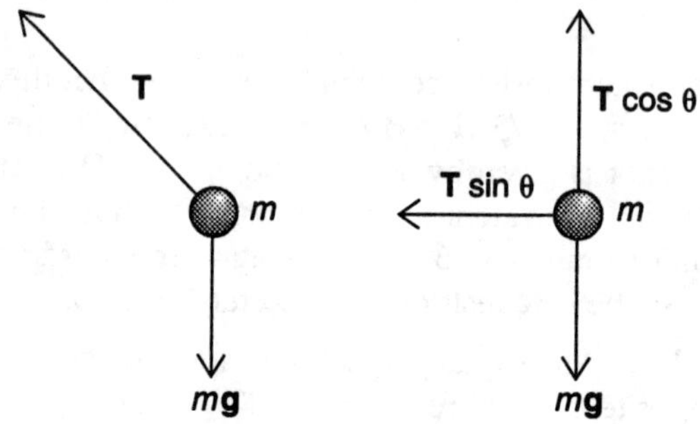

(b) We write that in equilibrium the sum of all vector forces is equal to zero. If there is acceleration, then the sum of all vector forces must equal $m\mathbf{a}$. In the vertical direction, we have equilibrium between the weight $m\mathbf{g}$ and the upward component of the tension:

$$\Sigma \mathbf{F}y = 0 = \mathbf{T} \cos \theta - m\mathbf{g}$$

This establishes that $\mathbf{T} = m\mathbf{g}/\cos \theta$.

In the x direction, the horizontal component of the tension is in the same line as the apparent centrifugal force acting on the ball since the ring stand is resting on the rotating platform. Thus, $\Sigma \mathbf{F}_x = \text{m}\mathbf{a} = \text{m}\,\boldsymbol{\omega}^2\text{r} = \mathbf{T} \sin \theta$. Substituting for the tension $\mathbf{T}$, we obtain

$$\tan \theta = \frac{\omega^2 r}{\mathbf{g}}$$

(c) To find θ, we first need to know that the value of $\boldsymbol{\omega} = 2\pi/\text{T} = 6.28/0.45 \text{ s} = 14$ rad/s. Making the other substitutions, we have $\tan \theta = \boldsymbol{\omega}^2 r/\mathbf{g} = (14 \text{ rad/s})^2(0.25 \text{ m})/(9.8 \text{ m/s}^2) = 5$. Thus, the angle is 78.7 degrees.

2. (a) The free body diagrams for both the top and bottom situations are given here.

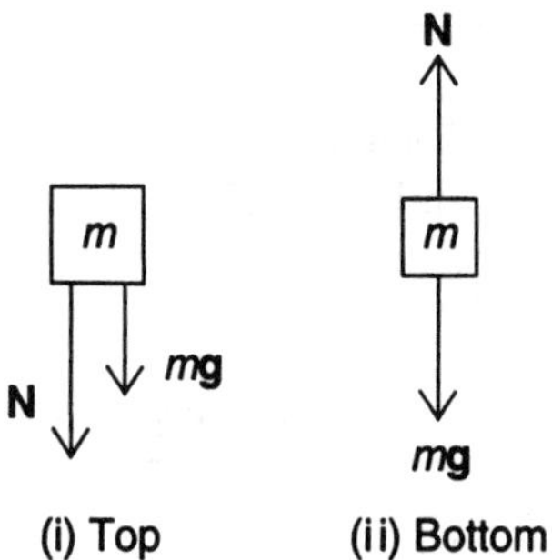

(b) For the top, $\Sigma\mathbf{F} = -m\mathbf{a} = -m\mathbf{v}^2/R = -\mathbf{N} - m\mathbf{g}$, which implies that $\mathbf{N} = m\mathbf{v}^2/R - m\mathbf{g}$. For the bottom, $\Sigma\mathbf{F} = m\mathbf{a} = m\mathbf{v}^2/R = \mathbf{N} - m\mathbf{g}$, which implies that $\mathbf{N} = m\mathbf{v}^2/R + m\mathbf{g}$.

(c) We want the reaction force **N** at the top to be equal to $m\mathbf{g}$. Thus,

$$\mathbf{N} = \frac{m\mathbf{v}^2}{R} - m\mathbf{g}$$

$$m\mathbf{g} = \frac{m\mathbf{v}^2}{R} - m\mathbf{g}$$

$$2\mathbf{g}R = \mathbf{v}^2$$

We now need to find the square of the velocity using the law of conservation of energy:

$$m\mathbf{g}h = m\mathbf{g}(2R) + \frac{1}{2}\, m\mathbf{v}^2$$

$$\mathbf{v}^2 = 2\mathbf{g}h - 4\mathbf{g}R$$

Substituting, we have $2\mathbf{g}R = 2\mathbf{g}h - 4\mathbf{g}R$, which implies that $h = 3R$.

3. (a) For diffraction our relationship is $n\lambda = d \sin \theta$. The information given is that $\lambda = 501.5$ nm, $n = 2$, and $\theta = 45$ degrees. Solving for d, we get $d = 1.433 \times 10^{-6}$ m. The reciprocal of d is the number of lines per meter. Thus, $1/d = 69{,}706.28$ lines per meter $= 698$ lines per millimeter.

(b) Using the value for d just obtained, we use $n = 1$ and $\lambda = 667.8$ nm to solve for θ. Direct substitution yields $\theta = 27.8$ degrees.

4. (a) $\Delta Q/\Delta t = kA/\Delta T/\Delta L$ is the equation for heat transfer, where k is the conductivity in W/(m-K) (watts per meter per Kelvin) and ΔL is the thickness in meters.

(b) The left-hand side is the rate of change in heat in J/s. We can perform the dimensional analysis on the right-hand side to verify the units:

$$\frac{[\mathrm{W/(m-K)}](\mathrm{m})(\mathrm{m})(\mathrm{K})}{\mathrm{m}} = \mathrm{W} = \mathrm{J/s}$$

(c) The area A is the total surface area. The dimensions in SI units are l = 0.5 m, h = 0.2 m, and w = 0.2 m. Each square side has an area of 0.04 m^2, and each rectangular side has an area of 0.1 m^2. Thus, A = 0.48 m^2. Given that ΔT = 30 K, we have

$$\frac{\Delta Q}{\Delta t} = \frac{[0.045\ \mathrm{W/(m\text{-}K)}](0.48\ \mathrm{m^2})(30\ \mathrm{K})}{0.03}\ \mathrm{m} = 21.6\ \mathrm{J/s}$$

In 1 h, we transfer 77,760 J, which is equal to 18,558.472 cal. Now, it takes 80 cal/g to melt ice at 0°C. Thus, we must have m = 232 g of ice melted in 1 h.

5. (a) The electron is initially projected into the field at some arbitrary position relative to the ground. We can treat this as the "zero" level and determine the total time of flight. From the theory of projectiles we know that the magnitude acceleration of the electron (downward) is $\mathbf{a} = e\mathbf{E}/m$. Substituting, we find that $\mathbf{a} = e\mathbf{E}/m = (1.6 \times 10^{-19}$ C)(600 N/C)/(9.1×10^{-31} kg) = 1.05×10^{14} m/s^2. Now the initial vertical velocity is given by $\mathbf{v}_{0y} = \mathbf{v}_0 \sin\theta$ = (300,000 m/s)(sin 35) = 172,073 m/s. The total time is equal to $t = 2\mathbf{v}_{0y}/\mathbf{a} = 3.28 \times 10^{-9}$ s.

(b) The maximum height achieved is given by $y = \frac{1}{2}\mathbf{a}t^2$ = (0.5)(1.05 × 10^{14} m/s^2)(1.64 × 10^{-9} s$)^2$ = 1.4 × 10^{-4} m, where we have used half the total time of flight and incorporated the maximum height at the equivalent free fall distance in this uniform electric field provided we have uniformly accelerated motion.

(c) Neglecting air resistance and gravity, the displacement in the total time of flight is given by the product of the total time and the horizontal velocity $\mathbf{v}_{0X} = \mathbf{v}_0 \cos\theta$ = 245,745.6 m/s. Thus, x = (245,745.6 m/s)(3.27 × 10^{-9} s) = 8 × 10^{-4} m.

6. (a) To find the mass defect we find the total constituent mass:

$$88\ \text{protons: } m = (88)(1.0078) = 88.6864\ \mathrm{u}$$

$$138\ \text{neutrons: } m = (138)(1.0087) = 139.2006\ \mathrm{u}$$

$$M_{\text{total}} = 227.8870\ \mathrm{u}$$

$$\text{Mass defect} = 227.8870\ \mathrm{u} - 226.0244\ \mathrm{u} = 1.8626\ \mathrm{u}$$

(b) BE = (931.5 MeV/u)(1.8626 u) = 1,735.0119 MeV. BE/nucleon = 1,735.0119/226 = 7.677 MeV/nucleon.

(c) The equivalent mass of Q = 226.0244 − 222.0165 − 4.0026 = 0.0053 u, and E = (931.5 MeV/u)(0.0053 u) = 4.94 MeV

PROBLEM SET 3

Exercises

1. A vector has the following components: $\mathbf{A}_x = 3$ units and $\mathbf{A}_y = 5$ units. What angle does this vector make with the positive x axis?

2. Two vectors **A** and **B** have components $\mathbf{A}_x = -2$, $\mathbf{A}_y = 3$, $\mathbf{B}_x = 1$, and $\mathbf{B}_y = 4$. What is the magnitude of the vector $\mathbf{A} + \mathbf{B}$?

3. How long will it take a rock, thrown downward from a height of 96 m with an initial velocity of -20 m/s, to hit the ground?

4. A car with a 500-N driver goes over a hill that has a radius of 60 m as shown. The velocity of the car is 20 m/s. What is the force that the car exerts on the driver?

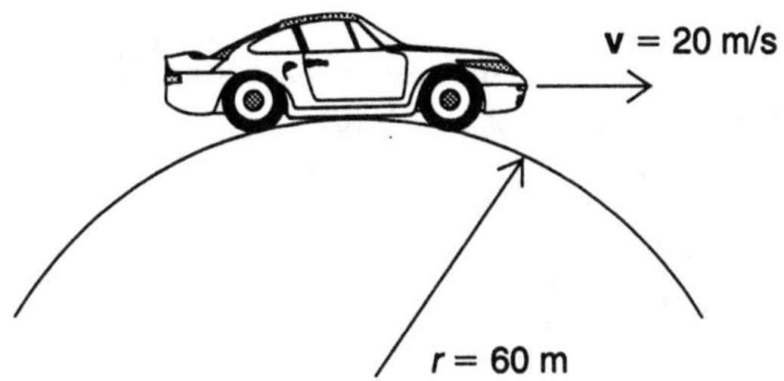

5. In the accompanying diagram, two masses are connected by a light string and slide over frictionless surfaces. If the angle of the incline is θ, what is the acceleration of both masses?

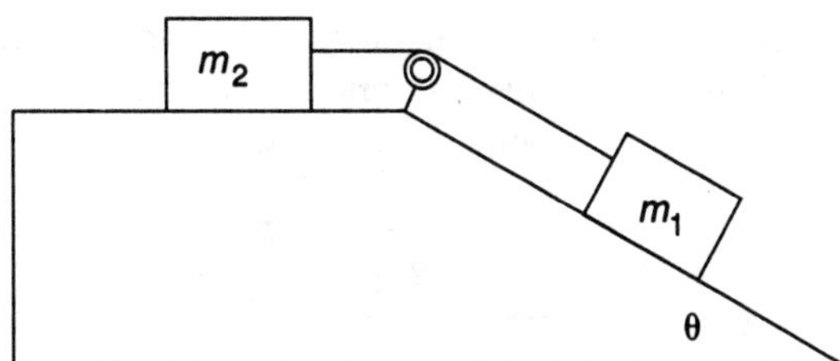

6. A boat moving due north crosses a 190-m-wide river with a velocity of 8 m/s relative to the water. The river flows east with a velocity of 4 m/s. How long will it take the boat to cross the river?

Questions 7 and 8 are based on the following information: A variable force acts on a 2-kg mass according to the graph shown.

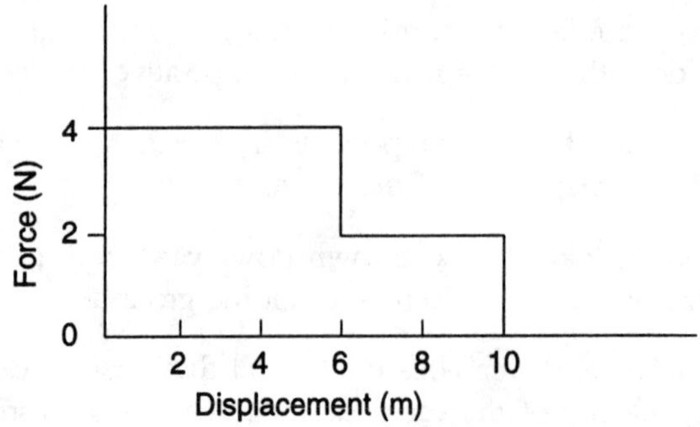

7. How much work was done to displace the mass 10 m?
8. What was the average force supplied to the mass for the entire 10-m displacement?
9. A block is pushed along a frictionless surface for a distance of 2.5 m. How much work has been done if the force of 10 N makes an angle of 60 degrees with the horizontal?

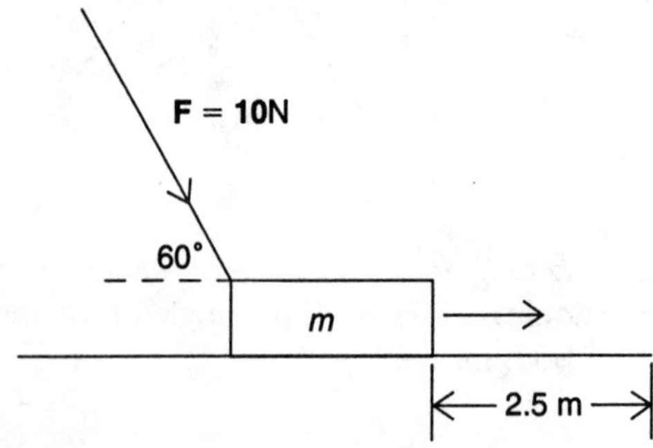

Questions 10 and 11 are based on the accompanying diagram. Two masses are connected by a light string. The horizontal mass of 2 kg is being pulled to the right with a force of 30 N along a frictionless surface.

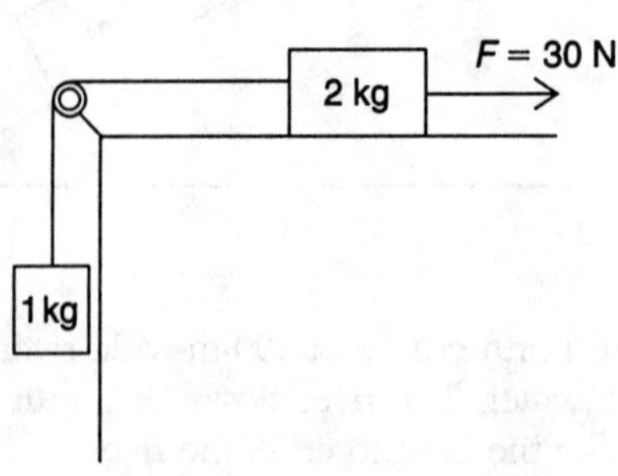

10. What is the magnitude of the tension in the connecting string?
11. What is the magnitude of the acceleration of both masses?
12. A 50-kg person is sitting on a see-saw 1.2 m from the balance point. On the other side, a 70-kg person is balanced (see the accompanying diagram). How far from the balance point is the second person sitting?

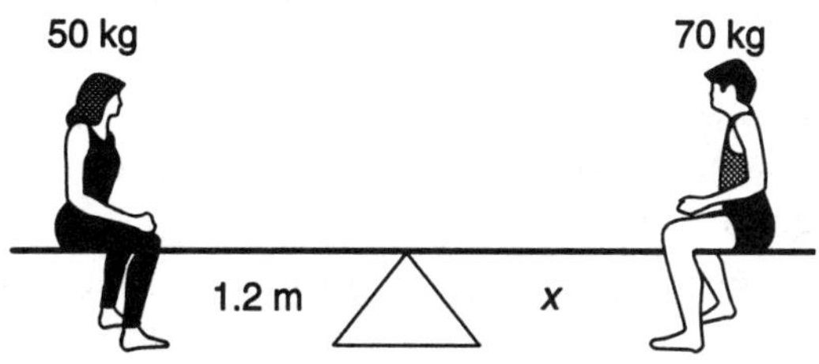

13. What is the net torque acting on the pivot supporting a 2-m-long, 10-kg beam as shown?

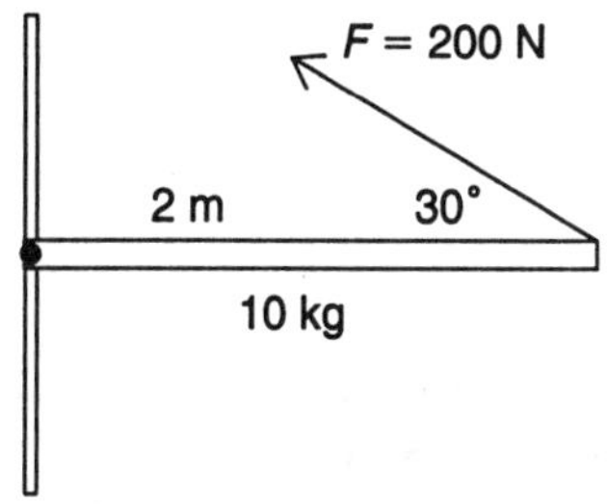

Questions 14 through 16 are based on the following information. A 0.3-kg sample of some material begins as a solid and absorbs energy according to the accompanying graph.

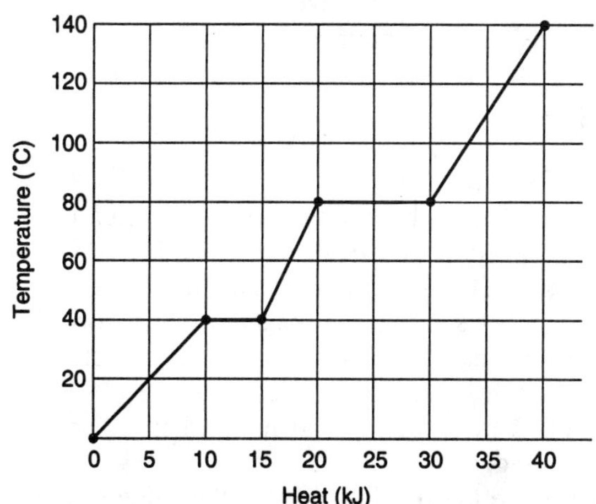

14. What is the melting point of this substance?
15. What is the latent heat of fusion for this substance?
16. What is the specific heat capacity of this substance in the liquid phase?
17. At 2 atm of pressure, 100 m^3 of an ideal gas at 50°C is heated until its pressure and absolute temperature are doubled. What is the new volume of the gas?
18. What is the electric force between two identical charges of magnitude +5 μC separated by a distance of 22 mm?
19. An electron is moving between two metal plates separated by 20 mm. The acceleration of the electron is 3×10^{12} m/s^2. What is the potential difference between the two plates?
20. A circuit consists of two 10-μF capacitors in series, which are then connected in parallel to a 5-μF capacitor. What is the equivalent capacitance of this circuit?

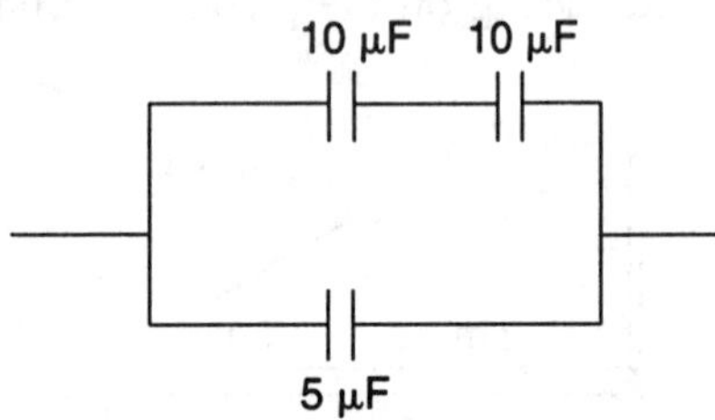

21. An electric motor operates from a 120-V source using 4 A of current. What is the power rating of this motor?
22. When is the back EMF of an electric motor the greatest?
23. A steel wire is stretched between two points. The length of the wire is 2.5 m, and the mass of the wire is 0.5 kg. A transverse pulse is sent down the wire and returns in 0.2 s. What is the tension in the wire?
24. What must be your velocity, relative to the speed of light **c**, such that your observed mass is 1.5 times your rest mass?
25. What is the DeBroglie wavelength for a 1,500-kg car moving with a velocity of 30 m/s?

Free-Response Problems

1. A cannon fires a shell with an initial velocity **v** and a launch angle θ from the ground. At a height *h* above the ground, fast winds will alter the trajectory of the shell. Under these conditions:
   (a) Derive an expression for the maximum range possible such that the projectile remains on course. Your expression should be in terms of **v**, **g**, and *h* only.
   (b) If *h* = 2,000 m and **v** = 500 m/s, what is the magnitude of the maximum range?
   (c) Using the same information, what is the value of the launch angle θ?

2. Two cars are separated by 25 m. Both are initially at rest. The car in front begins to accelerate uniformly at 2 m/s^2. Behind it the second car begins to accelerate at 3 m/s^2.
   (a) How long does it take the faster car to catch up with the slower car?
   (b) How far did the faster car go during that time?
   (c) What is the velocity of each car at the moment the second car meets the first car?
3. (a) Determine the rms velocities of hydrogen and oxygen molecules at 0°C.
   (b) What is the ratio of the rms velocity of hydrogen to the rms velocity of oxygen?
   (c) How does this ratio of rms velocities compare with the ratio of their molecular masses?
   (d) Write a formula expressing the relationship described in part (c) for any two molecules X and Y.
   (e) Use the formula from part (d) to determine the ratio of rms velocities of hydrogen and carbon dioxide at 0°C.
4. Determine the magnitude of each current in the network shown.

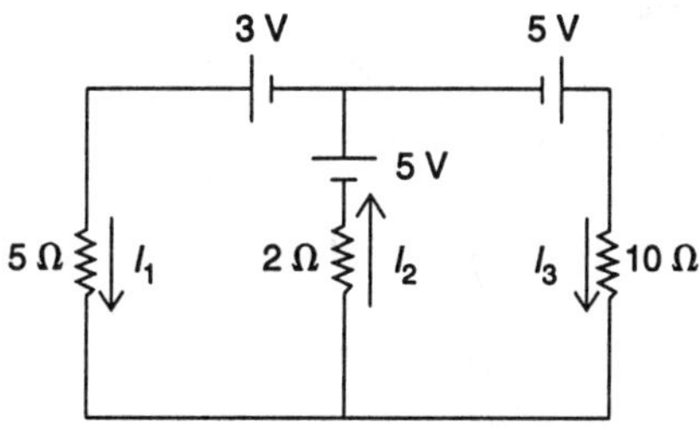

5. A person is looking at a fish in a tank of water from a near-vertical position. The actual depth of the fish is h, and its apparent refractive depth is h'. The distance to each fish from the normal is x.

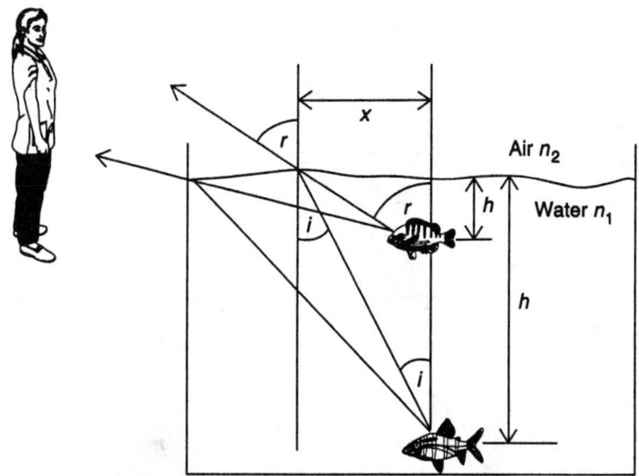

(a) Using a small-angle approximation, show that

$$\frac{\tan i}{\tan r} = \frac{h'}{h}$$

(b) Additionally, show that the expression in part (a) implies that

$$\frac{h'}{h} = \frac{n_2}{n_1}$$

where n_1 and n_2 are the absolute indices of refraction of the two media.

(c) If the absolute index of refraction for water is 1.33 and the actual depth of the fish is 1.5 m, what is the observed depth of the fish?

6. The half-life of $^{222}_{86}Rn$ is 3.8 days.
   (a) How many atoms are contained in 5 mg of radon?
   (b) What is the activity rate of this amount of radon?
   (c) What percentage of radon will remain after 8 days?

Solutions to Exercises

1. The angle is given by tan $\theta = \mathbf{a}_y/\mathbf{a}_x = \frac{5}{3} = 1.67$. Thus, $\theta = 59$ degrees.
2. The sum of two vectors can be found by adding up their components: $\mathbf{c}_x = \mathbf{a}_x + \mathbf{b}_x = -2 + 1 = -1$ and $\mathbf{c}_y = \mathbf{a}_y + \mathbf{b}_y = 3 + 4 = 7$. The magnitude of the sum is now given by the Pythagorean theorem:

$$|\mathbf{A} + \mathbf{B}| = \sqrt{(-1)^2 + (7)^2} = \sqrt{50} = 7.1$$

3. The formula for a downward displacement is given by

$$-y = -\mathbf{v}_{0y}t + \frac{1}{2}\mathbf{g}t^2$$
$$-96 = -20t - 4.9t^2$$

The time can be found using the quadratic formula and taking positive time:

$$t = \frac{-b \pm \sqrt{b^2 - 4ac}}{2a}$$
$$t = \frac{-20 \pm \sqrt{400 + (4)(4.9)(96)}}{(2)(4.9)} = 2.8 \text{ s}$$

4. The force that the car exerts on the driver is equal to the normal force **N**. If we write Newton's second law for the car at the top of the hill, we will have

$$\Sigma\mathbf{F}_y = m\mathbf{a} = -\frac{m\mathbf{v}^2}{r} = \mathbf{N} - m\mathbf{g}$$

Since $\mathbf{W} = m\mathbf{g} = 500$ N, we know that the mass is approximately 51 kg. Making the necessary substitutions for velocity and radius, we find that $\mathbf{N} = +160$ N (up).

5. From the diagram we can see that the force acting on the mass m_1 is the component of its weight down the incline: $\mathbf{F} = m_1\mathbf{g} \sin \theta$. However, this force must cause both masses to accelerate together since they are coupled: $\mathbf{F} = (m_1 + m_2)\mathbf{a}$. Equating these two expressions gives us the accleration: $\mathbf{a} = m_1 \mathbf{g} \sin \theta/(m_1 + m_2)$.
6. The two velocities are independent of each other. Thus, in order to find the time needed to cross the river, we consider only the velocity component heading across the river:

$$t = \frac{190 \text{ m}}{8 \text{ m/s}} = 24 \text{ s (actually 23.75 s)}$$

7. The work done by the variable force is equal to the total area under the curve. Using rectangles, this total area is equal to 32 J.
8. The average force applied will be equal to the total work done divided by the displacement: $\mathbf{F}_{avg} = 32$ J/10 m $= 3.2$ N.
9. The work done is equal to the product of the magnitude of the force component parallel to the direction of motion and the displacement. In this case, the force component is equal to (10 N)(cos 60) = 5 N. Thus, $\mathbf{W} = \mathbf{Fd} = (5 \text{ N})(2.5 \text{ m}) = 12.5$ J.
10. Using free body diagrams for both masses, we can set up the equations of motion for forces in the x and y directions: x direction, $30 - \mathbf{T} = m\mathbf{a} = 2\mathbf{a}$; y direction, $\mathbf{T} - M\mathbf{g} = \mathbf{T} - 9.8 = M\mathbf{a} = 1\mathbf{a}$. We need to solve two simultaneous equations for the tension $\mathbf{T}$:

$$30 - \mathbf{T} = 2\mathbf{a}$$

$$\mathbf{T} - 9.8 = \mathbf{a}$$

Eliminating $\mathbf{a}$ from both equations yields $\mathbf{T} = 16.5$ N.

11. Using the tension derived in 10, we substitute and find $\mathbf{a} = 6.75$ m/s^2.
12. We need to balance the torques on either side of the pivot: (50 kg)($\mathbf{g}$)(1.2 m) = (70 kg)($\mathbf{g}$)x implies that $x = 0.86$ m.
13. We take counterclockwise torques as being positive. Acting on the pivot are two forces contributing to a net torque. First, the component of the force $\mathbf{F}$ acting at right angles to the beam contributes a torque $\tau_1 = (200$ N)(sin 30)(2 m) = 200 N-m. The weight of the beam acts from its center of mass ($x = 1$ m since the beam is uniform). This torque will be clockwise and negative: $\tau_2 = -(98 \text{ N})(1 \text{ m}) = -98$ N. Thus, the net torque is -102 N-m.
14. The melting point is the temperature at which the substance changes from a solid to a liquid. This occurs at 40°C.

15. The heat of fusion is the number of kJ/kg required to change the state of the substance from a solid to a liquid. According to the graph, it takes 5 kJ to melt the 0.3 kg of material. Thus, the heat of fusion is 16.7 kJ/kg.

16. The specific heat capacity $c = \Delta Q/m/\Delta T$. As a liquid, we see that

$$c = \frac{5 \text{ kJ}}{(0.3 \text{ kg})(40°\text{C})} = 0.42 \text{ kJ/(kg°C)}$$

17. We use the ideal gas law by first converting 50°C to 323 K. Now we know that for an ideal gas, PV/T = constant. Thus, if both the pressure and the absolute temperature are doubled, the volume of gas will remain the same (100 m^3).

18. We use Coulomb's law to find the magnitude of this repulsive force:

$$\mathbf{F} = \frac{kQ_1Q_2}{d_2} = \frac{(9 \times 10^9 \text{ N-m}^2/\text{C}^2)(5 \times 10^{-6} \text{ C})(5 \times 10^{-6} \text{ C})}{(0.022 \text{ m})^2} = 465 \text{ N}$$

19. The force on the electron is given by $\mathbf{F} = m\mathbf{a} = (9.1 \times 10^{-31} \text{ kg})(3 \times 10^{12} \text{ m/s}^2) = 2.73 \times 10^{-18}$ N. The electric field experienced by the electron is given by

$$\mathbf{E} = \frac{\mathbf{F}}{Q} = \frac{2.73 \times 10^{-18} \text{ N}}{1.6 \times 10^{-19} \text{ C}} = 17 \text{ N/C}$$

Between two parallel plates, $V = \mathbf{E}d$ = (17 N/C)(0.02 m) = 0.34 V

20. The two series capacitors add up reciprocally: $1/C = 1/C_1 + 1/C_2 = \frac{1}{10}$ μF + 1/10 μF implies C = 5 μF. This capacitor is now in parallel with another 5-μF capacitor. The equivalent capacitance is equal to the sum of these two. Thus, C_{eq} = 10 μF.

21. Power is equal to potential difference times current. Thus, $P = VI$ = 480 W.

22. The back EMF of a motor is greatest when the velocity of the motor is greatest.

23. The formula for the velocity of wave propagation along a stretched wire is

$$\mathbf{v} = \sqrt{\frac{\mathbf{T}}{M/L}}$$

where the tension **T** is in newtons. In this example the velocity of the pulse is 25 m/s. Solving for the tension, we find that **T** = 125 N.

24. The formula for relativistic mass is

$$m = \frac{m_0}{\sqrt{1 - \mathbf{v}^2/\mathbf{c}^2}}$$

We want m to be equal to 1.5 times the rest mass. Making the substitution and solving for $\mathbf{v}$, we obtain $\mathbf{v} = 0.75\mathbf{c}$.

25. The DeBroglie wavelength is given by the formula $\lambda = h/m\mathbf{v}$. Using the given values,

$$\lambda = \frac{h}{m\mathbf{v}} = \frac{6.63 \times 10^{-34} \text{ J-s}}{(1{,}500 \text{ kg})(30 \text{ m/s})} = 1.5 \times 10^{-38} \text{ m}$$

Solutions to Free-Response Problems

1. (a) The general equations of motion for a projectile are

$$x = (\mathbf{v} \cos \theta)t$$

$$y = (\mathbf{v} \sin \theta)t + \frac{1}{2}\mathbf{g}t$$

Now the time to reach maximum height is related to the time it takes gravity to decelerate the initial upward component of velocity:

$$t' = t_{up} = \frac{\mathbf{v} \sin \theta}{\mathbf{g}}$$

Thus, y_{max} is the value of y when t' is used for the time. If we eliminate time from the equation, the formula for y_{max} will be $y_{max} = \mathbf{v}^2 \sin^2 \theta/2\mathbf{g}$. Based on our initial conditions, we want $y_{max} = h$. We can now solve for sin θ:

$$\sin \theta = \frac{\sqrt{2\mathbf{g}h}}{\mathbf{v}}$$

The formula for the range, independent of the time, is

$$R = \frac{2\mathbf{v}^2 \sin \theta \cos \theta}{\mathbf{g}}$$

We have solved for sin θ, and now we must find cos θ. Using the definition of the sine of an angle in a right triangle, we form the arrangement in the accompanying figure. From the triangle, we see that $\cos \theta = x/\mathbf{v}$ and $x^2 + 2\,\mathbf{g}h = \mathbf{v}^2$. Thus, $x = \sqrt{\mathbf{v}^2 - 2\mathbf{g}h}$ and $\cos \theta = \sqrt{\mathbf{v}^2 - 2\mathbf{g}h}/\mathbf{v}$.

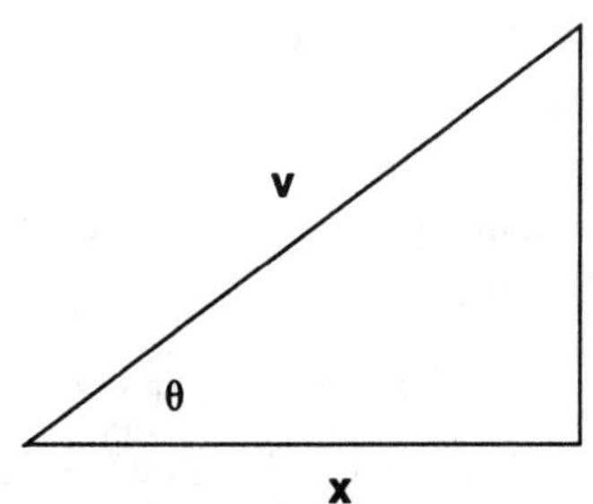

Finally, substituting for sin θ and cos θ in the range formula, we obtain

$$R = \frac{2\mathbf{v}^2\sqrt{2\mathbf{g}h/\mathbf{v}}\sqrt{\mathbf{v}^2 - 2\mathbf{g}h/\mathbf{v}}}{\mathbf{g}}$$

$$= \frac{2\sqrt{2\mathbf{g}h(\mathbf{v}^2 - 2\mathbf{g}h)}}{\mathbf{g}}$$

(b) To find the range, we use h = 2,000 m and $\mathbf{v}$ = 500 m/s in the preceding equation:

$$R = \frac{2\sqrt{2\mathbf{g}h(\mathbf{v}^2 - 2\mathbf{g}h)}}{\mathbf{g}} = \frac{2\sqrt{(39{,}200)(250{,}000 - 39{,}200)}}{0.8} = 18{,}551.6 \text{ m}$$

We can now find the value of θ using the formula derived for sin θ:

$$\sin\theta = \frac{\sqrt{2\mathbf{g}h}}{\mathbf{v}} = \frac{\sqrt{39{,}200}}{500} = 0.396$$

$$\theta = 23 \text{ degrees}$$

2. (a) At the beginning, car 1 is 25 m ahead of car 2. The initial velocities of both cars are zero. When they meet, the position of each car relative to the origin must be the same. Since each car undergoes uniformly accelerated motion from rest, we can write

$$\frac{1}{2}\mathbf{a}_2 t^2 = \frac{1}{2}\mathbf{a}_1 t^2 + 25 \text{ m}$$

$$\frac{1}{2}(3 \text{ m/s}^2)t^2 = \frac{1}{2}(2 \text{ m/s}^2)t^2 + 25 \text{ m}$$

$$t^2 = 50 \text{ s}^2$$

$$t = 7.07 \text{ s}$$

(b) In 7.07 s car 2, the faster car, goes a distance of $\frac{1}{2}\mathbf{a}_2 t^2 = (0.5)(3 \text{ m/s}^2)(7.07 \text{ s})^2 = 75$ m.

(c) Since each car has zero initial velocity, for any time t, $\mathbf{v} = \mathbf{a}t$:

Car 1: $\mathbf{v} = (2 \text{ m/s}^2)(7.07 \text{ s}) = 14.14 \text{ m/s}$

Car 2: $\mathbf{v} = (3 \text{ m/s}^2)(7.07 \text{ s}) = 21.21 \text{ m/s}$

3. (a) To find the root mean square velocity, we first need the mass of each molecule in kilograms:

H_2: $m = 2 \text{ u} = 2(1.66 \times 10^{-27} \text{ kg/u}) = 3.32 \times 10^{-27}$ kg

O_2: $m = 32 \text{ u} = 32(1.66 \times 10^{-27} \text{ kg/u}) = 5.312 \times 10^{-26}$ kg

The formula for the root mean square velocity is (where k is Boltzmann's constant)

$$H_2: \mathbf{v}_{rms} = \sqrt{\frac{3kT}{m}} = \sqrt{\frac{3(1.38 \times 10^{-23}\ \text{J/K})(273\ \text{K})}{3.32 \times 10^{-27}\ \text{kg}}} = 1{,}845\ \text{m/s}$$

$$O_2: \mathbf{v}_{rms} = \sqrt{\frac{3kT}{m}} = \sqrt{\frac{3(1.38 \times 10^{-23}\ \text{J/K})(273\ \text{K})}{5.312 \times 10^{-26}\ \text{kg}}} = 461.23\ \text{m/s}$$

(b) The ratio of the root mean square velocities is 4:1.

(c) The ratio of the molecular masses of oxygen and hydrogen is 16:1. Thus, the ratio of the root mean square velocities is equal to the inverse ratio of their molecular masses.

(d) Thus, for any two molecules X and Y,

$$\frac{\mathbf{v}_{rms(X)}}{\mathbf{v}_{rms(Y)}} = \sqrt{\frac{M_Y}{M_X}}$$

(e) The molecular mass of carbon dioxide is 44 u, and the molecular mass of hydrogen is 2 u. The ratio of the molecular masses is therefore $\frac{44}{2}$ or 22:1. This means that hydrogen molecules (at the same temperature) are faster than carbon dioxide molecules by a factor of $\sqrt{22}$, which is equal to about 4.7.

4. We write down Kirchoff's laws for the network:

$$I_2 = I_1 + I_3.$$

As we go around each loop clockwise, we examine the potential differences we cross and the resistors we meet.

$$-3 - 5 + 2I_2 + 5I_1 = 0$$
$$5 - 10I_3 - 2I_2 + 5 = 0$$

These two equations become

$$5I_1 + 2I_2 = 8$$
$$2I_2 + 10I_3 = 10$$

We eliminate I_2 from these equations using the first equation:

$$5I_1 + 2(I_1 + I_3) = 8$$
$$2(I_1 + I_3) + 10I_3 = 10$$

These equations now become

$$7I_1 + 2I_3 = 8$$
$$2I_1 + 12I_3 = 10$$

Solving for I_1 yields $I_1 = 0.95$ A. Substituting in for I_1 and solving for I_3 yields $I_3 = 0.675$ A. Since $I_2 = I_1 + I_3$, $I_2 = 1.625$ A.

5. (a) For small angles, $\sin\theta \approx \tan\theta$ (see Appendix A). We can therefore write

$$\frac{\sin i}{\sin r} \approx \frac{\tan i}{\tan r}$$

Now, from the diagram, we can see that $\tan i = x/h$ and $\tan r = x/h'$ and so

$$\frac{\tan i}{\tan r} = \frac{x/h}{x/h'} = \frac{h'}{h}$$

(b) From Snell's law, for an angle of incidence i and an angle of refraction r,

$$\frac{\sin i}{\sin r} = \frac{n_2}{n_1}$$

Therefore, using the results from part (a), we see that $h'/h = n_2/n_1$.

(c) If $h = 1.5$ m, then $h' = (1.5\text{ m})(1.00/1.33) = 1.13$ m is the apparent depth.

6. (a) To find out how many atoms are contained in 5 mg of radon, we note that the mass number 222 is approximately equal to the actual atomic mass of the nucleus. Therefore,

$$N_0 = \frac{5 \times 10^{-6}\text{ kg}}{(222\text{ u/atom})(1.66 \times 10^{-27}\text{ kg/u})} = 1.36 \times 10^{19}\text{ atoms}$$

(b) The activity is given by the formula

$$R = \frac{\Delta N}{\Delta t} = \frac{0.693}{T}N_0 = \frac{0.693}{(3.8\text{ days})(86{,}400\text{ s/day})}(1.36 \times 10^{19}\text{ atoms})$$
$$= 2.86 \times 10^{13}\text{ Bq}$$

Notice that for the activity, the half-life T is in units of seconds. The units of activity are events per second or becquerels.

(c) To find the percentage remaining after 8 days, we use the formula

$$\frac{N}{N_0} = \frac{1}{2^{t/T}}$$

In this case, $t = 8$ days and $T = 3.8$ days (the half-life). Thus, $t/T = 2.1$ and therefore

$$\frac{N}{N_0} = \frac{1}{2^{2.1}} = 0.233$$

The percentage left is therefore 23.3%.

Appendix A

List of Useful Constants

1 atomic mass unit	$1\ \text{u} = 1.66 \times 10^{-27}$ kg
Rest mass of the proton	$m_p = 1.673 \times 10^{-27}$ kg
Rest mass of the neutron	$m_n = 1.675 \times 10^{-27}$ kg
Rest mass of the electron	$m_e = 9.11 \times 10^{-31}$ kg
Magnitude of the electron charge	$e = 1.60 \times 10^{-19}$ C
Avogadro's number	$N_0 = 6.02 \times 10^{23}$/mole
Universal gas constant	$R = 8.32$ J/(mole = K)
Boltzmann's constant	$k_B = 1.38 \times 10^{-23}$ J/K
Speed of light	$\mathbf{c} = 2.998 \times 10^{8}$ m/s
Planck's constant	$h = 6.63 \times 10^{-34}$ J-s $= 4.14 \times 10^{-15}$ eV-s
1 electronvolt	$1\ \text{eV} = 1.6 \times 10^{-19}$ J
Vacuum permittivity	$\epsilon_0 = 8.85 \times 10^{-12}\ \text{C}^2/(\text{N-m}^2)$
Coulomb's law constant	$k = \frac{1}{4\pi\epsilon_0} = 8.998 \times 10^{9}\ \text{N-m}^2/\text{C}^2$
Vacuum permeability	$\mu_0 = 4\,\pi \times 10^{-7}$ Wb/ (A-m)
Magnetic constant	$k' = k/c^2 = \mu_0/4\,\pi = 10^{-7}$ Wb/(A-m)
Acceleration due to gravity at the earth's surface	$\mathbf{g} = 9.8\ \text{m/s}^2$
Universal gravitational constant	$G = 6.67 \times 10^{-11}\ \text{m}^3/(\text{kg-s}^2)$

Appendix B

Review of Mathematics

ALGEBRA

Equations and Relationships

Physical relationships are often expressed as mathematical equations. The techniques of algebra are often used to solve these equations as part of the process of analyzing the physical world. In general, the letters *u*, *v*, *w*, *x*, *y*, and *z* are used as variable unknown quantities (with *x*, *y*, and *z* the most popular choices). The letters *a*, *b*, *c*, *d*, *e*, . . . , are often used to represent constants or coefficients (with *a*, *b*, and *c* the most popular).

An equation of the form $y = ax + b$ is called a **linear** equation, and its graph is a straight line. The coefficient *a* is called the **slope** of the line, and *b* is called the ***y*-intercept**, which is the point at which the line crosses the *y* axis (using standard Cartesian coordinates). An equation of the form $y = ax^2 + bx + c$ is called a **quadratic** equation, and a graph of this relationship is a parabola. The **order** of the equation is the highest power of *x*. The variables *x* and *y* can represent any physical quantity being studied. For example, the following quadratic equation represents the displacement of a particle undergoing one-dimensional uniformly accelerated motion:

$$\mathbf{x} = x_0 + \mathbf{v}_0 t + \frac{1}{2}\mathbf{a}t^2$$

In this equation $\mathbf{x}_0 = \mathbf{v}_0$, and **a** are all constants, while the letters *x* and *t* are the variables (**x** representing the displacement, and *t* the time). While there is one solution to a linear equation, a quadratic equation has two solutions. However, in physics it is possible that only one solution is physically reasonable (e.g., there is no "negative time" in physics).

Solutions of Algebraic Equations

A set of points (x,y) is a solution of an algebraic equation if for each x there is one and only one y value (this is also the definition of a **function**) when the value of x is substituted into the equation and the subsequent arithmetical operations performed. In other words, the equation $3x + 2 = -4$ has only one solution since there is only one variable, x (the solution being $x = -2$). The equation $y = 3x + 2$ requires a pair of numbers (x,y) since there are two variables.

Quadratic equations can be solved using the **quadratic formula**. Given the quadratic equation $ax^2 + bx + c = 0$, we can write

$$x = \frac{-b \pm \sqrt{b^2 - 4ac}}{2a}$$

The quantity under the radical sign is called the *discriminant* and can be positive, negative, or zero. If the discriminant is negative, then the square root is an *imaginery number* which usually has no physical meaning for our study.

For example, suppose we wish to solve $x^2 + 3x + 2 = 0$ using the quadratic formula. In this equation $a = 1$, $b = 3$, and $c = 2$. Direct substitution and taking the necessary square roots give $x = -1$ or $x = -2$.

Often, equations are given in one form and need to be expressed in an alternative form. The rules of algebra allow us to manipulate the form of an equation. For example, given $x = \frac{1}{2}\mathbf{a}t^2$, solve for t.

First, we clear the fraction by multiplying by 2 to obtain $2x = \mathbf{a}t^2$. Now, divide both sides by $\mathbf{a}$ to get $2x/\mathbf{a} = t^2$. Finally, to solve for t, we take the square root of both sides. Mathematically, there are two solutions. However, in physics we must allow for the physical reality of a solution. Since this equation represents uniformly accelerated motion from rest, the concept of negative time is not realistic. Hence, we shall discard the negative square root solution and simply state that $t = \sqrt{2x/\mathbf{a}}$.

Exponents and Scientific Notation

Any number n can be written in the form of some base number B raised to a power a. The number a is called the exponent of the base number B. In other words, we can write that $n = B^a$. One common base number is the number 10. The use of products of numbers with powers of 10 is called **scientific notation**. Some examples of powers of 10 are $10 = 10^1$, $100 = 10 \times 10 = 10^2$, $1{,}000 = 10 \times 10 \times 10 = 10^3$, and so on. By definition, any number raised to the zero power is equal to 1. That is, $10^0 = 1$, by definition.

Numbers less than 1 have *negative exponents* since they are fractions of the powers of 10 discussed previously. Some examples are $0.1 = 10^{-1}$, $0.01 = 10^{-2}$, $0.001 = 10^{-3}$, and so on.

The use of negative exponents for small numbers comes from the law of division of exponents which is defined as

$$\frac{10^a}{10^b} = 10^{a-b}$$

Since a reciprocal means "1 over . . . " and 10 raised to the power of zero is equal to 1, this means that in the division example, if $a = 0$ and $b =$ any number, then we have negative exponents for fractions less than 1. The law of multiplication of exponents is expressed as

$$10^a 10^b = 10^{a+b}$$

$$(10^a)^b = 10^{ab}$$

Scientific notation involves the use of numbers and powers of 10 as products. For example, the number 200 can be expressed in scientific notation as 2.0×10^2. The number 3,450 can be expressed in scientific notation as 3.45×10^3. Finally, the number 0.045 can be expressed in scientific notation as 4.5×10^{-2}.

On most scientific calculators, scientific notation can be activated using the "exp" key. It is not necessary to push "× 10" (or "times 10") on the calculator. When you press the "exp" key, it automatically implies "times 10 to the . . . " in the notation.

The following prefixes often appear in physics:

10^{18}	*exa-*	10^{-1}	*deci-*
10^{15}	*peta-*	10^{-2}	*centi-*
10^{12}	*tera-*	10^{-3}	*milli-*
10^{9}	*giga-*	10^{-6}	*micro-*
10^{6}	*mega-*	10^{-9}	*nano-*
10^{3}	*kilo-*	10^{-12}	*pico-*
10^{2}	*hecto-*	10^{-15}	*femto-*
10^{1}	*deca-*	10^{-18}	*atto-*

Since any number n can be written as a base number B raised to a certain power a, there is a process opposite that of **exponentiation**. In this designation, $n = B^a$, we can also state that a is the **logarithm** of n with the base B: $a = \log_B n$. If we are using base 10, then we do not usually write the base number but simply use the designation "log" for what we call the *common logarithm*. For example, the logarithm of 100 is 2 (since 2 is the power 10 is raised to in order to make 100). That is, $\log 100 = 2$. Since logarithms are basically exponents, the rules of multiplication and division of exponents translate to the following:

$$\log(xy) = \log + \log y$$

$$\log\left(\frac{x}{y}\right) = \log x - \log y$$

$$\log x^n = n \log x$$

Another common base number is the irrational number e which is equal to 2.71828 . . . . . Logarithms to the base e are called *natural logarithms* and are designated by "ln". These logarithms follow the same rules as common logarithms. The use of scientific calculators makes extensive logarithm tables obsolete. Consult your own caluclator's instructions for using logarithms.

GEOMETRY

There are some common formulas from geometry that are of use in physics. We first review some of the more familiar geometric shapes and their equations:

Straight line with slope a and y-intercept b: $y = ax + b$
Circle of radius R centered at the origin: $x^2 + y^2 = R^2$
Parabola whose vertex is at $y = b$: $y = ax^2 + b$
Hyperbola: $xy =$ constant
Ellipse with semimajor axis a and semiminor axis b: $\frac{x^2}{a^2} + \frac{y^2}{b^2} = 1$

We now review some of the physical characteristics of some of these shapes.

Circle of radius R: area $= \pi R^2$; circumference $= 2\pi R$
Sphere of radius R: volume $= \frac{4}{3}\pi R^3$; surface area $= 4\pi R^2$
Cylinder of radius R and length l: volume $= \pi R^2 l$
Right circular cone of base radius R and height h: volume $= \frac{1}{3}\pi R^2 h$
Ellipse with semimajor axis a and semiminor axis b: area $= \pi ab$
Rectangle of length l and width w: area $= lw$
Triangle with base b and altitude h: area $= \frac{1}{2}bh$

TRIGONOMETRY

The branch of mathematics called **trigonometry** involves the algebraic relationships between angles in triangles. If we have a right triangle with hypotenuse c and sides a and b, then the relationship between these sides is given by the Pythagorean theorem $a^2 + b^2 = c^2$.

The ratios of the sides of the right triangle to its hypotenuse define the *trigonometric functions* sine and cosine.

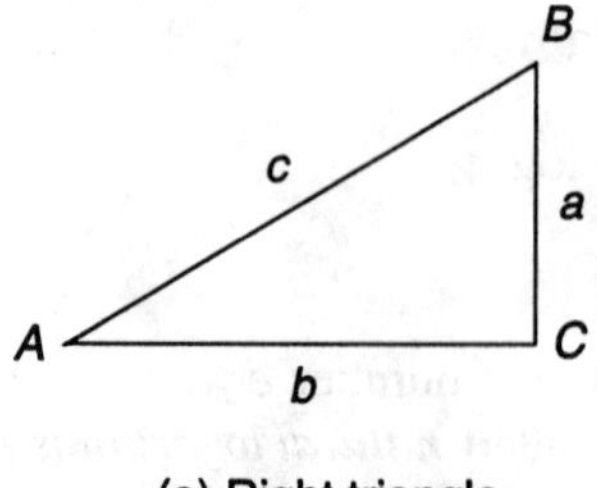

(a) Right triangle

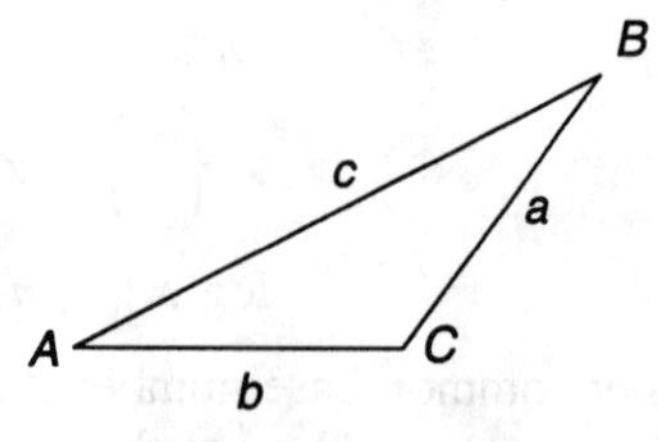

(b) Arbitrary triangle

$$\sin A = \frac{a}{c} \qquad \cos A = \frac{b}{c} \qquad \tan A = \frac{a}{b} \qquad \tan A = \frac{\sin A}{\cos A}$$

$$\sec A = \frac{c}{a} \qquad \csc A = \frac{c}{b} \qquad \cot A = \frac{b}{a}$$

The second set of functions are the reciprocals of the first, main set. For any angle θ in the right triangle, $\sin \theta = \cos (90 - \theta)$ and $\cos \theta = \sin (90 - \theta)$.

In the arbitrary general triangle, the sides of the triangle are related through the law of cosines and the law of sines. These formulas are extremely useful in the study of vectors:

$$\text{Law of cosines: } c^2 = a^2 + b^2 - 2ab \cos C$$

$$\text{Law of sines: } \frac{a}{\sin A} = \frac{b}{\sin B} = \frac{c}{\sin C}$$

For angles between 0 and 90 degrees, all the trigonometric functions have positive values. For angles between 90 and 180 degrees, only the sine function is positive. For angles between 180 and 270 degrees, only the tangent function is positive. Finally, between 270 and 360 (or 0) degrees, only the cosine function is positive.

Some useful identities (relationships) involving the various trigonometric functions that often appear in the the study of physics are listed here without proof:

$$\sin^2 \theta + \cos^2 \theta = 1$$
$$\sin 2\theta = 2 \sin \theta \cos \theta$$
$$\cos 2\theta = \cos^2 \theta - \sin^2 \theta$$
$$\sin (\theta + \phi) = \sin \theta \cos \phi + \cos \theta \sin \phi$$
$$\sin (\theta - \phi) = \sin \theta \cos \phi - \cos \theta \sin \phi$$
$$\cos (\theta + \phi) = \cos \theta \cos \phi - \sin \theta \sin \phi$$
$$\cos (\theta - \phi) = \cos \theta \cos \phi + \sin \theta \sin \phi$$

SERIES EXPANSIONS OF FUNCTIONS

Often it is useful to approximate a more complicated equation or function numerically. Usually this is because the magnitudes of quantities involved are extremely small (i.e., much less than 1). Without further discussion, three series expansions are presented for possible use during the course of reviewing and studying physics.

Binomial Expansion

If the number a is very small, such that $a << 1$, then the quantity

$$(1 + a)^n \approx 1 + na + \frac{1}{2} n(n - 1)a^2 + \cdots$$

can be further approximated (sometimes) to just $1 + na$.

Sine Function Expansion

If the angle θ is measured in radians, then we can state that

$$\sin\theta \approx \theta - \frac{\theta^3}{6} + \frac{\theta^5}{120} - \cdots \approx \theta - \frac{\theta^3}{3!} + \frac{\theta^5}{5!} - \cdots$$

The exclamation point means *factorial* such that $3! = 3 \times 2 \times 1 = 6$, for example. Further, if $\theta << 1$, then $\sin\theta \approx \theta$. Also if $\theta < < 1$, $\sin\theta \approx \tan\theta$.

Cosine Function Expansion

If the angle θ is measured in radians, then we can state that

$$\cos\theta \approx 1 - \frac{\theta^2}{2!} + \frac{\theta^4}{4!} - \cdots$$

Further, if $\theta << 1$, then we can state that $\cos\theta \approx 1$.

Appendix C

The Greek Alphabet

Alpha	Α	α	Nu	Ν	ν
Beta	Β	β	Xi	Ξ	ξ
Gamma	Γ	γ	Omicron	Ο	ο
Delta	Δ	δ	Pi	Π	π
Epsilon	Ε	ϵ	Rho	Ρ	ρ
Zeta	Ζ	ζ	Sigma	Σ	σ
Eta	Η	η	Tau	Τ	τ
Theta	Θ	θ	Upsilon	Υ	υ
Iota	Ι	ι	Phi	Φ	φ
Kappa	Κ	κ	Chi	Χ	χ
Lambda	Λ	λ	Psi	Ψ	ψ
Mu	Μ	μ	Omega	Ω	ω

Appendix D

Table of Conversions

Length
1 in = 2.54 cm
1 m = 100 cm = 39.37 in
1 ft = 12 in = 0.3048 m
1 yd = 3 ft = 0.9144 m
1 km = 1,000 m = 0.6214 mi
1 mi = 5,280 ft = 1.609 km
1 Å = 1×10^{-10} m
1 μ = 1×10^{-6} m

Area
1 m^2 = 10^4 cm^2 = 10.76 ft^2
1 ft^2 = 0.0929 m^2

Volume
1 m^3 = 10^6 cm^3 = 35.32 ft^3
1 ft^3 = 0.0283 m^3
1 liter = 1,000 cm^3
1 gal = 3.785 L

Mass
1,000 kg = 1 metric ton
1 u = 1.66×10^{-27} kg
1 slug (English mass) = 14.59 kg

Force
1 N = 0.2248 lb
1 lb = 4.448 N

Velocity
1 mi/h = 1.467 ft/s = 0.4470 m/s = 1.609 km/hr
1 m/s = 100 cm/s = 3.281 ft/s

Acceleration
1 m/s^2 = 100 cm/s^2 = 3.28 ft/s^2
1 ft/s^2 = 0.3048 m/s^2 = 30.48 cm/s^2

Pressure
1 Pa = 1 N/m^2
1 atm = 1.013×10^5 Pa

Time
1 year = 365.25 days = 3.16×10^7 s
1 day = 24 h = 86,400 s

Energy
1 J = 0.738 ft-lb
1 cal = 4.186 J
1 Btu = 252 cal = 1,054 J
1 eV = 1.6×10^{-19} J
931.5 MeV is equivalent to 1 u

Power
1 W = 1 J/s
1 hp = 746 W

GLOSSARY

Absolute index of refraction: In a transparent material, a number that represents the ratio of the speed of light in a vacuum to the speed of light in the material.

Absolute temperature: A measure of the average kinetic energy of the molecules in an object; the Kelvin temperature is equal to the Celsius temperature of an object plus 273.16.

Absolute zero: A temperature of −273°C, designated as 0 K.

Absorption spectrum: A continuous spectrum crossed by dark lines representing the absorption of particular wavelengths of radiation by a cooler medium.

Acceleration: A vector quantity representing the time rate of change in velocity.

Action: A force applied to an object that leads to an equal but opposite reaction; the product of total energy and time in units of J-s; the product of momentum and position, especially in the case of the Heisenberg uncertainty principle.

Activity: The number of radioactive decays of a radioactive atom per unit of time.

Adhesion: The force holding the surfaces of unlike substances together.

Adiabatic: In thermodynamics, the process that occurs when no heat is added to or subtracted from a system.

Alpha decay: The spontaneous emission of a helium nucleus from certain radioactive atoms.

Alpha particle: The name given to a helium nucleus ejected from a radioactive atom.

Alternating current: An electric current that varies its direction and magnitude according to a regular frequency.

Ammeter: A device for measuring the current in an electric circuit when placed in series; a galvanometer with a low-resistance coil placed across it.

Ampere: The SI unit of electric current equal to 1 C/s.

Amplitude: The maximum displacement of an oscillating particle, medium, or field relative to its rest position.

Angle of incidence: The angle between a ray of light and the normal to a surface that the ray intercepts.

Angle of reflection: The angle between a reflected light ray and the normal to a reflecting surface.

Angle of refraction: The angle between the emerging light ray in a transparent material and the normal to the surface.

Angle of shear: The relative angular distortion in an elastic solid due to the application of a shear stress.

Angular frequency: In units of rad/s, the rate at which a point mass undergoing uniform circular motion sweeps out a given angle each second; the product of 2 π and the frequency f.

Angular momentum: For a point mass undergoing uniform circular motion, a vector quantity equal to the product of the mass, velocity, and radius.

Angular velocity: For a point mass undergoing uniform circular motion, a vector quantity equal to the ratio of the tangential velocity and the radius (in rad/s).

Ångstrom: A unit of length measurement equal to 1×10^{-10} m.

Antinodal lines: A region of maximum displacement in a medium where waves interact with each other.

Aphelion: In the solar system, the point in a planet's orbit when it is farthest from the sun.

Archimede's principle: A law stating that an object immersed in a fluid will be buoyed up by a force equal to the weight of the fluid displaced by the object.

Armature: The rotating coil of wire in an electric motor.

Atomic mass unit: A unit of mass equal to $\frac{1}{12}$ the mass of a carbon-12 atom.

Atomic number: The number of protons in an atom's nucleus.

Atmosphere: A unit of pressure defined at sea level to be equal to 101 kPa.

Average speed: A scalar quantity equal to the ratio of the total distance to the total elapsed time.

Avogadro's number: The number of molecules in 1 mol of an ideal gas equal to 6.02×10^{23}; named for the Italian scientist Amadeo Avogadro.

Back EMF: The induced potential difference in the armature of an electric motor that opposes the applied potential difference.

Balmer series: In a hydrogen atom, the visible spectral emission lines that correspond to electron transitions from higher excited states to a lower level 2.

Battery: A combination of two or more electric cells.

Beats: The interference caused by two sets of sound waves with only a slight difference in frequency.

Becquerel: A unit of radioactive decay equal to one decay event per second.

Beta decay: The spontaneous ejection of an electron from the nuclei of certain radioactive atoms.

Beta particle: The name given to an electron that is spontaneously ejected by a radioactive nucleus.

Binding energy: The energy required to break apart an atomic nucleus; the energy equivalent of the mass defect.

Boltzmann's constant: In thermodynamics, a constant that is the ratio of the universal gas constant R to Avogadro's number N_a and is equal to 1.38×10^{-23} J/K; named for the German physicist Ludwig Boltzmann.

Boyle's law: A principle stating that for an ideal gas at constant temperature, the pressure of the gas varies inversely with its volume; named for the English scientist Robert Boyle.

Bouyant force: The upward force applied by a fluid on a solid placed in it.

Bright-line spectrum: The display of brightly colored lines on a screen or in a photograph indicating the discrete emission of radiation by a heated gas at low pressure.

Bulk modulus: The ratio of the applied stress to the relative change in volume.

Capacitance: The ratio of the total charge to the potential difference in a capacitor.

Capacitive reactance: In an ac circuit, the reciprocal of the product of the angular frequency ω and the capacitance *C* of a capacitor; designated X_C.

Capacitive time constant: In an ac circuit, the product of the resistance *R* and the capacitance *C*.

Capacitor: A pair of conducting plates, with either a vacuum or an insulator between them, used in an electric circuit to store charge.

Capillarity: The action of liquid rising through small tube openings.

Carnot cycle: In thermodynamics, a sequence of four phases experienced by an ideal gas confined in a cylinder with a movable piston (a Carnot engine); includes an isothermal expansion, an adiabatic expansion, an isothermal compression, and an adiabatic compression; named for the French physicist Sadi Carnot.

Cartesian coordinate system: A set of two or three mutually perpendicular reference lines called axes, usually designated *x*, *y*, and *z*, used to define the location of an object in a frame of reference; named for the French scientist Rene Descartes.

Cathode ray tube: An evacuated gas tube in which a beam of electrons is projected and their energy produces an image on a fluorescent screen; they are deflected by external electric or magnetic fields.

Celsius temperature scale: A metric temperature scale in which, at sea level, water freezes at 0°C and boils at 100°C.

Center of curvature: A point that is equidistant from all other points on the surface of a spherical mirror; a point equal to twice the focal length of a spherical mirror.

Center of mass: The weighted mean point where all the mass of an object can be considered to be located; the point at which if a single force is applied, purely translational motion will result.

Centripetal acceleration: The acceleration of mass moving uniformly in a circle at a constant speed directed radially inward toward the center of the circular path.

Centripetal force: The deflecting force, directed radially inward toward a given point, that causes an object to follow a circular path.

Chain reaction: In nuclear fission, the reaction of neutrons splitting uranium nuclei and creating more neutrons that continue the process self-sustained.

Characteristic time: In radioactive decay, the time required for the original mass to decay to a factor of $1/e$ of that value.

Charles's law: In thermodynamics, a principle stating that at constant pressure the volume of an ideal gas varies directly with its absolute temperature.

Chromatic aberration: In optics, the defect in a converging lens that causes the dispersion of white light into a continuous spectrum and results in the lens refracting the colors to different focal points.

Coefficient of friction: The ratio of the force of friction to the normal force when one surface is sliding (or attempting to slide) over another surface.

Coherent: Describing a set of waves that have the same wavelength, frequency, and phase.

Cohesion: The ability of liquid molecules to stick together.

Commutator: A split ring in a dc motor whose segments are connected to opposite terminals of a battery; as the armature turns, the split in the rings acts to reverse the direction of the current every half-revolution.

Component: One of two mutually perpendicular vectors that lie along the principal axes in a coordinate system and can be combined to form a given resultant vector.

Compressibility: The reciprocal of the bulk modulus.

Concave lens: A lens that causes parallel rays of light to emerge in such a way that they appear to diverge away from a focal point behind the lens.

Concave mirror: A spherical mirror that causes parallel rays of light to converge to a focal point in front of the mirror.

Concurrent forces: Two or more forces that act at the same point at the same time.

Conductor: A substance, usually metallic, that allows the relatively easy flow of electric charges.

Conservation of angular momentum: A principle stating that in the absence of external torques, the total angular momentum of an isolated system remains the same.

Conservation of energy: A principle stating that in an isolated system, the total energy of the system remains the same during all interactions within the system.

Conservation of mass-energy: A principle stating that in the conversion of mass into energy or energy into mass, the total mass-energy of the system remains the same.

Conservation of momentum: A principle stating that in the absence of any external forces, the total momentum of an isolated system remains the same.

Conservative force: A force whose work is independent of the path taken; when work is done by this force, the work can be recovered without any loss.

Constructive interference: The additive result of two or more waves interacting with the same phase relationship as they move through a medium.

Continuous spectrum: A continuous band of colors, consisting of red, orange, yellow, green, blue, and violet, formed by the dispersion or diffraction of white light.

Control rod: A device (usually made of cadmium) used in a nuclear reactor to control the rate of fission.

Converging lens: A lens that causes parallel rays of light incident on its surface to be refracted and converge to a focal point; a convex lens.

Converging mirror: A spherical mirror that causes parallel rays of light incident on its surface to reflect and converge to a focal point; a concave mirror.

Convex lens: A converging lens.

Convex mirror: A diverging mirror.

Coordinate system: A set of reference lines, not necessarily perpendicular, used to locate the position of an object within a frame of reference using the rules of analytic geometry.

Core: The interior of a solenoid electromagnet usually made of a ferromagnetic material; the part of a nuclear reactor where the fission reaction occurs.

Coulomb: The SI unit of electric charge defined to be the amount of charge 1 A of current delivers each second while passing a given point.

Coulomb's law: A principle stating that the electrostatic force between two point charges is directly proportional to the product of the charges and inversely proportional to the square of the distance separating them.

Critical angle of incidence: The angle of incidence with a transparent substance such that the angle of refraction is 90 degrees relative to the normal drawn to the surface.

Current: A scalar quantity that measures the amount of charge passing a given point in an electric circuit each second.

Current length: A relative measure of the magnetic field strength produced by a length of wire carrying current and equal to the product of the current and the length.

Cycle: One complete sequence of periodic events or oscillations.

Damping: The continuous decrease in the amplitude of mechanical oscillations due to a dissipative force.

Decay constant: The ratio of the decay rate of nuclei to the total number of nuclei available at a given time.

Decibel: A logarithmic unit of sound intensity defined such that 0 dB equals 1.0×10^{-12} W/m^2.

Deflecting force: Any force that acts to change the direction of motion of an object.

Derived unit: Any combination of fundamental physical units.

Destructive interference: The result produced by the interaction of two or more waves with opposite phase relationships as they move through a medium.

Deuterium: An isotope of hydrogen containing one proton and one neutron in its nucleus; heavy hydrogen (a component of heavy water).

Dielectric: An electric insulator placed between the plates of a capacitor to alter its capacitance.

Diffraction: The spreading of waves around or past obstacles.

Diffraction grating: A reflecting or transparent surface with many thousands of lines ruled on it which is used to diffract light into a spectrum.

Direct current: Electric current that is moving in only one direction around an electric circuit.

Dispersion: The separation of light into its component colors or spectrum.

Dispersive medium: Any medium that produces a dispersion of light; any medium in which the velocity of a wave depends on its frequency.

Displacement: A vector quantity that determines the change in position of an object by measuring the straight-line distance and direction from the starting point to the ending point.

Dissipative force: Any force that removes kinetic energy from a moving object (like friction); a nonconservative force.

Distance: A scalar quantity that measures the total length of a path taken by a moving object.

Diverging lens: A lens that causes parallel rays of light incident on its surface to be refracted and diverge away from a focal point on the incoming light side of the lens; a concave lens.

Diverging mirror: A spherical mirror that causes parallel rays of light incident on its surface to be reflected and diverge away from a focal point on the center-of-curvature side of the mirror.

Doppler effect: The apparent change in the wavelength or frequency of a wave as the source of the wave moves relative to an observer.

Dynamics: The branch of mechanics that studies the effect of forces on objects.

Effective current: For an ac circuit, a measure of the root mean square (i.e., the square root of the average of the square) of the time-varying current; equal to 0.707 times the maximum current.

Effective voltage: For an alternating current circuit, a measure of the root mean square of the time-varying voltage; equal to 0.707 times the maximum voltage.

Efficiency: In a machine, the ratio of work output to work input; in a transformer, the ratio of the power from the secondary coil to the power into the primary coil.

Elastic collision: A collision between two objects in which there is a rebounding and no loss of kinetic energy.

Elastic potential energy: The energy stored in a spring when work is done to stretch or compress it.

Electric cell: A chemical device for generating electricity.

Electric circuit: A closed conducting loop consisting of a source of potential difference, conducting wires, and other devices that operate on electricity.

Electric field: The region where an electric force is exerted on a charged object.

Electric field intensity: A vector quantity that measures the ratio of the magnitude of the force to the magnitude of the charge on an object.

Electric motor: A coil of wire carrying current that is caused to rotate in an external magnetic field because of the alternation of electric current.

Electrical ground: The passing of charges to or from the earth to establish a potential difference between two points.

Electromagnet: A coil of wire wrapped around a ferromagnetic core (usually made of iron) that generates a magnetic field when current is passed through it.

Electromagnetic field: The field produced by an electromagnet or moving electric charges.

Electromagnetic induction: The production of a potential difference in a conductor as a result of the relative motion between the conductor and an external magnetic field; the production of a potential difference in a conductor as a result of the change in an external magnetic flux near the conductor.

Electromagnetic spectrum: The range of frequencies covering the discrete emission of energy from oscillating electromagnetic fields that includes radio waves, microwaves, infrared waves, visible light, ultraviolet light, x rays, and gamma rays.

Electromagnetic wave: A wave generated by the oscillation of electric charges producing interacting electric and magnetic fields that oscillate in space and travel at the speed of light in a vacuum.

Electron: A negatively charged particle that orbits a nucleus in an atom; the fundamental carrier of negative electric charge.

Electron capture: The process in which an electron is captured by a nucleus possessing too many neutrons with respect to protons; K capture.

Electron cloud: A theoretical probability distribution of electrons around a nucleus due to the uncertainty principle, the most probable location for an electron is in the densest regions of the cloud.

Electronvolt: The energy associated with moving a one-electron charge through a potential difference of 1 V; equal to 1.6×10^{-19} J.

Electroscope: A device for detecting the presence of static charges on an object.

Elementary charge: The fundamental amount of charge of an electron.

EMF: Electromotive force; the potential difference caused by the conversion of different forms of energy into electric energy; the energy per unit charge.

Emission spectrum: The discrete set of colored lines representing electromagnetic energy produced when atomic compounds are excited into emitting light as a result of heat, sparks, or atomic collisions.

Energy: A scalar quantity representing the capacity to work.

Energy level: One of several regions around a nucleus where electrons are considered to reside.

Entropy: The degree of randomness or disorder in a thermodynamic system.

Equation of continuity: For an ideal fluid, an equation expressing the fact that the rate of flow remains constant for any change in the velocity of the fluid or in the cross-sectional area or height of the pipe it is contained in. (See *Rate of flow*)

Equilibrant: The force equal in magnitude and opposite in direction to the resultant of two or more forces that bring a system into equilibrium.

Equilibrium: The balancing of all external forces acting on a mass; the result of a zero vector sum of all forces acting on an object.

Escape velocity: The minimum initial velocity of an object away from a planet at which the object will not be pulled back toward the planet.

Excitation: The process by which

an atom absorbs energy and causes its orbiting electrons to move to higher energy levels.

Excited state: The situation in an atom in which its orbiting electrons are residing in higher energy levels.

Farad: A unit of capacitance equal to 1 C/V.

Faraday's law of electromagnetic induction: A principle stating that the magnitude of the induced EMF in a conductor is equal to the rate of change in the magnetic flux.

Ferromagnetic substances: Metals or compounds made of iron, cobalt, or nickel that produce very strong magnetic fields.

Field: A region characterized by the presence of a force on a test body, such as a unit mass in a gravitational field or a unit charge in an electric field.

Field intensity: A measure of the force exerted on a unit test body; force per unit mass; force per unit charge.

First law of thermodynamics: The law of conservation of energy as applied to thermodynamic systems.

Fission: The splitting of a nucleus into two smaller and more stable nuclei by means of a slow-moving neutron, which releases a large amount of energy.

Fluid: A form of matter which assumes the shape of its container; includes liquids and gases.

Flux: A measure of the product of the perpendicular component of a magnetic field crossing an area and the magnitude of the area (in webers).

Flux density (magnetic): A measure of the field intensity per unit area (in Wb/m^2).

Focal length: The distance along the principal axis from a lens or spherical mirror to the principal focus.

Focus: The point of convergence of light rays caused by a converging mirror or lens; either of two fixed points on an ellipse that determine its shape.

Force: A vector quantity that corresponds to any push or pull due to an interaction that changes the motion of an object

Forced vibration: Vibrations caused by the application of an external force.

Frame of reference: A point of view consisting of a coordinate system in which observations are made.

Free body diagram: A diagram that illustrates all the forces acting on a mass at any given time.

Frequency: The number of completed periodic cycles per second in an oscillation or wave motion.

Friction: A force that opposes the motion of an object as it slides over another surface.

Fuel rods: Rods packed with fissionable material which are inserted into the core of a nuclear reactor.

Fundamental unit: An arbitrary scale of measurement assigned to certain physical quantities such as length, time, mass, and charge which are considered to be the basis for all other measurements.

Fusion: The combination of two or more light nuclei which produces a more stable heavier nucleus with the release of energy.

Galvanometer: A device used to detect the presence of small electric currents when connected in series in a circuit.

Gamma radiation: High-energy photons emitted by certain radioactive substances.

Gravitation: The mutual force of attraction between two uncharged masses.

Gravitational field strength: A measure of the gravitational force per unit mass in a gravitational field.

Gravity: Another name given to gravitation or gravitational force; the tendency of objects to fall to the earth.

Ground state: The lowest energy level in an atom.

Half-life: The time it takes a radioactive material to decay to one half its original amount.

Heat: The net disordered kinetic energy associated with molecular motion in matter.

Heat of fusion: The amount of energy per kilogram needed to change the state of matter from a solid to a liquid or from a liquid to a solid.

Heat of vaporization: The amount of energy per kilogram needed to change the state of matter from a liquid to a gas or from a gas to a liquid.

Hertz: The SI unit of frequency equal to 1 s^{-1}; the number of cycles/s in a vibration.

Hooke's law: A principle stating that the stress applied to an elastic material is directly proportional to the strain produced; named for Robert Hooke, an English scientist.

Hysteresis: The response of a system to the repeated application and removal of stresses; in magnetism, it is the response of the magnetic field of a ferromagnetic material to an increase and decrease in magnetization from an external magnetic field.

Ideal gas: A gas for which the assumptions of the kinetic theory are valid.

Image: An optical reproduction of an object by means of a lens or mirror.

Impedence: In general, a measure of the ability of a system to allow for the transfer of energy from one part of the system to another; in an ac circuit, it is equal to the square root of the sum of the square of the resistance and the square of the difference between the inductive and capacitive reactances (designated Z).

Impedence matching: Arranging for the maximum transfer of energy or power from one part of a mechanical or electromagnetic system to another.

Impulse: A vector quantity equal to the product of the average force applied to a mass and the time interval over which the force acts; the area under a force versus time curve.

Induced potential difference: A potential difference created in a conductor as a result of its motion relative to an external magnetic field.

Inductance: In an electric circuit containing a solenoid, the ratio of the induced EMF in the solenoid to the rate of change in current; designated *L* and measured in henrys.

Induction coil: A transformer in which a variable potential difference is produced in a secondary coil when a direct current applied to the primary is turned on and off.

Inductive reactance: In an ac circuit, it is the product of the angular frequency ω and the inductance L; designated X_L.

Inductive time constant: In an ac circuit, the ratio of the inductance to the resistance.

Inelastic collision: A collision in which two masses interact with a loss of kinetic energy.

Inertia: The property of matter that resists the action of an applied force trying to change the motion of an object.

Inertial frame of reference: A frame of reference in which the law of inertia holds; a frame of reference moving with constant velocity relative to the earth.

Instantaneous velocity: The slope of a tangent line to a point on a displacement versus time graph; the velocity of a particle at any instant in time.

Insulator: A substance which is a nonconductor of electricity because of the absence of free electrons.

Interference: The interaction of two or more waves producing an enhanced or diminished amplitude at the point of interaction; the superposition of one wave on another.

Interference pattern: The pattern produced by the constructive and destructive interference of waves generated by two point sources.

Isobaric: In thermodynamics, a process in which the pressure of a gas remains the same.

Isochoric: In thermodynamics, a process in which the volume of a gas remains the same.

Isolated system: Two or more interacting objects that are not being acted on by an external force.

Isotopes: Atoms with the same number of protons but different numbers of neutrons.

Joule: The SI unit of work equal to 1 N-m; the SI unit of mechanical energy equal to 1 kg-m^2/s^2.

Junction: The point in an electric circuit where a parallel connection branches off.

K capture: Electron capture.

Kelvin: The unit of absolute temperature defined such that 0 K equals −273°C.

Kepler's first law: A principle stating that the orbital paths of all planets are elliptical with the sun at one focus.

Kepler's second law: A principle stating that a line from the sun to a planet sweeps out equal areas in equal times.

Kepler's third law: A principle stating that the ratio of the cube of the mean radius to the sun to the square of the period is a constant for all planets orbiting the sun.

Kilogram: The SI unit of mass.

Kilojoule: A number representing 1,000 J.

Kilopascal: A number representing 1,000 Pa.

Kinematics: In mechanics, the study of how objects move.

Kinetic energy: The energy possessed by a mass as a result of its motion relative to a frame of reference.

Kinetic friction: The friction induced by sliding one surface over another.

Kinetic theory of gases: A theory stating that all matter is made of molecules that are in a constant state of motion.

Kirchoff's first law: A principle stating that the algebraic sum of all currents at a junction in a circuit equals zero.

Kirchoff's second law: A principle stating that the algebraic sum of all potential drops around any closed loop in a circuit equals zero.

Laminar flow: Successive layers in a fluid moving smoothly past each other.

Laser: A device for producing an intense coherent beam of monochromatic light; an acronym for *l*ight *a*mplification by the *s*timulated *e*mission of *r*adiation.

Lens: A transparent substance with one or two curved surfaces used to direct rays of light by refraction.

Lenz's law: A principle stating that an induced current in a conductor is always in a direction such that its magnetic field opposes the change in the magnetic field that induced it.

Line of force: An imaginary line drawn in a gravitational electric, or magnetic field that indicates the direction a test particle takes when experiencing a force in that field.

Longitudinal wave: A wave in which the oscillating particles vibrate in a direction parallel to the direction of propagation.

Magnet: An object that exerts a force on ferrous materials or a current.

Magnetic field: A region surrounding a magnet in which ferrous materials or moving charged particles experience a force.

Magnetic field strength: The force on a unit current length in a magnetic field.

Magnetic flux density: The total magnetic flux per unit area in a magnetic field.

Magnetic pole: The region on a magnet where magnetic lines of force are the most concentrated.

Mass: The property of matter used to represent the inertia of an object; the ratio of the net force applied to an object to its subsequent acceleration as specified by Newton's second law of motion.

Mass defect: The difference between the actual mass of a nucleus and the sum of the masses of the protons and neutrons it contains.

Mass number: The total number of protons and neutrons in an atomic nucleus.

Mass spectrometer: A device that uses magnetic fields to cause nuclear ions to assume circular trajectories and thereby determines their masses based on their charge, the radius of the path, and the external field strength.

Mean radius: For a planet, the average distance to the sun.

Meter: The standard SI unit for

length: equal to the distance light travels in 1/299,792,458 s.

Moderator: A material (such as water or graphite) used to slow neutrons in fission reactions.

Modulus of elasticity: Another name for Young's modulus.

Mole: Amount of substance containing Avogadro's number of molecules.

Moment arm distance: A line drawn from a pivot point.

Moment of inertia: A measure of the relative distribution of matter about a rotating axis; rotational inertia.

Momentum: A vector quantity equal to the product of the mass and velocity of a moving object.

Monochromatic light: Light consisting of only one frequency.

Natural frequency: The frequency with which an elastic body will vibrate if disturbed.

Net force: The vector sum of all forces acting on a mass.

Neutron: A subatomic particle, residing in the nucleus, which has a mass comparable to that of a proton but is electrically neutral.

Newton: The SI unit of force equal to 1 kg-m/s^2.

Newton's first law of motion: A principle stating that an object at rest tends to stay at rest, and that an object in motion tends to stay in motion at a constant velocity, unless acted on by an external force; the law of inertia.

Newton's second law of motion: A principle stating that the acceleration of a body is directly proportional to the applied net force; $\mathbf{F} = m\mathbf{a}$.

Newton's third law of motion: A principle stating that for every action force there is an equal but opposite reaction force.

Nodal line: A line of minimum displacement occurring when two or more waves interfere.

Nonconservative force: A force, like friction, that decreases the amount of kinetic energy after work is done by the force.

Normal: A line perpendicular to a surface.

Normal force: A force that is directed perpendicular to a surface when two objects are in contact.

Nuclear force: The force between nucleons in a nucleus that opposes the Coulomb force repulsion of the protons; the strong force.

Nucleon: Either a proton or a neutron as it exists in a nucleus.

Ohm: The SI unit of electric resistance equal to 1 V/A.

Ohm's law: A principle stating that in a circuit at constant temperature, the ratio of the potential difference to the current is a constant (called the resistance).

Optical center: The geometric center of a converging or diverging lens.

Parallel circuit: A circuit in which two or more devices are connected across the same potential difference, providing two parallel paths for current flow.

Pascal: The SI unit of pressure equal to 1 N/m^2.

Pascal's principle: A law stating that in a confined static fluid, any external pressure applied is distributed uniformly throughout the fluid.

Perihelion: The point of closest approach of a planet to the sun.

Period: The time (in seconds) required to complete one cycle of a repetitive oscillation or uniform circular motion.

Permeability: The property of matter that affects the external magnetic field the material is placed in, causing it to align free electrons within the substance and hence magnetize it; a measure of a material's ability to become magnetized.

Phase: The fraction of a period or wavelength from a chosen reference point on the wave.

Phasor diagram: In ac electricity, a diagram in which rotating vectors are plotted to demonstrate the phase relationships among voltages, currents, and reactances in various circuit designs.

Photoelectric effect: The process by which surface electrons in a metal are freed through the incidence of electromagnetic radiation above a certain minimum frequency.

Photon: The quantum of electromagnetic energy.

Planck's constant: A universal constant representing the ratio of the energy of a photon to its frequency; the fundamental quantum of action having a value of 6.63×10^{-34} kg-m^2/s.

Polar coordinate system: A coordinate system in which the location of a point is determined by a vector from the origin of a Cartesian coordinate system and the angle it makes with the positive horizontal axis.

Polarization: The process by which the vibrations of a transverse wave are selected to lie in a preferred plane.

Polarized light: Light that has been polarized by passing it through a suitable polarizing filter.

Positron: A subatomic particle having the same mass as an electron except for a positive electric charge; an antielectron.

Potential difference: The work necessary to move a positive unit charge between any two points in an electric field.

Potential energy (gravitational): The energy possessed by a body as a result of its vertical position relative to the surface of the earth or any other arbitrarily chosen base level.

Power: The ratio of the work done to the time needed to complete the work.

Power factor: In an ac circuit, the cosine of the phase angle for the circuit; the ratio of the resistance in an ac circuit to the impedence Z.

Pressure: The force per unit area.

Primary coil: In a transformer, the coil connected to an alternating potential difference source.

Principal axis: An imaginary line passing through the center of curvature of a spherical mirror or the optical center of a lens.

Prism: A transparent triangular shape that disperses white light into a continuous spectrum.

Pulse: A localized disturbance in an elastic medium.

Quantum: A discrete packet of electromagnetic energy; a photon.

Radian: The angle that intercepts an arc length; a unit of angle measure based on the unit circle such that 360 degrees equals 2π rad.

Radioactive decay: The spontaneous disintegration of a nucleus as a result of its instability.

Radius of curvature: The distance from the center of curvature to the surface of a spherical shape.

Rate of flow: Volume flow per unit time; equal to the product of the velocity of the fluid and the cross-sectional area of the flowing fluid. (See *Equation of continuity.*)

Ray: A straight line used to indicate the direction of travel of a light wave.

Real image: An image formed by a converging mirror or lens that can be focused onto a screen.

Refraction: The bending of a light ray when it passes obliquely from one transparent substance to another in which its velocity is different.

Resistance: The ratio of the potential difference across a conductor to its current.

Resistivity: A property of matter that measures the ability of a substance to act as a resistor.

Resolution of forces: The process by which a given force is decomposed into a pair of perpendicular forces.

Resonance: Vibrations in a body at its natural vibrating frequency caused by the transfer of energy.

Rest mass: The mass of an object when the object is at rest with respect to the measuring instrument.

Rotational equilibrium: The point at which the vector sum of all torques acting on a rotating mass equals zero.

Rotational inertia: The property allowing an object to resist the action of a torque; moment of inertia.

Scalar: A physical quantity, like mass or speed, that is characterized by only magnitude or size.

Second: The SI unit of time.

Second law of thermodynamics: A principle stating that heat flows naturally only from a hot body to a cold one; according to this law, the entropy of the universe is always increasing and no process is possible from which the sole result is the complete conversion of heat from a source into work.

Secondary coil: In a transformer, a coil in which an alternating potential difference is induced.

Series circuit: A circuit in which electric devices are connected sequentially in a single conducting loop, allowing only one path for charge flow.

Shear modulus: The ratio of the shear stress to the angle of shear.

Shear stress: A measure of the longitudinal force per unit area that causes vertical layers of a solid to slide over one another.

Shunt: A low-resistance resistor connected parallel across a galvanometer, turning it into an ammeter.

Sliding friction: A force that resists the sliding motion of one surface over another; kinetic friction.

Snell's law: A principle stating that for light passing obliquely from one transparent substance to another, the ratio of the sine of the angle of incidence to the sine of the angle of refraction is equal to the relative index of refraction for the two media.

Solenoid: A coil of wire used for electromagnetic induction.

Specific heat: The amount of energy (in kilojoules) needed to change the temperature of 1 kg of a substance by 1°C.

Speed: A scalar quantity measuring the time rate of change in distance.

Spherical aberration: In a spherical mirror or a lens, the inability to properly focus parallel rays of light because of its shape.

Spring constant: The force needed to stretch or compress a spring a unit length; force constant; Young's modulus; modulus of elasticity.

Standard pressure: At sea level, 1 atm which is equal to 101.3 kPa.

Standing wave: A stationary wave pattern formed in a medium when two sets of waves with equal wavelength and amplitude pass through each other, usually after a reflection.

Starting friction: The force of friction overcome when two objects initially begin to slide over one another.

Static electricity: Stationary electric charges.

Static friction: The force that prevents one object from beginning to slide over another.

Statics: The study of the forces acting on an object that is at rest and unaccelerated relative to a frame of reference.

Strain: The resultant deformation in a solid due to the application of a stress.

Streamline flow: Laminar flow.

Stress: An force applied to a solid per unit area.

Superposition: When waves overlap, a condition in which the resultant wave is the algebraic sum of the individual waves.

Surface tension: In a liquid, the tendency of the top layers of the liquid to behave like a membrane under tension due to adhesive molecular forces.

Temperature: The relative measure of warmth or cold based on a standard; a measure of the average kinetic energy of molecules in matter.

Tesla: The SI unit of magnetic field strength equal to 1 W/m^2.

Test charge: A small, positively charged mass used to detect the presence of a force in an electric field.

Thermionic emission: The emission of electrons from a hot filament.

Thermometer: An instrument that makes a quantitative measurement of temperature based on an accepted scale.

Third law of thermodynamics: A principle stating that the temperature of any system can never be reduced to absolute zero.

Threshold frequency: The minimum frequency of electromagnetic radiation necessary to induce the photoelectric effect in a metal.

Torque: The application of a force at right angles to a designated line, from a pivot, that tends to produce circular motion; the product of a force and a perpendicular moment arm distance.

Total internal reflection: The process by which light is incident on an interface from a higher to a lower index of refraction medium in which its velocity increases such as to imply an angle of refraction greater than 90 degrees, reflecting the light back into the original medium.

Total mechanical energy: In a mechanical system, the sum of the kinetic and potential energies.

Transformer: Two coils used to change the alternating potential difference across one of them (the primary) to either a larger or smaller value in the secondary.

Transmutation: The process by which one atomic nucleus changes into another by radioactive decay or nuclear bombardment.

Transverse wave: A wave in which the vibrations of a medium or field are at right angles to the direction of propagation.

Uniform circular motion: Motion around a circle at a constant speed.

Uniform motion: Motion at a constant speed in a straight line.

Unit: An arbitrary scale assigned to a physical quantity for measurement and comparison.

Universal law of gravitation: A principle stating that the gravitational force between any two masses is directly proportional to the product of their masses and inversely proportional to the square of the distance between them.

Vector: A physical quantity characterized by having magnitude and direction; a directed arrow drawn in a Cartesian coordinate system used to represent a quantity such as force, velocity, or displacement.

Velocity: A vector quantity representing the time rate of change in displacement.

Virtual focus: The point at which the rays from a diverging lens would meet if they were traced back with straight lines.

Virtual image: An image formed by a mirror or lens that cannot be focused onto a screen.

Viscosity: In a fluid, the measure of the resistance adjacent vertical layers have a sliding over each other; the internal resistance of a fluid to flowing.

Visible light: The portion of the electromagnetic spectrum (ROYGBIV) that can be detected by the human eye.

Volt: The SI unit of potential difference equal to 1 J/C.

Volt per meter: The SI unit of electric field intensity equal to 1 N/C.

Voltmeter: A device used to measure potential difference between two points in a circuit when connected in parallel; a galvanometer with a large resistor placed in series with it.

Watt: The SI unit of power equal to 1 J/s.

Wave: A series of periodic disturbances in an elastic medium or field.

Wavelength: The distance between any two successive points in phase on a wave.

Weber: The SI unit of magnetic flux equal to 1 T times 1 m^2.

Weight: The force of gravity exerted on a mass at the surface of a planet.

Work: A scalar measure of the amount of change in mechanical energy; the product of the magnitude of the displacement of an object and the component of applied force in the same direction as the displacement; the area under a plot of force versus displacement.

Work function: The minimum amount of energy needed to free an electron from the surface of a metal using the photoelectric effect.

Young's modulus: The ratio of the applied stress to the linear strain in an elastic solid.

INDEX

BARRON'S COLLEGE REVIEW SERIES

Focus on Facts
Summarize Vital Topics
Excel in Your Course

Each title in *Barron's College Review Series* offers you an overview of a college-level course, and makes a fine supplement to your main textbook. You'll find topic summaries, lists of key terms, bibliographies, review questions, and more. **Each book: Paperback, $11.95, Can$15.95**

American Literature:
1930 to the Present
A. Bell, D. W. Heiney, and L. H. Downs
ISBN 1836-2

Biology
J. Snyder and C. L. Rodgers
ISBN 1862-1

English Literature
The Beginnings to 1800, 2nd Ed.
A. Bell and B. Grebanier
ISBN 1775-7

English Literature
1800 to 1900, 2nd Ed.
A. Bell and B. Grebanier
ISBN 1678-5

English Literature
1900 to the Present, 2nd Ed.
A. Bell, D. W. Heiney, and L. H. Downs
ISBN 1837-0

United States History
To 1877
N. Klose and R. F. Jones
ISBN 1834-6

United States History
Since 1863
N. Klose and C. Lader
ISBN 1835-4

World Literature
Early Origins to 1800, 2nd Ed.
A. Bell, V. P. Hopper, and B. Grebanier
ISBN 1811-7

World Literature
1800 to the Present, 2nd Ed.
A. Bell, V. P. Hopper, and B. Grebanier
ISBN 1812-5

Books may be purchased at your bookstore, or by mail from Barron's. Enclose check or money order for the total amount plus sales tax where applicable and 10% for postage and handling charge (minimum charge $3.75, Canada $4.00). Prices subject to change without notice.

Barron's Educational Series, Inc.
250 Wireless Blvd. • Hauppauge, NY 11788 • In Canada: Georgetown Book Warehouse • 34 Armstrong Ave., Georgetown, Ont. L7G 4R9

ISBN Prefix: 0-8120
$ = U.S. Dollars Can$ = Canadian Dollars

R 5/95 (#58)